Advanced Topics on Computer Vision, Control and Robotics in Mechatronics

Osslan Osiris Vergara Villegas
Manuel Nandayapa · Israel Soto
Editors

Advanced Topics on Computer Vision, Control and Robotics in Mechatronics

Springer

Editors
Osslan Osiris Vergara Villegas
Industrial and Manufacturing Engineering
Universidad Autónoma de Ciudad Juárez
Ciudad Juárez, Chihuahua
Mexico

Israel Soto
Industrial and Manufacturing Engineering
Universidad Autónoma de Ciudad Juárez
Ciudad Juárez, Chihuahua
Mexico

Manuel Nandayapa
Industrial and Manufacturing Engineering
Universidad Autónoma de Ciudad Juárez
Ciudad Juárez, Chihuahua
Mexico

ISBN 978-3-030-08535-3 ISBN 978-3-319-77770-2 (eBook)
https://doi.org/10.1007/978-3-319-77770-2

Printed on acid-free paper

This Springer imprint is published by the registered company Springer International Publishing AG part of Springer Nature
The registered company address is: Gewerbestrasse 11, 6330 Cham, Switzerland

Preface

The field of mechatronics, which is the synergistic combination of precision mechanical engineering, electronic control and thinking systems in the design of products and manufacturing processes, is gaining much attention in industries and academics. Most complex innovations in several industries are possible due to the existence of mechatronics systems.

From an exhaustive perusal and the experience gained from several years in the field, we detected that several disciplines such electronics, mechanics, control and computers are related to the design and building of mechatronic systems. However, computer vision, control and robotics are currently essential to achieve a better design and operation of intelligent mechatronic systems. Computer vision is the field of artificial intelligence devoted to acquiring, processing, analyzing and interpreting images from the real world with the goal of producing numerical information that can be treated by a computer. On the other hand, Control is a discipline that governs the physical laws of dynamic systems for variable regulations. Finally, Robotics is an interdisciplinary branch of engineering that deals with the design, construction, operation and application of robots.

This book is intended to present the recent advances in computer vision, control and robotics for the creation of mechatronics systems. Therefore, the book content is organized in three main parts: a) Computer Vision, b) Control, and c) Robotics, each one containing a set of five chapters.

Part I Computer Vision

In this part, the book reports efforts in developing computer vision systems implemented in different mechatronics industries including medical and automotive, and reviews in the field of pattern recognition, super-resolution and artificial neural networks.

Chapter 1 presents an implementation and comparison of five different denoising methods to reduce multiplicative noise in ultrasound medical images. The methods were implemented in the fixed-point DM6437 high-performance digital media processor (DSP).

In Chap. 2, a survey of the most recent advances concerning to morphological neural networks with dendritic processing (MNNDPs) is presented. The basics of each model and the correspondent training algorithm are discussed, and in some cases an example is presented to facilitate understanding.

The novel technology of augmented reality (AR) is addressed in Chap. 3. Particularly, a mobile AR prototype to support the process of manufacturing an all-terrain vehicle is discussed. The prototype was tested in a real automotive industry with satisfactory results.

Chapter 4 introduces the upcoming challenges of feature selection in pattern recognition. The paper particularizes in a new type of data known as chronologically linked, which is proposed to describe the value that a feature can acquire with respect to time in a finite range.

Finally, in Chap. 5 an overview of the most important single-image and multiple-image super-resolution techniques is given. The methods and its correspondent implementation and testing are showed. In addition, the main advantages and disadvantages of each methods were discussed.

Part II Control

The second part of the book related to control focused mainly into propose intelligent control strategies for helicopters, manipulators and robots.

Chapter 6 focuses on the field of cognitive robotics. Therefore, the simulations of an autonomous learning process of an artificial agent controlled by artificial action potential neural networks during an obstacle avoidance task are presented.

Chapter 7 analyzes and implements the hybrid force/position control using a fuzzy logic in a Mitsubishi PA10-7CE Robot Arm which is a seven degrees of freedom robot.

Chapter 8 reports the kinematic and dynamic models of the 6-3-PUS-type Hexapod parallel mechanism and also covers the motion control of the Hexapod. In addition, the chapter describes the implementation of two motion tracking controllers in a real Hexapod robot.

The application of a finite time-time nonlinear proportional–integral–derivative (PID) controller to a five-bar mechanism, for set-point controller, is presented in Chap. 9. The stability analysis of the closed-loop system shows global finite-time stability of the system.

Finally, Chap. 10 deals with the tracking control problem of three degrees of freedom helicopter. The control problem is solved using nonlinear H_∞ synthesis of time-varying systems. The proposed method considers external perturbations and parametric variations.

Part III Robotics

The final part of the book is devoted to the field of robotics implemented as mechatronics systems.The applications include rehabilitation systems, challenges in cognitive robotics, and applications of haptic systems.

Chapter 11 proposes a novel ankle rehabilitation parallel robot with two degrees of freedom consisting of two linear guides. Also, a serious game and a facial expression recognition system were added for entertainment and to improve patient engagement in the rehabilitation process.

Chapter 12 explains the new challenges in the area of cognitive robotics. In addition, two low-level cognitive tasks are modeled and implemented in an artificial agent. In the first experiment an agent learns its body map, while in the second experiment the agent acquires a distance-to-obstacles concept.

Chapter 13 covers a review of applications of two novel technologies known as haptic systems and virtual environments. The applications are divided in two categories including training and assistance. For each category the fields of education, medicine and industry are addressed.

The aerodynamic analysis of a bio-inspired three degrees of freedom articulated flat empennage is presented in Chap. 14. The proposal mimics the way that the tail of some birds moves.

Finally, the problem of performing different tasks with a group of mobile robots is addressed in Chap. 15. In order to cope with issues like regulation to a point or trajectory tracking, a consensus scheme is considered. The proposal was validated by a group of three differential mobile robots.

Also, we would like to thank all our book contributors and many other participants who submitted their chapters that cannot be included in the book, we value your effort enormously. Finally, we would like to thank the effort of our chapter reviewers that helped us sustain the high quality of the book.

Chihuahua, Mexico
April 2018

Osslan Osiris Vergara Villegas
Manuel Nandayapa
Israel Soto

Contents

Part I
Computer Vision

Chapter 1
Denoising of Ultrasound Medical Images Using the DM6437 High-Performance Digital Media Processor

Gerardo Adrián Martínez Medrano, Humberto de Jesús Ochoa Domínguez and Vicente García Jiménez

Abstract Medical ultrasound images are inherently contaminated by a multiplicative noise called speckle. The noise reduces the resolution and contrast, decreasing the capability of the visual evaluation of the image, and sometimes small speckles can mask ills in early stages. Therefore, denoising plays an important role in the diagnostic. Many investigations reported in the literature claim their performance. However, this is limited because the unclear indicators or sometimes the algorithms proposed are not suitable for implementations in hardware. In this chapter, the implementation of five methods, specifically designed to reduce multiplicative noise, in a digital signal processor is presented. The chapter includes performance evaluation of each method implemented in a fixed point, DM6437 digital signal processor (digital media processor) of Texas Instruments™. Results show that the performance of the Frost and Lee filters, with a local window of 5×5 pixels, is better to reduce high-variance speckle noise than the rest of the filters. For noise variance less than 0.1, the SRAD with 15 iterations has a higher performance. However, the Frost and SRAD filters take more time to yield a result.

Keywords Denoising · Ultrasound medical images · Digital signal processor
Filtering

G. A. Martínez Medrano · H. de Jesús Ochoa Domínguez (✉)
V. García Jiménez
Universidad Autónoma de Ciudad Juárez, Instituto de Ingeniería y Tecnología, Ciudad Juárez, Chihuahua, Mexico
e-mail: hochoa@uacj.mx

G. A. Martínez Medrano
e-mail: al131542@alumnos.uacj.mx

O. O. Vergara Villegas et al. (eds.), *Advanced Topics on Computer Vision, Control and Robotics in Mechatronics*,
https://doi.org/10.1007/978-3-319-77770-2_1

1.1 Introduction

Medical ultrasound (US) is a low cost, real-time, and noninvasive technique that requires processing signals at high speed (Adamo et al. 2013). This type of imaging modality has several advantages over computed tomography (CT), positron emission tomography (PET), and magnetic resonance imaging (MRI) especially in obstetric applications where radiation or the injection of a radiotracer can be harmful to the fetus. Besides, in medical US, the patient does not have to remain still. US images are inherently contaminated with speckle noise because it is a coherent imaging system.

In the past, several methods to denoise US medical images have been proposed. However, many of them apply strategies designed for additive Gaussian noise. Before filtering, the noisy image is transformed into an additive process by taking the logarithm of the image. Therefore, by assuming that the noise is an additive Gaussian process, a Wiener filter (Portilla et al. 2001) or a wavelet shrinkage method (Pizurica et al. 2003; Rizi et al. 2011; Tian and Chen 2011; Premaratne and Premaratne 2012; Fu et al. 2015) is applied to remove the noise component. Nevertheless, in (Oliver and Quegan 2004; Goodman 2007; Huang et al. 2012), the authors study the speckle noise and indicate that the suitable distribution for this type of noise is Gamma or Rayleigh.

The denoising methods are divided in spatial filtering (Lee 1980; Frost et al. 1982; Kuan et al. 1985), transform methods (Argenti and Alparone 2002; Xie et al. 2002; Pizurica et al. 2003; Rizi et al. 2011; Tian and Chen 2011; Premaratne and Premaratne 2012), and, more recently, regularization methods for image reconstruction and restoration (Aubert and Aujol 2008; Shi and Osher 2008; Huang et al. 2009; Nie et al. 2016a, b). Regularization methods are based on partial differential equations, and the first denoising filter for multiplicative noise was the total variation (TV) proposed in (Rudin et al. 2003). However, the problem of TV regularization method is that in smooth regions produces a stair-case effect. In other words, the texture features are not restored. Hence, other regularization methods introduce an extra term to the functional named prior to work with the TV and the data fidelity terms (Nie et al. 2016b) to overcome the piecewise constant of the smooth region.

Despite the results obtained in the transformed and the variational methods, the limitation for its implementation in real time is the computational burden, since the former need to change to a transform domain and after removing the noise returning to the spatial domain. The variational methods need several iterations to converge and it is usually very complicated their implementation in fixed-point processor.

In this chapter, a comparative analysis of the performances of several filters to reduce the speckle effect, in US medical images, is presented. The filters are especially designed for multiplicative noise, operate in the spatial domain and programmed in the DM6437 digital signal processor (DSP) of Texas Instruments™ (TI) to study their performance. This processor is also known as the digital media (DM) 6437.

The chapter is organized as follows: In Sect. 1.2, a literature review is given. In Sect. 1.3, the methods used in this research and the metrics to measure the performance are explained. In Sect. 1.4, the experimental results are presented. The chapter concludes in Sect. 1.5.

1.2 Literature Review

The aim of this section is to provide a brief and useful description of the hardware and techniques implemented in DSP for different image processing applications.

In Xuange et al. (2009), authors propose an online hydrological sediment detection system based on image processing. This consists on an image collection subsystem, an image transport subsystem, a network transmission subsystem, and an ARM-based processing subsystem based on PXA270 processor. The system acquires the image of mountain rivers using an online technique and performs the hydrological analysis of sediment. The denoising algorithm uses wavelet transformation. However, the overall performance is not reported.

In Bronstein (2011), the design of bilateral filter for noise removal is carried out for a parallel single instruction, multiple data (SIMD)-type architecture using a sliding window. For each pixel, in raster order, neighbor pixels within a window around it are taken and used to compute the filter output; the window is moved right by one pixel and so on. This implementation is optimized for windows sizes between 10 and 20 to keep low the complexity. However, it approximates the performance to the bilateral filter in terms of root mean square error (RMSE), and the proposed implementation can operate at real time.

In Lin et al. (2011), authors propose a novel restoration algorithm based on super-resolution concept using the wavelets decomposition implemented on the OMAP3530 platform performing the effectiveness of the images restoration. The architecture utilized is designed to provide good quality video, image, and graphics processing. To verify the execution time of the algorithm, they use four different methods: the Cortex-A 8 only implementation, the Cortex-A 8 + NEON implementation, the DSP only implementation, and the dual-core implementation. Method 2 shows the best performance. Method 3 or 4 did not have the best performance because the proposed algorithm involves heavy floating-point computation which is not supported by the fixed-point C64x + DSP. For the well-known Lena, Baboon, Barbara and Peppers images of size 256 $\times$ 256 report an execution time from 1.41 to 2.5 s with PSNRs of 32.78, 24.49, 25.36 and 31.43 dBs respectively, using a dual-core implementation, outperforming the bilinear and bicubic algorithms.

In Zoican (2011), the author develops an algorithm that reduces impulsive noise in still images that allows to reduce more than 90% of the noise. The algorithm presented is a median filter modification. The median filter is typically applied uniformly across the image. To avoid this and reduce the noise, the author uses the modified median filter, where impulse detection algorithm is used before filtering to control the pixel to be modified. The algorithm is non-parametric comparing with the progressive median algorithm that must be predetermined with four parameters. The performance of the new algorithm is evaluated by measurement of mean square error (MSE) and peak signal-to-noise ratio (PNSR). The results show the efficiency of the new algorithm comparing with median progressive algorithm while computational burden is similar. However, the proposal is for small images using the BF5xx (Analog Devices Inc.™) DSP family.

In Akdeniz and Tora (2012), authors present a study of the balanced contrast limited adaptive histogram equalization (BCLAHE) implementation for infrared images on an embedded platform to achieve a real-time performance for a target that uses a dual processor OMAP3530. The debug access port (DAP) and the advanced risk machine (ARM) are optimized to obtain a significant speed increase. The performance analysis is done over infrared images with different dynamic range. The performance reached a real-time processing at 28 FPS with 16-bit images.

In Dallai and Ricci (2014), the authors present a real-time implementation for a bilateral filter for the TMS320DM64x + DSPs. Real-time capability was achieved through code optimization and exploitation of the DSP architecture. The filter, tested on the ULA-OP scanner, processes images from 192 × 512 to 40 FPS. The images are obtained from a phantom and in vivo.

In Zhuang (2014), the author develops a system to enhance images using the dual-core TI DaVinci DM6467T with MontaVista Linux operating system running on the ARM subsystem to handle the I/O and the result of the DSP. The results show that the system provides equivalent capabilities to a X86 computer processing 25 FPS on D1 resolution (704 × 480 pixels).

Finally, in Fan et al. (2016), authors focus on the optimized implementation of the linear line detection system based on multiple image pre-processing methods and an efficient Hough transformation. To evaluate the performance of the real-time algorithm, the DSP TMS320C6678 was used. Lane detection takes up only a small portion of the processing time and should be implemented with a much higher performance than 25 frames per second (FPS) to make room for the rest of the system. The linear detection algorithm presented in this paper is faster-than-real-time, which achieves a high-speed performance with over 81 fps on a multicore DSP. They used C language to program the linear lane detection algorithm to achieve compatibility across multiple platforms especially for DSP to yield a much faster performance than real time. The processor has eight cores, and each core can run at 1.25 GHz. To develop a faster-than-real-time algorithm, they use optimize the DSP, such as restricted search area, an efficient Hough transform, and a better memory allocation. Also, with the purpose of reducing the Hough transformation accumulated noise and decreasing the processing time, Gaussian blur, edge thinning, and edge elimination are used.

1.3 Methods

This section introduces the methods used throughout this research, including the US image formation, the model of the image with speckle noise and the classic filtering strategies to remove it. Also, a brief description of the DSP as well as the metrics used is included.

1.3.1 Ultrasound Image Formation

US is described by some wave parameters such as pressure density, propagation, and particle displacement. It is a sound wave that transports energy and propagates through several means as a pulsating pressure wave with a frequency above 20 kHz (Suetens 2002). The modalities or formats of US are described following.

1.3.1.1 B-Mode

The B-mode or brightness-mode is currently the most used in US medical imaging. The B-mode image is produced by a small transducer array, arranged in a straight line. The most common are the linear, the convex, and the sector transducer. The image is builtup line by line as the beam progresses along the transducer array as shown in Fig. 1.1. The transducer is placed on the patient skin and sends a pulse traveling along a beam into the tissue. The reflected echo is amplified to form signal that is coherently summed to form the 2-D image.

Typically, the complete image is formed in a 1/30th second with negligible delay. Observe that in a linear transducer, the lines are perpendicular to the line of transducer elements. This allows to image superficial structures; in other words, it has a rectangular beam shape.

1.3.1.2 Speckle Noise

In US scans, the speckle noise is granular structures formed by the superposition of acoustical echoes with random phases and amplitudes from structures smaller than the spatial resolution of the medical US system (Wagner et al. 1983). Speckle is an inherent property of medical US imaging, and it generally tends to reduce the image resolution and contrast as well as blur image details, thereby reducing the diagnostic value of this imaging modality. Image processing methods for reducing the speckle noise (despeckling) have proven useful for enhancing the image quality and increasing the diagnostic potential of medical ultrasound (Abbott and Thurstone 1979; Ozcan et al. 2007; Ovireddy and Muthusamy 2014; Koundal et al. 2015;

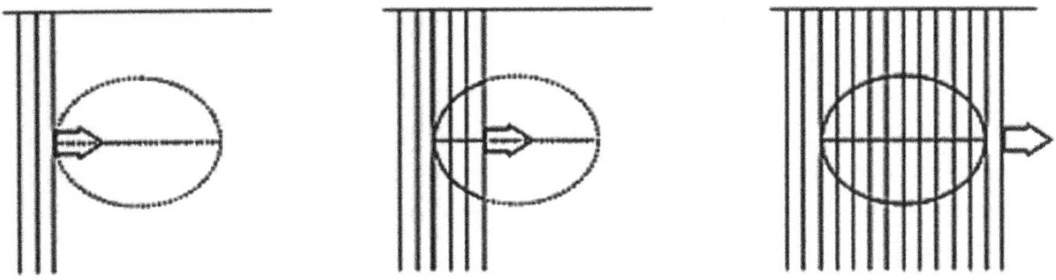

Fig. 1.1 Formation of a 2D B-mode image

Kang et al. 2016; Wen et al. 2016; Li et al. 2017; Singh et al. 2017). The multiplicative noise can be expressed as

$$g = f\,n + v \tag{1.1}$$

where g and f are the noisy and the noise-free images, respectively, $n(m, n)$ and v are the amount of multiplicative an additive noise component in the image. The effect of additive noise is considered smaller compared with that of multiplicative noise (coherent interface) $\|v\|^2 \ll \|n\|^2$ then Eq. (1.1) becomes,

$$g \approx f\,n \tag{1.2}$$

1.3.2 Despeckling Filters

Noise reduction without blurring the edges is a speckle noise reduction problem in US images. Speckle suppression in ultrasound images is usually done by techniques that are applied directly to the original image domain like median (Maini and Aggarwal 2009), Lee (1980), Frost et al. (1982), and Kuan et al. (1985) filters that achieve very good speckle reduction in homogeneous areas and ignore the speckle noise in areas close to edges and lines. Perona and Malik (1990) developed a method called anisotropic diffusion based on heat equation. It works well in homogenous areas with edge preservation for an image corrupted by additive noise, but the performance is poor for the speckle noise, which is a multiplicative noise; then, Yu and Acton (2002) introduced a method called speckle reduction anisotropic diffusion (SRAD). In this method, diffusion coefficient which defines the amount of smoothing is based on ratio of local standard deviation to mean and these are calculated using nearest neighbor window and it smoothens the edges and structural content in images. Median, Lee, Kuan, Frost, and SRAD filters were programmed in the DSP.

1.3.2.1 Median Filter

The median filter is a nonlinear image processing technique used to reduce impulsive noise from images and has the particularity of preserving the edges of the image. Hence, it produces a less blurred image. This spatial filtering operation applies a two-dimensional window mask to an image region and replaces its original center pixel value with the median intensity of the pixels contained within the window. The window is a sliding window that moves to the next image region, and the cycle is repeated until the entire image is processed. Hence, the median filter preserves the edges and reduces the blur in images. If the window length is $2k + 1$, the filtering is given by Eq. (1.3),

$$\hat{f}_n = \text{med}[g_{n-k}, \ldots, g_n, \ldots, g_{n+k}], \tag{1.3}$$

where med[•] is the median operator. To find the median value, it is necessary to sort all the intensities in a neighborhood into a numerical ascendant order. This is a computationally complex process due to the time needed to sort pixels to find the median value of the window.

1.3.2.2 Lee Filter

The Lee filter is popular in the image processing community for despeckling and enhancing SAR images. The Lee filter and other similar sigma filters reduce multiplicative noise while preserving image sharpness and details. It uses a sliding window that calculates a value with the neighbor pixels of the central window pixel and replaces it with the calculated value. Calculate the variance of the window and if the variance is low, smoothing will be performed. On the other hand, if the variance is high, assuming an edge, the smoothing will not be performed. Therefore,

$$\hat{f}_n = \bar{g} + \frac{\sigma_f^2 (g_n - \bar{g})}{\sigma_f^2 (\text{ENL} + 1)\bar{g}^2} \text{ENL}, \tag{1.4}$$

Equation (1.4) can be simplified as:

$$\hat{f}_n = \bar{g} + k(g_n - \bar{g}), \tag{1.5}$$

The Lee filter is a case of the Kuan filter (Kuan et al. 1985) without the term σ_f^2/ENL.

1.3.2.3 Kuan Filter

The Kuan filter (Kuan et al. 1985) is an adaptive noise smoothing filter that has a simple structure and does not require any information of the image. The filter considers the multiplicative noise model of Eq. (1.2) as an additive model of the form:

$$g = f + (n - 1)f, \tag{1.6}$$

Assuming unit mean noise, the estimated pixel value in the local window is:

$$\hat{f}_n = \bar{g} + \frac{\sigma_f^2(g_n - \bar{g})}{\sigma_f^2(\mathrm{ENL}+1)\bar{g}^2}\mathrm{ENL}, \tag{1.7}$$

with

$$\sigma_f^2 = \frac{\mathrm{ENL}\sigma_g^2 - \bar{g}^2}{\mathrm{ENL}+1}, \tag{1.8}$$

and

$$\mathrm{ENL} = \left(\frac{\mathrm{Mean}}{\mathrm{StDev}}\right)^2 = \left(\frac{\bar{g}}{\sigma_g^2}\right)^2, \tag{1.9}$$

The equivalent number of looks (ENL) estimates the noise level and is calculated in a uniform region of the image. One shortcoming of this filter is that the ENL parameter needs to be computed beforehand.

1.3.2.4 Frost Filter

The Frost filter (Frost et al. 1982) is an adaptive as well as exponential, based on weighted middling filter, that reduces the multiplicative noise while preserving edges. It works with a window that is $2k+1$ size replacing the central pixel with the sum of weighted exponential terms. The weighting factors depend on the distance to the central pixel, the damping factor, and the local variance. The more far the pixel from the central pixel the less the weight. Also, the weighting factors increase as variance in the window increases. The filter convolves the pixel values within the window with the exponential impulse response:

$$h_i = \mathrm{e}^{-Ka_g(i_0)|i|}, \tag{1.10}$$

where K is the filter parameter, i_0 is the window central pixel, and $|i|$ is the distance measured from the window central pixel. The coefficient of variation is defined as $a_g = \sigma_g/\bar{g}$, were $\bar{g}$ and σ_g are the local mean and standard deviation of the window, respectively.

1.3.2.5 SRAD Filter

SRAD is called speckle reducing anisotropic diffusion filter (Yu and Acton 2002), and it is obtained by rearranging Eq. (1.6) as:

$$\hat{f} = g + (1-k)(\bar{g} - g), \tag{1.11}$$

The term $(\bar{g} - g)$ approximate to the Laplacian operator (with $c = 1$) and then can be expressed as:

$$\hat{f} = g + k' \mathrm{div}(\Delta g). \tag{1.12}$$

This equation is an isotropic process. Hence, $\partial_t g = \mathrm{div}(c\Delta g)$ can be easily transformed into an anisotropic version by including only the c factor:

$$\partial_t g = \mathrm{div}(c\Delta g) = c\mathrm{div}(\Delta g) + \Delta c \Delta g. \tag{1.13}$$

The output image $g(x, y; t)$ is evolved according to the following partial derivative equation (PDE):

$$\begin{cases} \frac{\partial g(x,y;t)}{\partial t} = \mathrm{div}[c(q)\Delta g(x, y; t] \\ g(x, y; 0) = g(x, y) \\ \left.\frac{\partial g(x,y;t)}{\partial \vec{n}}\right|_{\partial\Omega} = 0 \end{cases}, \tag{1.14}$$

where $c(q)$ is the diffusion coefficient, and $g(x, y; t)$ is an edge detector. The last boundary condition states that the derivative of the function along the outer normal, at the image boundary, must vanish. This assures that the average brightness will be preserved.

1.3.3 Description of the TMS320DM6437 Digital Media Processor

1.3.3.1 DSP Core Description

The C64x + DSP core contains eight functional units (.M1, .L1, .D1, .S1, .M2, .L2, .D2, and .S2); each one can execute one instruction every clock cycle. The .M functional units perform multiply operations.

.M units can perform one of the following each clock cycle: one 32 × 32 bit multiply, one 16 × 16 bit multiply, two 16 × 16 bit multiplies, two 16 × 16 bit multiplies with add/subtract capabilities, four 8 × 8 bit multiplies with add operations, and four 16 × 16 multiplies with add/subtract capabilities also supports complex multiply (CMPY) instructions that take for 16-bit inputs and produces a 32-bit packed output that contains 16-bit real a 16-bit imaginary values. The 32 × 32 bit multiply instructions provide the extended precision necessary for audio and other high-precision algorithms on a variety of signed and unsigned 32-bit data types.

The .S and .L units perform a general set of arithmetic, logical, and branch functions. The .D units primarily load data from memory to the register file and store results from the register file into memory, also, two register files, and two data paths. There are two general-purpose register files (A and B), and each contains 32-bit registers for a total of 64 registers.

The .L or arithmetic logic units have the ability to do parallel add/subtract operations on a pair of common inputs. Versions of this instructions exist to work on 32-bit data or on pairs of 16-bit data performing dual 16-bit add–subtracts in parallel.

1.3.3.2 Evaluation Module

The DM6437 evaluation module (EVM) is a platform that allows to evaluate and develop applications for the TI DaVinci processors family. The EVM board includes a TI DM6437 processor operating up to 600 megahertz (MHz), one video decoder, supports composite or S-video, four video digital-to-analog converter (DAC) outputs—component, red, green, blue (RGB) composite, 128 megabytes (MB) of double data rate synchronous dynamic random-access memory (DDR2 DRAM), one universal asynchronous receiver-transmitter (UART) and a programmable input/output device for controller area network (CAN I/O), 16 MB of non-volatile flash memory, 64 MB of flash memory based on nand gates (NAND flash), 2 MB of static random-access memory (SRAM), a low power stereo codec (AIC33), inter-integrated circuit interface (I2C) with onboard electrically erasable programmable read-only memory (EEPROM) and expanders, 10/100 megabit per second (MBPS) Ethernet interface, configurable boot load options, embedded emulation interface known as joint test action group (JTAG), four user light emitting diodes (LEDs) and four position user switches, single voltage power supply (5 volts), expansion connectors for daughter card use, a full-duplex serial bus to perform transmit and receive operations separately for connecting to one or more external physical devices which are mapped to local physical address space and appear as if they are on the internal bus of the DM6437 processor, and one Sony/Philips digital interface format (S/PDIF) to transmit digital audio.

The EVM is designed to work with Code Composer Studio. Code Composer communicates with the board through the embedded emulator or an external JTAG emulator. Figure 1.2 shows the block diagram of the EVM.

The US images were loaded into the memory of the EVM using the JTAG emulator; after finishing the process, a copy of the clean image was sent to the computer and to the video port to be displayed in a monitor.

1.3.3.3 Memory Map

Figure 1.3 shows the memory of the address space of a DM6437, portions of memory can be remapped in software, the total amount of memory for data, program code, and video is 128 megabytes. In this work, the US images were allocated in DDR memory (unsigned char *) 0×80000000. This memory has a dedicated 32-bits bus.

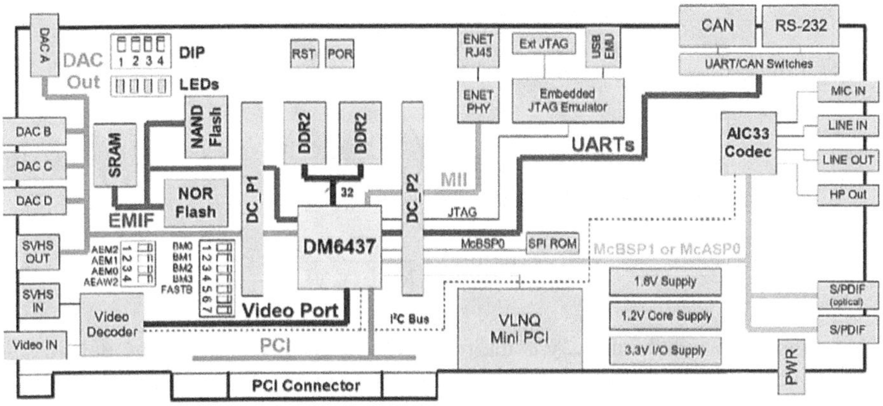

Fig. 1.2 Block Diagram of the EVM DM6437 (Texas Instruments 2006)

Address	DM6437 EVM
0x10800000	Cache/RAM
0x42000000	CS2
0x44000000	CS3
0x46000000	CS4
0x48000000	CS5
0x4C000000	VLNQ
0x80000000	DDR

Fig. 1.3 Memory map (Texas Instruments 2006)

For processing purposes, ten memory regions were allocated in DDR memory to store the results of filtered images. Before sending to the display, the images are reshaped to 480 × 720.

1.3.4 Metrics

In order to compare the restoration quantitatively, we use eight error measures including the MSE of Eq. (1.15), PSNR of Eq. (1.16), the signal-to-noise ratio (SNR) of Eq. (1.17), the structural similarity index (SSIM) (Wang et al. 2004) of Eq. (1.18), mean structural similarity index (MSSIM) of Eq. (1.19), the contrast to background contrast (CBC) of Eq. (1.20), perceptual sharpness index (PSI) (Blanchet and Moisan 2012), and Pratt figure of merit (FOM) (Pratt 2001), which are widely used in the image processing literature.

$$\mathrm{MSE} = \frac{1}{MN}\sum_{i=0}^{M-1}\sum_{j=0}^{N-1}\|x(i,j) - y(i,j)\|^2, \tag{1.15}$$

$$\mathrm{PSNR} = 10\log_{10}\left(\frac{v_{\max}^2}{\mathrm{MSE}(x,y)}\right), \tag{1.16}$$

$$\mathrm{SNR} = 10\log_{10}\left(\frac{\sigma_y^2}{\mathrm{MSE}(x,y)}\right), \tag{1.17}$$

where x is the original image, y is the recovered image after denoising, and $v_{\max}$ is the maximum possible value in the range of the signals. The SSIM factor (Wang et al. 2004) is calculated as,

$$\mathrm{SSIM}(x,y) = \frac{(2\mu_x\mu_y + c_1)(2\sigma_{xy} + c_2)}{\left(\mu_x^2 + \mu_y^2 + c_1\right)\left(\sigma_x^2 + \sigma_y^2 + c_2\right)}, \tag{1.18}$$

where μ_x and μ_y are the mean value of x and y, σ_x^2, σ_y^2 and σ_{xy} are the variance and covariance of x and y; and c_1 and c_2 are constants terms. Another metric derived from the SSIM is the MSSIM of Eq. (1.19),

$$\mathrm{MSSIM} = \frac{1}{M}\sum_{j=1}^{M}\mathrm{SSIM}(x_j, y_j), \tag{1.19}$$

where M is the number of the areas being compared.

$$CB = \frac{|\mu_{\mathrm{FR}} - \mu_{\mathrm{BR}}|}{|\mu_{\mathrm{FP}} + \mu_{\mathrm{BP}}|}, \tag{1.20}$$

where μ_{FR} and μ_{BR} are the mean value of the foreground and the background of the recovered image, and μ_{FP} and μ_{BP} are the mean value of the foreground and the background of the synthetic image. Both obtained on homogeneous regions.

1.4 Results

In this section, performance evaluation of the filters on synthetic data and on real data is obtained. The phantom of a fetus (Center for Fast Ultrasound Imaging 2017) was contaminated with speckle noise with different variances and uploaded to the memory of the board. Then, a filtering process is applied to the noisy image. The resulting clean image is sent back to the computer. Different metrics were calculated using the clean phantom as a reference.

Fig. 1.4 System configuration: code composer software, interface to the DM6437 EVM and the display unit

Figure 1.4 shows the system configuration to process the US images. The code composer is used to program the processor and to upload the image to the DDR memory. The interface connects the computer to module, and the module sends the image to a display and to the computer for visualization and performance evaluation purposes, respectively. In the next sections, the results or synthetic and real data are presented.

1.4.1 Experiments on Synthetic Data

The synthetic images (Center for Fast Ultrasound Imaging 2017) and speckle noise model are considered for the experiments, and different metrics to evaluate the noise are used to compare objectively several methods. Figure 1.5 shows the original image and the affected images with different speckle values.

In this experiment, the synthetic image of Fig. 1.5 (Center for Fast Ultrasound Imaging 2017) was corrupted with different levels of noise. The synthetic image (phantom) was modified according to the national television system committee (NTSC) standard to 8-bit image of 480 × 720 pixels for display purposes. The speckle noise process, applied to the synthetic image, follows the model of Eq. (1.2). Seven different levels of noise variance were tested by setting $\sigma = \{0.02, 0.05, 0.1, 0.15, 0.2, 0.25, 0.3\}$. To assess denoising methods, the previous metrics defined in Sect. 1.3.4 were computed between the synthetic and the reconstructed image. Quantitative results are shown in Tables 1.1, 1.2, 1.3, 1.4, 1.5, 1.6, and 1.7.

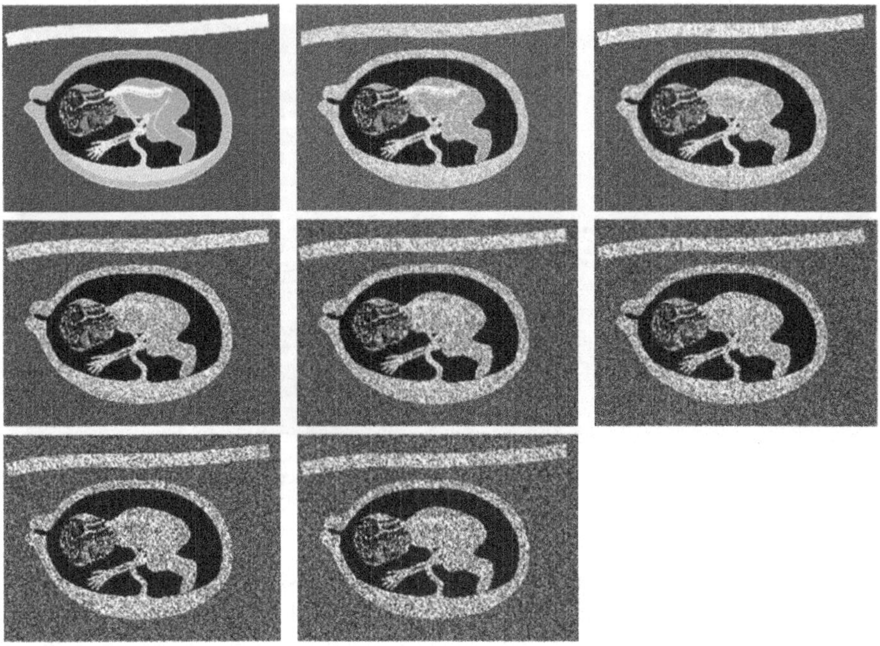

Fig. 1.5 Synthetic images, from left to right. First row shows the original image and the images contaminated with a speckle noise variance of 0.02 and 0.05, respectively. Second row shows the original image contaminated with a speckle noise variance of 0.1, 0.15, and 0.2, respectively, and the third row shows the original image contaminated with a speckle noise variance of 0.25 and 0.3, respectively

Table 1.1 Results with filters applied to the affected image with 0.02 of noise variance

Filter	MSE	PSNR	SNR	CB	SSIM	MSSIM	SI	FOM
Speckle 0.02	489.099	21.236	10.756	1.116	0.3621	0.4766	296.593	0.7797
Median 3 × 3	647.448	20.018	9.538	1.119	0.3192	0.5623	626.425	0.7257
Median 5 × 5	859.717	18.787	8.307	1.119	0.2751	0.6035	569.047	0.7166
Median 7 × 7	1082.408	17.786	7.306	1.118	0.2503	0.6104	556.798	0.6665
Lee 3 × 3	845.326	18.860	8.380	1.112	0.2755	0.5747	111.047	0.8868
Lee 5 × 5	1300.727	16.988	6.508	1.016	0.1855	0.4729	19.122	0.7730
Lee 7 × 7	1391.133	16.697	6.217	1.107	0.1384	0.5112	18.385	0.6531
Kuan (15 it.)	2106.461	14.895	4.415	1.065	0.1319	0.5213	363.414	0.5547
Frost 5 × 5	812.249	19.033	8.553	1.109	0.2552	0.6142	147.586	0.8072
SRAD (15 it.)	287.326	23.547	13.067	1.113	0.3830	0.7129	1205.296	0.8352

Table 1.2 Results with filters applied to the affected image with 0.2 of noise variance

Filter	MSE	PSNR	SNR	CB	SSIM	MSSIM	SI	FOM
Speckle 0.2	2622.640	13.943	3.463	0.982	0.2960	0.3126	29.194	0.3072
Median 3 × 3	1428.633	16.58	6.101	0.976	0.2212	0.3293	145.724	0.3524
Median 5 × 5	1410.025	16.638	6.158	0.976	0.1843	0.3864	124.430	0.4327
Median 7 × 7	1578.258	16.149	5.669	0.981	0.1504	0.4269	89.081	0.5420
Lee 3 × 3	1410.097	16.638	6.158	0.985	0.2230	0.3616	32.443	0.4055
Lee 5 × 5	1418.131	16.613	6.133	0.978	0.1724	0.4465	16.415	0.7803
Lee 7 × 7	1627.629	16.015	5.535	0.976	0.1206	0.4561	9.122	0.6280
Kuan (15 it.)	2357.738	14.405	3.925	0.952	0.0923	0.4601	6.631	0.5881
Frost 5 × 5	1082.135	17.787	7.307	0.978	0.2840	0.5181	97.837	0.5566
SRAD (15 it.)	1949.973	15.230	4.750	0.993	0.2809	0.3371	68.303	0.3434

Table 1.3 Results with filters applied to the affected image with 0.05 noise variance

Filter	MSE	PSNR	SNR	CB	SSIM	MSSIM	SI	FOM
Speckle 0.05	721.608	19.547	9.067	0.9979	0.4229	0.4590	126.939	0.6122
Median 3 × 3	686.310	19.765	9.285	0.9904	0.2997	0.4897	371.883	0.7023
Median 5 × 5	858.621	18.792	8.312	0.9938	0.2629	0.5532	321.724	0.7218
Median 7 × 7	1072.914	17.825	7.345	0.9950	0.2259	0.5690	297.163	0.6402
Lee 3 × 3	819.340	18.996	8.516	0.9953	0.2804	0.5266	75.003	0.7893
Lee 5 × 5	1300.727	16.988	6.508	1.0161	0.1855	0.4729	19.122	0.7730
Lee 7 × 7	1331.404	16.887	6.407	0.9935	0.1359	0.5022	14.161	0.6541
Kuan (15 it.)	1685.911	15.862	5.382	0.9014	0.1148	0.4746	7.368	0.6057
Frost 5 × 5	758.378	19.331	8.851	0.9948	0.3025	0.6060	118.401	0.8178
SRAD (15 it.)	374.178	22.400	11.920	1.003	0.4720	0.6642	712.479	0.7405

Table 1.4 Results with filters applied to the affected image with 0.1 of noise variance

Filter	MSE	PSNR	SNR	CB	SSIM	MSSIM	SI	FOM
Speckle 0.1	1358.284	16.800	6.320	1.003	0.3552	0.3778	59.144	0.3888
Median 3 × 3	953.5335	18.337	7.857	0.9982	0.2647	0.4089	245.035	0.5166
Median 5 × 5	1071.67	17.830	7.350	1.003	0.2252	0.4734	198.070	0.6694
Median 7 × 7	1280.756	17.056	6.576	1.003	0.1825	0.4992	171.538	0.5598
Lee 3 × 3	1104.697	17.698	7.218	1.018	0.2440	0.4153	44.087	0.4482
Lee 5 × 5	1300.727	16.988	6.508	1.016	0.1855	0.4729	19.122	0.7730
Lee 7 × 7	1426.965	16.586	6.106	0.9962	0.1294	0.4840	11.755	0.6529
Kuan (15 it.)	1872.912	15.405	4.925	0.9081	0.1075	0.4694	7.120	0.5955
Frost 5 × 5	864.236	18.764	8.284	0.9980	0.2974	0.5723	107.896	0.8182
SRAD (15 it.)	880.621	18.682	8.202	1.009	0.3574	0.4489	242.935	0.5477

Table 1.5 Results with filters applied to the affected image with 0.15 of noise variance

Filter	MSE	PSNR	SNR	CB	SSIM	MSSIM	SI	FOM
Speckle 0.15	1998.636	15.123	4.643	1.024	0.3212	0.3392	37.895	0.3967
Median 3 × 3	1214.233	17.287	6.807	1.018	0.2381	0.3574	182.118	0.3902
Median 5 × 5	1256.460	17.139	6.659	1.021	0.2033	0.4203	158.164	0.5545
Median 7 × 7	1434.514	16.563	6.083	1.020	0.1645	0.4565	125.176	0.5472
Lee 3 × 3	1104.697	17.698	7.218	1.018	0.2440	0.4153	44.087	0.4482
Lee 5 × 5	1300.727	16.988	6.508	1.016	0.1855	0.4729	19.122	0.7730
Lee 7 × 7	1518.448	16.316	5.836	1.013	0.1267	0.4700	10.422	0.6528
Kuan (15 it.)	2123.588	14.860	4.380	0.9359	0.0993	0.4647	7.133	0.5973
Frost 5 × 5	970.406	18.261	7.781	1.016	0.2914	0.5424	106.232	0.7643
SRAD (15 it.)	1417.497	16.615	6.135	1.032	0.3100	0.3762	114.514	0.3921

Table 1.6 Results with filters applied to the affected image with 0.25 of noise variance

Filter	MSE	PSNR	SNR	CB	SSIM	MSSIM	SI	FOM
Speckle 0.25	3250.787	13.010	2.530	0.9861	0.2771	0.2935	22.871	0.2969
Median 3 × 3	1671.314	15.900	5.420	0.9806	0.2047	0.3030	137.273	0.3312
Median 5 × 5	1579.811	16.148	5.664	0.9878	0.1727	0.3588	129.235	0.4253
Median 7 × 7	1729.038	15.752	5.272	0.9898	0.1445	0.4069	92.049	0.5062
Lee 3 × 3	1410.097	16.638	6.158	0.9859	0.2230	0.3616	32.443	0.4055
Lee 5 × 5	1300.727	16.988	6.508	1.016	0.1855	0.4729	19.122	0.7730
Lee 7 × 7	1743.332	15.717	5.237	0.9833	0.1250	0.4475	8.2734	0.6560
Kuan (15 it.)	2624.992	13.939	3.459	0.9752	0.0854	0.4560	6.736	0.5838
Frost 5 × 5	1208.513	17.308	6.828	0.9830	0.2776	0.4937	99.0494	0.5461
SRAD (15 it.)	2474.478	14.195	3.715	0.9995	0.2602	0.3124	49.905	0.3159

Table 1.7 Results with filters applied to the affected image with 0.3 of noise variance

Filter	MSE	PSNR	SNR	CB	SSIM	MSSIM	SI	FOM
Speckle 0.3	3871.467	12.252	1.772	0.9866	0.2607	0.2780	22.102	0.2825
Median 3 × 3	1911.848	15.316	4.836	0.9774	0.1944	0.2877	128.782	0.2993
Median 5 × 5	1731.528	15.746	5.266	0.9775	0.1633	0.3392	116.175	0.3503
Median 7 × 7	1881.839	15.384	4.90	0.9749	0.1292	0.3824	76.382	0.4697
Lee 3 × 3	1104.697	17.698	7.218	1.018	0.2440	0.4153	44.087	0.4482
Lee 5 × 5	5605.530	10.644	0.1646	0.9925	0.0782	0.1372	22.024	0.2347
Lee 7 × 7	6554.056	9.965	0.5142	0.9896	0.0392	0.1009	21.828	0.2022
Kuan (15 it.)	2902.858	13.502	3.022	0.9885	0.0772	0.4484	6.597	0.5759
Frost 5 × 5	1335.729	16.873	6.393	0.9848	0.2722	0.4705	96.415	0.4828
SRAD (15 it.)	2985.113	13.381	2.90	1.001	0.2454	0.2943	42.202	0.2985

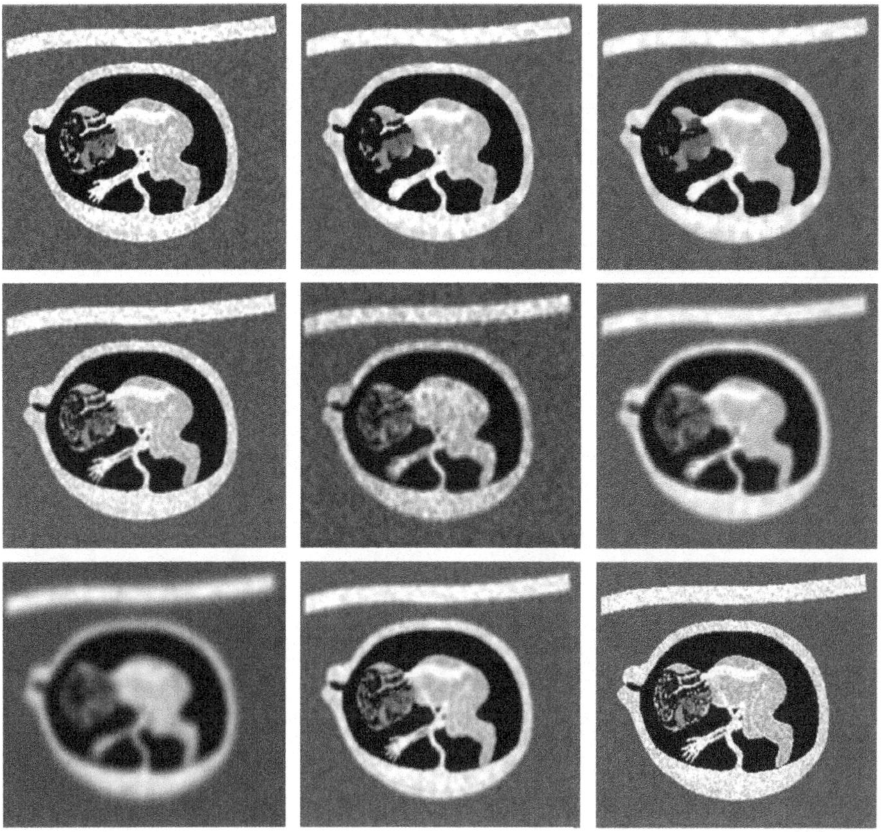

Fig. 1.6 Synthetic images after filtering process to remove a noise variance of 0.02, from left to right. First row shows the filtered image using median filter with window sizes of 3 × 3, 5 × 5, and 7 × 7, respectively. Second row shows the filtered image using the Lee filter with window sizes of 3 × 3, 5 × 5, and 7 × 7, respectively. Third row shows the filtered image using the Kuan, Frost, and SRAD filters, respectively

Figure 1.6 shows the synthetic images after applying the filtering process to remove the noise. The speckle noise variance was 0.02. From left to right, the first row shows the filtered image using a median filter with window sizes of 3 × 3 (median 3 × 3), 5 × 5 (median 5 × 5) and 7 × 7 (median 7 × 7) respectively. The second row shows the filtered image using the Lee filter with window sizes of 3 × 3, 5 × 5, and 7 × 7, respectively, and the third row shows the filtered images using the Kuan, the Frost, and the SRAD (15 iterations) filters, respectively. Notice that the SRAD filter yields the best visual results.

Median filter of 7 × 7 (median 7 × 7) yields a clean image. However, the regions of the fingers are mixed, and the same happens in the image processed with the median filter of 5 × 5 (median 5 × 5). Also, Lee of 3 × 3 and SRAD filters yield a better image. The quantitative evaluation is summarized in Table 1.1. The best FOM was obtained by using the Lee 3 × 3 filter followed by the performance of the SRAD. However, SRAD yielded the best PSNR, SSIM, MSSIM, and SI.

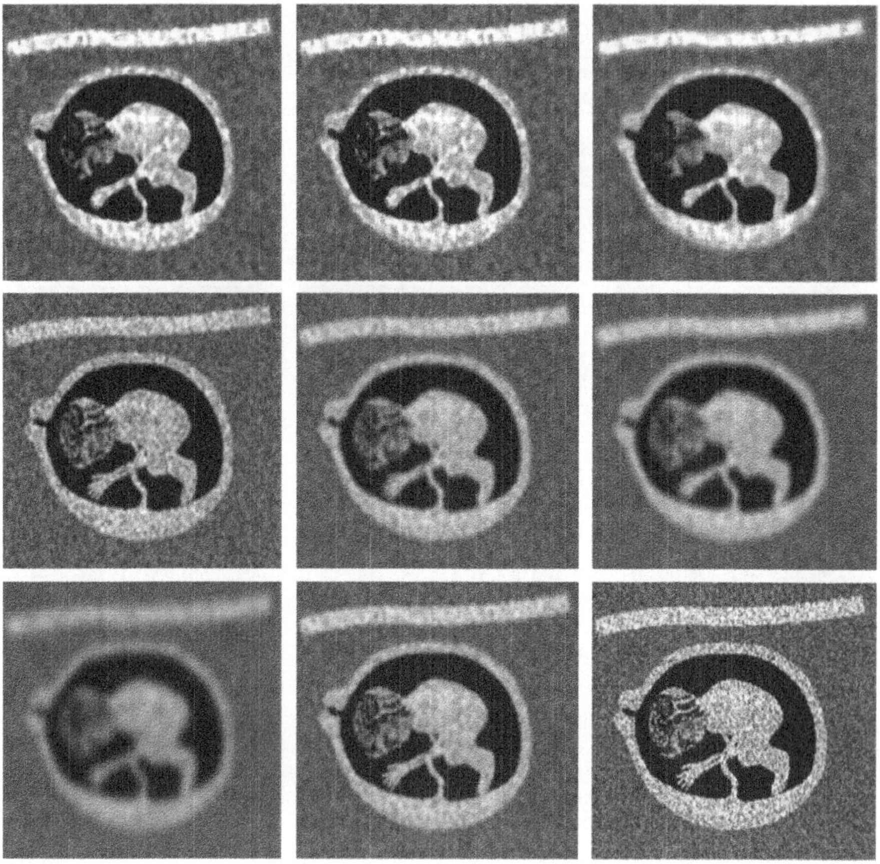

Fig. 1.7 Synthetic images after filtering process to remove a noise variance of 0.2, from left to right. First row shows the filtered image using median filter with window sizes of 3 × 3, 5 × 5, and 7 × 7, respectively. Second row shows the filtered image using the Lee filter with window sizes of 3 × 3, 5 × 5, and 7 × 7, respectively. Third row shows the filtered image using the Kuan, Frost, and SRAD filters, respectively

Figure 1.7 shows the synthetic images after applying the filtering process. The speckle noise variance was 0.2. From left to right, the first row shows the filtered image using a median filter with window sizes of 3 × 3, 5 × 5, and 7 × 7, respectively. The second row shows the filtered image using the Lee filter with window sizes of 3 × 3, 5 × 5, and 7 × 7, respectively, and the third row shows the filtered images using the Kuan, the Frost, and the SRAD (15 iterations) filters, respectively. Notice that the Lee of 3 × 3, Frost, and SRAD filters preserve most of the image details in spite of the noise.

Table 1.8 Comparison of performance in time

Algorithm	Execution time (s)
Median 3 × 3	0.81
Median 5 × 5	2.37
Median 7 × 7	4.7
Lee 3 × 3	0.62
Lee 5 × 5	1.43
Lee 7 × 7	2.61
Kuan (15 it.)	12.3
Frost 5 × 5	50.55
SRAD (15 it.)	50.43

The quantitative evaluation is summarized in Table 1.2 for Fig. 1.7. The best FOM was obtained by using the Lee 5 × 5 filter in spite of the blur image followed by the Lee 7 × 7 and Kuan after 15 iterations. However, the Frost 5 × 5 yielded the best PSNR, SNR, SSIM, MSSIM. However, the SRAD gives a better CBC because it also produces a piecewise effect in smooth areas.

Tables 1.4, 1.5, 1.6, and 1.7 show the performance of the filters for different noise powers. For example, when the noise variance is 0.1, Frost, SRAD, and median 3 × 3 filters yield the best PSNR results and Lee 3 × 3 preserves better the contrast. Frost 5 × 5 yields the best SI, MSSIM, and FOM. SRAD (15 it) yields the best SSIM and median 3 × 3 the best SI.

The PSNR and MSE values from the tables show that the filters have good noise smoothing, especially in the SRAD and Frost filters that have a higher PSNR in most of the cases. The SRAD and Lee filters yield the best results in contrast CB and FOM, meaning that these filters reduce the noise and preserve the contrast.

The results show the effectiveness of the Frost filter with the highest score of MSSIM in most of the cases. We note that the MSSIM approximates the perceived visual quality of an image better than PSNR. The median and SRAD filters reached the highest values of sharpness index that means that these filters can restore more image details.

The time reached by the filters is shown in Table 1.8, highlighting that the SRAD and Kuan filters have 15 iterations for better quality, and the rest of the algorithms is only in a single iteration.

In these results, it is shown that the algorithms can be suitable for the DM6437 DSP reaching good results in the metrics and acceptable processing times. The more performance in restoring image the more time consumed. However, classical filters are suitable for implementation using fixed-point hardware.

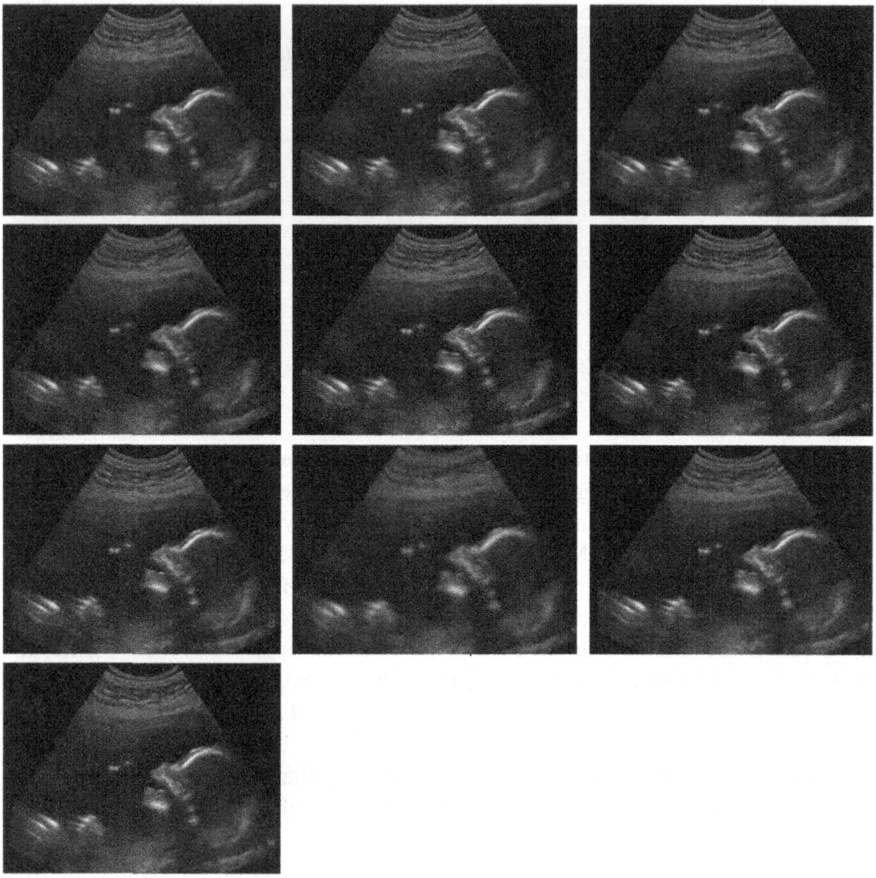

Fig. 1.8 Real images from left to right. First row, original obstetric image, image restored using a median 3 × 3 and a median 5 × 5 filter. Second row, image restored using a median 7 × 7, a Lee 3 × 3, and a Lee 5 × 5 filter. The third row, image restored using a Lee 7 × 7 a Kuan with 15 iterations and the Frost filter, and the fourth row shows the SRAD filter with 15 iterations

1.4.2 Experiments on Real Data

The algorithms show a good visual performance in the synthetic image, and now they are going to be tested in real data. For this experiment, it was used an US obstetric image. After processing, the image was adjusted to be displayed in a display unit as shown in Fig. 1.4. Figure 1.8 shows the original and the denoised images using the different algorithms implemented.

Speckle noise in US images has very complex statistical properties which depend on several factors. Experimental results show that the edge preservation of the Lee and SRAD filter is visible on the removed noise image.

The benefits of using the DM6437 processor (C6000 family) are its capabilities of instruction scheduling to ensure the full utilization of the pipeline, parallel processing, and high throughput. These proficiencies make the selected DSP suitable for computation-intensive real-time applications. The TI's C6000 core utilizes the very long instruction word (VLIW) architecture to achieve this performance and affords lower space and power footprints to implement compared to superscalar architectures. The eight functional units are highly independents and include six 32-bits and 40-bits arithmetic logic units (ALUs), and 64 general-purpose registers of 32 bits (Texas Instruments 2006). In this research, a sample was represented in Q-format as Q9.7, meaning a gap of only 0.0078125 between adjacent non-integer numbers and a maximum decimal number of 0.9921875. As it can be seen, the effect of the granular noise introduced by this quantization process is negligible. Nevertheless, the speed gain is high (about 1.67 ns per instruction cycle) (Texas Instruments 2006) compared to a floating-point processor.

1.5 Conclusions

The existence of speckle noise in US images is undesirable since it reduces the image quality by affecting edges and details between interest data that is the most interesting part for diagnostics. In this chapter, the performance of different strategies to remove speckle noise using the fixed-point, DM6437 digital signal processor was analyzed. The performance of the filters in synthetic images, with different noise variance, and images acquired with a real US-scanner were compared. Measurements of reconstruction quality and performance in time were carried out. It is noted that the median, the Lee, and the Kuan filters perform very fast. However, Frost and SRAD filters provide the best reconstruction quality even with images severely affected by noise, but their performance in time is less than the previous filters.

As future directions, we are working on a framework to include stages such as filtering, zooming, cropping, and segmentation of regions using active contours (Chan and Vese 2001).

References

Abbott, J., & Thurstone, F. (1979). Acoustic speckle: Theory and experimental analysis. *Ultrasonic Imaging, 1*(4), 303–324.

Adamo, F., Andria, G., Attivissimo, F., Lucia, A., & Spadavecchia, M. (2013). A comparative study on mother wavelet selection in ultrasound image denoising. *Measurement, 46*(8), 2447–2456.

Akdeniz, N., & Tora, H. (2012). Real time infrared image enhancement. In *Proceedings of the 20th Signal Processing and Communications Applications Conference (SIU)*, Mugla, Turkey (vol. 1, pp. 1–4).

Argenti, F., & Alparone, L. (2002). Speckle removal from SAR images in the undecimated wavelet domain. *IEEE Transactions on Geoscience and Remote Sensing, 40*(11), 2363–2374.

Aubert, G., & Aujol, J. (2008). A variational approach to removing multiplicative noise. *SIAM Journal on Applied Mathematics, 68*(4), 925–946.

Blanchet, G., & Moisan, L. (2012) An explicit sharpness index related to global phase coherence. In *Proceedings of 2012 IEEE International Conference on Acoustics, Speech and Signal Processing (ICASSP)*, Kyoto, Japan (vol. 1, pp. 1065–1068).

Bronstein, M. (2011). Lazy sliding window implementation of the bilateral filter on parallel architectures. *IEEE Transactions on Image Processing, 20*(6), 1751–1756.

Center for Fast Ultrasound Imaging. (September, 2017). Field II Simulation Program. [online] Available at: http://field-ii.dk/?examples/fetus_example/fetus_example.html.

Chan, T., & Vese, L. (2001). Active contours without edges. *IEEE Transactions on Image Processing, 10*(2), 266–277.

Dallai, A., & Ricci, S. (2014). Real-time bilateral filtering of ultrasound images through highly optimized DSP implementation. In *Proceedings of 6th European Embedded Design in Education and Research Conference (EDERC)*, Milano, Italy (vol. 1, pp. 278–281).

Fan, R., Prokhorov, V., & Dahnoun, N. (2016). Faster-than-real-time linear lane detection implementation using SoC DSP TMS320C6678. In *Proceedings of the IEEE International Conference on Imaging Systems and Techniques (IST)* (Chania, Greece, vol. 1, pp. 306–311).

Frost, V., Abbott, J., Shanmugan, K., & Holtzman, J. (1982). A model for radar images and its application to adaptive digital filtering of multiplicative noise. *IEEE Transactions on Pattern Analysis and Machine Intelligence, 4*(2), 157–166.

Fu, X., Wang, Y., Chen, L., & Dai, Y. (2015). Quantum-inspired hybrid medical ultrasound images despeckling method. *Electronic Letters, 51*(4), 321–323.

Goodman, J. (2007). *Speckle phenomena in optics: Theory and applications* (1st ed.). Englewood, Colorado, USA: Roberts and Company Publishers.

Huang, Y., Ng, M., & Wen, Y. (2009). A new total variation method for multiplicative noise removal. *SIAM Journal on Imaging Sciences, 2*(1), 20–40.

Huang, Y., Moisan, L., Ng, M., & Zeng, T. (2012). Multiplicative noise removal via a learned dictionary. *IEEE Transactions on Image Processing, 21*(11), 4534–4543.

Kang, J., Youn, J., & Yoo, Y. (2016). A new feature-enhanced speckle reduction method based on multiscale analysis for ultrasound b-mode imaging. *IEEE Transactions on Biomedical Engineering, 63*(6), 1178–1191.

Koundal, D., Gupta, S., & Singh, S. (2015). Nakagami-based total variation method for speckle reduction in thyroid ultrasound images. *Journal of Engineering in Medicine, 230,* 97–110.

Kuan, D., Sawchuk, A., Strand, T., & Chavel, P. (1985). Adaptive noise smoothing filter for images with signal-dependent noise. *IEEE Transactions on Pattern Analysis and Machine Intelligence, 7*(2), 165–177.

Lee, J. (1980). Digital image enhancement and noise filtering by use of local statistics. *IEEE Transactions on Pattern Analysis and Machine Intelligence, 2*(2), 165–168.

Li, H., Wu, J., Miao, A., Yu, P., Chen, J., & Zhang, Y. (2017). Rayleigh-maximum-likelihood bilateral filter for ultrasound image enhancement. *Biomedical Engineering Online, 16*(46), 1–22.

Lin, R., Su, B., Wu, X., & Xu, F. (2011). Image super resolution technique based on wavelet decomposition implemented on OMAP3530 platform. In *Proceedings of Third International Conference on Multimedia Information Networking and Security (MINES)* (Shanghai, China, vol. 1, pp. 69–72).

Maini, R., & Aggarwal, H. (2009). Performance evaluation of various speckle noise reduction filters on medical images. *International Journal of Recent Trends in Engineering, 2*(4), 22–25.

Nie, X., Zhang, B., Chen, Y., & Qiao, H. (2016a). A new algorithm for optimizing TV-based Pol-SAR despeckling model. *IEEE Signal Processing Letters, 23*(10), 1409–1413.

Nie, X., Qiao, H., Zhang, B., & Huang, X. (2016b). A nonlocal TV-based variational method for PolSAR data speckle reduction. *IEEE Transactions on Image Processing, 25*(6), 2620–2634.

Oliver, C., & Quegan, S. (2004). *Understanding synthetic aperture radar images* (1st ed.). Raleigh, North Carolina, USA: SciTech Publishing, Inc.

Ovireddy, S., & Muthusamy, E. (2014). Speckle suppressing anisotropic diffusion filter for medical ultrasound images. *Ultrasonic Imaging, 36*(2), 112–132.

Ozcan, A., Bielnca, A., Desjardins, A., Bouma, B., & Tearney, G. (2007). Speckle reduction in optical coherence tomography images using digital filtering. *Journal of the Optical Society of America A, 24*(7), 1901–1910.

Perona, P., & Malik, J. (1990). Scale-space and edge detection using anisotropic diffusion. *IEEE Transactions on Pattern Analysis and Machine Intelligence, 12*(7), 629–639.

Pizurica, A., Philips, W., Lemahieu, I., & Acheroy, M. (2003). A versatile wavelet domain noise filtration technique for medical imaging. *IEEE Transactions on Medical Imaging, 22*(3), 323–331.

Portilla, J., Strela, V., Wainwright, M., & Simoncelli, E. (2001). Adaptive Wiener denoising using a Gaussian scale mixture model in the wavelet domain. In *Proceedings of the International Conference on Image Processing (ICIP)* (Thessaloniki, Greece, vol. 2, pp. 37–40).

Pratt, W. (2001). *Digital image processing* (4th ed.). Hoboken, New Jersey, USA: Wiley.

Premaratne, P., & Premaratne, M. (2012). Image similarity index based on moment invariants of approximation level of discrete wavelet transform. *Electronic Letters, 48*(23), 465–1467.

Rizi, F., Noubari, H., & Setarehdan, S. (2011). Wavelet-based ultrasound image de-noising: Performance analysis and comparison. In *Proceedings of the 2011 Annual International Conference of the IEEE Engineering in Medicine and Biology Society (EMBC)* (Boston, Massachusetts, USA, vol. 1, pp. 3917–3920).

Rudin, L., Lions, P., & Osher, S. (2003). Multiplicative denoising and deblurring: Theory and algorithms. In Geometric Level Set Methods in Imaging, Vision, and Graphics. New York, USA: Springer.

Shi, J., & Osher, S. (2008). A nonlinear inverse scale space method for a convex multiplicative noise model. *SIAM Journal on Imaging Sciences, 1*(3), 294–321.

Singh, K., Ranade, S., & Singh, C. (2017). A hybrid algorithm for speckle noise reduction of ultrasound images. *Computer Methods and Programs in Biomedicine, 148*, 55–69.

Suetens, P. (2002). *Fundamentals of medical Imaging* (2nd ed.). Cambridge, United Kingdom: Cambridge University Press.

Texas Instruments. (2006). TMS320DM6437 Digital Media Processor, SPRS345D, Rev. D.

Tian, J., & Chen, L. (2011). Image despeckling using a non-parametric statistical model of wavelet coefficients. *Biomedical Signal Processing and Control, 6*(4), 432–437.

Wagner, R., Smith, S., Sandrik, J., & Lopez, H. (1983). Statistics of speckle in ultrasound B-Scans. *IEEE Transactions on Sonics and Ultrasonics, 30*(3), 156–163.

Wang, Z., Bovik, A., Sheikh, H., & Simoncelli, E. (2004). Image quality assessment: from error visibility to structural similarity. *IEEE Transactions on Image Processing, 13*(4), 600–612.

Wen, T., Gu, J., Li, L., Qin, W., & Xie, Y. (2016). Nonlocal total-variation-based speckle filtering for ultrasound images. *Ultrasonic Imaging, 38*(4), 254–275.

Xie, H., Pierce, L., & Ulaby, L. (2002). SAR speckle reduction using wavelet denoising and Markov random field modeling. *IEEE Transactions on Geoscience and Remote Sensing, 40*(10), 2196–2212.

Xuange, P., Ming, L., Bing, Z., Chunying, H., & Xuyan, Z. (2009). The online hydrological sediment detection system based on image process. In *Proceedings of 4th IEEE Conference on Industrial Electronics and Applications (ICIEA)* (Xi'an, China, vol. 1, pp. 3761–3764).

Yu, Y., & Acton, S. (2002). Speckle reducing anisotropic diffusion. *IEEE Transactions on Image Processing, 11*(11), 1260–1270.

Zhuang, L. (2014). Realization of a single image haze removal system based on DaVinci DM6467T processor. In *Proceedings of SPIE 9273, Optoelectronic Imaging and Multimedia Technology III* (Beijing, China, vol. 9273, pp. 1–7).

Zoican, S. (2011). Adaptive algorithm for impulse noise suppression from still images and its real time implementation. In *Proceedings of 10th International Conference on Telecommunication in Modern Satellite Cable and Broadcasting Services (TELSIKS)* (Nis, Serbia, vol. 1, pp. 337–340).

Chapter 2
Morphological Neural Networks with Dendritic Processing for Pattern Classification

Humberto Sossa, Fernando Arce, Erik Zamora and Elizabeth Guevara

Abstract Morphological neural networks, in particular, those with dendritic processing (MNNDPs), have shown to be a very promising tool for pattern classification. In this chapter, we present a survey of the most recent advances concerning MNNDPs. We provide the basics of each model and training algorithm; in some cases, we present simple examples to facilitate the understanding of the material. In all cases, we compare the described models with some of the state-of-the-art counterparts to demonstrate the advantages and disadvantages. In the end, we present a summary and a series of conclusions and trends for present and further research.

Keywords Morphological neural networks with dendritic processing
Pattern classification · Artificial intelligence

2.1 Introduction

Pattern classification is recognized to be one of the most important problems in many areas. One of these areas is artificial intelligence and robotics. Pattern classification allows a robot, for example, to recognize surrounding objects (things, sounds, and so on). This way the robot can navigate inside its environment in an

---

H. Sossa (✉) · F. Arce
Instituto Politécnico Nacional (CIC), Av. Juan de Dios Batiz S/N, Col. Nueva Industrial Vallejo, 07738 Gustavo A. Madero, Ciudad de México, Mexico
e-mail: hsossa@cic.ipn.mx

E. Zamora
Instituto Politécnico Nacional (UPIITA), Av. Instituto Politécnico Nacional 2580, Col. Barrio la Laguna, 07340 Ticoman, Ciudad de México, Mexico

E. Guevara
Universidad Anáhuac México Sur, Av. de las Torres 131, Col. Olivar de los Padres, 01780 Torres de Potrero, Ciudad de México, Mexico

O. O. Vergara Villegas et al. (eds.), *Advanced Topics on Computer Vision, Control and Robotics in Mechatronics*,
https://doi.org/10.1007/978-3-319-77770-2_2

autonomous way in such a way that it can avoid hitting those objects, obeying orders, locate and grasp them to perform a given task.

The pattern classification problem can be stated as follows: Given a pattern X in vector form composed of or of n features as follows: $X = [x_1, x_2, \ldots, x_n]^T$, determine its corresponding class $C^k, k = 1, 2, \ldots, p$. Several approaches were developed during the last decades to provide different solutions to this problem; among them are the statistical approach, the syntactical or structural approach, and the artificial neural approach.

The artificial neural approach is based on the fact that many small processing units (the neurons) combine their capabilities to determine the class $C^k, k = 1, 2, \ldots, p$ given an input pattern: $X = [x_1, x_2, \ldots, x_n]^T$. Considering that an artificial neural network is a mapping between X and the set of labels: $K = \{1, 2, \ldots, p\}$; if this mapping is defined as M then: $X \rightarrow M \rightarrow K$.

Several artificial neural network (ANN) models have been reported in literature, since the very old threshold logic unit (TLU) model introduced to the world during the 40s by McCulloch and Pitts (1943), the well-known Perceptron developed by Rosenblatt during the 50s (Rosenblatt 1958, 1962), the radial basis function neural network (RBFNN) proposed by Broomhead and Lowe (1988a, b), the elegant support vector machine (SVM) introduced to the world by Cortes and Vapnik in the 90s (Cortes and Vapnik 1995), the extreme learning machine (ELM) model proposed by Guang et al. (2006) and Huang et al. (2015), among other.

A kind of ANNs not very well known by the scientific community that has demonstrated very promising and competitive pattern classification results is the so-called morphological neural network with dendritic processing (MNNDP) model (Ritter et al. 2003; Ritter and Urcid 2007).

Instead of using the standard multiplications $(\times)$ and additions $(+)$ to obtain the values used by the activation functions of the computing unities in classical models, MNNDPs combine additions $(+)$ and **max** $(\vee)$ or **min** $(\wedge)$ operations. As we will see along this chapter, this change will modify the way separating among pattern classes; instead of using decision surfaces integrated by a combination of separating hyperplanes, MNNDPs combine hyper-boxes to perform the same task: Divide classes to find the class to which a given input pattern $X = [x_1, x_2, \ldots, x_n]^T$ should be put.

The rest of this chapter is organized as follows. Section 2.2 is oriented to present to the reader the basics of MNNDP. Section 2.3, on the other hand, is focused to explain the operation of the most popular and useful training algorithms. When necessary, a simple numerical example is provided to help the reader to easily grasp the idea of the operation of the training algorithm. In Sect. 2.4, we compare the performance of the presented models as well as the training algorithms in respect to other artificial neural networks models. Finally, in Sect. 2.5, we conclude and give some directives for present and future research.

2.2 Basics on MNNDPs

Due to some of the important discoveries in the biophysics of computation, in (Ritter et al. 2003; Ritter and Schmalz 2006; Ritter and Urcid 2007), the authors present an improvement over the so-called morphological perceptron (MP) (Ritter and Beaver 1999). This improvement consists in adding a dendritic structure to the neuron to enhance its computing capabilities. This new model named morphological perceptron with dendritic processing (MPDP) allows generating decision boundaries formed by combinations of rectangular regions in the plane (hyper-boxes in n-dimensional space). Thus, in n dimensions, a dendrite represents a hyper-box; a combination of hyper-boxes allows grouping classes.

Dendrites have played an important role in previously proposed training methods. The most common approach consists in enclosing patterns by one or more hyper-boxes assigning a label to each group of hyper-boxes.

MPDPs computations are based on lattice algebra. More information can be found in (Ritter et al. 2003; Ritter and Schmalz 2006; Ritter and Urcid 2007).

Morphological processing involves additions $(+)$ and **min** $(\wedge)$ or **max** $(\vee)$ operations; min and max operators allow generating piecewise boundaries for classification problems. These operations can be easily implemented in logic devices because computational processing is based on comparative operators instead of products.

Several MPDPs can be arranged into a net called MNNDP. Usually, a MNNDP has a layer of several MPDPs. Without loss of generality, let us consider the case of a MNNDP with one MPDP.

In what follows an incoming pattern is firstly processed by all the dendrites of the MPDP (For an example, refer to Fig. 2.1a). Once this is done, the dendrite with the biggest value is chosen by means of a selecting function. Figure 2.1a shows the architecture of a MPDP with K dendrites and an example of a hyper-box (Fig. 2.1b) generated by kth dendrite in terms of its weights $w^0_{i,k}$ and $w^1_{i,k}$ in two dimensions.

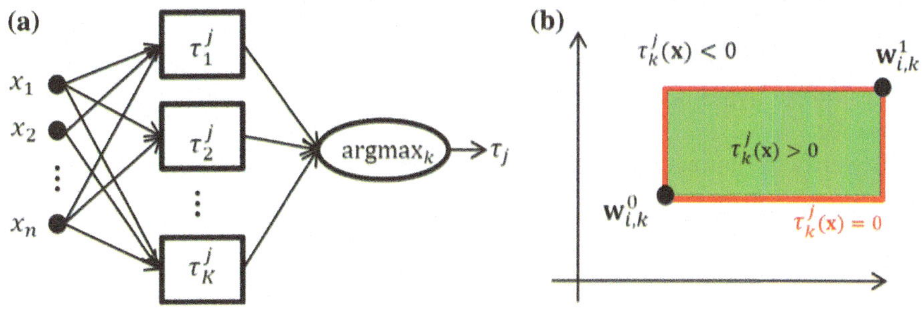

Fig. 2.1 **a** Typical MPDP and **b** example of hyper-box in 2D generated by kth dendrite

The computation τ_k^j performed by the kth dendrite for the jth class in 2D can be expressed as follows:

$$\tau_k^j = \wedge_{i=1}^n \left(x_i + w_{ik}^1\right) \wedge -\left(x_i + w_{ik}^0\right) \tag{2.1}$$

where x_i is the ith component of input vector $X = [x_1, x_2, \ldots, x_n]^{\mathrm{T}}$; n is the vector dimensionality of X, $i \in I$; and $I \in \{1, \ldots, n\}$ represents the set of all input neurons with terminal fibers that synapsing kth dendrite; w_{ik}^0 and w_{ik}^1 are the synaptic weights corresponding to the set of terminal fibers of the ith neuron that synapse on the kth dendrite; w_{ik}^1 represent an activation terminal fiber while w_{ik}^0 an inhibition terminal fiber.

If $\tau_k^j > 0$, X is inside the hyper-box; if $\tau_k^j = 0$, X is over the hyper-box boundary, and if $\tau_k^j < 0$, X is outside the hyper-box.

On the other hand, the output value of the MPDP: τ_j is obtained by computing the argument of the maximum over all the computations obtained by the set of dendrites connected to the MPDP as follows:

$$\tau_j = \operatorname{argmax}_k\left(\tau_k^j\right) \tag{2.2}$$

From Eq. (2.2), we can see that the argmax function selects only one of the dendrites values, and the result is a scalar. This argmax function permits a MPDP classifying patterns that are outside the hyper-boxes. It also allows building more complex decision boundaries by combining the actions of several hyper-boxes. If Eq. (2.2) produces more than one output, the argmax function selects the first maximum argument as the index class to which the input pattern is classified.

In order to explain how a dendrite computation is performed for a MPDP, let us refer to Fig. 2.2a displaying two hyper-boxes that could be generated by any MPDP trained with any training algorithm covering all the patterns (green crosses and blue dots). In the example, blue dot points belong to class C^1 while green crosses belong to class C^2. Figure 2.2b presents the MPDP that allows generating the two aforementioned boxes. As can be appreciated, the input pattern values x_1 and x_2 are connected to the output neuron via the dendrites. The geometrical calculation explanation executed by the dendrites is that each of these determines a box in two dimensions (a hyper-box in n dimensions) which can be represented by its weight values w_{ij}.

To verify the correct operation of the MPDP shown in Fig. 2.2b, let us consider the following two noisy patterns: $\tilde{x}_1 = \begin{bmatrix} 3 \\ 0 \end{bmatrix}$ which is supposed to belong to class C^1 and $\tilde{x}_2 = \begin{bmatrix} 7 \\ 3 \end{bmatrix}$ to class C^2.

According to Eq. (2.1), the following dendrite computations for both patterns can be obtained:

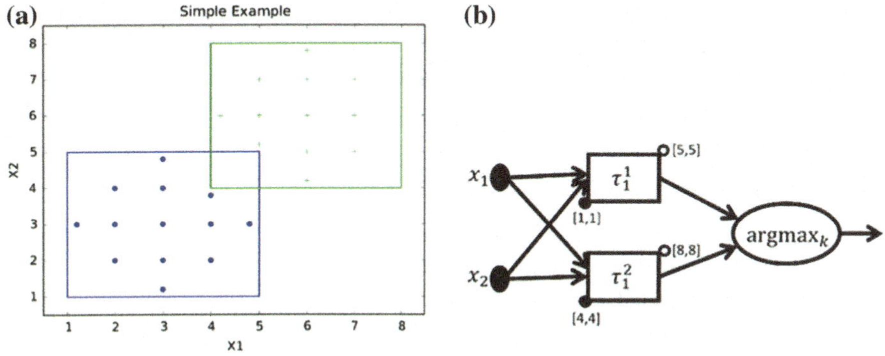

Fig. 2.2 Simple example of a DMNN. **a** Two boxes that cover all the patterns and **b** MPDP based on the two boxes (black circles denote excitatory connections and white circles inhibitory connections)

$$\tau_1^1(\tilde{x}_1) = \tau_1^1\begin{bmatrix}3\\0\end{bmatrix} = [(3-1)\wedge-(3-5)]\wedge[(0-1)\wedge-(0-5)] = [2\wedge-1] = -1.$$

$$\tau_1^2(\tilde{x}_1) = \tau_1^2\begin{bmatrix}3\\0\end{bmatrix} = [(3-4)\wedge-(3-8)]\wedge[(0-4)\wedge-(0-8)] = [-1\wedge-4] = -4.$$

$$\tau_1^1(\tilde{x}_2) = \tau_1^1\begin{bmatrix}7\\3\end{bmatrix} = [(7-1)\wedge-(7-5)]\wedge[(3-1)\wedge-(3-5)] = [-2\wedge 2] = -2.$$

$$\tau_1^2(\tilde{x}_2) = \tau_1^2\begin{bmatrix}7\\3\end{bmatrix} = [(7-4)\wedge-(7-8)]\wedge[(3-4)\wedge-(3-8)] = [1\wedge-1] = -1.$$

Now, by applying Eq. (2.2) to these four values:

$$\tau = \text{argmax}_k\left(\tau_1^1(\tilde{x}_1), \tau_1^2(\tilde{x}_1)\right) = \text{argmax}_k(-1, -4) = 1.$$
$$\tau = \text{argmax}_k\left(\tau_1^1(\tilde{x}_2), \tau_1^2(\tilde{x}_2)\right) = \text{argmax}_k(-2, -1) = 2.$$

As expected, the MPDP maps out $\tilde{x}_1$ to C^1 and $\tilde{x}_2$ to C^2 correctly.

2.3 Training Algorithms

It is well known that to be useful, any ANN has to be trained. In the case of MNNDPs, several training methods have been reported in the literature. Most of these methods utilize a sort of heuristic and do not make use of an optimization technique to tune the interconnection parameters.

In this section, we describe some of the most useful methods reported in the literature to train a MNNDP. Without loss of generality, let us consider the case of a MNNDP composed of just one neuron, i.e., a MPDP.

2.3.1 Elimination and Merging Methods

According to Ritter et al. (2003), a MPDP can be trained in two different ways. The first is based on iteratively eliminating boxes, the second one on merging boxes. The principle of operation of both approaches is described in the following two subsections.

2.3.1.1 Elimination Method

This method was originally designed to work for one morphological perceptron applied to two-class problems. The method first builds a hyper-box that encloses all the patterns from the first class and possibly patterns of the second class. For an example, refer to Fig. 2.3a. As can be appreciated, the hyper-box generated contains patterns of both classes.

The elimination method then, in an iterative way, generates boxes containing patterns of the second class, carving the first hyper-box producing a polygonal region containing, at each iteration, more patterns of the first class. The elimination method continues this way until all patterns of second class are eliminated from the original hyper-box. Figure 2.3b and c illustrates this process.

2.3.1.2 Merging Method

This method begins by generating several hyper-boxes around groups of patterns belonging to the same class. A hyper-box is generated in such a way that only patterns of the first class are enclosed by it. Next, in an iterative way, the generated hyper-boxes are merged (unified) generating at the end a polygonal region enclosing only patterns of the first class. Figure 2.4a, b, and c shows the operation of the aforementioned procedure for the two-class problem of Fig. 2.3.

Although these two training methods for single MPDP were originally designed to work with two-class problems, they can be easily extended to the case of multi-class problems. The main difference would be the form of the decision boundaries and the number of generated hyper-boxes.

Complete details about the operation of two methods can be found in (Ritter et al. 2003).

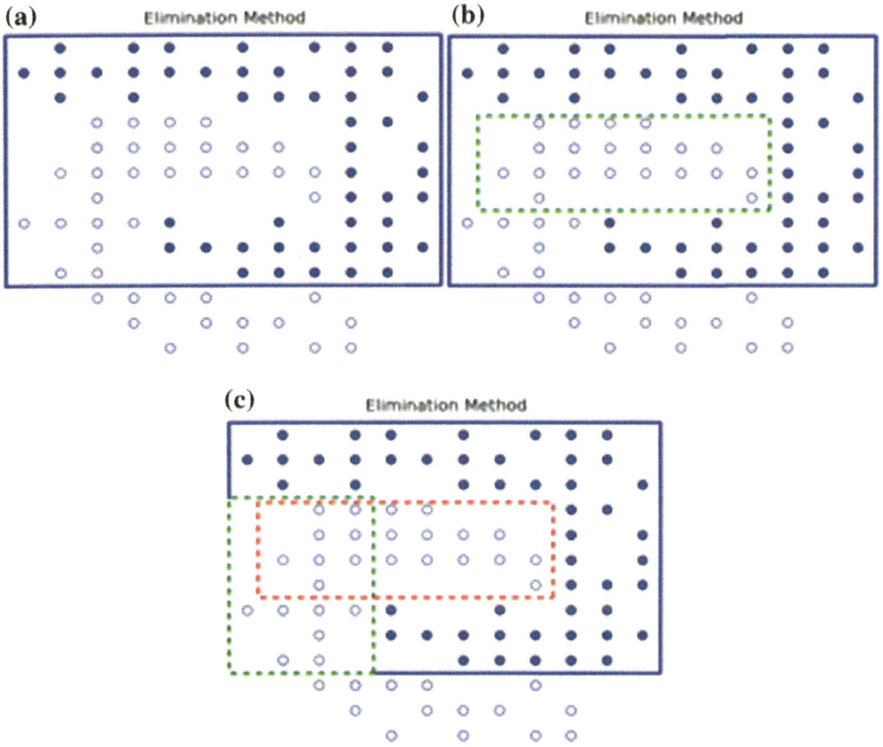

Fig. 2.3 Illustration of the operation of the elimination method. **a** A two-class problem, **b** and **c** consecutive steps until the resulting region only encloses patterns of the first class

2.3.2 *Divide and Conquer Methods*

The second set of methods we are going to describe first takes a set of patterns, divided into classes producing a clustering. They then utilize the generated clustering to obtain the weights of the corresponding dendrites. We present two methods, one of exponential complexity and one improvement of linear complexity.

2.3.2.1 Divide and Conquer Method

In Sossa and Guevara (2014), the authors introduce the so-called divide and conquer method (DCM) for training MPDP. The main idea behind this training method is to first group the patterns of classes into clusters (one cluster for each class of patterns), then to use this clustering to obtain the weights of dendrites of the morphological perceptron.

For purposes of explaining the functioning of the algorithm, a simple example of three classes with two attributes will be used. Figure 2.5a shows the whole set of

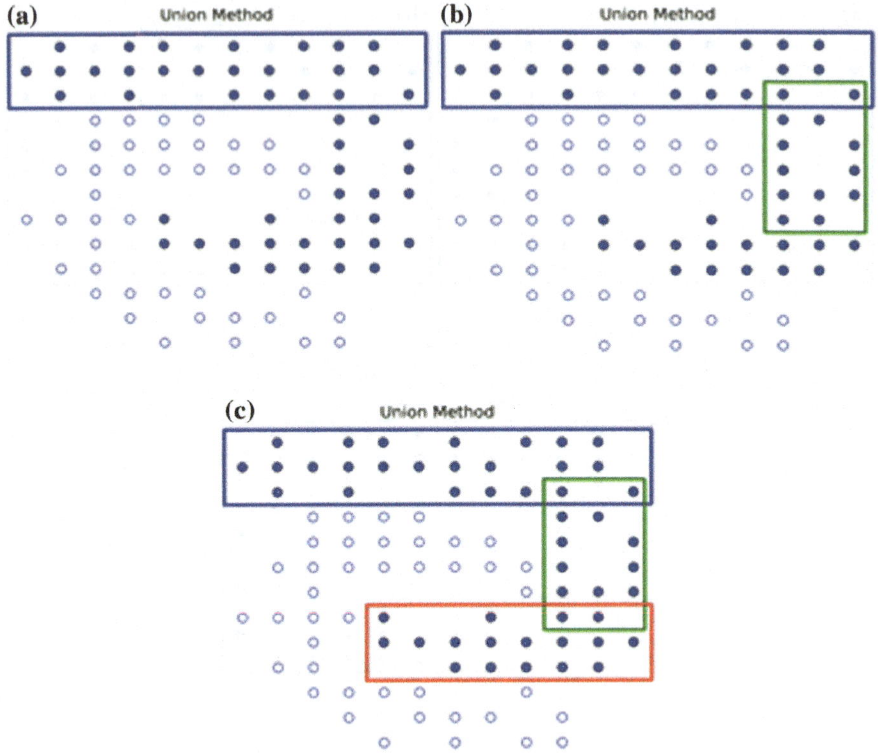

Fig. 2.4 Illustration of the operation of the merging method. **a** A two-class problem, **b** and **c** consecutive steps of the merging process until only one region is obtained

patterns as points—points belonging to class C^1 as red dots, patterns of class C^2 as green dots and points of class C^3 as blue dots.

Given p classes of patterns, $C^k, K = 1, 2, \ldots, p$, with n the dimensionality of each pattern, the algorithm applies the following steps:

Algorithm DCM:

(1) Generate a n-dimensional hyper-box H^n that encloses all pattern of all p classes. Hyper-box H^n can be a minimal one (one that matches the coordinates of the limit coordinates of the class boundaries). Let this box be $H^n_{\min}$. For having better tolerance to noise, at moment, of pattern classification the enclosing box can be a little greater, with a margin M to each side of the box. This margin is obtained as a function of the size T of box $H^n_{\min}$. If, for example, $M = 0.1T$ then the new hyper-box will extend that ratio to every side of $H^n_{\min}$. Figure 2.5a presents the box that covers the patterns of all the three classes with $M = 0.1T$.

(2) If hyper-box H^n contains patterns of different classes, being the case for the provided example, divide it into 2^n smaller hyper-boxes. If each generated smaller hyper-box encloses patterns from only one class, label each hyper-box

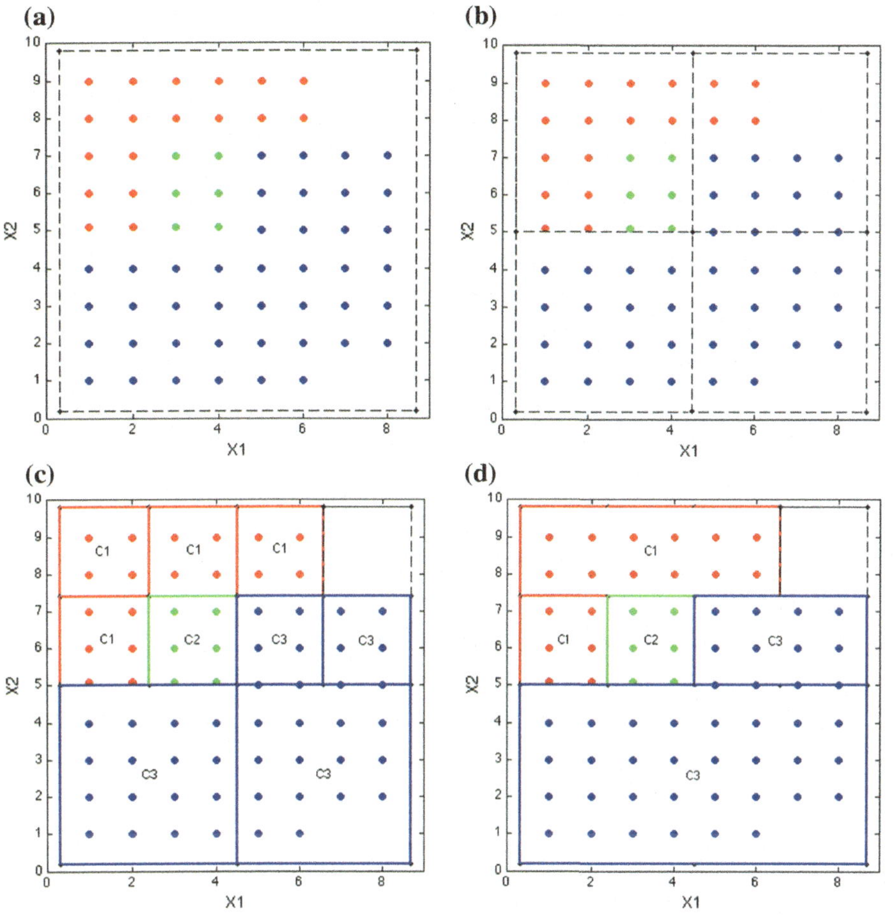

Fig. 2.5 Illustration of the operation of the DCM. **a** Three-class problem and hyper-box enclosing all patterns of all classes, **b** first division when step 2 of the DCM is applied, **c** consequent divisions generated when step 3(a) is applied, and **d** simplification step, resulting in five dendrites

with the label of the corresponding class, stop the learning process and proceed to step 4. For example, the first division of the box is presented in Fig. 2.5b.

(3) This step is divided into two stages as follows:

  (a) If at least one of the generated hyper-cubes H^n has patterns of more than one class, divide H^n into 2^n smaller hyper-boxes. Iteratively repeat the verification division process onto each smaller hyper-box until the stopping criterion is satisfied. Figure 2.5c shows all the boxes generated by the training algorithm.

  (b) Once all the hyper-boxes are generated, if two or more of them of the same class share a common side, group them into one region. Figure 2.5d presents the application of this simplification procedure that automatically reduces the number of hyper-box.

(4) Based on the coordinates on each axis, compute the weights for each hyper-box that encloses patterns belonging to C^k. By taking into account only those hyper-boxes enclosing patterns belonging to C^k, at this moment the MNNDP is designed. The final MNNDP has five dendrites as depicted in Table 2.1. Each input x_1 and x_2 are connected to each dendrite. These connections are excitatory (w_{ik}^1) or inhibitory (w_{ik}^0) as depicted in columns 2–5 in Table 2.1. This MNNDP allows separating among the three classes: C^1, C^2, and C^3. The geometrical interpretation of the computations performed by a dendrite is that every single dendrite determines a hyper-box defined by a single dendrite via its weight values w_{ij} as the example shows.

To verify the correct operation of the MPDP whose dendrites weight values are shown in Table 2.1, let us consider the two noisy versions: $\tilde{x}_1 = \begin{bmatrix} 4.5 \\ 8.5 \end{bmatrix}$ and $\tilde{x}_2 = \begin{bmatrix} 4 \\ 3.5 \end{bmatrix}$, which are supposed to belong, respectively, to the classes C^1 and C^3.

The reader can easily verify that when Eq. (2.1) is applied to the first dendrite, the following results for each test pattern can be obtained:

$$\tau_1^1(\tilde{x}_1) = \tau_1^1\begin{bmatrix} 4.5 \\ 8.5 \end{bmatrix} = [(4.5 - 0.3) \wedge -(4.5 - 6.6)] \wedge [(8.5 - 7.4) \wedge -(8.5 - 9.8)]$$
$$= [2.1 \wedge 1.1] = 1.1.$$
$$\tau_1^1(\tilde{x}_2) = \tau_1^1\begin{bmatrix} 4 \\ 3.5 \end{bmatrix} = [(4 - 0.3) \wedge -(4 - 6.6)] \wedge [(3.5 - 7.4) \wedge -(3.5 - 9.8)]$$
$$= [2.6 \wedge -3.9] = -3.9.$$

In the same way, all the other calculations for:

$$\tilde{x}_1 : \tau_2^1 = -2.1, \tau_3^2 = -1.1, \tau_4^3 = -1.1, \tau_5^3 = -3.5.$$
$$\tilde{x}_2 : \tau_2^1 = -1.6, \tau_3^2 = -1.5, \tau_4^3 = -1.5, \tau_5^3 = 1.5.$$

Table 2.1 Excitatory and inhibitory weights for the five dendrites of the morphological neural network for example

Dendrite	w_{ik}^1	w_{ik}^0	w_{ik}^1	w_{ik}^0
τ_1^1	−0.3	−6.6	−7.4	−9.8
τ_2^1	−0.3	−2.4	−5.0	−7.4
τ_3^2	−2.4	−4.5	−5.0	−7.4
τ_4^3	−4.5	−8.7	−5.0	−7.4
τ_5^3	−0.3	−8.7	−0.2	−5.0

With these values by means of Eq. (2.2), the two patterns are classified as follows:

$$\begin{aligned}\tau(\tilde{x}_1) &= \operatorname{argmax}_k\left(\tau_1^1(\tilde{x}_1), \tau_2^1(\tilde{x}_1), \tau_3^2(\tilde{x}_1), \tau_4^3(\tilde{x}_1), \tau_5^3(\tilde{x}_1)\right) \\ &= \operatorname{argmax}_k(1.1, -2.1, -1.1, -1.1, -3.5) = 1.\end{aligned}$$

Thus, $\tilde{x}_1$ is put in class C^1 as expected. In the same way, for pattern $\tilde{x}_2$:

$$\begin{aligned}\tau(\tilde{x}_2) &= \operatorname{argmax}_k\left(\tau_1^1(\tilde{x}_2), \tau_2^1(\tilde{x}_2), \tau_3^2(\tilde{x}_2), \tau_4^3(\tilde{x}_2), \tau_5^3(\tilde{x}_2)\right) \\ &= \operatorname{argmax}_k(-3.9, -1.6, -1.5, -1.5, 1.5) = 3.\end{aligned}$$

Thus, $\tilde{x}_2$ is put in class C^3 as expected.

Successful applications of the DCM in pattern classification and object and pattern recognition can be found in (Sossa and Guevara 2013a, b; Vega et al. 2013, 2015; Sossa et al. 2014; Ojeda et al. 2015).

2.3.2.2 Linear Divide and Conquer Method

A main problem of the DCM introduced in (Sossa and Guevara 2014) and explained in the last section is its exponential complexity. Each time a hyper-box is divided, 2^n computations are required. This could be very restrictive in most sequential platforms.

In this section, we briefly present the operation of a substantial improvement of the DCM that operates in linear time. We call this new method the LDCM. Instead of generating all the 2^n hyper-boxes at each iteration, the new method generates only the necessary hyper-boxes directly from the data by analyzing it in a linear way. The method operates in a recursive way. The steps that the method uses for training a LDCM are explained as follows.

Given m patterns belonging to p classes, with n the dimensionality of each pattern:

Algorithm LDCM:

(1) Enclose all the m patterns inside a first hyper-box denoted as H_0 in $\mathbb{R}^n$. Again, to have better tolerance to noise, add a margin M to each side of H_0.

(2) Divide the length of each dimension of H_k, including H_0 half: $\frac{d_i}{2} = \frac{\max\{x_i\} - \min\{x_i\}}{2}$, obtaining two parts at each dimension $\min\{x_i\} \leq h_{i1} \leq \max\{x_i\} + \frac{d_i}{2}$, $\min\{x_i\} + \frac{d_i}{2} \leq h_{i2} \leq \max\{x_i\}$. Determine inside which intervals the first sample is found and generate the corresponding hyper-box. Let us call this box H_1.

(3) Take each sample pattern (in the provided example) from left to right and up and down. If a pattern is out of the generated box, generate a new box for each of these patterns. Repeat this step for the whole set of patterns. Let us designate these boxes as: H_2, H_3,…, one for each first point outside the preceding box.

(4) Verify all the generated boxes generated in step 3, and apply one of the following steps:

    (a) If the patterns inside a box belong to the same class, label this box with the label of the corresponding class. If all boxes contain patterns of the same class, stop training and go to step 5.
    (b) If at least one box contains pattern of different classes, then iterate between steps 3 and 4(a) until the stopping criterion is satisfied (until each generated box contains patterns of the same class).

(5) If two or more generated boxes share dimensions, merge those regions, as it is done with the DCM.
(6) By taking into account the coordinates, select the weights for each box (dendrite).

To illustrate the operation of LDCM, let us consider the two-class problem shown in Fig. 2.6a. Figure 2.6b shows also the corresponding box H_0 generated when the first step of the LDCM is applied. Figure 2.6c shows the box generated by the application of the second step of the LDCM. Figures 2.6d–f depicts the three boxes generated when the third step of the LDCM is applied over the sample points. Each generated box is labeled with the black dot at the left upper over the first pattern outside the previous box. As can be appreciated from Fig. 2.6f, only the first two boxes contain patterns of the first class (red dots) and second class (green dots), and the other two boxes contain patterns of both classes; thus, step 4(b) is applied over these two boxes, giving as a result the subdivision of boxes shown in Fig. 2.6g–i. Finally, Fig. 2.6j depicts the simplification of the boxes provided by the application of the fifth step of the LDCM. As can be seen, the optimized final neuron will have seven dendrites, four for the first class and three for the second class. The weights of dendrite should be calculated in terms of the limit coordinates of each box (step 6 of the LDCM). One important result concerning the LDCM is that it produces exactly the same result as if the DCM was applied. The proof can be found in (Guevara 2016). More details concerning the LDCM can be found in (Guevara 2016).

2.3.3 Evolutionary-Based Methods

This kind of methods makes use of so-called evolutionary techniques to find the weights of the dendrites of a MPDP. Recently in (Arce et al. 2016, 2017), the authors describe a method that utilizes so-called evolutionary computation to optimally find the weights of the dendrites of a MPDP. The method utilizes differential evolution to evolve the weights. Let us call this method the differential evolution method (DEM).

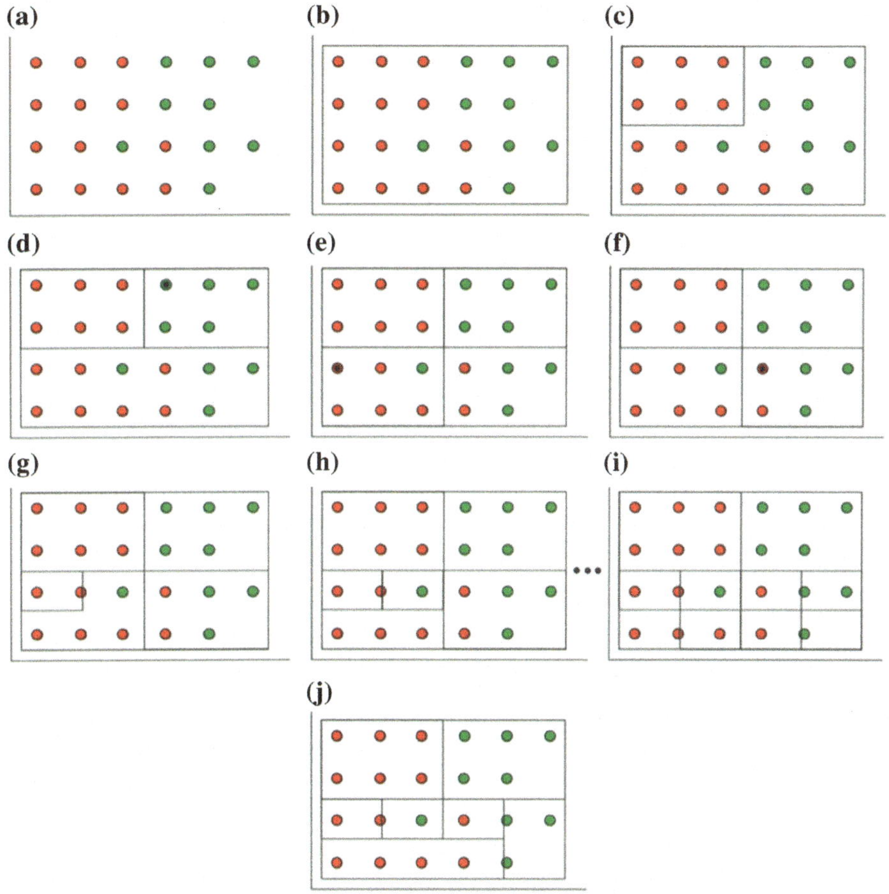

Fig. 2.6 Illustration of the operation of the LDCM. **a** Two-class problem, **b** first step (box H_0), **c** box generated by the second step, **d**, **e**, **f** boxes generated by third step, **g**, **h**, **i** iterative subdivision of the boxes by the application of step 4(b), and **j** simplification of the boxes by the application of the fifth step

The DEM proceeds in three steps as follows. Given m patterns belonging to p classes, with n the dimensionality of each pattern:

Algorithm DEM:

(1) Initialization.
(2) Application of DE as in (Ardia et al. 2011) to the actual boxes.
(3) Select the set of boxes that produces the smallest error. Algorithm 1 shows the pseudo-code of DE applied to the training of a MNNDP.

Algorithm 1: Pseudo-code of DE applied to MNNDP

```
Begin
Generate initial population of solutions.
For d = 1  to q
    Repeat:
        For the entire population, calculate the fitness value.
        For each parent, select two solutions at random and the best parent.
        Create one offspring using DE operators.
        If the offspring is better than the parents:
            Replace parent by the offspring.
    Until a stop condition is satisfied.
End
```

In (Arce et al. 2016, 2017), the authors present two initialization methods. Here, we describe the operation of one of these methods. In general, the so-called HBd initialization method proceeds in two steps as follows:

(1) For each class C^j, open a hyper-box that encloses all its patterns.
(2) Divide each hyper-box into smaller hyper-boxes along the first axis on equal terms by a factor d, for $d\mathbb{Z}^+$ until q divisions have been carried out.

In order to explain the better, understand the operation of the HBd initialization algorithm, a straightforward example of four classes with two features is next presented. Figure 2.7a illustrates the problem to be solved. Blue dots belong to C^1, black crosses to C^2, green stars to C^3, and red diamonds to C^4.

As can be seen from Fig. 2.7b, during the first step, the patterns of each class are enclosed a box. The blue box encloses the patterns from C^1, the black box encloses those from C^2, then green box encloses those from C^3, and the red encloses those from C^4.

During the second step, Fig. 2.7b and c shows how each box is divided by a factor d ($d = 1$ in the first case, $d = 2$ in the second case), while Fig. 2.7d shows how DE is applied to the resultant boxes from Fig. 2.7c. This is best placement of boxes for $d = 2$ by the application of DE.

2.3.4 *Other Related Works*

In 2006, Barmpoutis and Ritter (2006) modified the dendritic model by rotating the orthonormal axes of each hyper-box. In this work, the authors create hyper-boxes with a different orientation of their respective coordinate axes.

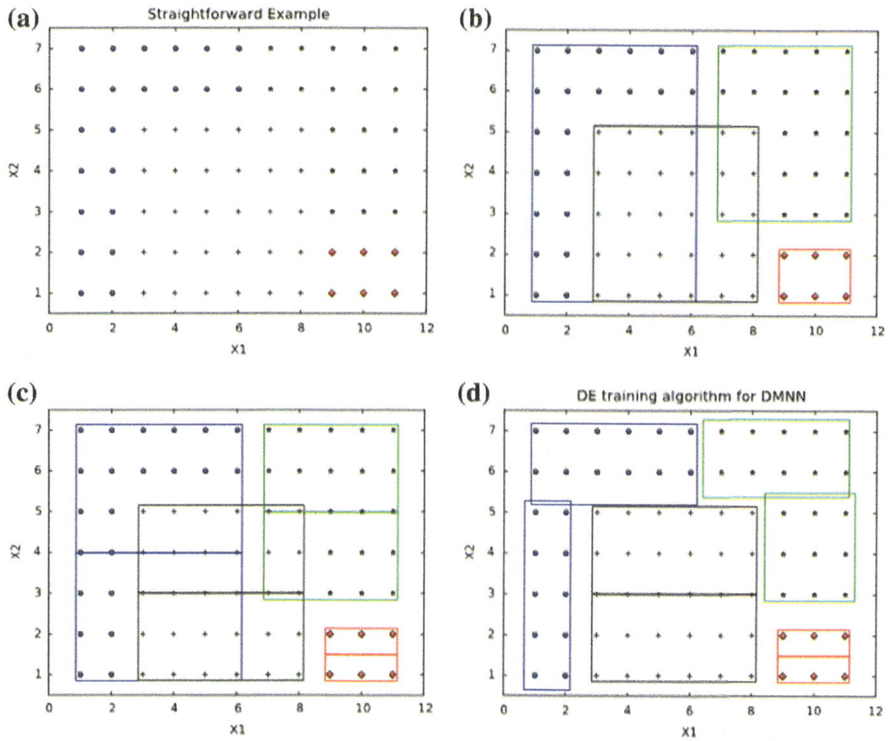

Fig. 2.7 Illustration of the operation of the HBd initialization method. **a** Simple example with four classes, **b** division with $d = 1$, **c** division with $d = 2$, and **d** optimal placement by the application of DE of the boxes for $d = 2$

Later in 2007, Ritter and Urcid (2007) present a generalization of the single-layer morphological perceptron introduced by Ritter et al. (2003) and Ritter and Schmalz (2006).

In 2009, Sussner and Esmi (2009) describe so-called morphological perceptron with competitive output neurons. In this paper, the authors describe a training algorithm that automatically adds neurons to the hidden layer.

Two years later, in 2011, Sussner and Esmi (2011) add competitive neurons to the MP by proposing using the maximum argument at the output neuron.

In short, in 2014, Ritter et al. (2014) extend the elimination training algorithm proposed in (Ritter and Schmalz 2006; Ritter and Urcid 2007) by using two lattice metrics to the SLLP, L_1 and L_∞. This algorithm allows creating more complex decision boundaries, known as polytopes.

In short, in (Zamora and Sossa 2016, 2017), the authors show how the gradient decent principle can be successfully applied to obtain the weights of the dendrites of a morphological perceptron.

2.4 Comparison

In this section, we present a comparison of the described models and training algorithms. We compare the performance among three training methods for MPDP: Ritter's method (RM), DCM, and the DEM. We compare also the described models with other ANNs models, the multilayered perceptron (MLP), the RBFNN, and SVM.

2.4.1 *Results with Synthetic Data*

To compare among the training methods for MPDP, we use three synthetic databases shown in Fig. 2.8a and b, the spiral with laps depicted in Fig. 2.8c and a, spiral with 10 laps (not shown).

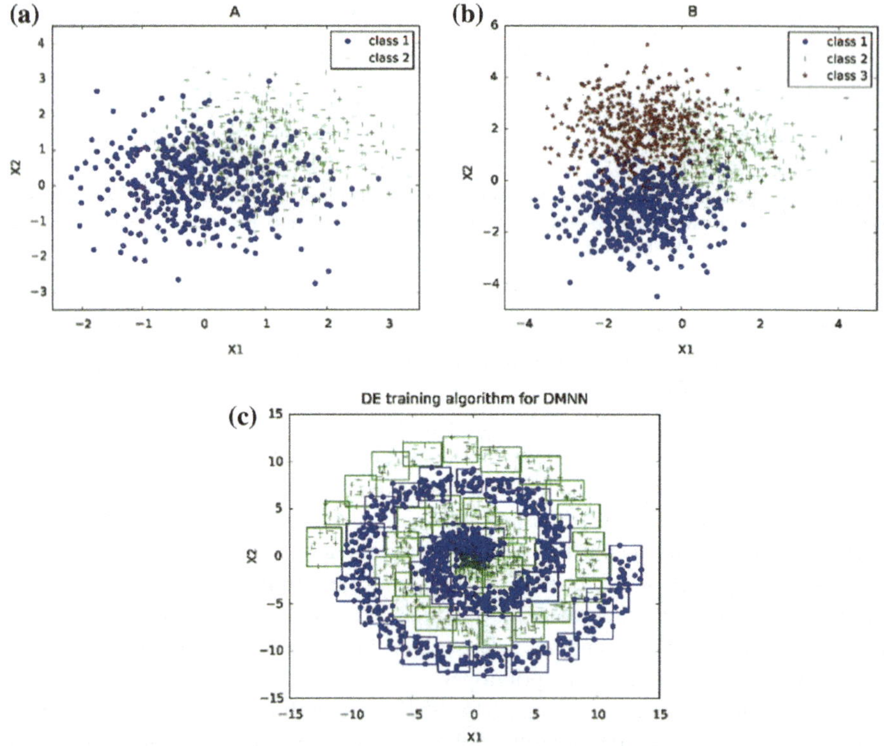

Fig. 2.8 Three of the synthetic databases to test the performance of three of the methods for training MNNDP. **a** Two-class synthetic problem, **b** Three-class synthetic problem, and **c** two-class spiral synthetic problem

Table 2.2 Comparison between RM, DCM, and DEM methods for four synthetic databases

Dataset	RM		DCM			DEM		
	ND	e_{test}	*ND*	e_{train}	e_{test}	*ND*	e_{train}	e_{test}
A	194	28.0	419	0.0	25.0	**2**	21.7	**20.5**
B	161	50.3	505	0.0	20.3	**3**	16.6	**15.2**
Spiral 2	160	8.6	356	0.0	7.2	**60**	7.3	**6.4**
Spiral 10	**200**	26.7	1648	0.0	10.6	1094	1.8	**6.3**

Bold indicates DEM method offers a better e_{test} error; for the first three databases (A, B and Spiral 2) it requires a less number of dendrites ND to produce the result; only in the case of Spiral 10 DEM method requires a greater number of dendrites (1094) compared to the 200 required by the RM method

Table 2.2 shows a comparison with these four databases among the methods reported in (Ritter et al. 2003; Sossa and Guevara 2014; Arce et al. 2016), respectively. In all three cases, *ND* is the number of dendrites generated by the training method, e_{train} is the training error, and e_{test} is the error during testing.

As can be appreciated, in all four cases, the DE-based algorithm provides the best testing errors. Note also that although the DE-based method obtains a training error, it provides the best testing error. We can say that this method generalizes better than the DCM that obtains a 0% for training error e_{train}. In the first two problems, we can see that the DEM needs a reduced number of dendrites to provide a reduced training error. In all four experiments, 80% of the data were used for training and 20% for testing.

2.4.2 Results with Real Data

In this section, we first compare the three training methods for MPDP with 11 databases taken from the UCI Machine Learning Repository (Asuncion 2007). Table 2.3 presents the performance results. As we can appreciate from Table 2.3, in all cases, the DE-based training method provides the smallest number of dendrites to solve the problem as well as the smallest testing error.

Because the DEM provides the best results among the three training methods for MNNDP, we now compare its performance against three well-known neural network-based classifiers: a MLP, a SVM, and a RBFNN. We do it with the same 11 databases from UCI. Table 2.4 presents the performance results; as we appreciate from this table, in most of the cases, the DEM provides the best testing errors.

Table 2.3 Comparison between RM, DCM, and DEM methods for 11 databases from UCI Machine Learning Repository

Dataset	RM		DCM			DEM		
	ND	e_{test}	*ND*	e_{train}	e_{test}	*ND*	e_{train}	e_{test}
Iris	5	6.7	28	0.0	3.3	**3**	3.3	**0.0**
Mammographic mass	51	14.4	26	0.0	19.2	**8**	15.8	**10.4**
Liver disorders	41	42.0	183	0.0	35.5	**12**	37.6	**31.1**
Glass identification	60	36.7	82	0.0	31.8	**12**	4.7	**13.6**
Wine quality	120	51.0	841	0.0	42.1	**60**	42.1	**40.0**
Mice protein expression	77	18.9	809	0.0	5.0	**32**	6.6	**4.5**
Abalone	835	88.2	3026	0.0	80.6	**27**	77.1	**78.2**
Dermatology	192	57.8	222	0.0	15.5	**12**	4.8	**4.2**
Hepatitis	19	53.3	49	0.0	46.7	**9**	9.4	**33.3**
Pima Indians diabetes	180	70.6	380	0.0	31.4	**2**	23.8	**23.5**
Ionosphere	238	10.0	203	0.0	35.7	**2**	2.8	**2.8**

Bold indicates DEM method requires a less number of dendrites to produce the result; in all cases it also offers a better e_{test} error

Table 2.4 Comparison between the MLP, the SVM, the RBFNN, and the DEM methods for 11 databases from UCI Machine Learning Repository

Dataset	MLP		SVM		RBFNN		DEM	
	e_{train}	e_{test}	e_{train}	e_{test}	e_{train}	e_{test}	e_{train}	e_{test}
Iris	1.7	**0.0**	4.2	**0.0**	4.2	**0.0**	3.3	**0.0**
Mammographic mass	15.7	11.2	18.4	11.2	17.9	16.0	15.8	**10.4**
Liver disorders	40.3	40.6	40.0	40.2	29.0	37.8	37.6	**31.1**
Glass identification	14.1	20.4	12.3	18.2	0.0	20.4	4.7	**13.6**
Wine quality	34.0	**39.0**	40.6	43.0	41.5	44.3	42.1	40.0
Mice protein expression	0.0	0.6	0.1	**0.5**	11.4	13.9	6.6	4.5
Abalone	75.0	**75.0**	73.1	**75.0**	72.0	76.0	77.1	78.2
Dermatology	0.0	**0.0**	1.4	1.4	1.0	2.8	4.8	4.2
Hepatitis	1.6	40.0	15.6	**33.3**	15.6	**33.3**	9.4	**33.3**
Pima Indians diabetes	15.5	29.4	22.3	24.8	22.3	24.8	23.8	**23.5**
Ionosphere	0.3	7.1	6.8	6.8	6.4	8.6	2.8	**2.8**

Bold indicates DEM wins when compared to other standard classification methods. In some cases it losses, for example, for the Dermatology problems MLP methods obtains the best results with 0 e_{train} and e_{test} errors

2.5 Summary, Conclusions, and Present and Future Research

In this chapter, we have presented the basic operation of so-called morphological perceptron with dendritic processing. We have also described the fundamental functioning of several of the most popular training for these kind of perceptrons. In some cases, we have provided simple numerical examples to facilitate the explanation to the reader.

We have compared the performance of Ritter's approach, the so-called divide and conquer method, and the differential evolution-based technique with four synthetic and 11 real databases. As can be appreciated from Table 2.3, in all cases, the DEM provides the best results, that is way, we compared it with three of the most well-known neural networks classifiers: the MLP, the RBFNN, and SVM. From the experimental results, we can conclude that the DEM provided very competitive results, resulting thus in a competitive alternative to solve pattern classification problems.

Nowadays, we are trying parallel implementations of MNNDP in GPU platforms. We are also replacing the MLP part in a convolutional neural network to verify if it can obtain better performances.

Acknowledgements E. Zamora and H. Sossa would like to acknowledge UPIITA-IPN and CIC-IPN for the support to carry out this research. This work was economically supported by SIP-IPN (grant numbers 20160945, 20170836 and 20161116, 20170693, 20180730, 20180180), and CONACYT (grant number 155014 (Basic Research) and grant number 65 (Frontiers of Science). F. Arce acknowledges CONACYT for the scholarship granted toward pursuing his PhD studies.

References

Arce, F., Zamora, E., Sossa, H., & Barrón, R. (2016). Dendrite morphological neural networks trained by differential evolution. In *Proceedings of 2016 IEEE Symposium Series on Computational Intelligence (SSCI)*, Athens, Greece (vol. 1, pp. 1–8).

Arce, F., Zamora, E., Sossa, H., & Barrón, R. (2017). Differential evolution training algorithm for dendrite morphological neural networks. Under Review in *Applied Soft Computing*.

Ardia, D., Boudt, K., Carl, P., Mullen, K., & Peterson, B. (2011). Differential evolution with DEoptim: An application to non-convex portfolio optimization. *The R Journal, 3*(1), 27–34.

Asuncion, D. (2007). UCI machine learning repository. [online] Available at: http://archive.ics.uci.edu/ml/index.php.

Barmpoutis, A., & Ritter, G. (2006). Orthonormal basis lattice neural networks. In *Proceedings of the IEEE International Conference on Fuzzy Systems*, Vancouver, British Columbia, Canada (vol. 1, pp. 331–336).

Broomhead, D., & Lowe, D. (1988a). Radial basis functions, multi-variable functional interpolation and adaptive networks. (Technical Report). RSRE. 4148.

Broomhead, D., & Lowe, D. (1988b). Multivariable functional interpolation and adaptive networks. *Complex Systems, 2*(3), 321–355.

Cortes, C., & Vapnik, V. (1995). Support vector networks. *Machine Learning, 20*(3), 273–297.

Guevara, E. (2016). *Method for training morphological neural networks with dendritic processing*. Ph.D. Thesis. Center for Computing Research. National Polytechnic Institute.

Guang, G., Zhu, Q., & Siew, Ch. (2006). Extreme learning machine: theory and applications. *Neurocomputing, 70*(1–3), 489–501.

Huang, G., Huang, G., Song, S., & You, K. (2015). Trends in extreme learning machines: A review. *Neural Networks, 61,* 32–48.

McCulloch, W., & Pitts, W. (1943). A logical calculus of the ideas immanent in nervous activity. *Bulletin of Mathematical Biophysics, 5,* 115–133.

Ojeda, L., Vega, R., Falcon, L., Sanchez-Ante, G., Sossa, H., & Antelis, J. (2015). Classification of hand movements from non-invasive brain signals using lattice neural networks with dendritic processing. In *Proceedings of the 7th Mexican Conference on Pattern Recognition (MCPR)* LNCS 9116, Springer Verlag (pp. 23–32).

Ritter, G., & Beaver, T. (1999). Morphological perceptrons. In *Proceedings of the International Joint Conference on Neural Networks (IJCNN)*. Washington, DC, USA (vol. 1, pp. 605–610).

Ritter, G., Iancu, L., & Urcid, G. (2003). Morphological perceptrons with dendritic structure. In *Proceedings of the 12th IEEE International Conference in Fuzzy Systems (FUZZ)*, Saint Louis, Missouri, USA (vol. 2, pp. 1296–1301).

Ritter, G., & Schmalz, M. (2006). Learning in lattice neural networks that employ dendritic computing. In *Proceedings of the 2006 IEEE International Conference on Fuzzy Systems (FUZZ)*, Vancouver, British Columbia, Canada (vol. 1, pp. 7–13).

Ritter, G., & Urcid, G. (2007). Learning in lattice neural networks that employ dendritic computing. *Computational Intelligence Based on Lattice Theory, 67,* 25–44.

Ritter, G., Urcid, G., & Valdiviezo, J. (2014). Two lattice metrics dendritic computing for pattern recognition. In *Proceedings of the 2014 IEEE International Conference on Fuzzy Systems (FUZZ)*, Beijing, China (pp. 45–52).

Rosenblatt, F. (1958). The perceptron: A probabilistic model for information storage and organization in the brain. *Psychological Review, 65*(6), 386–408.

Rosenblatt, F. (1962). *Principles of neurodynamics: Perceptron and theory of brain mechanisms* (1st ed.). Washington, DC, USA: Spartan Books.

Sossa, H., & Guevara, E. (2013a). Modified dendrite morphological neural network applied to 3D object recognition. In *Proceedings of the Mexican Conference on Pattern Recognition (MCPR)*, LNCS (vol. 7914, pp. 314–324).

Sossa, H., & Guevara, E. (2013b). Modified dendrite morphological neural network applied to 3D object recognition on RGB-D data. In *Proceedings of the 8th International Conference on Hybrid Artificial Intelligence Systems (HAIS)*, LNAI (vol. 8073, pp. 304–313).

Sossa, H., & Guevara, E. (2014). Efficient training for dendrite morphological neural networks. *Neurocomputing, 131,* 132–142.

Sossa, H., Cortés, G., & Guevara, E. (2014). New radial basis function neural network architecture for pattern classification: First results. In *Proceedings of the 19th Iberoamerican Congress on Pattern Recognition (CIARP)*, Puerto Vallarta, México, LNCS (vol. 8827, pp. 706–713).

Sussner, P., & Esmi, E., (2009). An introduction to morphological perceptrons with competitive learning. In *Proceedings of the 2009 International Joint Conference on Neural Networks (IJCNN)*, Atlanta, Georgia, USA (pp. 3024–3031).

Sussner, P., & Esmi, E. (2011). Morphological perceptrons with competitive learning: Lattice-theoretical framework and constructive learning algorithm. *Information Sciences, 181* (10), 1929–1950.

Vega, R., Guevara, E., Falcon, L., Sanchez, G., & Sossa, H. (2013). Blood vessel segmentation in retinal images using lattice neural networks. In *Proceedings of the 12th Mexican International Conference on Artificial Intelligence (MICAI)*, LNAI (vol. 8265, pp. 529–540).

Vega, R., Sánchez, G., Falcón, L., Sossa, H., & Guevara, E. (2015). Retinal vessel extraction using lattice neural networks with dendritic processing. *Computers in Biology and Medicine, 58,* 20–30.

Zamora, E., & Sossa, H. (2016). Dendrite morphological neurons trained by stochastic gradient descent. In *Proceedings of the 2016 IEEE Symposium Series on Computational Intelligence (SSCI 2016)*, Athens, Greece (pp. 1–8).

Zamora, E., & Sossa, H. (2017). Dendrite morphological neurons trained by stochastic gradient descent. *Neurocomputing, 260,* 420–431.

Chapter 3
Mobile Augmented Reality Prototype for the Manufacturing of an All-Terrain Vehicle

Erick Daniel Nava Orihuela, Osslan Osiris Vergara Villegas, Vianey Guadalupe Cruz Sánchez, Ramón Iván Barraza Castillo and Juan Gabriel López Solorzano

Abstract In this chapter, a mobile augmented reality prototype to support the process of manufacturing an all-terrain vehicle (ATV) is presented. The main goal is assisting the automotive industry in the manufacturing process regarding vehicle design and new model's introduction; in addition, the activities of training and quality control can be supported. The prototype is composed of three main stages: (a) welding inspection, (b) measuring of critical dimensions inspection, and (c) mounting of virtual accessories in the chassis. A set of 3D models and 2D objects was used as virtual elements related to augmented reality. The prototype was tested regarding usability in a real industrial stage by measuring the scope of markers' detection and by means of a survey. The results obtained demonstrated that the prototype is useful for the manufacturing of an ATV.

Keywords Mobile augmented reality · Automotive manufacturing
All-terrain vehicle · Android OS · Unity 3D · Vuforia

3.1 Introduction

Mechatronics is the science of intelligent machines, since its dissemination has been useful for the development of several industries such as manufacturing, robotics, and automotive (Bradley et al. 2015). Particularly, most complex innovations in the automotive industry are highly integrated mechatronics systems that include electronic, mechanical, computer, and control structures (Bradley 2010).

E. D. Nava Orihuela · O. O. Vergara Villegas (✉) · V. G. Cruz Sánchez
R. I. Barraza Castillo · J. G. López Solorzano
Instituto de Ingeniería y Tecnología, Universidad Autónoma
de Ciudad Juárez, Ciudad Juárez, Chihuahua, Mexico
e-mail: overgara@uacj.mx

O. O. Vergara Villegas et al. (eds.), *Advanced Topics on Computer Vision, Control and Robotics in Mechatronics*,
https://doi.org/10.1007/978-3-319-77770-2_3

The automotive industry is related with the design, development, manufacturing, marketing, and selling of motor vehicles and is considered as one of the main drivers for the development of a nation's industrial economy (Schoner 2004). Companies in the automotive industry are divided into two segments: car manufacturers and car parts' manufacturers. China and USA are considered the largest automobile markets worldwide, in terms of production and sales. Only in 2017, global sales of passenger cars were estimated at 77.7 million vehicles. However, the process of manufacturing a car is complex and involves many parts and electronics (Liu et al. 2015).

A particular branch in the automotive industry is the related with the manufacturing of all-terrain vehicles (ATV) introduced in the USA in 1971 (Benham et al. 2017). In this industry, aesthetic and functional changes in the design of the existed ATVs happened frequently. The changes impact not only the design but also the manufacturing of the vehicle. All the changes are subjected to rigorous quality controls and must comply with strict security measures. Therefore, the process of personal training for design and manufacture of this kind of vehicles is complex and needs to be done rapidly.

In recent years, technologies such as virtual reality (VR) and augmented reality (AR) have been used in training scenarios with promising results (Gavish et al. 2015). VR consists of a 3D computer-generated environment updated in real time that allows human interaction through various input/output devices, while AR technology refers to the inclusion of virtual elements in views of actual physical environments, in order to create a mixed reality in real time. It supplements and enhances the perceptions humans gain through their senses in the real world (Mota et al. 2017).

With the recent studies, it has been proved that AR offers competitive advantages over VR because it offers a more natural interface and enriches the reality (Gavish et al. 2015; Westefield et al. 2015). Moreover, mobile AR (MAR) attracted interest from industry and academy because it supplements the real world of a mobile user with computer-generated virtual contents (Chatzopoulos et al. 2017). In addition to the mobility offered to the user, MAR uses the sensors included in the device such as accelerometers, gyroscope, and global positioning system (GPS), which offer additional attractive features against computer-based AR.

Based on the potential offered by MAR, in this chapter, we propose the design and the construction of a MAR prototype to support the operations of manufacturing an ATV that includes welding inspection, measuring of critical dimensions and accessories mounting.

The rest of the chapter is organized as follows. In Sect. 3.2, a brief introduction to the manufacturing of ATVs is presented. In Sect. 3.3, a literature review of the works that uses AR in manufacturing processes is showed. The proposed methodology to create the MAR prototype is presented in Sect. 3.4. In Sect. 3.5, the results obtained from experiments and its correspondent discussion are shown. Finally, Sect. 3.6 presents the conclusions obtained with this research, followed by the further works.

3.2 All-Terrain Vehicles

An ATV is a four-wheel motorized transport fueled by gasoline or diesel, which is also known as a quad-bike or four-wheeler (Azman et al. 2014). As can be observed in Fig. 3.1, an ATV is composed by an engine part, an electric system, a brake system, a power terrain (engine), a transmission, a chassis, a steering system (handlebar), a seat stride, and the tires (Aras et al. 2015).

An ATV is operated by only one people, and because of large size and low pressure of its tires, it can be driven in different terrains including sand, stony ground, road, and mud (Fleming 2010). However, the use of ATVs whether for work or recreational use continues to be major contributors to fatal and serious injuries worldwide including shocks, rollovers, and falls. Therefore, the processes of design and manufacturing an ATV including quality control and safety considerations are a very important topic (Williams et al. 2014).

The chassis is the most important structure when an ATV is manufactured; it holds the engine, bodywork, and other subsystems. The structure is composed of several components made of steel alloy that are prefab by manufacturing processes such as lamination, extrusion, die cutting, and bending to obtain the required physical characteristics. After that, the components are joined (fused) in subassemblies by means of metal inert gas (MIG) welding. The correct positioning of the components to build a subassembly with the desired specifications is achieved by means of scantlings. Posteriorly, the individual subassemblies are joined by welds to finally obtain the complete structure of the chassis.

In parallel, the steel accessories of the chassis are fabricated. After that, the accessories and the chassis obtained their aesthetic features by means of a painting process. Finally, the accessories are added to the chassis structure using screws. As aforementioned, the metal mechanic manufacturing process to create the complete chassis is complex and claimant. It is very important to ensure that the complete structural system is strong, solid, and impact resistant, so that when an accident happens to avoid breakage or detachment of chassis components.

The key aspects to review by the people in charge of quality include: (1) the critical weldings to observe the importance (critical or non-critical), trajectory type,

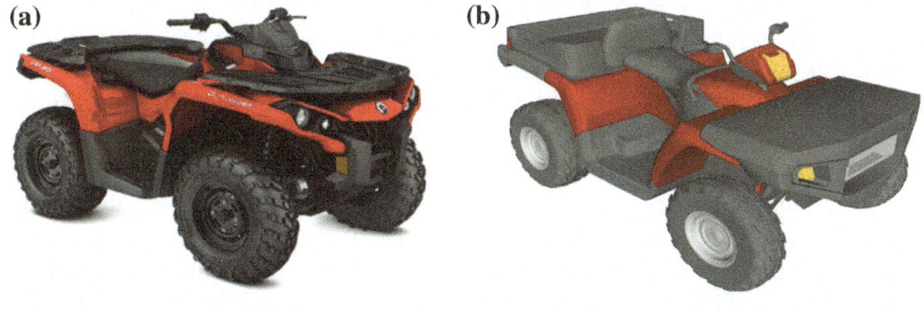

Fig. 3.1 Example of an ATV. **a** The real ATV and **b** the 3D model of an ATV

fusion, and no porosity; (2) specific dimensions, for example, the distance from one blast-hole to another, or the distance between two components; and (3) the correct mounting of the accessories to build the complete ATV structure. The measures and welding are determined by the design plans, and if they are not correct, the correspondent assembly cannot be carried out.

The difficulties encountered in the assembly lines are due to human errors or by bad weldings and out of specification dimensions, which causes that the accessories cannot be mounted properly. Therefore, it is important to create a system to support the processes of welding inspection and measure critical dimensions and accessories mounting.

3.3 Literature Review

The use of AR applications for industrial uses is increasingly common as was stated in the works of Odhental et al. (2012), Nee et al. (2012), Elia et al. (2016), Syberfeldt et al. (2017), Palmarini et al. (2018). However, in the literature, a limited number of papers have been presented which showed the ability of AR to support processes in the manufacturing industry with promising results. Following some of the papers detected in the perusal of current literature are briefly discussed.

A typical problem in operations and maintenance (O&M) practice is the collection of various types of data to locate the target equipment and facilities and to properly diagnose them at the site. In the paper of Lee and Akin (2011), an AR-based interface for improving O&M information in terms of time spent and steps taken to complete work orders was developed. The BACnet protocol was used to get sensor-derived operation data in real time from building automation system (BAS). A series of experiments was conducted to quantitatively measure improvement in equipment O&M fieldwork efficiency by using a software prototype of the application. Two research and educational facilities and their heating, ventilating, and air conditioning (HVAC) systems were used for tests: a ventilation system and a mullion system in one facility, and an air-handling unit (AHU) in the other facility. The verification tests consist of retrieval of operation data from HVAC systems in real time and superimposition of the 3D model of the mullion system. The results obtained show that with the proposal the subjects saved, on average, 51% of time spent at the task when they located target areas, and 8% of the time at task while obtaining sensor-based performance data from BAS.

The use of robots in the processes of a manufacturing plant is increasingly common for handling tasks, for example, in assembly operations. The paper of Fang et al. (2012) developed an AR system (RPAR-II) to facilitate robot programming and trajectory planning considering the dynamic constraints of the robots. The users are able to preview the simulated motion, perceive any possible overshoot, and resolve discrepancies between the planned and simulated paths prior to the execution of a task. A virtual robot model, which is a replicate of a real robot, was used to perform and simulate the task planning process. A hand-held device, which is attached with a marker-cube, was used for human–robot interaction in the task and path planning processes. By means of a pick-and-place simulation, the performance of the trajectory planning and the fitness of the selection of the robot controller model/parameters in the robot programming process can be visually evaluated.

Because maintenance and assembly tasks can be very complex, training technicians to efficiently perform new skills is challenging. Therefore, the paper of Webel et al. (2013), presented an AR platform that directly links instructions on how to perform the service tasks to the machine parts that require processing. The platform allows showing in real time the step-by-step instructions to realize a specific task and, as a result, accelerating the technician's acquisition of new maintenance procedures. The experimental task was composed of 25 steps grouped into six subtasks to assemble an electro-mechanical actuator. Twenty technicians with at least 2 years of experience on field assembly/disassembly operations served as participants. The sample was divided into two groups of ten participants: the control group executes the task by watching videos and the second group using AR. The execution time of the task was enhanced in 5%, and the affectivity rate obtained was 77% using AR.

Maintenance is crucial in prolonging the serviceability and lifespan of the equipment. The work of Ong and Zhu (2013) presented an AR real-time equipment maintenance system including: (1) context-aware information to the technicians, (2) a mobile user interface that allows the technicians to interact with the virtual information rendered, (3) a remote collaboration mechanism that allows the expert to create and provide AR-based visual instructions to the technicians, and (4) a bidirectional content creation tool that allows dynamic AR maintenance contents creation offline and on-site. The system was used to assist the machinist and maintenance engineers in conducting preventive and corrective computer maintenance activities. From the studies conducted, it was found that providing context-aware information to the technicians using AR technology can facilitate the maintenance workflow. In addition, allowing the remote expert to create and use AR-based visual interactions effectively enables more efficient and less error prone remote maintenance.

For decades, machine tools have been widely used to manufacture parts for various industries including automotive, electronics and aerospace. Due to the pursuit of mechanical precision and structural rigidity, one of the main drawbacks in machine tool industry is the use of traditional media, such as video and direct mail advertising instructional materials. In order to solve this, the machine tools augmented reality (MTAR) system for viewing machine tools from different angles with 3D demonstrations was developed by Hsien et al. (2014). Based on markerless AR, the system can integrate real and virtual spaces using different platforms, such as a webcam, smartphone, or tablet device without extra power or demonstration space. The clients can project the virtual information to a real field and learn the features of the machine form different angles and aspects. The technology also provides information for area planning.

AR is a technology that has contributed to the development of the Industry 4.0 due to its flexibility and adaptability to the production systems. The works of Gattullo et al. (2015a, b) designed a solution to the crucial problem of legibility of text observed through a head-worn display (HWD) in industrial environments. Legibility depends mainly on background, display technology (see-through optical or video HWDs), and text style (plain text, outline, or billboard). Furthermore, there are constraints to consider in industrial environments, such as standard color-coding practices and workplace lighting. The results suggest that enhancing text contrast via software, along with using the outline or billboard style, is an effective practice to improve legibility in many situations. If one text style is needed for both types of HWD, then colored billboards are effective. When color coding is not mandatory, white text and blue billboard are more effective than other styles.

The paper of Yew et al. (2016) describes an AR manufacturing system that aims to improve the information perception of the different types of workers in a manufacturing facility and to make interaction with manufacturing software natural and efficient. The traditionally paper-based and computer-based tasks are augmented to the workers' interactions in the environment. The system was distributed and modular as the different functions of CAD/CAM software are provided by individual physical or virtual objects such as CNC machines and CAD designs in the environment or by a combination of them working cooperatively. For testing purposes, a scenario following the interactions of two engineers with their AR environment was proposed obtaining good results.

AR was used in automotive industry for service training and assistance. The work of Lima et al. (2017) presented a complete markerless tracking solution to the development of AR applications for the automotive industry. The case study consists of accurate tracking of vehicle components in dynamic and sometimes noisy environments. The first scenario comprised tracking a rotating vehicle, and the second scenario involved capturing and tracking of different parts of a real Volkswagen Golf™. For both scenarios, third tasks were defined including tracking the engine part, tracking the interior from driver's seat, and tracking of the trunk (tracking from a bright into a dark environment). The proposed system allows automatic markerless model generation without the need of a CAD model, also the model covered several parts of the entire vehicle, unlike other systems that focus only on specific parts. The main positive aspect is that regular users are able to track the vehicle exterior and identify its parts.

Finally, the use of a projector-based spatial augmented reality system to highlight spot-weld locations on vehicle panels for manual welding operators was presented by Doshi et al. (2017). The goal of the work was to improve the precision and accuracy of manual spot-weld placements with the aid of visual cues in the automotive industry. Production trials were conducted, and techniques developed to analyze and validate the precision and accuracy of spot-welds both with and without the visual cues. A reduction of 52% of the standard deviation of manual spot-weld placement was observed when using AR visual cues. All welds were within the required specification, and panels evaluated in this study were used as the final product made available to consumers.

As can be observed from the literature review, most of the works use markers as the core to show the AR, only one work implemented markerless AR. None of the papers addressed the measuring of critical dimensions and mounting accessories for ATV manufacturing. However, the welding inspection for automotive purposes was addressed by the work of Doshi et al. (2017). It should be noted that the work of Doshi et al. (2017) checks the welds only in plane panels, unlike our work which checks welds even on irregular surfaces. On the other hand, none of the papers reviewed included a usability study such as the one presented in our chapter. The study is important to measure if the system complies with the initial goal and if it is ease of use.

The most observed applications focused on maintenance and training operations in different industries, including two works for automotive. It is important to note that in all the works revised the ability of AR to enhance some task is always highlighted. Motivated from the revision above, in the following section, the proposal of a methodology to create a MAR prototype to support the manufacturing of an ATV is proposed.

3.4 Proposed Methodology

The methodology for building the MAR prototype, as it is shown in Fig. 3.2, comprises five main stages: (1) selection of development tools, (2) selection and design of 3D models, (3) markers design, (4) development of the MAR application, and (5) graphical user interface (GUI) design. The individual stages of the methodology are deeply explained in the following subsections.

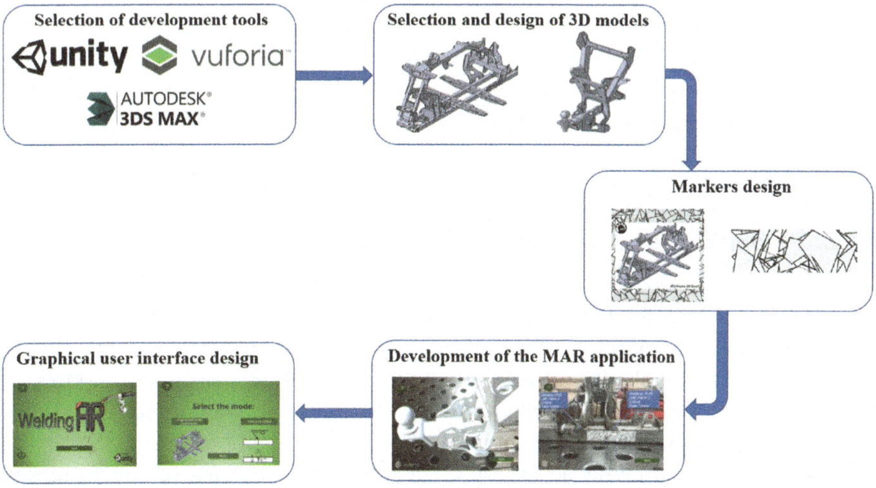

Fig. 3.2 Stages of building the MAR prototype

3.4.1 Selection of Development Tools

Three software packages were used to build the core of the MAR prototype. The selection was made by an exhaustive analysis of the commercial software for AR development. At the end, the softwares selected were Autodesk 3DS Max, Vuforia, and Unity 3D, all of them in its free or educational versions.

3DS Max is software for graphics creation and 3D modeling developed by Autodesk and contains integrating tools for 3D modeling, animation, and rendering. One of the advantages is that 3DS has an educational version that includes the same functionalities of the professional version. The software was used for the creation of all the 3D models and animations of the MAR prototype (Autodesk 2017).

Vuforia software developer kit (SDK) was selected because it is a powerful platform that contains the necessary libraries to carry out the tasks related to AR including the markers detection, recognition and tracking, and the computations for object superimposition. Nowadays, Vuforia is the world's most widely deployed AR platform (PTC Inc. 2017).

Unity is a multiplatform game engine created by Unity Technologies that offer the possibility of building 3D environments. It was selected because of the facility of having control of the content of the mobile device in a local way. In addition, the visual environment of the platform provides a transparent integration with Vuforia. The language C# was used to create script programming which includes all the logical operations of the MAR prototype Unity and includes an integrated system that allows the creation of a GUI for execution at different platforms including iPhone operating system (iOS), Android, and universal windows platform (UWP). It is compatible with 3D graphics and animations created by 3DS such as *.max, *.3ds, *.fbx, *.dae, *.obj, among others (Unity Technologies 2017).

The integration of Unity and Vuforia is explained in Fig. 3.3. The MAR application is fully designed in Unity including all the programming logic related to system navigation and 3D model's behavior. The necessary resources to create AR are taken from Vuforia that includes administration (local, remote), detection, recognition, and tracking of all the markers. Finally, the developer defines the 3D models and animations associated with each marker.

3.4.2 Selection and Design of 3D Models

Two different ATVs models known as short chassis ATV and large chassis ATV were selected as the core for 3D modeling purposes. The selection was mainly due to the associated complexity of fabrication and assembly, because both models are the most sold in the company where the MAR prototype was implemented. The short and large ATV chassis are shown in Fig. 3.4.

In addition, six different accessories, including (a) the arms of the front suspension, (b) the arms of the rear suspension, (c) the tail structure (seat support),

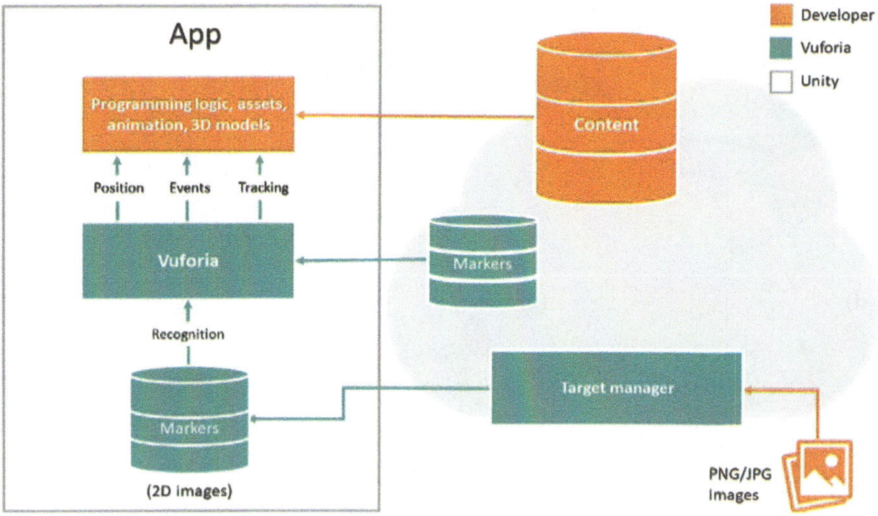

Fig. 3.3 Unity and Vuforia integration scheme to develop the MAR prototype

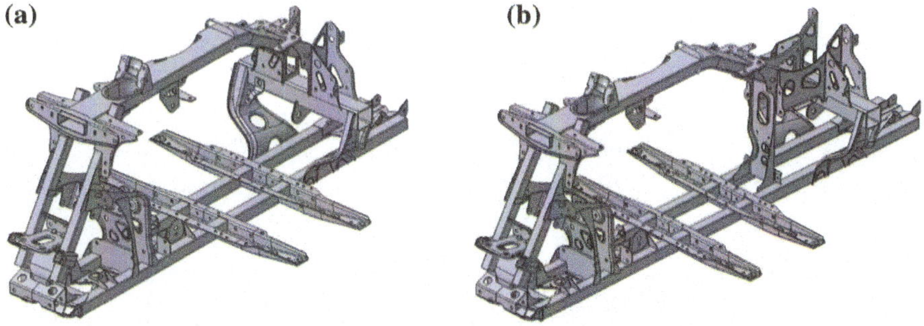

Fig. 3.4 Two versions of the ATV chassis. **a** Short, and **b** large

(d) the front bumper, (e) the rear loading structure, and (f) the steering column, were selected and 3D modeled. In the real process, the accessories are added to the chassis by means of temporal mechanical joints (screws) to shape the final ATV structure. The main idea is to make a montage of the 3D models of the accessories over the physical chassis, to observe the critical dimensions, and the weldings that will be inspected and controlled in the ATV manufacturing process. The 3D models of the six accessories selected are shown in Fig. 3.5.

The original 3D models of the chassis and the six accessories were originally designed by the manufacturing company in the computer-aided three-dimensional interactive application (CATIA) software, with a file extension *.CATPart. Therefore, the models were converted from CATPart to STEP format. Finally, STEP format was opened in 3DS Max and saved as *.max file, which is compatible with Vuforia, and

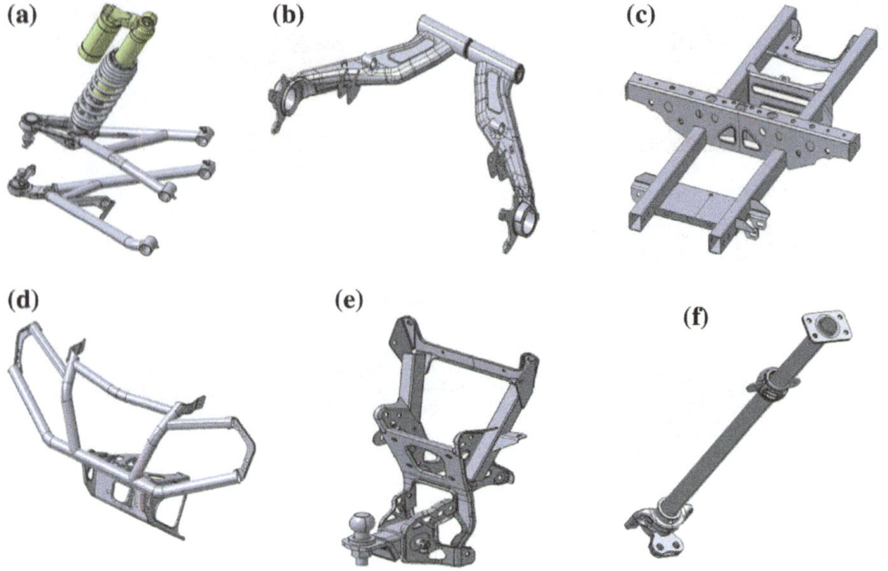

Fig. 3.5 3D models of the accessories selected. **a** The arms of the front suspension, **b** the arms of the rear suspension, **c** the tail structure, **d** the front bumper, **e** the rear loading structure, and **f** the steering column

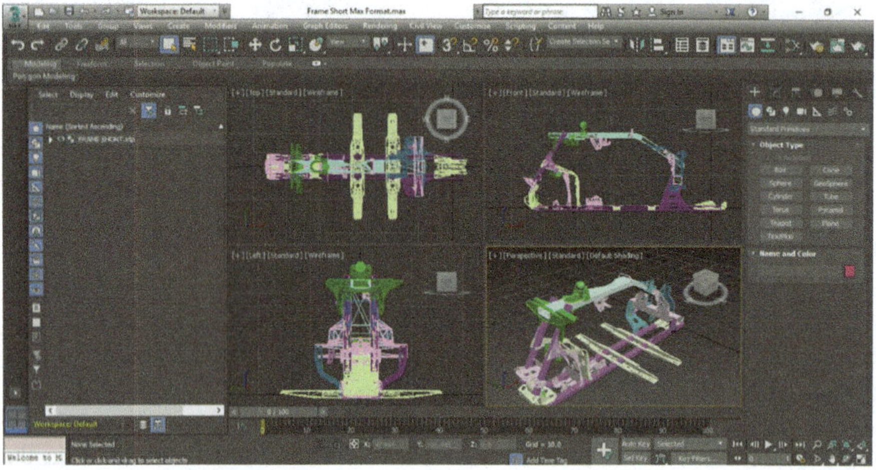

Fig. 3.6 Short chassis ATV modeled in 3DS Max

this was to allow model manipulation in the MAR prototype. The file in 3DS preserves the original model geometries and creates the necessary meshes with graphics features to be projected in an AR application. In Fig. 3.6, the model of the short chassis ATV represented in 3DS is shown.

3.4.3 Markers Design

Markers in conjunction with the programming scripts for detection and tracking are one of the main parts of the MAR prototype. The design of the markers associated with the 3D models designed was made with the AR marker generator Brosvision (2017). The generator uses an algorithm for the creation of images with predefined patterns composed of lines, triangles, and rectangles. A unique image is created randomly in full color or in gray scale.

In this stage, nine markers with different sizes were created for the MAR prototype. The size of a marker was defined in accordance with the physical space which will be located. The first two markers named Short and Max were associated with the two chassis sizes and allow to virtually observe the particular size of the chassis (short or large) as shown in Fig. 3.7.

The seven remaining markers were associated with the six selected accessories mentioned in Sect. 3.4.2 and will be mounted in the real chassis to execute the superimposition of the associated 3D models. It is important to mention that in the experimentation process, it was detected that for the case of the arms of the front suspension, the use of only one marker was not sufficient to observe the entire details. Therefore, an additional marker was created; one was used on left and the other on right side of the ATV, obtaining the total quantity of seven. The 3D model of the arms of the front suspension associated with the additional marker is just a mirror of the original ones. The set of seven markers associated with accessories are shown in Fig. 3.8.

It should be noted that markers have a scale of 1:2 and with the adequate proportion to be collocated in strategic parts of the chassis. The markers shown in Fig. 3.8a, b will be located at left and right lower tubes of the frontal suspension,

(a) (b)

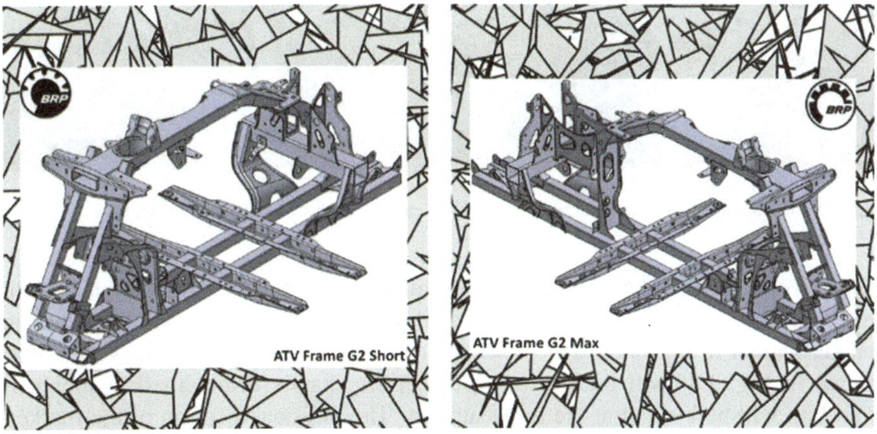

Fig. 3.7 Markers associated with ATV chassis. **a** Short, and **b** Max

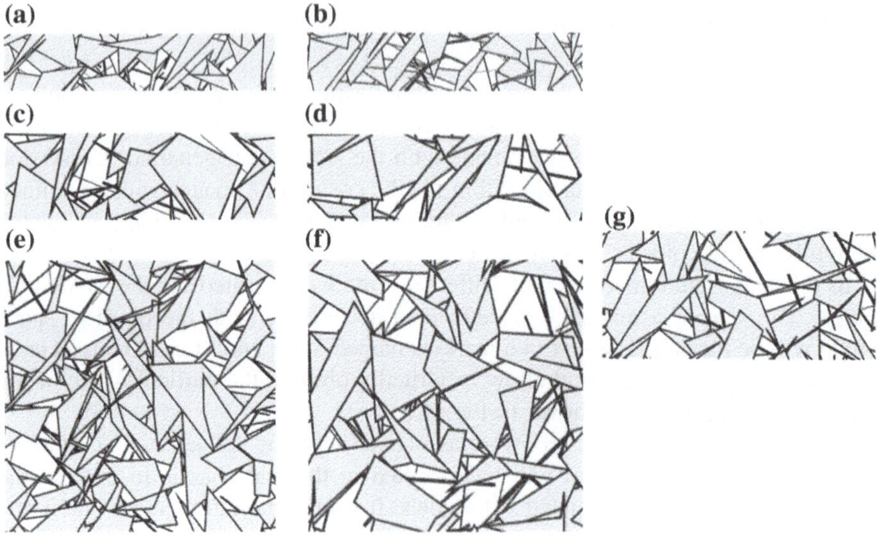

Fig. 3.8 Seven markers. **a** Front_susp_left, **b** Front_susp_right, **c** Rear_arm_suspension, **d** Rear_structure, **e** Steering_column, **f** Tail_structure, and **g** Front_bumper

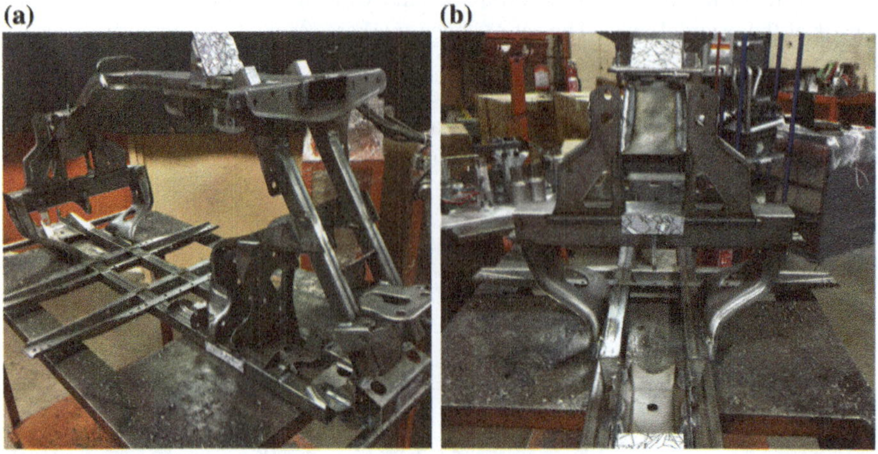

Fig. 3.9 Location of the markers in a short ATV. **a** Front section, and **b** rear section

respectively; marker shown in Fig. 3.8c will be located at the support tube of the rear suspension; marker shown in Fig. 3.8d will be located at the end of the chassis; marker shown in Fig. 3.8e will be located at the steering column bracket; marker shown in Fig. 3.8f will be located at the upper chassis beam; marker shown in Fig. 3.8g will be located at the front support. The physical location of the markers in a short chassis ATV can be observed in Fig. 3.9.

3.4.4 Development of the MAR Application

In this stage, the MAR application was developed and it is important to note that Android operating system (OS) was selected for deployment. In the first step, it is necessary to import the markers created with Brosvision to Vuforia by means of the *target manager*. To do this, a database where the markers will be stored was created. The store location can be remotely (cloud) or directly in the mobile device; for this chapter and because of facility, the last one was selected. After that, the markers were added to the database in joint photographic experts group (JPEG) format.

Once the database was created, each marker was subjected to an evaluation performed by Vuforia to measure the ability of detection and tracking. Vuforia uses a set of algorithms to detect and track the features that are present in an image (marker) recognizing them by comparing these features against a local database. A star rating is assigned for each image that is uploaded to the system. The star rating reflects how well the image can be detected and tracked, and can vary among 0–5. The higher rating of an image target, the stronger the detection and tracking ability it contains. A rating of zero explains that a target would not be tracked at all, while an image given a rating of 5 would be easily tracked by the AR system. The developers recommend to only using image targets that result in 3 stars an above. The rating of stars obtained for each of the nine markers used in the MAR prototype is shown in Fig. 3.10.

It should be noted from Fig. 3.10 that all the markers used in the MAR prototype obtained at least a rating of three stars, which means that are appropriate for AR purposes. In addition, every single marker was analyzed in a detailed way to observe the set of traceable points (fingerprints) as it is shown in Fig. 3.11.

After the process of marker rating, the creation of AR scenes is carried out using the AR camera prefab offered by Vuforia. The camera was included by dragging it toward the utilities tree. The configuration of the AR camera includes the license and the definition of the maximum number of markers to track and detect.

In a similar way than AR camera, the markers must be added to the utilities tree. In this part, the database that contains the markers was selected, and the respective markers to detect were defined. At this time, a Unity scene is ready and able to detect and track the markers and display the related 3D models. The 3D models also must be imported by dragging it from its location in a local directory to Unity interface. The models can be observed in the assets menu. Afterward, each model was associated with a particular marker by a dragging action similar to the previously explained.

Once that the main AR functionality was explained, the three experiences related to welding inspection, measuring of critical dimensions, and accessories mounting were developed.

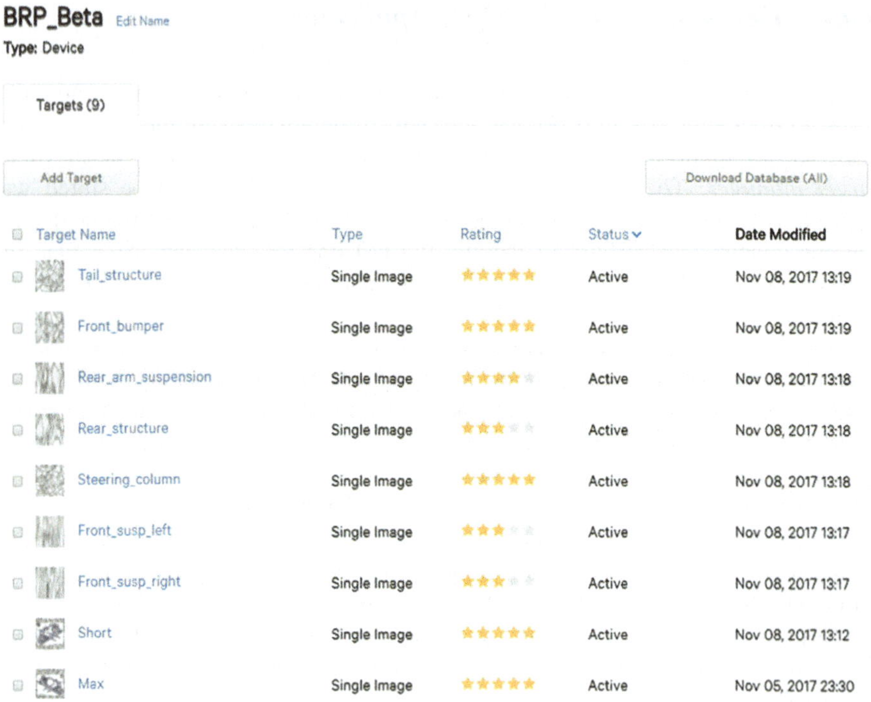

Fig. 3.10 Rating of each marker obtained by Vuforia

3.4.4.1 Welding Inspection

Quality control in the manufacturing of an ATV is very important to ensure the vehicle safety. One of the most crucial stages consists on to verify the welding features against the specifications defined by the quality department. In other words, the minimum requirements that a weld must comply are established and checked.

The welding inspection AR scene contains two 2D components to show the information regarding a particular weld. The first component displays information inside a square related to the weld number assigned by the metallurgy department, the location of the workstation (cell) where the weld was made, the importance of the weld (critical or non-critical), and the weld trajectory type (linear, circular, oval). The second component consists of predefined virtual arrows that are included to the AR scene to indicate the welds that are inspected. An example of the final scene for welding inspection at different points of the ATV chassis is shown in Fig. 3.12. As can be observed, the camera of the mobile device is pointed out in front of one of the seven markers, and the information regarding weldings in that particular locations superimposed inside the real scene, generating the AR.

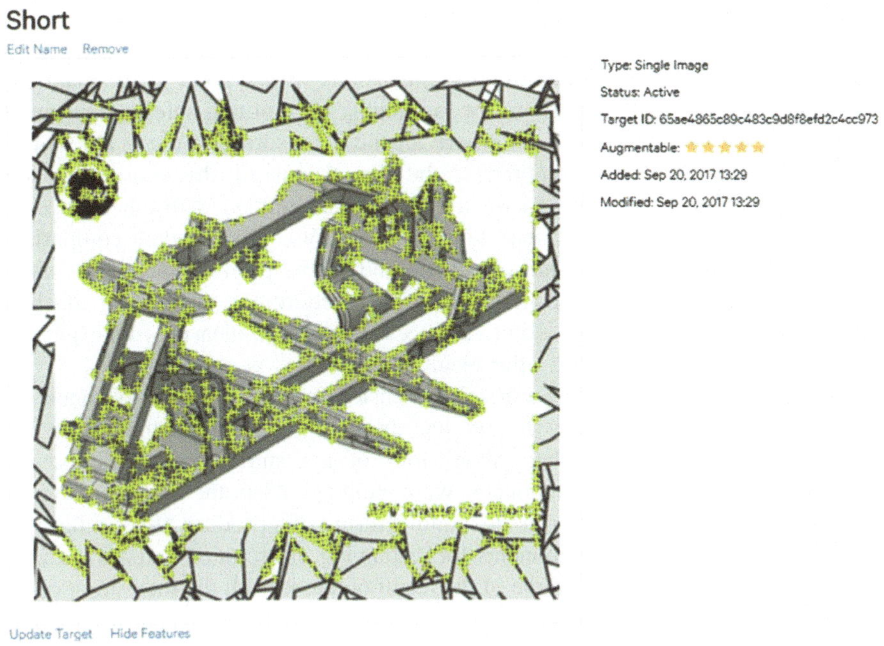

Fig. 3.11 Fingerprints obtained by Vuforia for short marker

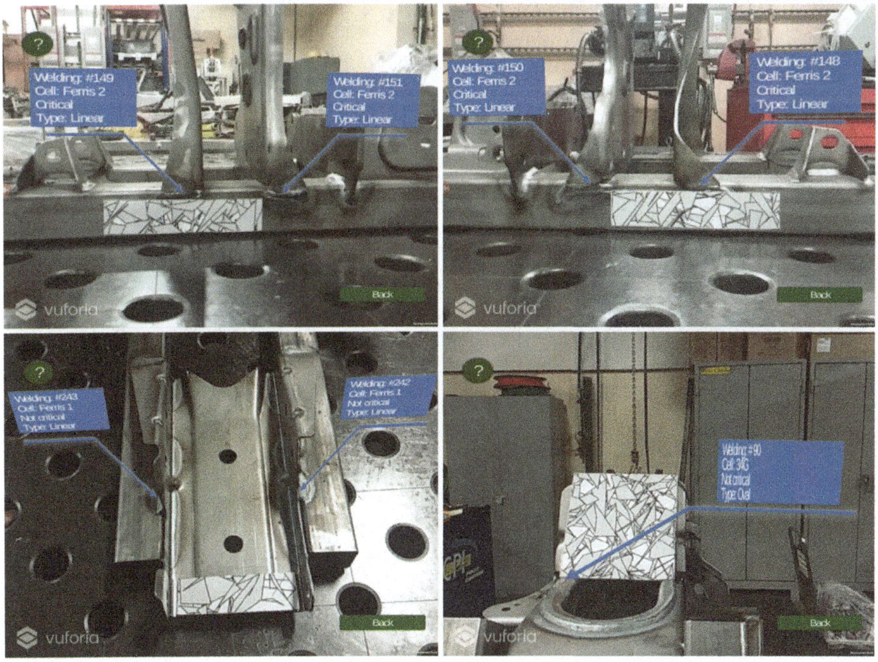

Fig. 3.12 Example of the AR scenes for welding inspection

3.4.4.2 Measuring Critical Dimensions

The possibility of reviewing the product dimensions with respect to the manufacturing plans is an important activity for the quality control and safety department. The chassis is the base component where all the accessories of the ATV will be mounted and assembled. Therefore, if the dimensions of the chassis are not according to the manufacturing plans, it cannot be assembled with other pieces.

Currently, the revision of critical dimensions and its correspondent comparison with the nominal value is carried out at the end of the production process using gauges. In addition, specialized machines such as coordinate measuring machine (CMM) are used. However, it is not easy to see full-scale measures with gauges, and the time taken by CMM to offer results of measures is quite long.

The AR tool for measuring critical dimensions offers a guide to check the measures that impact the quality of a product. In the final prototype, the dimensions from one component to another, dimensions of a manufacturing process, and dimensions from individual components were included. Also, the tool can serve for fast training of people that labor in the manufacturing of a product. In a similar way to welding inspection, 2D components for displaying information and arrows were inserted into the scene. An example of the result obtained with the measuring of the critical dimensions tool is shown in Fig. 3.13. It should be noted that information

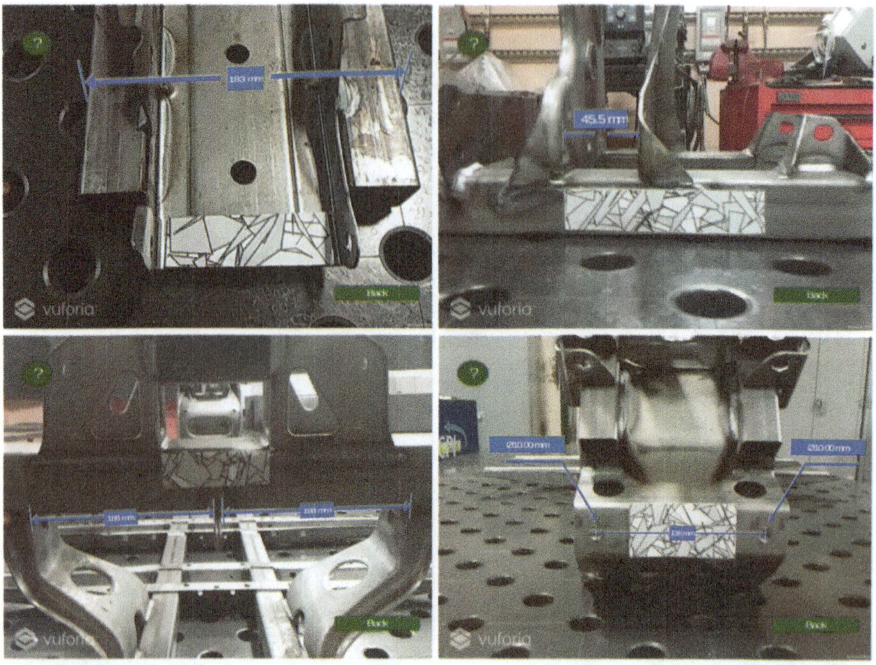

Fig. 3.13 Example of the AR scenes for measuring critical dimensions

related to a particular dimension is displayed when the camera of the mobile device is pointed out to one of the seven markers.

3.4.4.3 Accessories Mounting

The main goal of this stage is creating an AR experience to show the real place in the chassis where the accessories will be mounted. The seven markers of the accessories were placed in the chassis structure corresponding to the real location of a particular component. This tool is very important for helping in the process of training people that in the future it will construct the ATV. In this stage, unlike welding inspection and measuring of critical dimensions where only boxes and arrows were used, the transformation properties of the virtual 3D models inserted in the scene must be adjusted to determine the proper position and scale according to the size of the real accessories. The good determination of the transformation properties inside a Unity scene helps the final perspective observed by the user when the application is running on the mobile device.

The transformation properties obtained for each 3D model are shown in Table 3.1. The values obtained include the position in *X*-, *Y*-, and *Z*-axis, with the respective values in scale and orientation.

Table 3.1 Transformation properties of the 3D models

3D model	Property	*X*	*Y*	*Z*
Tail_structure	Position	−3.7	−5	0
	Rotation	−90	0	0
	Scale	0.27	0.27	0.27
Front_bumper	Position	0	−13.5	−0.4
	Rotation	0	0	90
	Scale	0.27	0.27	0.27
Rear_structure	Position	0	0	1
	Rotation	−90	−90	0
	Scale	0.27	0.27	0.27
Steering_column	Position	0	12.57	3.2
	Rotation	27.8	0	−90
	Scale	0.27	0.27	0.27
Front_susp_left	Position	0	−0.3	0.4
	Rotation	−90	0	0
	Scale	0.27	0.27	0.27
Front_susp_right	Position	0.2	0.3	0.4
	Rotation	90	0	0
	Scale	0.27	0.27	0.27
Rear_arm_suspension	Position	0	0.5	2.37
	Rotation	0	180	−90
	Scale	0.27	0.27	0.27

Fig. 3.14 AR scene for showing the accessories mounting

An example of the results obtained for mounting accessories AR scene is shown in Fig. 3.14.

3.4.5 *GUI Design*

The name of the MAR prototype is "*Welding AR*" due to the welding metalworking process for ATVs chassis manufacturing. The complete GUI structure can be observed in Fig. 3.15, and each block corresponds to one individual scene designed in Unity.

3.4.5.1 Scene Creation in Unity

The first scene created was the main screen that is displayed when the icon of the Welding AR is taped in the mobile device as shownin Fig. 3.16. The scene includes buttons to display the prototype help, for closing the application, and to start the main menu.

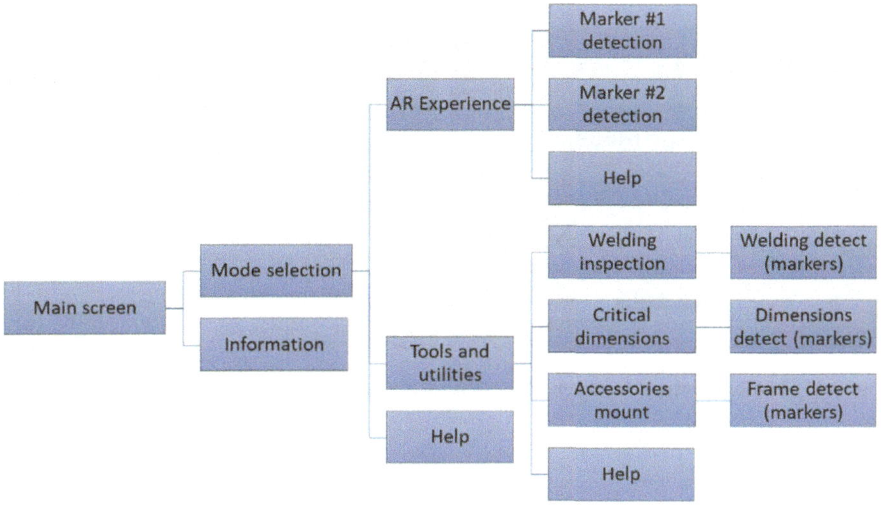

Fig. 3.15 Complete structure of the GUI

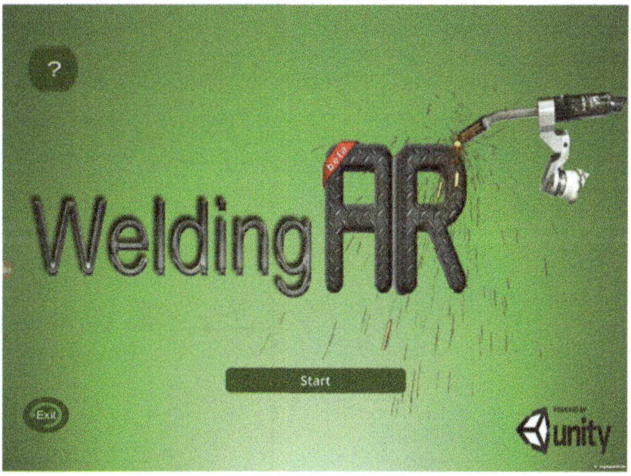

Fig. 3.16 Main scene of MAR prototype

After main scene creation, eight additional scenes were created regarding to: (1) mode selection, (2) information, (3) AR experience (observing the short and large chassis ATV), (4) tools and utilities menu, (5) Help for all the scenes, (6) welding inspection, (7) measuring critical dimensions, and (8) accessories mounting. All the scenes include buttons to follow the flow of the application and to return to the previous scene. Figure 3.17 shows the scenes of mode selection and tools and utilities.

Figure 3.18 shows the flow diagram to understand the function of the prototype regarding AR experience. The diagram was used in the MAR prototype for welding inspection, measuring critical dimensions, and accessories mounting.

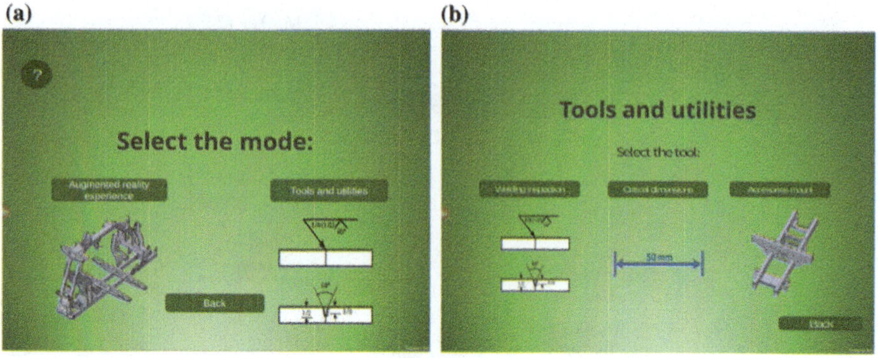

Fig. 3.17 MAR prototype scenes. **a** Mode selection, and **b** tools and utilities

Fig. 3.18 Flow diagram for AR experience

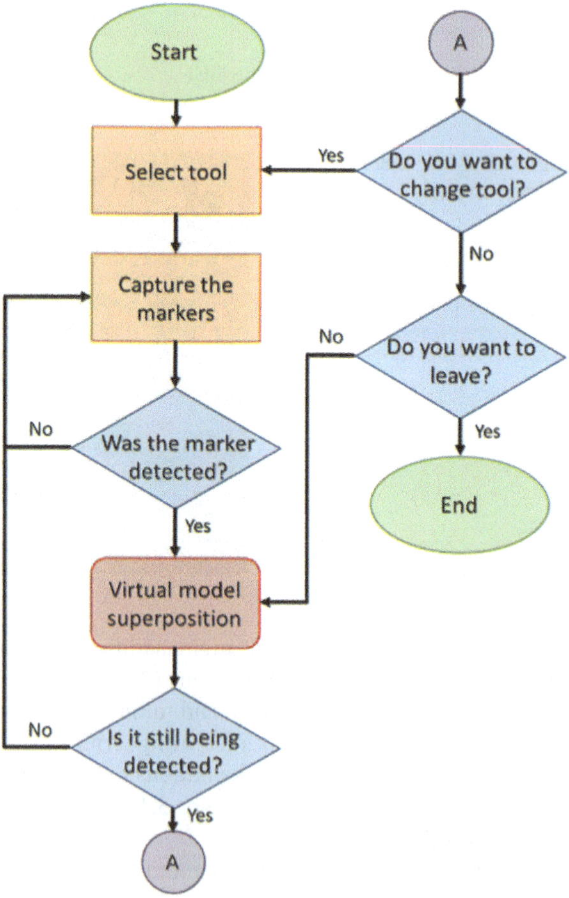

The resulting application has the *.apk extension which can be shared in a virtual store such as Google Play. Once the application is downloaded, it is deployed to the mobile device to be used.

3.5 Experimental Results

Two different tests were executed in order to measure and demonstrate the performance of the MAR prototype inside a real manufacturing industry. Both experiments are explained in the following subsections.

3.5.1 Scope of Markers Detection

The first test consists in reviewing the scope of marker detection and the behavior of the whole prototype. In this test, the measures were obtained in the real industrial environment where the ATV is manufactured, with a constant illumination of 300 lumens. By using a Bosch GLM 40 laser, and a typical flexometer, the minimal and maximal distances in centimeters to detect the 7 markers of accessories were calculated. The specifications of the two mobile devices used for testing are shown in Table 3.2.

The test was carried out by approaching the camera of the mobile device to the marker as close as possible, and after that, moving the device away until the point that the marker cannot be detected. The results obtained from the test are shown in Table 3.3.

It should be observed from Table 3.3 that the general range to detect markers is wide. In addition, even when the Galaxy S6 has better characteristics, the detection range is greater with Tab S2, concluding that this was the device with the better performance. In addition, the area covered by the marker is important for good recognition. In Table 3.4, the information about the area covered by each marker is shown.

It should be noted from Table 3.4 that the detection abilities are influenced by the area covered by the marker. For example, the marker of steering_columns is the biggest; therefore, the scope distance range is greater than the others. In conclusion,

Table 3.2 Technical specifications of the mobile devices used for tests

Brand	Model	Operating system	RAM	Camera (MP)
Samsung	Galaxy Tab S2 8.0 (SM-T713)	Android 6.0.1 (Marshmallow)	3 GB LPDDR3	8
Samsung	Galaxy S6 (SM-G920V)	Android 6.0.1 (Marshmallow)	3 GB LPDDR4	16

Table 3.3 Minimal and maximal distance to detect markers

Marker	Minimum distance (cm)		Maximum distance (cm)	
	Tab S2	Galaxy S6	Tab S2	Galaxy S6
Tail_structure	6	12.5	178	163
Steering_column	5	14	212	80
Rear_arm_suspension	9.5	17	59	48
Rear_structure	9.5	28	68	62
Front_susp_left	9.5	21.5	94	43
Front_susp_right	9.5	21.5	94	43
Front_bumper	6.5	21	147	117

Table 3.4 Area covered by the markers

Marker	Width (cm)	Height (cm)	Area (cm^2)
Tail_structure	8.8	8.8	77.44
Steering_column	8.9	8.9	79.21
Rear_arm_suspension	10.1	3.4	34.34
Rear_structure	9.1	2.8	25.48
Front_susp_left	12	2.6	31.2
Front_susp_right	12	2.6	31.2
Front_bumper	8.4	4.5	37.8

the MAR prototype allows working inside a real manufacturing scenario with different devices and different distances for pointing out the markers with good detection and tracking.

3.5.2 Survey for Users

In the second test, a questionnaire was designed for measuring the user satisfaction when using the MAR prototype inside the real manufacturing environment. Ten subjects participated in the survey, with an age ranged from 22 to 60 years. Nine subjects were men, while one was women, all of them employees of the ATV manufacturing company. In the sample, three subjects were technicians, two group chiefs, two welding engineers, one quality engineer, one supervisor, and one welder. The survey is shown in Table 3.5.

In the Likert scale used, the 1 means totally disagree, while a 10 means totally agree. Each participant received an explanation about the purpose of the survey; after that, both devices were used to test the MAR prototype. Each user takes around 15 min for testing the prototype, and after that, the survey was filled.

The results obtained for questions 1–7 are shown in Fig. 3.19, while the results obtained for question 10 are shown in Fig. 3.20. For the case of question 8, 80% of

Table 3.5 Questions of the survey

Question	Scale									
	1	2	3	4	5	6	7	8	9	10
1. Welding AR application is easy to understand and use?										
2. Welding AR application is physically or mentally demanding?										
3. Welding AR application interface is ease of use?										
4. Using Welding AR application, it's frustrating?										
5. Welding AR application will help for improving quality control?										
6. Welding AR application will be helpful to be used in manufacturing processes?										
7. Welding AR application will allow that the introduction of new chassis or modifications of current models takes less time for its manufacturing process?										
8. Welding AR application has opportunities for improvement? Which?	Yes or no, and open question									
9. Do you consider that the process of training in the department of quality control using Welding AR application will be easier and faster?	Yes or no									
10. Which kind of employees will exploit better the Welding AR application? (a) Welder, (b) trainer, (c) group chief, (d) production supervisor, (e) programmer, (f) maintenance technician, (g) quality technician, (h) engineer technician, (i) welder engineer, (j) quality engineer, (k) administrative	Multiple option selection									

the participants responded yes. The comments include augmenting the number of weldings inspected, augmenting the distance in which the prototype can detect the markers, and augmenting the number of critical dimensions measured. Most of the participants comment the benefits that could be obtained if the prototype will be installed in AR lenses such Microsoft Hololens. Finally, for the case of question 9, 100% of the participants expressed that it will be easy and fast the process of training a new employee using the MAR prototype, and this is mainly due to its visual and ease of use interface.

3.5.3 *Discussion*

By observing the results obtained for both experiments, it should be noted that the prototype is useful for supporting the ATV manufacturing process including the training stage. It is important to highlight that the users demonstrate interest in using the application and enthusiasm to include it in the dairy work. Effectively, the

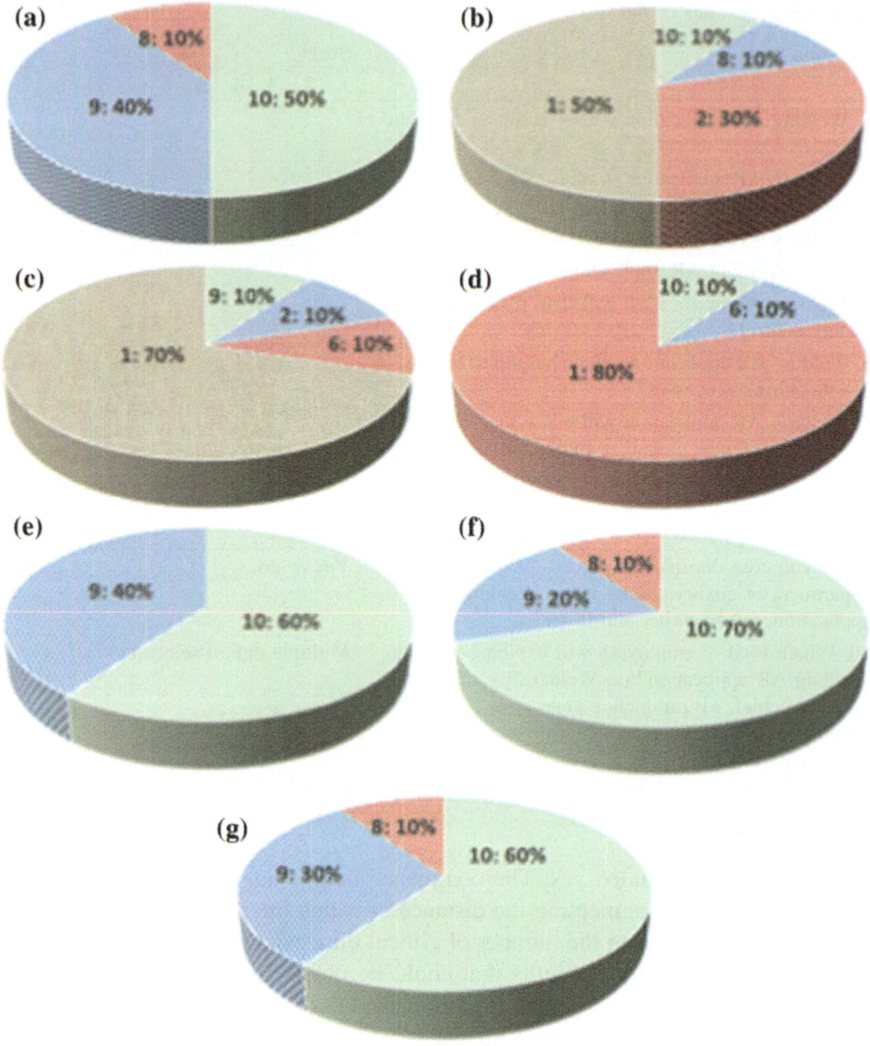

Fig. 3.19 Results obtained for questions 1–7. **a** Question 1, **b** question 2, **c** question 3, **d** question 4, **e** question 5, **f** question 6, and **g** question 7

prototype helps in the task programmed with the use of AR that includes welding inspection, measuring critical dimensions and mounting accessories.

Regarding the ability to detect the markers, it should be noted that a wide range of distances could be handled, which will help the user to observe the superimposed models at different sizes and orientations. When a detailed view is necessary, then the user approaches the device in a very short distance, if a macrovision is needed, then the user moves away from the markers.

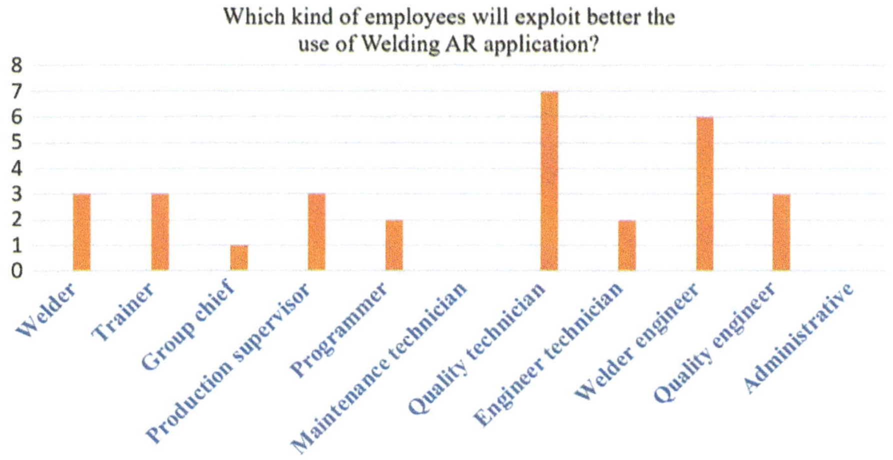

Fig. 3.20 Results obtained for question 10

It is important to highlight the ability of the prototype to be useful inside the real manufacturing environment, where changes of illumination, noisy environment, and eventually occlusions happened almost all the time.

With respect to the results obtained from the survey, we confirmed that users are interested in using the application. Nevertheless, the comments offered about improvement opportunities were very valuable to enhance the prototype in the future. At the end, the experiments allow confirming the premise that AR is a valuable technological tool that can be used to support the process of manufacturing an ATV.

3.6 Conclusions

In this chapter, a MAR prototype (Welding AR) to support the manufacturing of an ATV was presented. Particularly, the prototype serves in three crucial stages of manufacturing such as welding inspection, measuring of the critical dimensions, and accessories mounting. After the results obtained from experiments, we conclude that the prototype fully complies its function of visually showing the process of manufacturing an ATV.

The prototype can help experimented and novice users. The application executes well on mobile devices with different architectures based on Android OS. The use of this kind of technological tools is essential to finally reach the real explosion of the 4.0 industry that includes cyber-physical systems, the Internet of things (IoT), cloud computing, and cognitive computing.

Future work will be directed toward on implementing the prototype in other operating systems such as iOS. Also, it will be important to carry out a set of tests using AR glasses such as ORA Optinvent or Microsoft Hololens, which will provide the user the total mobility of the hands. It will be important to increase the number of 3D models and include more types of ATV models. It is also necessary to increase the number of welds inspected and the number of critical dimensions to measure. Finally, it will be desirable to change the functionality of the prototype from marker-based AR to a markerless system which will offer a more natural interface.

References

Aras, M., Shahrieel, M., Zambri, M., Khairi, M., Rashid, A., Zamzuri, M., et al. (2015). Dynamic mathematical design and modelling of autonomous control of all-terrain vehicles (ATV) using system identification technique based on pitch and yaw stability. *International Review of Automatic Control (IREACO), 8*(2), 140–148.

Autodesk. (2017, September). *3D modeling with Autodesk*, [On Line]. Available: https://www.autodesk.com/solutions/3d-modeling-software.

Azman, M., Tamaldin, N., Redza, F., Nizam, M., & Mohamed, A. (2014). Analysis of the chassis and components of all-terrain vehicle (ATV). *Applied Mechanics and Materials, 660,* 753–757.

Benham, E., Ross, S., Mavilia, M., Fescher, P., Britton, A., & Sing, R. (2017). Injuries from all-terrain vehicles: An opportunity for injury prevention. *The American Journal of Surgery, 214*(2), 211–216.

Bradley, D. (2010). Mechatronics—More questions than answers. *Mechatronics, 20,* 827–841.

Bradley, D., Russell, D., Ferguson, I., Isaacs, J., MacLeod, A., & White, R. (2015). The Internet of Things—The future or the end of mechatronics. *Mechatronics, 27,* 57–74.

Brosvision. (2017, September). *Augmented reality marker generator* [On Line]. Available: http://www.brosvision.com/ar-marker-generator/.

Chatzopoulos, D., Bermejo, C., Huang, Z., & Hui, P. (2017). Mobile augmented reality survey: From where we are to where we go. *IEEE Access, 5,* 6917–6950.

Doshi, A., Smith, R., Thomas, B., & Bouras, C. (2017). Use of projector based augmented reality to improve manual spot-welding precision and accuracy for automotive manufacturing. *The International Journal of Advanced Manufacturing Technology, 89*(5–8), 1279–1293.

Elia, V., Grazia, M., & Lanzilotto, A. (2016). Evaluating the application of augmented reality devices in manufacturing from a process point of view: An AHP based model. *Expert Systems with Applications, 63,* 187–197.

Fang, H., Ong, S., & Nee, A. (2012). Interactive robot trajectory planning and simulation using augmented reality. *Robotics and Computer-Integrated Manufacturing, 28*(2), 227–237.

Fleming, S. (2010). *All-terrain vehicles: How they are used, crashes, and sales of adult-sized vehicles for children's use* (1st ed.). Washington D.C., USA: Diane Publishing Co.

Gattullo, M., Uva, A., Fiorentino, M., & Gabbard, J. (2015a). Legibility in industrial AR: Text style, color coding, and illuminance. *IEEE Computer Graphics and Applications, 35*(2), 52–61.

Gattullo, M., Uva, A., Fiorentino, M., & Monno, G. (2015b). Effect of text outline and contrast polarity on AR text readability in industrial lighting. *IEEE Transactions on Visualization and Computer Graphics, 21*(5), 638–651.

Gavish, N., Gutiérrez, T., Webel, S., Rodríguez, J., Peveri, M., Bockholt, U., et al. (2015). Evaluating virtual reality and augmented reality training for industrial maintenance and assembly tasks. *Interactive Learning Environments, 23*(6), 778–798.

Hsien, Y., Lee, M., Luo, T., & Liao, C. (2014). Toward smart machine tools in Taiwan. *IT Professional, 16*(6), 63–65.

Lee, S., & Akin, O. (2011). Augmented reality-based computational fieldwork support for equipment operations and maintenance. *Automation in Construction, 20*(4), 338–352.

Lima, J., Robert, R., Simoes, F., Almeida, M., Figueiredo, L., Teixeira, J., et al. (2017). Markerless tracking system for augmented reality in the automotive industry. *Expert Systems with Applications, 82,* 100–114.

Liu, Y., Liu, Y., & Chen, J. (2015). The impact of the Chinese automotive industry: Scenarios based on the national environmental goals. *Journal of Cleaner Production, 96,* 102–109.

Mota, J., Ruiz-Rube, I., Dodero, J., & Arnedillo-Sánchez, I. (2017). Augmented reality mobile app development for all. *Computers and Electrical Engineering*, article in press.

Nee, A., Ong, S., Chryssolouris, G., & Mourtzis, D. (2012). Augmented reality applications in design and manufacturing. *CIRP Annals—Manufacturing Technology, 61*, 657–679.

Odenthal, B., Mayer, M., KabuB, W., & Schlick, C. (2012). A comparative study of head-mounted and table-mounted augmented vision systems for assembly error detection. *Human Factors and Ergonomics in Manufacturing & Service Industries, 24*(1), 105–123.

Ong, S., & Zhu, J., (2013). A novel maintenance system for equipment serviceability improvement. *CIRP Annals—Manufacturing Technology, 62*(1), 39–42.

Palmarini, R., Ahmet, J., Roy, R., & Torabmostaedi, H. (2018). A systematic review of augmented reality applications in maintenance. *Robotics and Computer-Integrated Manufacturing, 49,* 215–228.

PTC Inc. (2017, September). *Vuforia*, [On Line]. Available: https://www.vuforia.com/.

Schoner, H. (2004). Automotive mechatronics. *Control Engineering Practice, 12*(11), 1343–1351.

Syberfelt, A., Danielsson, O., & Gustavson, P. (2017). Augmented reality smart glasses in the smart factory: Product evaluation guidelines and review of available products. *IEEE Access, 5,* 9118–9130.

Unity Technologies. (2017, September). *Unity-products*, [On Line]. Available: https://unity3d.com/es/unity.

Webel, S., Bockholt, U., Engelke, T., Gavish, N., Olbrich, M., & Preusche, C. (2013). An augmented reality training platform for assembly and maintenance skills. *Robotics and Autonomous Systems, 61*(4), 398–403.

Westerfield, G., Mitrovic, A., & Billinghurst, M. (2015). Intelligent augmented reality training for motherboard assembly. *International Journal of Artificial Intelligence in Education, 25*(1), 157–172.

Williams, A., Oesch, S., McCartt, A., Teoh, E., & Sims, L. (2014). On-road all-terrain vehicle (ATV) fatalities in the United States. *Journal of Safety Research, 50,* 117–123.

Yew, A., Ong, S., & Nee, A. (2016). Towards a griddable distributed manufacturing system with augmented reality interfaces. *Robotics and Computer-Integrated Manufacturing, 39,* 43–55.

Chapter 4
Feature Selection for Pattern Recognition: Upcoming Challenges

Marilu Cervantes Salgado and Raúl Pinto Elías

Abstract Pattern recognition is not a new field, but the challenges are coming on the data format. Today's technological devices provide a huge amount of data with extensive detail evolving the classical pattern recognition approaches for dealing with them. Given the size of and quantity of descriptors data possess, traditional pattern recognition techniques have to draw on feature selection to handle problems like the excess of computer resources and dimensionality. Feature selection techniques are evolving, as well, for data related reasons. Chronologically linked data brings new challenges to the field. In the present chapter, we expose the gap in feature selection research to handle this type of data, as well as give suggestions of how to perform or pursue an approach to chronologically linked data feature selection.

Keywords Feature selection · Pattern recognition · Chronologically linked data

4.1 Introduction

Today, in the big data era, data are coming at large formats, from high-definition video to interaction posts on social media, making it hard for pattern recognition algorithms to process and make decisions with them. Pattern recognition aims to search into data for regularities that can automatically classify (among other tasks) the data into different classes.

Let's look at this last statement with a specific problem: one of the latest smartphone tackles unlocking function with facial recognition, since this task has to differentiate a face from non-face image, it has to search for facial characteristics, thus pattern recognitions is involved.

M. Cervantes Salgado · R. Pinto Elías (✉)
Centro Nacional de Investigación y Desarrollo Tecnológico (CENIDET),
Cuernavaca, Morelos, Mexico
e-mail: rpinto@cenidet.edu.mx

O. O. Vergara Villegas et al. (eds.), *Advanced Topics on Computer Vision, Control and Robotics in Mechatronics*,
https://doi.org/10.1007/978-3-319-77770-2_4

Differentiating facial images from non-facial images requires training. Training is done by feeding a pattern recognition algorithm with representative examples of images with faces and images with no faces, until the algorithm can differentiate between one and another. This kind of procedure (training/learning) can be implemented in many fields and applications which has done pattern recognition and machine learning to undergo an extensive development over the past decades (Bishop 2006).

During the last decade (2007–2017), there has been a consider amount of research related to feature selection, over 2000 results were found in Google Scholar by looking the keywords "feature selection" and at least one of the words: "machine," "learning," "data," and "mining." which tell us that the research in this area is evolving and there are, still, unsolved matters. In Fig. 4.1, a graph to support this information is presented.

Using the example of facial recognition on a smartphone, we characterize the different approaches which pattern recognition has for the training step:

- *Supervised learning*: we supply different examples previously label to the learning algorithm; i.e., face images and examples of images without faces, telling it which is which, so it learns to recognize what a face looks like.
- *Unsupervised learning*: we introduce in the algorithm images of two types, let say faces and trees, without saying which is which. Thus, the algorithm has to find the pattern of each class due to the significant characteristics of each class.
- *Semi-supervised learning*: we provide examples of faces and some other classes but not all of them have the label of being or not faces.

Providing examples sounds easy task, but sometimes these examples could overcome the learning algorithm, either with excess of variables (what definitions make a face to look like a face) or with definitions that do not contribute with enough meaning to the classification process and could, at the end, confuse the learning/training task. Here is where feature selection techniques are crucial.

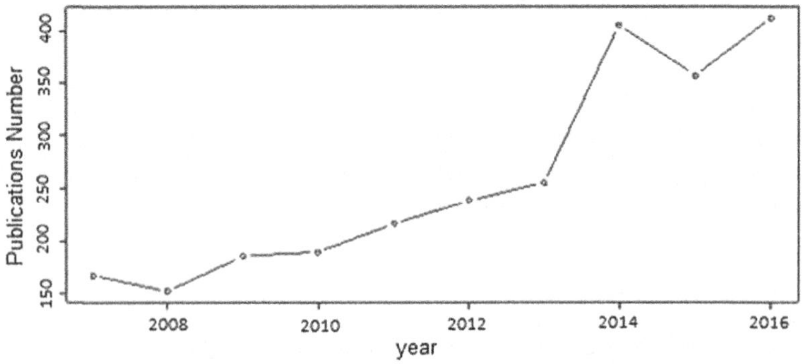

Fig. 4.1 Number of publications by year related to feature selection in the last decade

Feature selection aims to select a subset of characteristics[1] which is reduced on number compared with the original one (set of characteristics), but still keeps the separability of classes. For face recognition, the number of eyes is an irrelevant characteristic if the goal is to differentiate between the owner of the phone and somebody else so that description can be ignored from the feature list.

Characterization of feature selection methods by the way they interact with the classification algorithms is very common (Webb and Copsey 2011). This characterization organized the methods in three categories:

1. *Filter methods*: these methods use the statistical description to evaluate the relevance of one attribute with another without taking in consideration the algorithm to be used for further classification.
2. *Wrapper methods*: this approach is classifier dependent due to its evaluating performance tide to the learning algorithm.
3. *Embedded methods*: it aims to include the feature selection into the training process. Thus, this method is classifier dependent as well.

Approaches for feature selection, according to data, have been categorized in Li et al. (2018). In this work, a wider categorization is worked up into just two: (1) punctual data and (2) chronologically linked data.

Motivation for ordering the approaches into these categories comes from the evolution of data recollection. Many authors treat data in the classical way (He et al. 2006; Murty and Devi 2011; Pedregosa et al. 2011; Sathya and Aramudhan 2014; Sridevi and Murugan 2014; Hancer et al. 2015; Zou et al. 2015) where data is given in a format as shown in Fig. 4.2 with only one dimension per observation. Into our example of facial recognition, the training stage is given by single images of faces or non-faces. In the other hand, chronologically liked data is given by multiple observations per face; there is movement that represents action, in this case the recognition will be done by the owner of the phone saying, for instance, the word "hello."

A description of punctual data and chronologically linked data is shown in Fig. 4.2. Also, in Fig. 4.2b, it can be seen that chronologically linked data is not cardinal through the items, it has to preserve the order in the sample, and it has a time variable.

The rest of the chapter is organized as follows: In Sect. 4.2, a brief state of the art that covers the essential key points that will lead us to discover gaps for chronologically linked data feature selection is presented. In Sect. 4.3, a more condense review of the algorithms presented in Sect. 4.2 is stated, aiming leads to future research work. In Sect. 4.4, we present the characteristics of chronologically linked data which convert this type of data in a challenge for feature selection. Finally, conclusion will be given in Sect. 4.5.

[1]Definition, characteristic, variable, and description are all used as synonyms in this chapter.

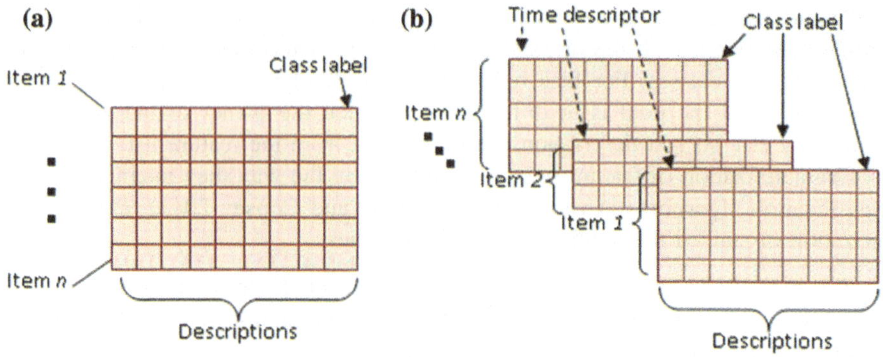

Fig. 4.2 Visual representation of **a** classic punctual data, and **b** chronologically linked data

4.2 Theoretical Context and State of the Art

Feature selection has had an increase amount of research because every day technological devices need to perform some kind of machine learning/pattern recognition task. The applications are many, and the algorithms to deal with this demanding activity are evolving; from punctual data approaches to chronologically linked data methodologies. A state-of-the-art review will be presented; dividing it on the methodologies, the studies are based on to perform feature selection. This survey of the state of the art is not intended to be a complete guide to perform feature selection but to search for gaps in order to offer a big panorama of how the upcoming challenges could be tackled.

4.2.1 Statistical-Based Methods for Feature Selection

Descriptive statistics make use of measures to describe data; these measures have the intention to respond, among others, questions like: what kind of data we are dealing with? How spread is data? Location measurements intend to give an idea of the kind of data within an attribute, and dispersion measurements describe how spreads are the values within an attribute (Ayala 2015). This information is the baseline of statistical feature selection methods. Statistical feature selection compares the attributes without considering the classification algorithm, so the majority of these methods are considered to be filter feature selection methods (Chandrashekar and Sahin 2014).

4.2.1.1 Statistical Measures

There are some concepts to go through in order to facilitate the understanding of the rest of the section. These measurements are: *mean, variance,* and *standard deviation*.

- Mean: It is one of the most common location measurements used in statistics. It helps us to localize our data. To obtain the mean value of a variable, we use

$$\widetilde{x} = \sum_{i=1}^{n} \frac{x_i}{n} \tag{4.1}$$

  The term *mean* is often interchanged by *average*, which gives a clear explanation of the equation.
- Variance: This measurement is used in order to know the spread of the values within a variable. Thus, the variance measures how much the values of a variable differs from the mean (Myatt and Johnson 2014), and the variance formula is

$$s^2 = \frac{\sum_{i=1}^{n} (x_i - \widetilde{x})^2}{n} \tag{4.2}$$

  where n represents the total of observations, and it differs when the formula is used to calculate a sample.
- Standard deviation: The standard deviation is the square root of the variance and is given by

$$s = \sqrt{s^2} = \sqrt{\frac{\sum_{i=1}^{n} (x_i - \widetilde{x})^2}{n}} \tag{4.3}$$

where x_i is the actual data value, $\widetilde{x}$ is the mean of the variable, and n is the number of observation. The higher the value of s, the most broadly scattered the variable's data values are from and toward the mean.

4.2.1.2 Statistical-Based Methods: State of the Art

Other than giving a complete tutorial of how each method is done, we pretend to dig into the methodology in order to catch any possibility of using the method with chronologically linked data.

An algorithm for feature selection on chronologically linked data (Somuano 2016).

A very ingenious statistical feature selection approach was done at Somuano (2016) where the measures reviewed in last section take a key point on discrimination. The author describes the feature selection algorithm with the following steps:

1. Calculate variation coefficients, from original data, with the following equation

$$\mathrm{CV}_{n=1,\ldots N} = \frac{\sigma_n}{\mu_n} \quad (4.4)$$

where CV is a vector that saves N values of variance, N is the total number of variables, μ is a vector of N dimension that contains the mean of each variable, σ is a N dimension vector containing the standard deviation of each variables.
2. Reorder data taking the highest CV value in the vector.
3. Obtain a discriminated matrix, by using

$$\mathrm{MVD} = \begin{cases} 1 & \text{if } r(n,n') = \frac{\mathrm{cov}(n,n')}{\mathrm{CV}} \leq 0.8 \\ 0 & \text{otherwise} \end{cases} \quad (4.5)$$

where cov(n, n') is the covariance of two variables (n, n') and is defined by

$$\mathrm{cov}(n,n') = \frac{1}{L \cdot t} \sum_{i=1}^{L \cdot t} \left(x_{i,n} - \mu_n\right)\left(x_{i,n'} - \mu_{n'}\right) \quad (4.6)$$

where L is the total number of objects, t is the total number of observations, $x_{i,n}$ is the data values in row i column n, μ_n is the mean of the n feature for the first object. $x_{i,n}$ are the values in row i column n', and μ_n is the mean of the n' feature for the second object.
Note: this approach was constructed for chronologically linked data, which means there are several observations per object, find an example in Table 4.1, where object1 has three entries and object2 has three entries as well.
4. Find the Basic Matrix. The basic matrix is Boolean and is made of only basic rows from the matrix of differences (MD). One t row is basic only if in the MD there is not any p row that is a sub-row of t. If p and t are two rows on the MD, it is said that p is a sub-row of t if and only if:

$$\forall j(a_{p,j} = 1 \Rightarrow a_{tj=1}) \text{ and } \exists k\left(a_{tk} = 1 \wedge a_{pk} = 1\right) \quad (4.7)$$

5. Using the bottom-top algorithm, subsets of variables are codified. The algorithm is able to compute the subsets of variables as a vector of n dimension and Boolean values, where 0 denotes that the associated variable is not going to be included and number 1 indicates that the variable is included.

Table 4.1 Example of a chronologically linked data used in He et al. (2006)

Object	var1	var2	TimeStamp
O1	$x_{1,1}$	$x_{1,2}$	1
O1	$x_{2,1}$	$x_{2,2}$	2
O1	$x_{3,1}$	$x_{3,2}$	3
O2	$x_{1,1}$	$x_{1,2}$	1
O2	$x_{2,1}$	$x_{2,2}$	2
O2	$x_{3,1}$	$x_{3,2}$	3

We can notice this method uses chronologically linked data, as we are looking for; however, it seems to lose the sequence of the time stamp, for that reason we will continue exploring more possibilities.

A simple low-variance approach (Pedregosa et al. 2011).

Feature selection can be achieved using simpler descriptive statistics. Tools have been developed that perform feature selection removing variables with low variance whose value does not meet some threshold (Pedregosa et al. 2011). In the mentioned work, the authors approach Boolean variables as Bernoulli random variables, where the variance is giving by

$$\mathrm{var}[X] = p(1 - p) \tag{4.8}$$

where p is the percentage of ones (or zeros), and the discrimination can be done by setting a threshold, if the variance does not meet the threshold then it will be removed.

We can expect, in this type of approach, that features with zero variance value are the first options to remove because this implies that the variable or feature has the same value through all the samples. As we can see, this kind of discrimination has nothing to do with the classification task.

4.2.2 *Information Theory-Based Methods*

Consider a discrete random variable (let say x) and then think about the amount of information that is received when we select a specific value within it, this is call "degree of surprise" (Bishop 2006). We start this section presenting some basic concepts of information theory before exploring different developments that make used of it to search for a subset of features that preserve reparability (feature selection).

4.2.2.1 Basic Concepts

Correlation, a quantity measuring the amount of interdependence of variable quantities, is one of the simplest approaches for feature selection (Webb and Copsey 2011). Starting with a simple correlation value: Pearson correlation; Pearson correlation measures the degree of linear correlation between two variables. Pearson correlation is given by

$$\rho(X, Y) = \frac{\sum_i (x_i - \widetilde{x})(y_i - \widetilde{y})}{\left[\sum_i (x_i - \widetilde{x})^2 \sum_i (y_i - \widetilde{y})^2\right]^{\frac{1}{2}}} \tag{4.9}$$

where X and Y are two variables and $\tilde{x}$ and $\tilde{y}$ their respective means. If two variables are entirely correlated, $\rho\pm=$, then one is redundant; thus, it could be eliminated.

Mutual information is a nonlinear correlation measure. One of its key concepts is entropy, and it is defined by

$$(X) = -\sum_{x} p(x)\log_2(p(x)) \tag{4.10}$$

Here, variable X must have a discrete set of values, an associated probability $p(x)$. The probability of a value is given by the frequency of that value in the sample.

The entropy of X given Y is defined as

$$H(X|Y) = -\sum_{y} p(y)\sum_{x} p(x|y)\log_2(p(x|y)) \tag{4.11}$$

Mutual information is the additional information about X provided by Y and is the decrease in the entropy of X, defined by

$$\mathrm{MI}(X|Y) = H(X) - H(X|Y) \tag{4.12}$$

This could be used to categorize features to further discrimination (Webb and Copsey 2011). A feature X is more correlated than a feature Z to the class variable Y if

$$\mathrm{MI}(X|Y) > \mathrm{MI}(Z|Y) \tag{4.13}$$

Information and concepts presented above will give us an idea of how these families of methods work.

4.2.2.2 Information Theory-Based Methods: State of the Art

Having a dataset that describe images of faces or non-faces with hundreds of attributes will require that pattern recognition algorithms work extra and even get confused with some of them. Using information theory methods, we can reduce the number of features that we use to train the pattern recognition algorithm. Let's explore how they are being used and the type of data they can handle.

Feature selection based on information theory for pattern classification (Sathya and Aramudhan 2014).

In this research, the authors present a feature selection algorithm and then they create mutual information criteria to achieve feature selection. In order to create the criteria, they made use of entropy estimation methods.

The algorithm is listed in the following steps:

1. Calculate mutual information.
2. Calculate the information gain ratio, which measures the correlation between the class and a feature. The information gain ratio is given by

$$\mathrm{IG}_R(C, F_i) = \frac{I(C, F_i)}{H(F_i)} \tag{4.14}$$

where $FS = \{x_1, x_2, x_3, \ldots, x_n\}$ is the subset of size n and C represent the class label. $H(F)$ is the entropy and can be calculated with Eq. (4.10).
3. Decompose FS using the chain rule of mutual information given by

$$I(\mathrm{FS}, C) = \sum_{i=1}^{n} I(X_i, C) - \sum_{i=2}^{n} I(X_i, \mathrm{FS}_{1,i-1} - I(X_i, \mathrm{FS}_{1,i-1}|C) \tag{4.15}$$

4. Remove the features that do not provide any added information about the class.

The significance of the complete estimation of mutual information is discussed when employed as a feature selection criterion. Nevertheless, this looks like a simple task, the authors only deploy it with a dataset of 500 samples, and conclude that if the number of samples increase, the computational time will increase as well.

Feature selection based on correlation (Sridevi and Murugan 2014).

Finding a proper classification for breast cancer diagnosis is achieved using correlation-rough set feature selection joint method.

The proposed method combines two methods in a two-step feature selection algorithm. The first step selects a subset of features, and this is done by applying the rough set feature selection algorithm to discrete data. The resultant set *R1* is set with the attribute with the highest correlation value, then with an attribute with average correlation (*R2*), and for the third time (*R3*) is done with the lowest correlated attribute. The rough set feature selection algorithm relays, for this work, in the QuickReduct algorithm, and it can be consulted at Hassanien et al. (2007). The algorithm can be visualized at Fig. 4.3.

Finally, the second step consists of a reselection of the *R1*–*R3* subsets using correlation feature selection. As a conclusion, the author affirm that this joint algorithm achieves the 85% of classification accuracy.

In order to pursue a chronologically linked data feature selection with, this family of methods, implies the use of probabilistic techniques such as Markov models.

4.2.3 Similarity-Based Methods

Similarity-based methods select a feature subset where the pairwise similarity can be preserved (Liu et al. 2014). Within these methods, there are two key concepts:

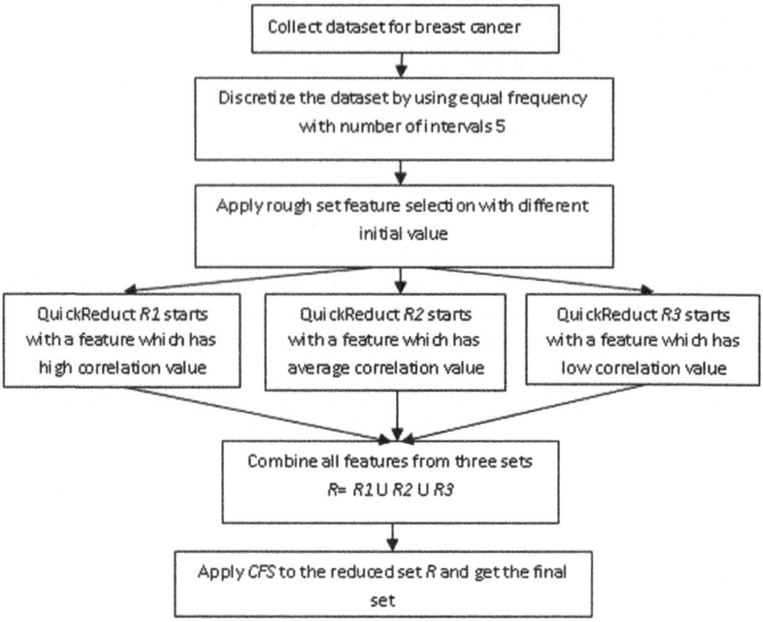

Fig. 4.3 Visual description of correlation-rough set feature selection joint method proposed by Sridevi and Murugan (2014)

(1) pairwise sample similarity and (2) local geometric structure of data. These concepts and their theoretical support are described in the next section.

4.2.3.1 Similarity-Basic Concepts

As indicated in the last section, similarity-based methods relay in two important concepts: pairwise similarity and geometric structure. Next, the concepts are explained for better understanding of the research works that follows the present section.

Pairwise sample similarity. It refers to the similarity between every two samples in a training data set. It can be computed with a similarity measure. Similarity gives us the estimation of how close or how alike two samples are from each other. Many distance/similarity measures are available in the literature to compare two data samples; a list of common distance measures will be presented next.

Euclidean Distance. The Euclidean distance between the *i*th and *j*th objects, as presented in University of Pennsylvania State (2017), is

$$d_E(i,j) = \left(\sum_{k=1}^{p} \left(x_{ik} - x_{jk}\right)^2\right)^{\frac{1}{2}} \tag{4.16}$$

for every pair (i, j) of observations.

In University of Pennsylvania State (2017), the Minkowski distance is denoted as

$$d_M(i,j) = \left(\sum_{k=1}^{p} |x_{ik} - x_{jk}|^{\lambda}\right)^{\frac{1}{\lambda}} \tag{4.17}$$

where $\lambda \geq 1$. It can be called the L_λ metric given

$\lambda = 1$ L_1 metric, Manhattan or city-block distance.
$\lambda = 2$ L_2 metric, Euclidean distance.
$\lambda \to \infty$ L_∞ metric, supremum distance. where

$$\lim \lambda \to \infty = \left(\sum_{k=1}^{p} |x_{ik} - x_{jk}|^{\lambda}\right)^{\frac{1}{\lambda}} = \max\left(|x_{i1} - xj_1|, \ldots, |x_{ip} - x_{jp}|\right) \tag{4.18}$$

Similarity between two binary variables. The last three distance definitions are suitable for continuous variables. But for binary variables, we need a different approach. First, we have to converge in the notation, let p and q be the two binary samples, and Table 4.2 shows all the possible combinations to their values that will be used to find the distance between them.

Then, to find the distance between samples p and q, there are two measures (University of Pennsylvania State 2017): the simple matching and the Jaccard coefficient.

Simple matching, SMC, coefficient is given by

$$\text{SMC} = \frac{(n_{1,1} + n_{0,0})}{(n_{1,1} + n_{1,0} + n_{0,1} + n_{0,0})} \tag{4.19}$$

Jaccard coefficient, J, is given by

$$J = \frac{n_{1,1}}{(x_{1,1} + x_{1,0} + x_{0,1})} \tag{4.20}$$

Local geometric structure of data. Often in the literature (He et al. 2006; Liu et al. 2014), geometric structure of data is done with a graph, where each sample is treated as a node and an edge is placed between two samples if they are neighbors. To find out if two nodes are neighbors, we can either use the label information (class feature for supervised learning) or we could use k-nearest neighbor (kNN) algorithm (University of Pennsylvania State 2017). Using the k-NN algorithm, we put an edge between nodes i and j if x_i and x_j are "close."

Table 4.2 Co-occurrence table for binary variables

	$q = 1$	$q = 0$
$p = 1$	$n_{1,1}$	$n_{1,0}$
$p = 0$	$n_{0,1}$	$n_{0,0}$

Let's take a look at the example given at Murty and Devi (2011) and its resulted graph. Let the training set consists of two variables, eighteen samples, and three classes, as shown in Table 4.3.

Using the label information of attribute *class* in the data set, we obtain a graph that looks like the one in Fig. 4.4, where connected nodes (samples) belong to the same class.

So far, we have covered some basics of the similarity approach without doing any feature selection. Until this point, the similarity concepts that were presented just show how close are the observations from each other. Thus, the research work that follows will give an idea of how they can be used for the purpose that is intended.

Table 4.3 Training set for practical example (Murty and Devi 2011)

	var1	var2	Class		var1	var2	Class		var1	var2	Class
*x*1	0.8	0.8	1	*x*2	1	1	1	*x*3	1.2	0.8	1
*x*4	0.8	1.2	1	*x*5	1.2	1.2	1	*x*6	4	3	2
*x*7	3.8	2.8	2	*x*8	4.2	2.8	2	*x*9	3.8	3.2	2
*x*10	4.2	3.2	2	*x*11	4.4	2.8	2	*x*12	4.4	3.2	2
*x*13	3.2	0.4	3	*x*14	3.2	0.7	3	*x*15	3.8	0.5	3
*x*16	3.5	1	3	*x*17	4	1	3	*x*18	4	0.7	3

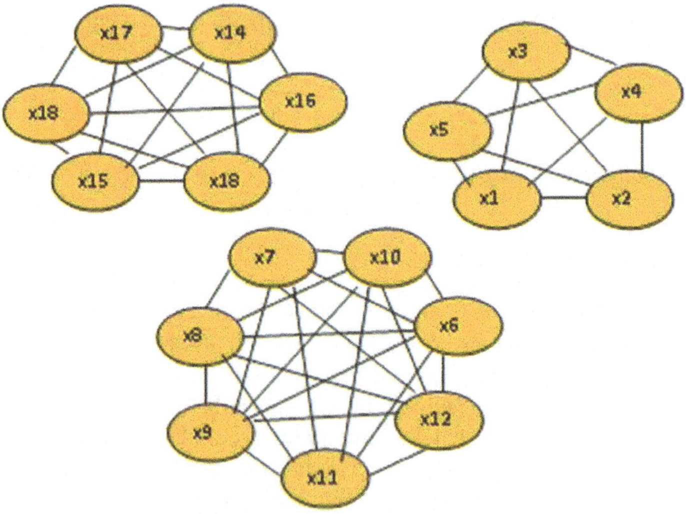

Fig. 4.4 Geometric structure from the example training set

4.2.3.2 Similarity-Based Methods: State of the Art

The methods in this family assess the importance of features by their ability to preserve data similarity between two samples (Li et al. 2018). It means, they have the characteristic of selecting a subset of attributes where the similarity for pairwise comparison can be preserved (He et al. 2006). For the face recognition example, samples of face images have close values when there is a face in the image and as long there is a face, in other way the values would be different. Research of the methods belonging to this type will be described next.

A binary ABC algorithm based on advanced similarity scheme for feature selection (Hancer et al. 2015).

The main goal of this research was to propose a variant of the discrete binary ABC (DisABC) algorithm for feature selection. The variant consists of introducing differential evolution (DE)-based neighborhood mechanism into the similarity-based search of DisABC. The main steps of the research are summarized in the next list:

1. Pick three neighbor samples and call them X_{r1}, X_{r2}, and X_{r3}.
2. Compute $\phi \times$ Dissimilarity(X_{r2}, X_{r3}) using

$$\text{Dissimilarity}(X_i, X_k) = 1 - \text{Similarity}(X_i, X_k) \tag{4.21}$$

   where similarity(X_i, X_k) represents the Jaccard coefficient, which was defined in the previous section, and ϕ is a positive random scaling factor.
3. Solve the equation

$$\min\left\{1 - \frac{M_{11}}{M_{11} + M_{10} + M_{01}} - \Phi \times \text{Dissimilarity}\ (X_{r2}, X_{r3})\right\} \tag{4.22}$$

   to obtain M values between Ω_i and X_i
4. Making use of the obtained M values, apply random selection to generate Ω_i.
5. Set recombination between Ω_i and X_i to get a new solution U_i, using equation

$$u_{\text{id}} = \begin{cases} \omega_{\text{id}}, & \text{if } r\text{and}(d) \leq \text{CR} \\ x_{\text{id}}, & \text{otherwise} \end{cases} \tag{4.23}$$

   where CR is the crossover rate and ω_{id} represents the dth dimension of Ω_i
6. Pick a better solution between X_i and U_i.

According to the authors and the results shown in this study, the joint of DE-based similarity search mechanism into the DisABC algorithm improve the ability of the algorithm in feature selection due to its ability to remove redundant features. The study was performed with different datasets at Asuncion (2007) where data is punctual.

Laplacian Score for Feature Selection (He et al. 2006).

Laplacian score is an algorithm that can be used either for unsupervised or supervised feature selection. This algorithm selects features that can preserve the data separability of classes based on the observation that data from the same class are often close to each other. The importance of a feature is evaluated by its power of locality preservation.

The algorithm stands on the following assertions: L_r denotes the Laplacian score of the *r*th feature. f_{ri} denotes the *i*th sample of the *r*th feature, $i = 1,\ldots, m$, and it consists of the following steps:

1. Build a nearest neighbor graph G with m nodes. The G graph can be built as shown in Sect. 4.2.3.1.
2. Obtain matrix S; if nodes i and j are connected, set

$$S_{ij} = \mathrm{e}^{-\frac{\left\|x_i - x_j\right\|^2}{t}} \tag{4.24}$$

where t is a suitable constant. Otherwise, $S_{ij} = 0$.

3. Define for the *r*th feature:

$$f_r = [f_{r1}, f_{r2}, \ldots, f_{rm}]^T, D = \mathrm{diag}(S1), 1 = [1, \ldots, 1]^T, L = D - S \tag{4.25}$$

4. Obtain the Laplacian score of the *r*th feature by

$$L_r = \frac{\tilde{f}_r^T L \tilde{f}_r}{\tilde{f}_r^T D \tilde{f}_r} \tag{4.26}$$

where $\tilde{f}_r$ is given by

$$\tilde{f}_r = f_r - \frac{f_r^T D1}{1^T D1} 1 \tag{4.27}$$

This method is a filter approach and was tested using two datasets formed of punctual data. As said before, it can be used for supervised or unsupervised approaches, and as conclusion, it is based on the observation that local geometric structure is crucial for discrimination. Similarity-based feature selection methods can be applied to supervised or unsupervised learning.

4.2.4 Neural Networks-Based Feature Selection

With the crescent advances on technological devices that capture high-resolution video/images, pattern recognition algorithms have challenged due to the curse of dimensionality (Bishop 2006). So far, we have presented three classifications that can tackle feature selection; however, not all of them can handle chronologically linked data. Now, it is the turn of neural networks; later we will discuss if they can reach the goal of feature selection on chronologically linked data.

4.2.4.1 Basic Concepts

This section provides a brief review of the multi-layer perceptron (MLP) and deep belief networks (DBNs) without intending to be an extend guide (see references for complete information). Artificial neural networks (ANNs) are believed to handle a bigger amount of data without compromising too much of the resources (Bishop 2006). At the end of this quick review of ANN approaches, a state of the art will be presented.

The MLP is a basic ANN structure that can have many numbers of layers; its configuration lies on the idea of having the outputs of one layer connected to the inputs to the next layer, and in between having a nonlinear differentiable activation function. MLP ANNs are trained using several backpropagation methods of reinforcement learning (Curtis et al. 2016). MLP is a supervised learning algorithm that learns a function $f(\cdot)$: $R^m \rightarrow R^o$ by training a dataset, where m is the number of dimensions for input and o is the number of dimensions for output. Given a set of features $X = x_1, x_2, \ldots, x_m$ and a target y, it can learn a nonlinear function for either classification or regression. Figure 4.5 shows a one-hidden layer MLP (Pedregosa et al. 2011).

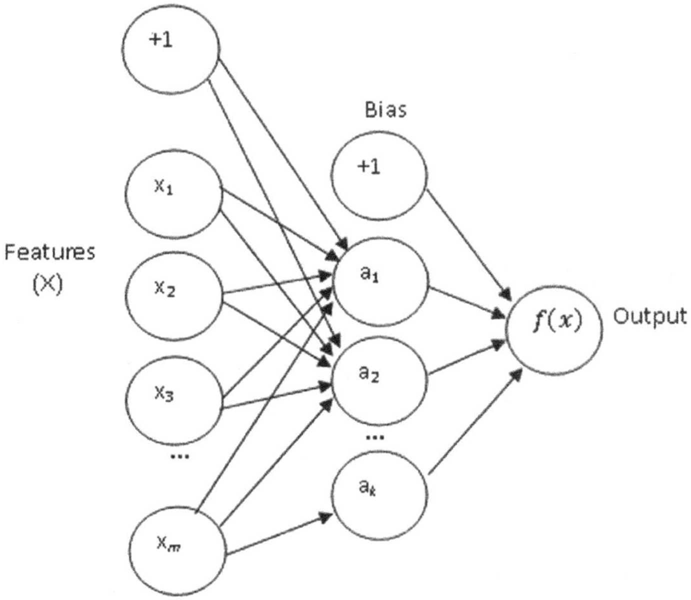

Fig. 4.5 One hidden layer MLP (Pedregosa et al. 2011)

In Fig. 4.5, the features are represented in the left side. The hidden layer (middle one) transforms the values from the left layer with a weighted linear summation $w_1x_1 + w_2x_2 + \cdots + w_mx_m$ followed by a nonlinear activation function $g(\cdot)$: $R \rightarrow R$ (i.e., hyperbolic function). The output layer receives the values from the last hidden layer and transforms them into output values (Pedregosa et al. 2011).

To model a classification task using a MLP, the ANN will consist of an output neuron for each class, where a successful classification produces a much higher activation level for the corresponding neuron class (Curtis et al. 2016).

Deep learning is a relatively new technique and has attracted wide attention (Zou et al. 2015), it uses artificial intelligence techniques of which we will refer here in particular to deep belief networks (DBNs). The deep-learning procedure of the DBN consists of two steps: Layer-wise feature abstracting and reconstruction weight fine-tuning (Hinton 2006). In the first step, the DBN make used of a restricted Boltzmann machine (RBM) to calculate the reconstruction of weights. During the second step, DBN performs a backpropagation to achieve the desired weights obtained from the first step (Hinton 2006).

To stand in solid ground, let consider v as the left layer (visual layer) and h the middle layer (hidden). In the DBN, all nodes are binary variables (to satisfy the Boltzmann distribution). In a RBM, there is a concept called "energy" which is a joint configuration of the visible and hidden layers and is defined as follows:

$$E(v, h; \theta) = -\sum_{i,j} W_{ij} v_i h_j - \sum_i b_i v_i - \sum_j a_j h_j \tag{4.28}$$

where θ denotes the parameters (i.e., W, a, b); W denotes the weights between visible and hidden nodes; and a and b denote de bias of the hidden and visible layers. The joint probability of the configuration can be defined as

$$P_\theta(v, h) = \frac{1}{Z(\theta)} \exp(-E(v, h; \theta)) \tag{4.29}$$

where $Z(\theta) = \sum_{v,h} (\exp(-E(v, h; \theta))$ is the normalization factor. Combining the last two equations (Zou et al. 2015), we have

$$P_\theta(v, h) = \frac{1}{Z(\theta)} \exp\left(\sum_{i,j} W_{ij} v_i h_j + \sum_i b_i v_i + \sum_j a_j h_j\right) \tag{4.30}$$

In the RBMs, the visible and hidden nodes are conditionally independent to each other, that is why, the marginal distribution of v respect h can be defined as

$$P_\theta(v) = \frac{1}{Z(\theta)} \exp\left[v^T W h + a^T h + b^T v\right] \tag{4.31}$$

In the second step of the DBN, a backpropagation is applied on all the layers to fine-tune the weight obtained from the first step.

4.2.4.2 Artificial Neural Networks Methods for Feature Selection: State of the Art

Having in mind that the survey done in this chapter has to lead us to the techniques that can handle the chronologically linked data feature selection, we present how recent research is dealing with feature selection using ANN. At Sect. 4.3, we will present a summary of the methods and personalized opinion of the approaches that could handle chronologically liked data.

Deep learning-based feature selection for remote sensing scene classification (Zou et al. 2015).

According to the authors, feature selection can be achieved by making use of the most reconstructible features due to its characteristic of holding the feature intrinsic. They proposed a method based in DBNs with two main steps. The steps are: interactive feature learning and feature selection. The details of this method are presented next.

1. Iterative feature learning. In this step, the goal is to obtain reconstruction weights, and that can be done removing the feature outliers. The feature outliers are those with larger reconstruction errors. They can be identified by analyzing the distribution of the reconstruction errors as output of the following algorithm: (1) as inputs enter:

$$V = \{v_i | i = 1, 2, \ldots, n\} \tag{4.32}$$

   which is the original input feature vector, η = the ratio of feature outliers, η as the stop criteria, $n\mathrm{It}$ = the max number of iterations.
2. Iterate from $j = 1$ to n It times only if

$$\widehat{e}_{j-1} - \widehat{e}_j < \varepsilon \tag{4.33}$$

   meanwhile get the weight matrix and the average error. Also, filter the features.
3. Finally, as output get M, the final reconstruction weight matrix.

The reconstruction error of feature v_i is defined as $e = \left\| v_i' - v_i \right\|$ and the average error as

$$\widehat{e} = \frac{1}{n} \sum_i^n e_i \tag{4.34}$$

Feature selection. At this step, the weight matrix is used to choose the better features since the outliers were eliminated in first step. Suppose M is the reconstruction weight matrix obtained in the last step, I is an image (since this method was intended for feature selection on images) in the testing data set,

$$V_N^I = \{v_i^I | i = 1, 2, \ldots, N\} \tag{4.35}$$

where N is the number of features extracted from I. As mention before, the purpose of this research is to select the feature with smaller reconstruction error which is given by

$$V^I = \{v_i^I | e_i^I < \mathrm{T}^\mathrm{I}, e_i^I \in E_N^I\} \tag{4.36}$$

Refer to Zou (2015) for complete set of equations. It can be seen that the main idea is to get rid of features that after going through the ANN, in this case a DBN, are considered having a greater error, giving us a different way to use the ANN, not as classifiers but as selectors.

Artificial Neural Networks (Murty and Devi 2011).

A less complicated technique is presented in Murty and Devi (2011). Using a multi-layer feed-forward network with a backpropagation learning algorithm, the authors propose the extraction of the more discriminative feature subset. First, it is proposed to set a larger network, then start the training, and as it goes trim some nodes being careful to adjust the remaining weights in such way that the network performance does not become worse over training process.

The criteria to remove a node, on this approach, are given by

- A node will be removed after analyzing the increase of the error caused by removing that specific node.

The pruning problem is formulated in terms of solving a system of linear equations using the optimization technique. Same as last study (Zou et al. 2015), data set consider for this specific study is punctual. Within this and the last studies, no chronologically linked data was used or consider during tests.

4.2.5 Sparse Learning Methods for Feature Selection

Given its nature, sparse learning is very suitable for feature selection. For a sparse statistical model just a relatively small number of features are important for the manifold of data (Hastie et al. 2015). They are said to handle linked data or multi-view data (Wang et al. 2013; Tang and Liu 2014), for that reason they will be presented in this survey.

4.2.5.1 Introductory Concepts

Sparse learning aims to find a simpler model out of data, as said in Hastie et al. (2015) simplicity could be a synonymous of sparsity. In order to comprehend this theme, we have to introduce important concepts.

In a linear regression problem, let N be the number of observations of an outcome variable y_i, and p associated predictor variables (or features) $x_i = (x_{i,1},\ldots x_{i,p})^T$. The solution to the problem will be to predict the outcome from the features, and this is done to get two goals: prediction with future data and to find the features that actually are important for separability. A linear regression model assumes that

$$y_i = \beta_0 + \sum_{j=1}^{p} x_{ij}\beta_j + e_i \tag{4.37}$$

where β_O and $\beta = (\beta_1,\beta_2,\ldots\beta_p)$ are unknown parameter and e_i is an error term. Usually, the method of least squares is used to obtain the parameters coefficients by

$$\text{minimize}_{\beta_0,\beta} = \sum_{i=1}^{N}\left(y_i - \beta_0 - \sum_{j=1}^{p} x_{ij}\beta_j\right)^2 \tag{4.38}$$

Due to the estimates of last equation, which typically will be nonzero, the interpretation of the model will be hard if p is large. Thus, in *lasso* or l_1-regularized regression a regulation is introduced, and the problem is solved as follows

$$\text{minimize}_{\beta_0,\beta}\sum_{i=1}^{N}\left(y_i - \beta_0 - \sum_{j=1}^{p} x_{ij}\beta_j\right)^2 \text{ subject to } \parallel \beta \parallel_1 \leq t \tag{4.39}$$

where $\|\beta\|_1 = \sum_{j=1}^{p} |\beta_j|$ is the l_1 norm of β, and t is parameter establish to find the within parameters.

So far, we show the introductory basis of sparse learning, next we present a pair of studies that will give us an understanding of how this method is been used.

4.2.5.2 Sparse Learning-Based Methods: State of the Art

Sparse model aims to push the feature coefficients close to zero, then these features can be eliminated. Sparse models are suspect of a lot of research in recent years (Li et al. 2018). We present recent studies that will lead us to find an appropriate method to handle chronologically linked data.

Feature Selection for social media data (Tang and Liu 2014).

This is a study of how social media data present new challenges to feature selection approaches. Since data in social media has multi-dimension, some approaches for feature discrimination will not perform well. Data in social media, i.e., Twitter, have a morphological representation as seen in Fig. 4.6 where the authors explain the interaction user-post and user-user. Users in social media have two behaviors: (1) following other users, represented in Fig. 4.6 as μ_i, and (2) generating some posts (post is a generalization of tweets, blogs or pictures) represented as p_i

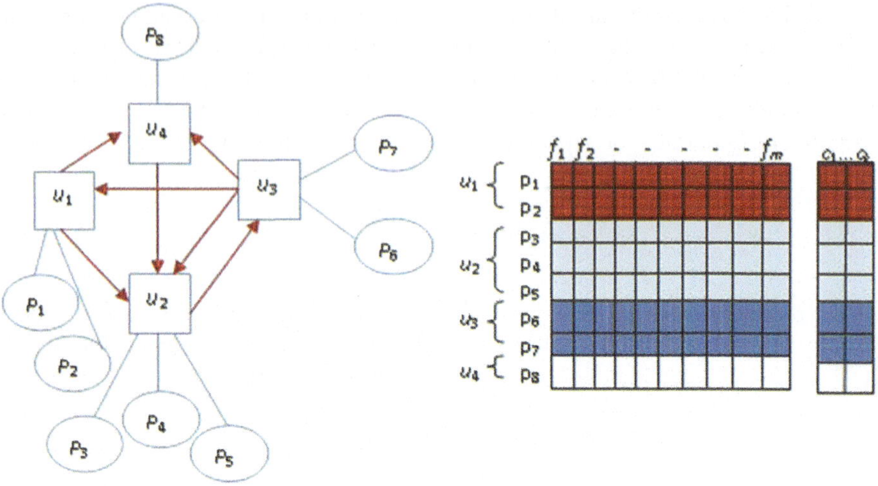

Fig. 4.6 Visual representation of social media data and its matrix illustration as shown in Tang and Liu (2014)

To model the hypotheses, the authors first introduce feature selection for punctual data based on $l_{2,1}$-norm regulation, which selects features across data points using

$$\min_w \| X^T W - Y \|_F^2 + \alpha \| W \|_{2,1} \tag{4.40}$$

where $\|\cdot\|_F$ denotes the Frobenius norm of a matrix and the parameter α controls the sparseness of W in rows. $W \in R^{m\times k}$ and $\|W\|_{2,1}$ is the $l_{2,1}$-norm of W and is define by

$$\| W \|_{2,1} = \sum_{i=1}^{m} \sqrt{\sum_{j=1}^{k} W^2(i,j)} \tag{4.41}$$

Then, the authors propose to add a regularization term with equation in first step that forces the hypothesis that the class labels of posts by the same user are similar. And this is given by

$$\min_w \|X^T W - Y\|_F^2 + \alpha \|W\|_{2,1} + \beta \sum_{u\in u} \sum_{f_i, f_j \in F_u} \|T(f_i) - T(f_j)\|_2^2 \tag{4.42}$$

The above hypothesis assumes that posts by the same user are of similar topics. In other words, the posts from same user are more similar, in terms of topics, than randomly selected posts.

Since the authors in this study work with linked data, it could be assumed that this method could handle the chronology linked data. To reinforce this thought, we present another work related to sparse learning. Until here, we can make a summary of the methods and the type of data they used and if there is a possibility to expand the model to accept and perform well with chronologically linked data.

4.3 Summary of Methods and Their Capability to Handle Chronologically Linked Data

In this section, we present a summary of feature selection methods presented in Sect. 4.2 in order to confirm or not the possible inclusion of chronologically linked data. As the main goal of the chapter is to present the challenges that feature selection faces as data characteristics change.

We presented methods based in different approaches that accomplish feature selection. Now, we congregate the information to give the reader a perspective of how those methods could be used or not for feature selection with chronologically linked data.

Statistical-based methods: as exposed in Sect. 4.2.1, the research in Somuano (2016) treats chronologically link data; however, it seems not to preserve the sequence of it. The work from Pedregosa et al. (2011) uses the lowest variance of the set of features to determine discrimination. With this technique, features selection in chronologically linked data cannot be achieved unless the objects are compare individually.

Information Theory-Based Methods: listed in Sect. 4.2.2, there are two approaches that both make use of punctual data. Probability theory has techniques that deal with time sequences, i.e., Markov models (Bishop 2006); thus, future work could raise from these techniques.

Similarity-based methods: as shown in Sect. 4.2.3, similarity-based methods rely in metric measurements. In order to accommodate similarity-based methods to feature selection on chronologically linked data measures of distances between groups of objects or distributions are needed (Webb and Copsey 2011).

Neural Networks based feature selection: this family of methods, presented in Sect. 4.2.4, has had used punctual data in the literature. However, a deeper research is needed to explore all the possibilities these methods offer.

Sparse learning methods for feature selection: shows to be leading toward the solution to feature selection on chronologically linked data due to the vast research that has been done with linked and multi-view data (Tang and Liu 2014).

In Table 4.4, we present the condensed information to have a better perspective of the challenge that chronologically linked data represent for feature selection approaches.

We could imply, from the analysis done in this study that chronologically linked data represent an unsolved problem and further research can be done. Also, we give some guidance on how to move toward the unsolved issue.

Table 4.4 Summary of all the approaches and the possibilities to adapt with chronologically linked data

Reference	Based method	Data type used	Comments
He et al. (2006)	Similarity	Punctual	Potential to deal with chronologically linked data introducing measures of distances between groups of objects or distributions
Murty and Devi (2011)	ANN	Punctual	There is not available literature that suggests the possible adjustment to use chronologically linked data
Pedregosa et al. (2011)	Statistical	Punctual	Simple approach that cannot deal with chronologically linked data
Sathya and Aramudhan (2014)	Information theory	Punctual	Potential to deal with chronologically linked data using, i.e., Markov models
Sridevi and Murugan (2014)	Information theory	Punctual	Potential to deal with chronologically linked data using, i.e., Markov models
Tang and Liu (2014)	Sparse learning	Linked	Advance research on multiple sample objects, suitable for chronologically linked data
Hancer et al. (2015)	Similarity	Punctual	Potential to deal with chronologically linked data introducing measures of distances between groups of objects or distributions
Zou et al. (2015)	ANN	Punctual	There is not available literature that suggests the possible adjustment to use chronologically linked data
Somuano (2016)	Statistical	Chronologically linked	It uses chronologically linked data nevertheless needs to improve the preservation of sequence

4.4 Upcoming Challenges

In Sects. 4.2 and 4.3, we presented the possibilities that different methods contain to work or to be adapted for chronologically linked data. Here, we present the challenges that this data type represents.

4.4.1 Characteristics of Chronologically Linked Data

In order to clarify the necessity of feature selection on chronologically linked data, some points have to be clear about the characteristics of it. Next, we summarized these characteristics:

- Multi-sample objects.
- Different cardinality between objects.
- Same number of attributes for all the objects.
- Time stamp available on the set.

Multi-sample objects are one of the principal characteristics, since data in this category contain multiple samples per object and every object could contain different number or samples. Refer to Fig. 4.2b for visual explanation.

4.4.2 *Challenges for Feature Selection Algorithms*

Today's data availability and utilization bring new challenges for pattern recognition algorithms. Feature selection aims to facilitate data visualization, prediction performance, reduce storage space, and reduce training time while keeping separability of classes.

In contrast to static concept, events have a dynamic characteristic which is represented by chronologically linked data.

As we presented in Sects. 4.2 and 4.3, there is work that needs to be done to tackle feature selection with chronologically linked data with the following requirements:

- Supervised learning.
- Samples with different cardinality.
- Conserve the sequence of the data.

Having supervised learning scenario will help to evaluate the quality of the selected subset. It is very important to preserve the sequence of the data since it might experience future processing.

4.5 Conclusions

The goal of this study was to find a gap in research to justify a proposal of chronologically linked data for feature selection investigation. We reviewed the importance of feature selection for pattern recognition algorithms and the work done until now with punctual and linked data. We provided a quick guide of basic concepts to introduce readers into the feature selection techniques. Within the introduction of concepts, we showed a representative state of the art using of the categorization used to organize the content and studies in Sect. 4.2 (state of the art). After summarizing and analyzing the state of the art, we found that chronologically linked data problem continues unsolved. We gave some guidance of the appropriated mathematical methods to handle the type of data mentioned. Finally, we presented the challenges of feature selection techniques given the dynamic nature of events (chronologically linked data).

References

Asuncion, D. (2007). *UCI machine learning repository*. [online] Available at: http://archive.ics.uci.edu/ml/index.php.

Ayala, G. (2015). *Estadística Básica* (1st ed.). Valencia, España: Universidad de Valencia.

Bishop, C. (2006). *Pattern recognition and machine learning* (1st ed.). New York, USA: Springer.

Chandrashekar, G., & Sahin, F. (2014). A survey on feature selection methods. *Computers & Electrical Engineering, 40*(1), 16–28.

Curtis, P., Harb, M., Abielmona, R., & Petriu, E. (2016). Feature selection and neural network architecture evaluation for real-time video object classification. In *Proceedings of 2016 IEEE Congress on Evolutionary Computation (CEC)* (pp. 1038–1045) Vancouver, British Columbia, Canada.

Hancer, E., Xue, B., Karaboga, D., & Zhang, M. (2015). A binary ABC algorithm based on advanced similarity scheme for feature selection. *Applied Soft Computing, 36*, 334–348.

Hassanien, A., Suraj, Z., Slezak, D., & Lingras, P. (2007). *Rough computing: Theories, technologies and applications* (1st ed.). Hershey, Pennsylvania, USA: IGI Global.

Hastie, T., Tibshirani, R., & Wainwright, M. (2015). *Statistical learning with sparsity: The lasso and generalizations* (1st ed.). Boca Raton, Florida, USA: CRC Press.

He, X., Cai, D., & Niyogi, P. (2006). Laplacian score for feature selection. In *Proceedings of the 18th International Conference on Neural Information Processing Systems* (Vol. 1, pp. 507–514). Vancouver, British Columbia, Canada.

Hinton, G., Osindero, S., & Teh, Y. (2006). A fast learning algorithm for deep belief nets. *Neural Computation, 18*(7), 1527–1554.

Li, J., Cheng, K., Wang, S., Morstatter, F., Trevino, R., Tang, J., et al. (2018). Feature selection: A data perspective. *ACM Computing Surveys, 50*(6), 1–45.

Liu, X., Wang, L., Zhang, J., Yin, J., & Liu, H. (2014). Global and local structure preservation for feature selection. *IEEE Transactions on Neural Networks and Learning Systems, 25*(6), 1083–1095.

Murty, N., & Devi, S. (2011). *Pattern recognition: An algorithmic approach*. London, United Kingdom: Springer Science & Business Media.

Myatt, G., & Johnson, W. (2014). *Making sense of data I: A practical guide to exploratory data analysis and data mining*. London, United Kingdom: Wiley.

Pedregosa, F., Varoquaux, G., Gramfort, A., Michel, V., Thirion, B., Grisel, O., et al. (2011). Scikit-learn: Machine learning in Python. *Journal of Machine Learning Research, 12*, 2825–2830.

Sathya, R., & Aramudhan M. (2014). Feature selection based on information theory for pattern classification. In *Proceedings of 2014 International Conference on Control, Instrumentation, Communication and Computational Technologies (ICCICCT)* (Vol. 1, pp. 1233–1236). Kanyakumari, India.

Somuano, J. (2016). *Algoritmo para la selección de variables en descripciones crono-valuadas*, Ms.C. Thesis, Centro Nacional de Investigación y Desarrollo Tecnológico (CENIDET).

Sridevi, T., & Murugan, A. (2014). A novel feature selection method for effective breast cancer diagnosis and prognosis. *International Journal of Computer Applications, 88*(11), 28–33.

Tang, J., & Liu, H. (2014). Feature selection for social media data. *ACM Transactions on Knowledge Discovery from Data, 8*(4), 1–27.

University of Pennsylvania State. (2017). *Applied data mining and statistical learning*. [Online]. Available: https://onlinecourses.science.psu.edu/stat857/node/3.

Wang, H., Nie, F., & Huang, H. (2013). Multi-view clustering and feature learning via structured sparsity. In *Proceedings of the 30th International Conference on Machine Learning (ICML)* (Vol. 28, pp. 352–360). Atlanta, Georgia, USA, 28.

Webb, A., & Copsey, K. (2011). *Statistical pattern recognition* (3rd ed.). London, United Kingdom: Wiley.

Zou, Q., Ni, L., Zhang, T., & Wang, Q. (2015). Deep learning based feature selection for remote sensing scene classification. *IEEE Geoscience and Remote Sensing Letters, 12*(11), 2321–2325.

Chapter 5
Overview of Super-resolution Techniques

Leandro Morera-Delfín, Raúl Pinto-Elías and Humberto-de-Jesús Ochoa-Domínguez

Abstract In the last three decades, multi-frame and single-frame super-resolution and reconstruction techniques have been receiving increasing attention because of the large number of applications that many areas have found when increasing the resolution of their images. For example, in high-definition television, high-definition displays have reached a new level and resolution enhancement cannot be ignored; in some remote sensing applications, the pixel size is a limitation; and in medical imaging, the details are important for a more accurate diagnostic or acquiring high-resolution images while reducing the time of radiation to a patient. Some of the problems faced in this area, that in the future require dealing more effectively, are the inadequate representation of edges, inaccurate motion estimation between images, sub-pixel registration, and computational complexity among others. In this chapter, an overview of the most important methods classified into two taxonomies, multiple- and single-image super-resolution, is given. Moreover, two new techniques for single-image SR are proposed.

Keywords Super-resolution · Frequency domain · Spatial domain
Total variation

5.1 Introduction

Image super-resolution (SR) refers to the process of creating clear and high-resolution (HR) images from a single low-resolution (LR) image or from a sequence of low-resolution observations (Schultz and Stevenson 1994). In this chapter, the most important SR techniques are explained.

L. Morera-Delfín (✉) · R. Pinto-Elías
Centro Nacional de Investigación y Desarrollo Tecnológico
(CENIDET), Cuernavaca, Morelos, Mexico
e-mail: lmorera@cenidet.edu.mx

H.-d.-J. Ochoa-Domínguez
Universidad Autónoma de Ciudad Juárez, Instituto de Ingeniería
y Tecnología, Ciudad Juárez, Chihuahua, Mexico

O. O. Vergara Villegas et al. (eds.), *Advanced Topics on Computer Vision, Control and Robotics in Mechatronics*,
https://doi.org/10.1007/978-3-319-77770-2_5

The methods for SR are addressed including the definition of each method. We address big topics of work in super-resolution as: the pure interpolation with high scales of amplification, the use of dictionaries, the variational procedures and the exploiting of gradients sharpening. Each section in this chapter yields a guide for the technical comprehension of each procedure. The technical procedures of the cited articles are not fully reproduced but neither is a superficial description made without ideas for a practical realization.

The first main separation between the SR methods is determined by the resources to employ in the process. In the first case, a group of LR images are used. These procedures refer to the first publications about the topic. In the second case, due to practical situations, the SR is carried out by using only the input image of low resolution. Figures 5.1 and 5.2 show the taxonomies of the more evident classification of the methods, multiple-image SR or single-image SR.

In the second class of methods, we refer to the domain of application, spatial domain or frequency domain. The following proposed differentiation between SR methods is based on the mathematical models in order to reach the high resolution. Transformations, probabilistic prediction, direct projection, learning dictionaries, reduction of dimension and reconstruction models under minimization procedures and residual priors are discussed. A common goal is the incorporation of the lost

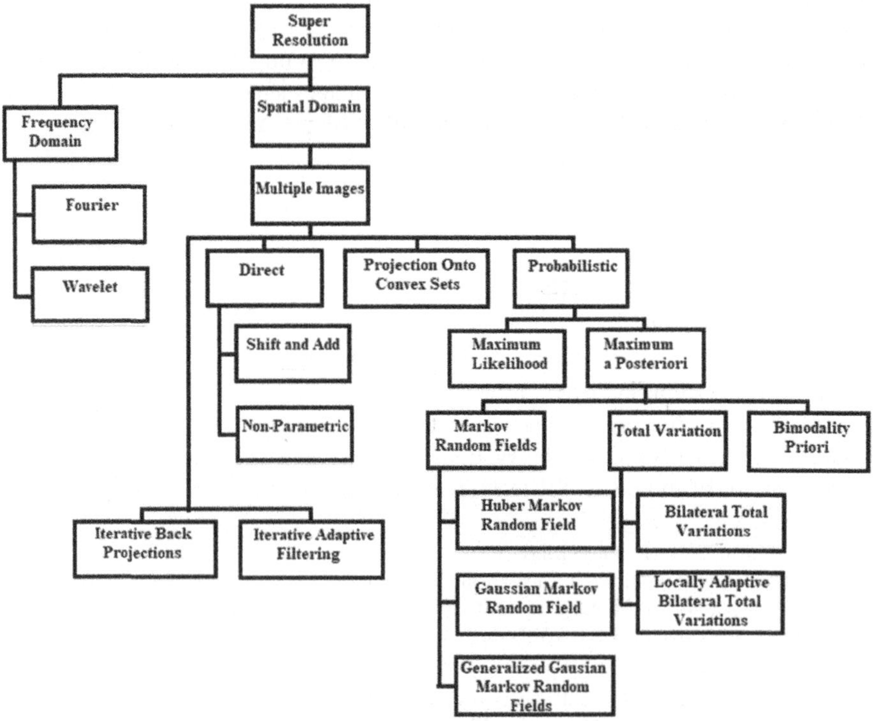

Fig. 5.1 Taxonomy of multiple-image super-resolution

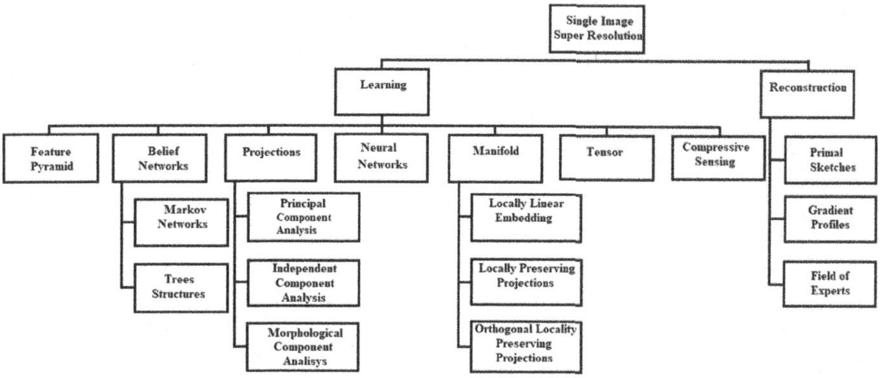

Fig. 5.2 Taxonomy of single-image super-resolution

high-frequency details. Finally, we propose two new methods for single-image SR. The first one is based on gradient control, and the second one is a hybrid method based on gradient control and total variation.

The rest of the chapter is organized as follows: In Sect. 5.2, the methods are explained. In Sect. 5.3, the results of the proposed methods are presented. In Sect. 5.4, the metrics used to characterize the methods are presented. Finally, the chapter concludes in Sect. 5.5.

5.2 Methods

The accuracy in the estimation of the HR image is result of a right selection of mathematical tools and signal processing procedures as transformations, learning models, minimization techniques, and others for reaching the major content of high spatial frequencies or details in the output image. In this section, current methods as well as the proposed SR procedures are explained. SR models for single image and multiple image are considered.

5.2.1 Image Model

Down-sampling and warping are two processes in consideration for a more realistic representation of the image at low resolution. In the first process, the image is averaged over equal areas of size $q \times q$ as can be seen from Eq. (2.1). In the warping process, the image is shifted along x and y directions, and the distances a and b are in pixels. Also, a rotation Θ is assumed on the image (Irani and Peleg 1990; Schultz and Stevenson 1994) as can be observed in Eq. (2.2).

$$g(m,n) = \frac{1}{q^2} \sum_{x=qm}^{(q+1)m-1} \sum_{y=qm}^{(q+1)n-1} f(x,y) \tag{5.1}$$

$$w\begin{bmatrix} x \\ y \\ 1 \end{bmatrix} = \left(\begin{bmatrix} 1 & 0 & a \\ 0 & 1 & b \\ 0 & 0 & 1 \end{bmatrix} \times \begin{bmatrix} \cos\theta & \sin\theta & 0 \\ -\sin\theta & \cos\theta & 0 \\ 0 & 0 & 1 \end{bmatrix} \right)^{-1} \begin{bmatrix} m \\ n \\ 1 \end{bmatrix} \tag{5.2}$$

In a SR algorithm, the model of degradation is fundamental for comparative purposes and evaluation of the effectivity of the algorithm. Equation (5.1) considers the blurring and down-sampling processes, and Eq. (5.2) represents the warping operation. For a number k of LR images with noise added, the model becomes (Irani and Peleg 1990),

$$g_k(m,n) = d(h_k(w_k(f(x,y)))) + \eta_k(m,n) \tag{5.3}$$

Equation (5.3) incorporates the distortions that yield a LR image, d is the down-sampling operator, h_k is the blurring operator, w_k is the warping operator, and $f(x, y)$ is the HR image. Furthermore, the blurring process can be composed by distortions due to the displacement h_d, the lens h_l, and the sensors h_s. The result is a convolution operation. The transformations are shown in Fig. 5.3.

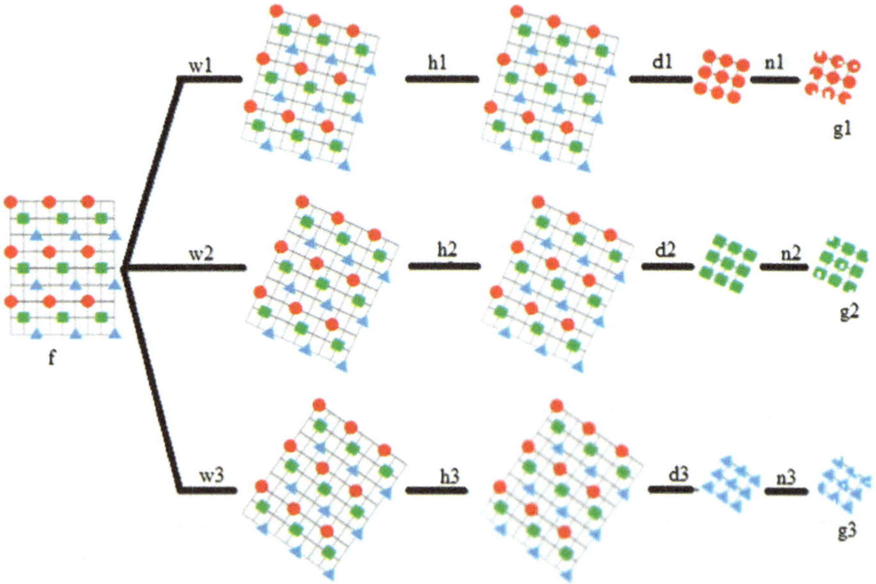

Fig. 5.3 Steps to form three LR images g_1, g_2, and g_3 from a HR image f. Each branch represents a different acquisition process

5.2.2 Image Registration

Translation and rotation relationship, between a LR and a HR image, is calculated using Eq. (5.4) (Keren et al. 1998),

$$\begin{aligned} x &= x_k^t + q_x m \cos\theta_k - q_y n \sin\theta_k \\ y &= y_k^t + q_x m \sin\theta_k + q_y n \cos\theta_k \end{aligned}, \tag{5.4}$$

where x_k^t and y_k^t are the displacements, q_x and q_y the sampling rates, and θ the rotation angle. Two acquisitions g_1 and g_2 with rotation and displacements can be related using the following Eq. (5.5).

$$g_2(m,n) = g_1(m\cos\theta - n\sin\theta + a, n\cos\theta + m\sin\theta + b) \tag{5.5}$$

5.2.3 Approximation Between Acquisitions

The approximation to this parameter has been solved using the Taylor series representation. In the first step, sin θ and cos θ are expressed in series expansion using the first two terms.

$$g_2(m,n) = g_1\left(m + a - n\theta - \frac{m\theta^2}{2}, n + b + m\theta - \frac{n\theta^2}{2}\right).$$

Then, the function g_1 can be expanded with Taylor series,

$$g_2(m,n) = g_1(m,n) + \left(a - n\theta - \frac{m\theta^2}{2}\right)\frac{\partial g_1}{\partial m} + \left(b + m\theta - \frac{n\theta^2}{2}\right)\frac{\partial g_1}{\partial n} \tag{5.6}$$

$$E(a,b,\theta) = \sum\left(g_1(m,n) + \left(a - n\theta - \frac{m\theta^2}{2}\right)\frac{\partial g_1}{\partial m} + \left(b + m\theta - \frac{n\theta^2}{2}\right)\frac{\partial g_1}{\partial n} - g_2(m,n)\right)^2 \tag{5.7}$$

Finally, the parameters *a*, *b*, and θ of Eq. (5.7) are determined using partial derivatives on the final expansion and solving the equation system.

5.2.4 Frequency Domain

The models in frequency domain consider the sampling theory. There, a 2D array of Dirac deltas (D_T) performs the sampler function. The array has the same form in time and frequency domains (2D impulse train). The acquisition process multiplies

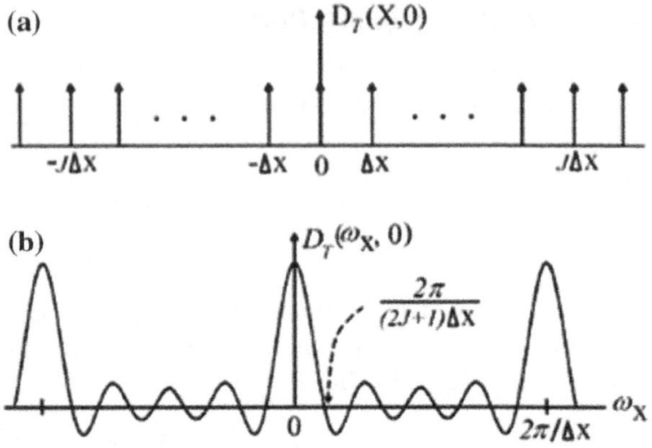

Fig. 5.4 Sampling array in **a** space domain and **b** frequency domain

the array of D_T with the image in the spatial domain point by point. This operation in frequency domain becomes a convolution operation. The advantage is that the resolution of the convolution kernel (sampling array in the frequency domain in the interval of $[-\pi, \pi]$) can be increased for optimal scales of amplification, checking the high-frequency content at the output of the process. The Fourier transform of the sampling is shown in Eq. (5.8),

$$D_T(\omega'_x, \omega'_y) = \frac{\sin\left\{\omega'_x\left(\frac{M-1}{2}\right)\Delta x\right\}\sin\left\{\omega'_y\left(\frac{L-1}{2}\right)\Delta y\right\}}{\sin\left\{\omega'_x\frac{\Delta x}{2}\right\}\sin\left\{\omega'_y\frac{\Delta y}{2}\right\}}, \tag{5.8}$$

and the convolution with the image can be expressed as in Eq. (5.9),

$$\begin{aligned} S_{\text{amp}}(j_1, j_2) = \sum_{n_x=-L/2}^{L/2}\sum_{m_x=-M/2}^{M/2} S(n_x\Delta\omega_x, n_y\Delta\omega_y)\cdot \\ D_T(j_1 - n_x\Delta\omega_x + M_{cx}, j_2 - n_y\Delta\omega_y + L_{cx}; j_1, j_2) \end{aligned} \tag{5.9}$$

The high-frequency content in S_{amp} must be maximized. This strategy has been used in (Morera 2015). Figure 5.4 shows a 1D sampling array in space and frequency domains.

5.2.5 *Wavelet Transform*

The wavelet transform introduces the analysis of the image generally in four fields of information. The common decomposition brings directional information of fluctuation of the image signal. The coefficients of the transformation are present in four groups. The low-frequency coefficients which are a coarse representation of the image, the horizontal, the vertical and the diagonal coefficients which represent details of directional variations of the image. The most common strategy for SR using wavelets applies a non-sub-sampled wavelet or static wavelet before a wavelet reconstruction, and the first step produces a decomposition of four images with the same dimension as the input. Then, the wavelet reconstruction produces an amplified image with scale factor 2, this strategy is employed in (Morera 2014).

5.2.6 Multiple-Image SR

The main goal in this group of techniques is the simulation of the process of formation of the image in order to reject the aliasing effects due to the down-sampling effect. A group of acquisitions of the same scene in LR is required for estimation of the HR image.

5.2.6.1 Iterative Back-Projection

Iterative back-projection (IBP) methods were the first methods developed for spatial-based SR. IBP algorithm yields the desired image that satisfies that the reconstruction error is close to zero. In other words, the IBP is convergent. Having defined the imaging model like the one given in Eq. (5.3), the distance $\|Af - g\|_2^2$ is minimized, where matrix A includes the blur, down-sampling and warping operations, f is the original HR image, and g is the observed image. The HR estimated image is generated and afterward refined. Such a guess can be obtained by registering the LR images over a HR grid and then averaged them (Irani and Peleg 1990, 1991, 1992, 1993). The iterative model given in Eq. (5.10) is used to refine the set of the available LR observations. Then, the error between the LR images and the observed ones is obtained and back-projected to the coordinates of the HR image to improve the initial estimation (Irani and Peleg 1993). The Richardson iteration is commonly used in these techniques.

$$f^{(t+1)}(x,y) = f^{(t)}(x,y) + \frac{1}{K}\sum_{k=1}^{K} w_k^{-1}\left(\left(\left(g_k - g_k^{(t)}\right)\dot{d}\right) * \dot{h}\right), \tag{5.10}$$

where w_k^{-1} is the inverse of the warping operator, $\dot{d}$ is the up-sampling operator, $\dot{h}$ is a deblurring kernel, $k = 1 \ldots K$ is the number of LR acquisitions, $f^{(t+1)}(x,y)$ is the reconstructed SR image in the $(t + 1)$th iteration, and $f^{(t)}(x,y)$ is the reconstructed SR image in the previous (t)th iteration. The shortcoming of this algorithm is that produces artifacts along salient edges.

5.2.6.2 Maximum Likelihood

The noise term in the imaging model given in Eq. (5.3) is assumed to be additive white Gaussian noise (AWGN) with zero mean and variance σ^2. Assuming the measurements are independent and the error between images is uncorrelated, the likelihood function of an observed LR image g_k for an estimated HR image $\hat{f}$ (Cheeseman et al. 1994; Capel and Zisserman 1998; Elad and Hel-Or 2001; Farsiu et al. 2004; Pickup et al. 2006; Pickup 2007; Prendergast and Nguyen 2008; Jung et al. 2011a) is,

$$p\left(g_k|\hat{f}\right) = \prod_{\forall m,n} \frac{1}{\sqrt{2\pi\sigma^2}} \exp\left(-\frac{\left(\hat{g}_k - g_k\right)^2}{2\sigma^2} \right). \tag{5.11}$$

The log-likelihood transforms the product into a summation. Therefore, Eq. (5.11) becomes the summation of a term C that does not depend on f and the summation of the exponents of the exponential function as shown in Eq. (5.12),

$$L(g_k) = C - \frac{1}{2\sigma^2} \sum_{\forall m,n} \left(\hat{g}_k - g_k\right). \tag{5.12}$$

The maximum likelihood (ML) solution (Woods and Galatsanos 2005) seeks a super-resolved image $\hat{f}_{\mathrm{ML}}$ which maximizes the log-likelihood for all observations. Notice that after maximization the constant term vanishes. Therefore, the super-resolved images can be obtained by maximizing Eq. (5.12) or, equivalently, by minimizing the distance between $\hat{g}_k$ and g_k as,

$$\hat{f}_{\mathrm{ML}} = \arg\max_f \left(\sum_{\forall m,n} L(g_k) \right) = \arg\min_f \left(\left\|\hat{g}_k - g_k\right\|_2^2 \right). \tag{5.13}$$

5.2.6.3 Maximum a Posteriori

Given the LR images g_k, the maximum a posteriori (MAP) method (Cheeseman et al. 1994) finds an estimate $\hat{f}_{\mathrm{MAP}}$ of the HR image by using the Bayes rule in Eq. (5.14),

$$p\left(\hat{f}|g_1, g_2, \ldots, g_k\right) = \frac{p(g_1, g_2, \ldots, g_k|f)p(f)}{p(g_1, g_2, \ldots, g_k)} \propto p(g_1, g_2, \ldots, g_k|f)p(f) \tag{5.14}$$

The estimate can be found by maximizing log of Eq. (5.14). Notice that the denominator is a constant term that normalizes the probability conditional. This term is going to be zero after maximization then,

$$\hat{f}_{\mathrm{MAP}} = \arg\max_{f}(\log(p(g_1, g_2, \ldots, g_k|f)) + \log p(f))). \tag{5.15}$$

Applying statistical independence between the images g_k, Eq. (2.15) can be written as,

$$\hat{f}_{\mathrm{MAP}} = \arg\max_{f}\left(\sum_{k=1}^{K}\log(p(g_k|f)) + \log(p(f))\right), \tag{5.16}$$

where

$$p(g_k|f) \propto \exp\left(-\frac{\left\|\hat{g}_k - g_k\right\|_2^2}{2\sigma_k^2}\right)$$

The probability $p(g_k|f)$ is named the regularization term. This term has been modeled in many different forms; some cases are:

1. Natural image prior (Tappen et al. 2003; Kim and Kwon 2008, 2010).
2. Stationary simultaneous autoregression (SAR) (Villena et al. 2004), which applies uniform smoothness to all the locations in the image.
3. Non-stationary SAR (Woods and Galatsanos 2005) in which the variance of the SAR prediction can be different from one location in the image to another.
4. Soft edge smoothness a priori, which estimates the average length of all level lines in an intensity image (Dai et al. 2007, 2009).
5. Double-exponential Markov random field, which is simply the absolute value of each pixel value (Debes et al. 2007).
6. Potts–Strauss MRF (Martins et al. 2007).
7. Non-local graph-based regularization (Peyre et al. 2008).
8. Corner and edge preservation regularization term (Shao and Wei 2008).
9. Multi-channel smoothness a priori which considers the smoothness between frames (temporal residual) and within frames (spatial residual) of a video sequence (Belekos et al. 2010).
10. Non-local self-similarity (Dong et al. 2011).
11. Total subset variation, which is a convex generalization of the total variation (TV) regularization strategy (Kumar and Nguyen 2010).
12. Mumford–Shah regularization term (Jung et al. 2011b).
13. Morphological-based regularization (Purkait and Chanda 2012).
14. Wavelet-based (Li et al. 2008; Mallat and Yu 2010).

5.2.7 Single-Image SR

Single-image SR problem is a very ill-posed problem. It is necessary an effective knowledge about the HR image to obtain a well-posed HR estimation. The algorithms are designed for one acquisition of low resolution of the image. Some of the strategies proposed are summarized following,

1. Pure interpolation using estimation of the unknown pixels in the HR image, modification of the kernel of interpolation, and checking the high-frequency content in the estimated output HR image.
2. Learning the HR information from external databases. In this case, many strategies of concentration of the information of the image and clustering are used. Then, the image is divided into overlapping patches and this information is mapped over a dictionary of LR–HR pairs of patches of external images.
3. Manage the information of gradients in the image.
4. Hybrid models used to reconstruct the image with a minimization procedure in which some prior knowledge about the estimation error is included.

5.2.7.1 Geometric Duality

The concept of geometric duality is one of the most useful tools in the parametric SR with least-square estimation for interpolation, and one of the most cited algorithm in comparison with SR method is the new edge-directed interpolation (NEDI) (Li and Orchad 2001).

The idea behind is that each low-resolution pixel also exists in the HR image and the neighbor pixels are unknown. Hence, with two orthogonal pairs of directions around the low-resolution pixel in the HR image (horizontal, vertical, and diagonal directions), a least-square estimation can be used in each pair. The equation system is constructed in the LR image, and then, the coefficients are used to estimate pixels in the HR initial image. The first estimation is made by using Eq. (5.17),

$$\hat{Y}_{2i+1,2j+1} = \sum_{k=0}^{1} \sum_{l=0}^{1} \alpha_{2k+l} Y_{2(i+k),2(j+l)} \tag{5.17}$$

where the coefficients are obtained in the same configuration as in the LR image. In this case, the unknown pixels between LR pixels that exist in the HR image (in vertical and horizontal directions) are estimated. In the next step, the unknown pixels between LR pixels that exist in the HR image (in diagonal directions) are estimated. The pixels of each category are shown in Fig. 5.5.

In (Zhang and Wu 2008), take advantage of NEDI. There, a new restriction is applied including the estimated pixels in the second step, and the minimum square estimation is made using the 8-connected pixels around a central pixel in the diamond configuration shown in Fig. 5.6. They define a 2D piecewise

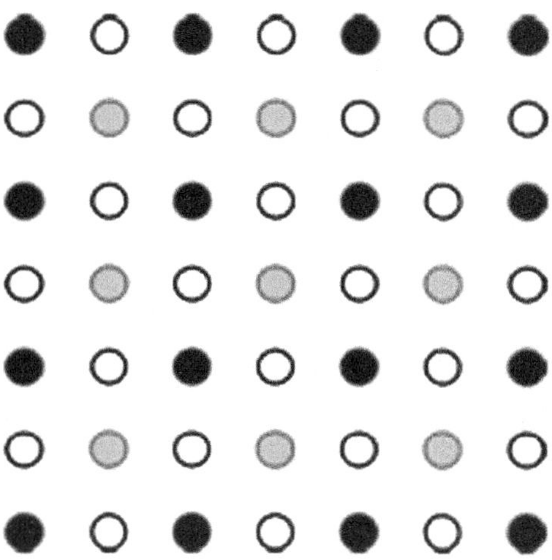

Fig. 5.5 Array of pixels in the initial HR image for NEDI interpolation. Black pixels are the LR pixels used to calculate the HR gray pixels. The white pixels are calculated using the white and the black pixels

autoregressive (PAR) image model of parameters $a_k, k \in [1, \ldots, 4]$ to characterize the diagonal correlations in a local diamond window W and the extra parameters $b_k, k \in [1, \ldots, 4]$ to impose horizontal and vertical correlations in the LR image as shown in Fig. 5.6b. The parameters are obtained using a linear least-square estimator using four 4-connected neighbors for b_k (horizontal and vertical), and four 8-connected diagonal neighbors, available in the LR image for a_k.

To interpolate the missing HR pixel in the window, the least-square strategy of Eq. (5.18) is carried out.

$$\hat{\mathbf{y}} = \arg\min_{\mathbf{y}} \left\{ \sum_{i \in W} \left\| y_i - \sum_{1 \le k \le 4} a_k x_{i \diamond k}^{(8)} \right\| + \sum_{i \in W} \left\| x_i - \sum_{1 \le k \le 4} a_k y_{i \diamond k}^{(8)} \right\| \right\}, \quad (5.18)$$

where x_i and y_i are the LR and the HR pixels, respectively, $x_{i \diamond k}^{(8)}$ are the four 8-connected LR neighbors available for a missing y_i pixel and for a x_i pixel, and $y_{i \diamond k}^{(8)}$ denotes its HR missing four 8-connected pixels.

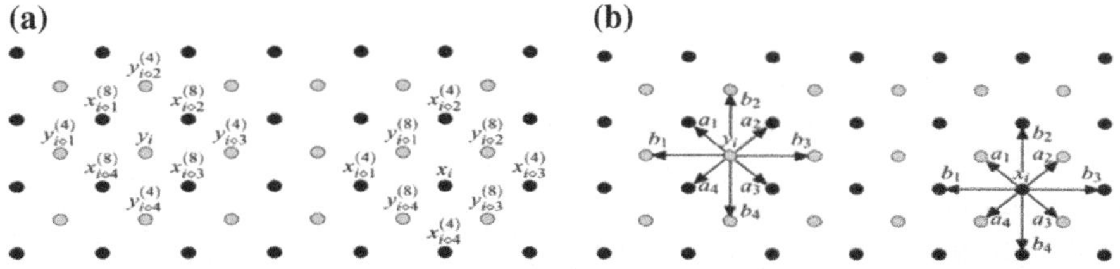

Fig. 5.6 **a** Spatial configuration for the known and missing pixels and **b** the parameters used to characterize the diagonal, horizontal, and the vertical correlations (Zhang and Wu 2008)

Other approach for NEDI algorithms (Ren et al. 2006; Hung and Siu 2012) uses a weighting matrix **W** to assign different influence of the neighbor pixels on the pixel under estimation. The correlation is affected by the distance between pixels. The diagonal correlation model parameter is estimated by using a weighted least-square strategy.

$$\mathbf{A} = \left(\mathbf{L}_{LA}^{T}\mathbf{W}\mathbf{L}_{LA}\right)^{-1}\mathbf{L}_{LA}^{T}\mathbf{W}\mathbf{L}, \tag{5.19}$$

where $\mathbf{A} \in R^{4\times1}$ is the diagonal correlation model parameter, $\mathbf{L} \in R^{64\times1}$ is a vector of the LR pixels, $\mathbf{L}_{LA} \in R^{64\times4}$ are the neighbors of **L**, and $\mathbf{W} \in R^{64\times64}$ is the weighting matrix of Eq. (5.20).

$$\mathbf{W}_{i,i} = \exp\left(-2\left(\|\mathbf{L}_c - \mathbf{L}_{LAi}\|_p/\sigma_1 + \|\mathbf{V}_c - \mathbf{V}_{LAi}\|_p/\sigma_2\right)\right), \tag{5.20}$$

where σ_1 and σ_2 are global filter parameters, $\mathbf{L}_c \in R^{4\times1}$ is the HR geometric structure, $\mathbf{L}_{LAi} \in R^{4\times1}$ is the *i*th LR geometric structure, $\|\cdot\|_p$ denotes the *p*-norm (1 or 2), and $\mathbf{V}_{LAi}, \mathbf{V}_c \in R^{2\times1}$ are the coordinates of L_i and L_c. The all-rounded correlation model parameter $\mathbf{B} \in R^{8\times1}$ is given by,

$$\mathbf{B} = \left(\mathbf{L}_{LB}^{T}\mathbf{W}\mathbf{L}_{LB}\right)^{-1}\mathbf{L}_{LB}^{T}\mathbf{W}\mathbf{L}, \tag{5.21}$$

where $\mathbf{L}_{LB} \in R^{64\times8}$ are the neighbor's positions in **L**.

5.2.7.2 Learning-Based SR Algorithms

In these algorithms, the relationship between some HR and LR examples (from a specific class like face images, animals) is learned. The training database as example shown in Fig. 5.8 needs to have proper characteristics (Kong et al. 2006). The learned knowledge is a priori term for the reconstruction. The measure of these two factors of sufficiency and predictability is explained in (Kong et al. 2006). In general, a larger database yields better results, but a larger number of irrelevant examples only increase the computational time of search and can disturb the results. The content-based classification of image patches (like codebook) during the training is suggested as alternative in (Li et al. 2009).

The dictionaries to be used can be a self-learned or an external-based dictionary. Typically, some techniques like the *K*-means are used for clustering of *n* observations into *k* clusters and the principal component analysis (PCA) algorithm is employed to reduce the dimensionality. In dictionary learning, it is important to reduce dimensionality of the data. Figure 5.7 shows a typical model of projection for dictionary learning for SR. Figure 5.8 shows an example of a LR image input and a pair of LR–HR dictionary images.

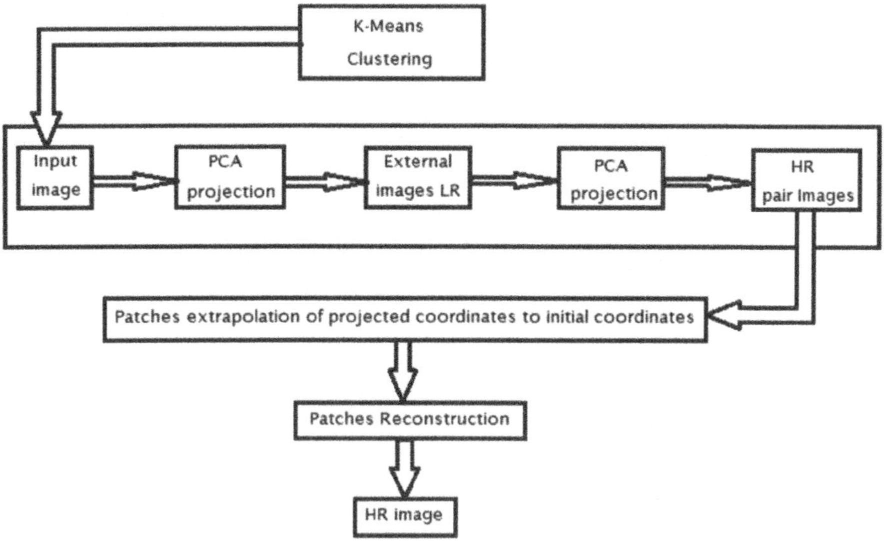

Fig. 5.7 Projection of an input image using two external LR–HR dictionaries

Fig. 5.8 Low-resolution input image and a pair of LR–HR dictionary images

The projection PCA is based on finding the eigenvectors and eigenvalues of an autocorrelation. This can be expressed as,

$$\mathbf{\Omega} = \mathbf{\Psi}\mathbf{\Lambda}\mathbf{\Psi}^T, \tag{5.22}$$

where $\mathbf{\Omega}$ is the autocorrelation matrix of the input data $\mathbf{U}$, $\mathbf{\Psi}$ is the matrix of eigenvectors, and $\mathbf{\Lambda}$ is a diagonal matrix containing the eigenvalues. The eigenspace is the projection of $\mathbf{U}$ into the eigenvectors. The data at high and low resolution $\mathbf{U}_h$ and $\mathbf{U}_l$ are used to find the minimum distance in a projection over the found eigenspace.

$$\boldsymbol{\Delta}_h = \mathbf{U}_h^k \boldsymbol{\Psi}_h. \tag{5.23}$$

In dictionary search, the patches represent rows or columns of the data matrix $\mathbf{U}^h$ or $\mathbf{U}^l$. The strategy is to find the position of the patch at HR with a minimum distance respect to the projection of a LR patch in the eigenspace of HR.

$$ph(\text{pos}) = \min_{v,l,j} \left\| \Delta_h^T \hat{\mathbf{U}}_k^h - \Delta_h^T \hat{v}_{l,j} \right\|_2, \quad \left| \hat{v}_{l,j} \in \hat{\mathbf{U}}_k^l. \right. \tag{5.24}$$

5.2.7.3 Diffusive SR

Perona and Malik (1990) developed a method that employs a diffusion equation for the reconstruction of the image. The local context of the image is processed using a function to restore the edges.

$$\text{div}\,(c\nabla\mathbf{I}) = \frac{\partial}{\partial x}(c\mathbf{I}_x) \tag{5.25}$$

where c is a function to control the diffusivity; for example if $c = 1$, the process is linear isotropic and homogeneous, and if c is a function that depends on $\mathbf{I}_x$, i.e., $c = c(\mathbf{I}_x)$, the process becomes a nonlinear diffusion; however, if c is a matrix-valued diffusivity, the process is called anisotropic and it will lead to a process where the diffusion is different for different directions. The image is differentiated in cardinal directions, and a group of coefficients are obtained in each point using the information of the gradient.

$$\nabla_N I_{i,j} \equiv I_{i-1,j} - I_{i,j},\ \nabla_S I_{i,j} \equiv I_{i+1,j} - I_{i,j},\ \nabla_E I_{i,j} \equiv I_{i,j+1} - I_{i,j},\ \nabla_W I_{i,j} \equiv I_{i,j-1} - I_{i,j} \tag{5.26}$$

$$c_{N\,i,j}^t = g\left(\left\|(\nabla\mathbf{I})_{i+(1/2),j}^t\right\|\right),\ c_{S\,i,j}^t = g\left(\left\|(\nabla\mathbf{I})_{i-(1/2),j}^t\right\|\right), c_{E\,i,j}^t = g\left(\left\|(\nabla\mathbf{I})_{i,j+(1/2)}^t\right\|\right)$$
$$c_{W\,i,j}^t = g\left(\left\|(\nabla\mathbf{I})_{i,j-(1/2)}^t\right\|\right). \tag{5.27}$$

Finally, the image is reconstructed by adding the variations in the iterative process of Eq. (5.28).

$$\mathbf{I}_{i,j}^{(t+1)} = \mathbf{I}_{i,j}^{(t)} + \lambda[c_N \nabla_N \mathbf{I} + c_S \nabla_S \mathbf{I} + c_E \nabla_E \mathbf{I} + c_W \nabla_W \mathbf{I}]_{i,j}^{(t)} \tag{5.28}$$

This principle has been a guide for local in time processing over the image used in image processing algorithms for adaptation to a local context in the image.

5.2.7.4 TFOCS

The reconstruction methods require powerful mathematical tools for minimization of the error in the estimation. A resource commonly used is the split-Bregman model in which the norm L_1 is employed. In (becker et al. 2011), the library templates for first-order conic solvers (TFOCS) were designed to facilitate the construction of first-order methods for a variety of convex optimization problems. Its development was motivated by its authors' interest in compressed sensing, sparse recovery, and low-rank matrix completion. In a general form, this tool let us solve x for the inverse problem:

$$\min \phi(x) \triangleq f(\mathrm{A}(x) + b) + h(x), \tag{5.29}$$

where the function f is smooth and convex, h is convex, A is a lineal operator, and b a bias vector. The function h also must be prox-capable; in other words, it must be inexpensive to compute its proximity operator of Eq. (5.30)

$$\min \Phi_h(x, t) = \arg\min_z \; h(z) + \frac{1}{2} t^{-1} \langle z - x, z - x \rangle \tag{5.30}$$

The following is an example of solution with TFOCS; consider the following problem,

$$\min \frac{1}{2} \|Ax - b\|_2^2, s.t. \|x\|_1 \le \tau \tag{5.31}$$

this problem can be written as:

$$\min \frac{1}{2} \|Ax - b\|_2^2 + h(x) \tag{5.32}$$

where $h(x) = 0$ if $\|x\|_1 \le \tau$ and $+\infty$ otherwise. Translated to a single line of code:

$$\mathrm{x} = \mathrm{tfocs}(\mathrm{smooth\_quad}, \{\mathrm{A}, -\mathrm{b}\}, \mathrm{proj\_l1}(\mathrm{tau}));$$

The library was employed in Ren et al. (2017) for the minimization of a function of estimation in which two priors are employed: the first a differential respect to a new estimation based on TV of a central patch respect to a window of search adaptive high-dimensional non-local total variation (AHNLTV) and the second a weighted adaptive geometric duality (AGD). Figure 5.9 shows the visual comparison between bicubic interpolation and AHNLTV-AGD method after HR image estimation.

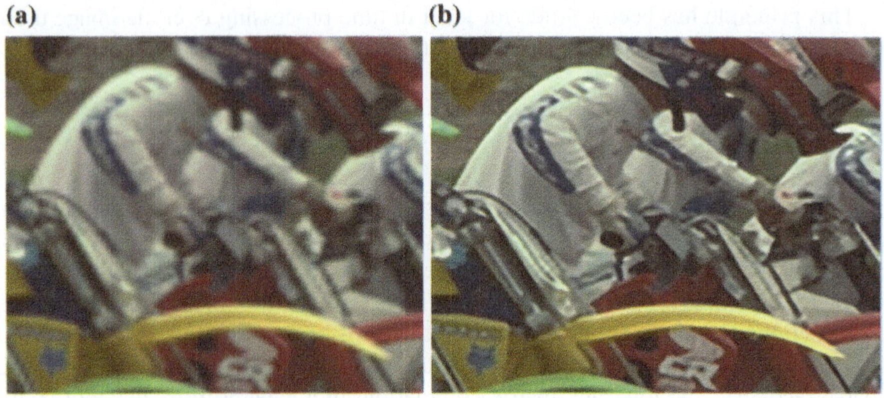

Fig. 5.9 Visual comparison of the HR image using **a** bicubic interpolation and **b** AHNLTV-AGD method

5.2.7.5 Total Variation

Total variation (TV) uses a regularization term as in MAP formulation. It applies similar penalties for a smooth and a step edge, and it preserves edges and avoids ringing effects; Eq. (5.33) is the term of TV,

$$\rho(f) = \|\nabla f\|_1 \tag{5.33}$$

where ∇ is the gradient operator. The TV term can be weighted with an adaptive spatial algorithm based on differences in the curvature. For example, the bilateral total variation (BTV) (Farsiu et al. 2003) is used to approximate TV, and it is defined in Eq. (5.34),

$$\rho(f) = \sum_{k=0}^{P}\sum_{l=0}^{P} \alpha^{l+1}\left\|f - S_x^k S_y^l f\right\|_1 \tag{5.34}$$

where S_x^k and S_y^l shift f by k and l pixels in the x and y directions to present several scales of derivatives, $0 < \alpha < 1$ imposes a spatial decay on the results (Farsiu et al. 2003), and P is the scale at which the derivatives are calculated (so it calculates derivatives at multiple scales of resolution (Farsiu et al. 2006). In (Wang et al. 2008), the authors discuss that an a priori term generates saturated data if it is applied to unmanned aerial vehicle data. Therefore, it has been suggested to combine it with the Hubert function, resulting in the BTV Hubert of Eq. (5.35),

$$\rho(|x|) = \begin{cases} \frac{|\nabla x|^2}{2} & ,\text{if } A < \alpha \\ \frac{\partial A}{\partial x} & \text{othewise} \end{cases}, \tag{5.35}$$

where A is the BTV regularization term and α is obtained as α = median [|A − median|A|·|]. This term keeps the smoothness of the continuous regions and preserves edges in discontinuous regions (Wang et al. 2008). In (Li et al. 2010), a locally adaptive version of BTV, called LABTV, has been introduced to provide a balance between the suppression of noise and the preservation of image details (Li et al. 2010). To do so, instead of the L_1 norm an Lp norm is used. The value of p for every pixel is defined based on the difference between the pixel and its surroundings. In smooth regions, where the noise reduction is important, p is set to a large value, close to two, and in non-smooth regions, where edge preservation is important, p is set to small values, close to one. The same idea of adaptive norms, but using different methods for obtaining the weights, has been employed in (Omer and Tanaka 2010; Song et al. 2010; Huang et al. 2011; Liu and Sun 2011; Mochizuki et al. 2011).

5.2.7.6 Gradient Management

The gradients are a topic of interest in SR. The changes in the image are a fundamental evidence of the resolution, and a high-frequency content brings the maximal changes of values between consecutive pixels in the image. The management of gradient has been addressed in two forms: first, by using a dictionary of external gradients of HR and second, by working directly on the LR image and reconstructing the HR gradients with the context of the image and regularization terms.

In these methods (Sun et al. 2008; Wang et al. 2013), a relationship is established in order to sharp the edges. In the first case, the gradients of an external database of HR are analyzed, and with a dictionary technique, the gradients of the LR input image are reconstructed. In the second case, the technique does not require external dictionaries, the procedure is guided by the second derivative of the same LR images amplified using pure interpolation, then a gradient scale factor is incorporated extracted from the local characteristics of the image.

In this chapter, we propose a new algorithm of gradient management and the application for a novel procedure of SR. For example, a bidirectional and orthogonal gradient field is employed. In our algorithm, two new procedures are proposed; in the first, the gradient field employed is calculated as:

$$l\nabla_{Th}^{u} = \mathbf{I}_{h}^{u}\begin{bmatrix} -1 & -1 & 1 \\ -1 & 0 & 1 \\ -1 & 1 & 1 \end{bmatrix}\frac{1}{2} \tag{5.36}$$

Then, the procedure is integrated as shown in Fig. 5.10; for deeper understanding, refer to (Wang et al. 2013).

The second form of our procedure is the application of the gradient field with independence. That is, the gradient fields are calculated by convolving the image with discrete gradient operators of Eq. (5.37) to obtain the differences along diagonal directions. The resulting model is shown in Fig. 5.11.

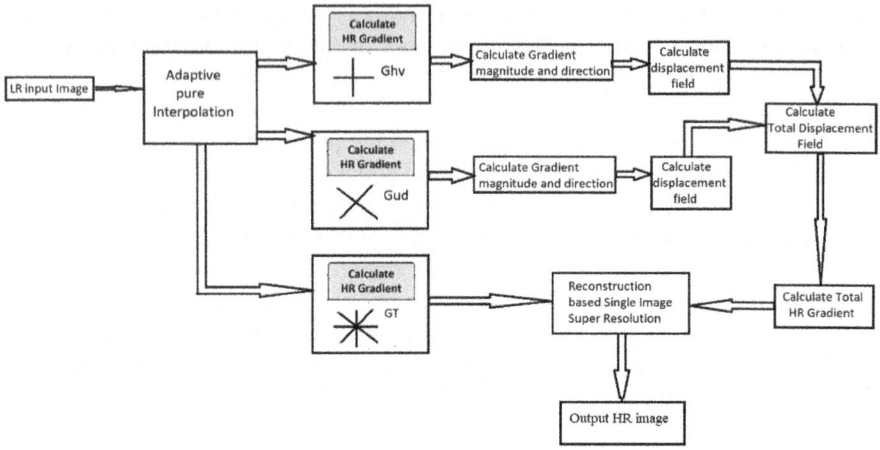

Fig. 5.10 Overview of the proposed SR algorithm. First, two orthogonal and directional HR gradients as well as a displacement field

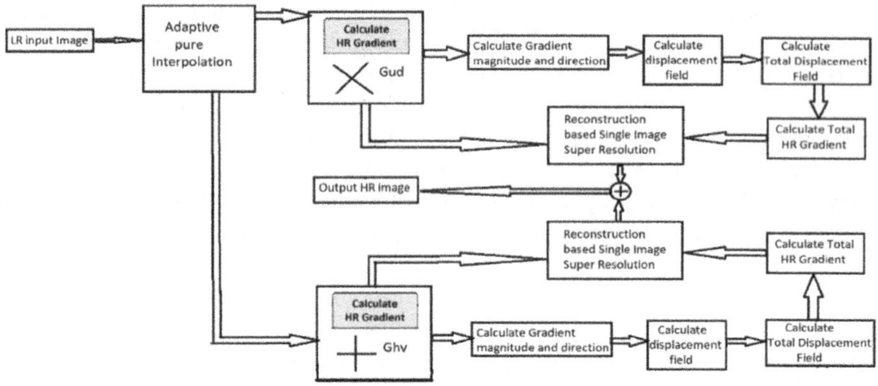

Fig. 5.11 Bidirectional and orthogonal gradient management with independent branches

$$\begin{bmatrix} -1 & 0 & 0 \\ 0 & 0 & 0 \\ 0 & 0 & 1 \end{bmatrix} \frac{1}{2} \text{ and } \begin{bmatrix} 0 & 0 & -1 \\ 0 & 0 & 0 \\ 1 & 0 & 0 \end{bmatrix} \frac{1}{2} \tag{5.37}$$

5.2.7.7 Hybrid BTV and Gradient Management

This section proposes the integration of two powerful tools for SR, the TV and gradient control. In the proposed case, the gradient regularization is applied first using the proposed model of Sect. 5.2.7.6. The technique produces some artifacts

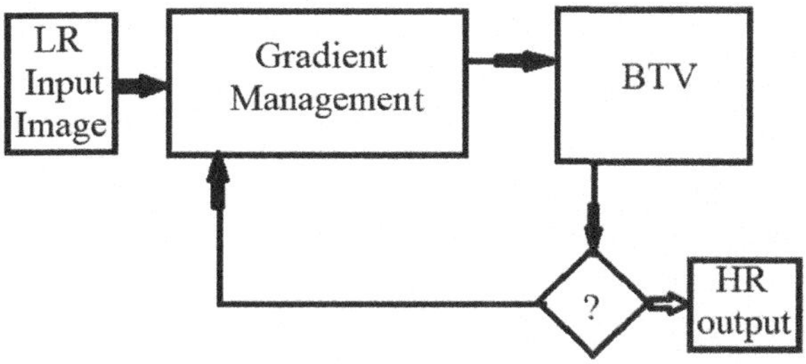

Fig. 5.12 Hybrid model for collaborative SR. The model combines the gradient control and BTV strategies

when the amplification scale is high, and the regularization term takes high values. The first problem is addressed by TV also exposed previously. This algorithm brings an average of similar pixels around the image for estimation of the high resolution. Here, two characteristics can collaborate for a better result.

The general procedure of the proposed method is shown in Fig. 5.12, and the visual comparison between the LR image and the HR image is exposed in Fig. 5.16. The proposed new algorithm is named orthogonal and directional gradient management and bilateral total variation (ODGM-BTV). It is only an illustration of the multiple possibilities for the creation of SR algorithms.

5.3 Results of the Proposed Methods

In this section, the results of the proposed methods are illustrated. Experiments on test and real images are presented with scaling factors of 2, 3, and 4. The objective metrics used were peak signal-to-noise ratio (PSNR) and the structural similarity (SSIM), and results are given in Tables 5.1, 5.2, and 5.3. Subjective performance of our SR schemes is evaluated in Figs. 5.13, 5.14, 5.15, and 5.16.

Table 5.1 PSNR/SSIM comparison for multi-directional, diagonal, horizontal, and vertical gradient management with a scale of 2

Method	Decoupled total gradient	Diagonal gradient	Horiz–vert gradient	Total gradient
Bike	**0.6509/22.399**	0.6547/22.321	0.6346/21.988	0.6507/22.026
Butterfly	**0.7714/22.360**	0.7714/22.580	0.7564/22.177	0.7634/22.237
Comic	0.5515/20.694	**0.5540/20.628**	0.5366/20.348	0.5633/20.697
Flower	**0.7603/23.964**	0.7553/26.221	0.7467/25.225	0.7691/24.710
Hat	**0.8165/26.025**	0.8163/25.888	0.8126/25.964	0.8168/26.284
Parrot	**0.8648/27.378**	0.8604/26.70	0.8602/26.594	0.8641/25.876
Parthenon	0.6585/20.632	**0.6593/20.593**	0.6451/20.371	0.6616/21.540
Plants	**0.8480/23.298**	0.8467/23.521	0.8418/28.645	0.8546/26.458

Table 5.2 PSNR/SSIM comparison for multi-directional, diagonal, horizontal, and vertical gradient management with a scale of 3

Method	Decoupled total gradient	Diagonal gradient	Horiz–vert gradient	Total gradient
Bike	**0.6103/21.83**	0.5950/21.229	0.5956/21.405	0.5854/21.068
Butterfly	**0.7345/22.23**	0.7174/21.121	0.7199/21.124	0.7065/20.714
Comic	**0.5074/20.12**	0.4812/18.789	0.4985/19.859	0.4999/19.925
Flower	**0.7275/24.37**	0.7079/25.200	0.7185/25.007	0.7242/24.678
Hat	**0.7996/26.43**	0.7939/26.757	0.7965/25.967	0.7904/26.129
Parrot	**0.8491/26.73**	0.8363/25.901	0.8486/26.307	0.8371/24.253
Parthenon	**0.6243/21.70**	0.6091/21.777	0.6151/20.868	0.6137/22.001
Plants	**0.8256/23.19**	0.8103/23.372	0.8252/28.664	0.8248/26.432

Table 5.3 PSNR/SSIM comparison for multi-directional, diagonal, horizontal, and vertical gradient management with scale of 4

Method	Decoupled total gradient	Diagonal gradient	Horiz–vert gradient	Total gradient
Bike	**0.5498/20.9911**	0.5197/19.3258	0.5368/20.6564	0.5325/20.7127
Butterfly	**0.6806/20.7812**	0.6532/19.0710	0.6552/18.0756	0.6613/19.9920
Comic	**0.4451/19.4540**	0.4156/18.0424	0.4363/19.0990	0.4417/19.3933
Flower	**0.6713/24.5320**	0.6440/24.0778	0.6644/24.5192	0.6697/24.0886
Hat	**0.7723/26.4887**	0.7633/26.2439	0.7684/26.2650	0.7673/26.1693
Parrot	**0.8214/25.5286**	0.8026/24.5686	0.8209/24.8905	0.8115/23.5459
Parthenon	**0.5844/22.4052**	0.5626/21.9254	0.5745/21.5580	0.5762/21.7595
Plants	**0.7837/22.3791**	0.7599/22.5480	0.7905/27.7839	0.7901/26.5721

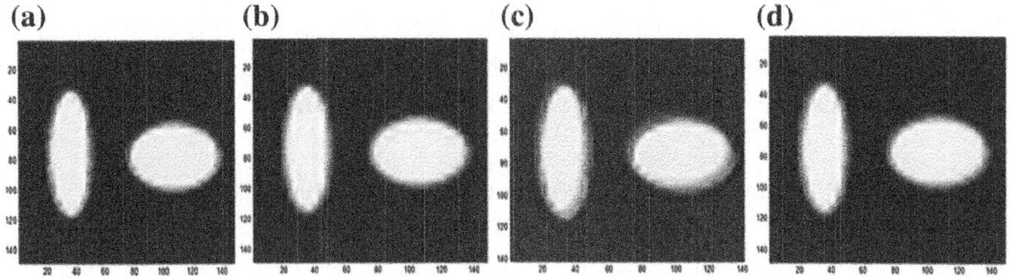

Fig. 5.13 4× amplification factor using a test image with **a** diagonal, **b** horiz–vert, **c** coupled, and **d** decoupled gradients

5.3.1 Gradient Management

In these experiments, the group of images shown in Fig. 5.15, included in the BSDS500 database, was used. The amplification factors were 2, 3, and 4. Tables 5.1, 5.2, and 5.3 show the increment in PSNR and SSIM of the second

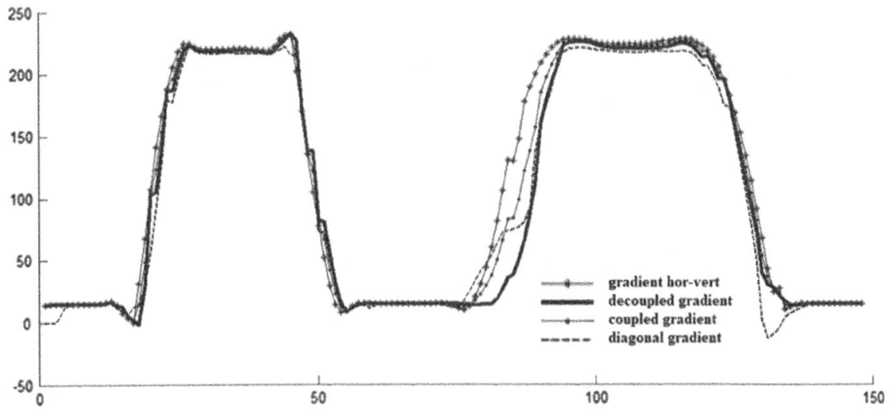

Fig. 5.14 Slopes of the estimated HR image (row 60 of the test image in Fig. 5.13). The image was processed using the two proposed algorithms with two orthogonal directions of the slopes independently

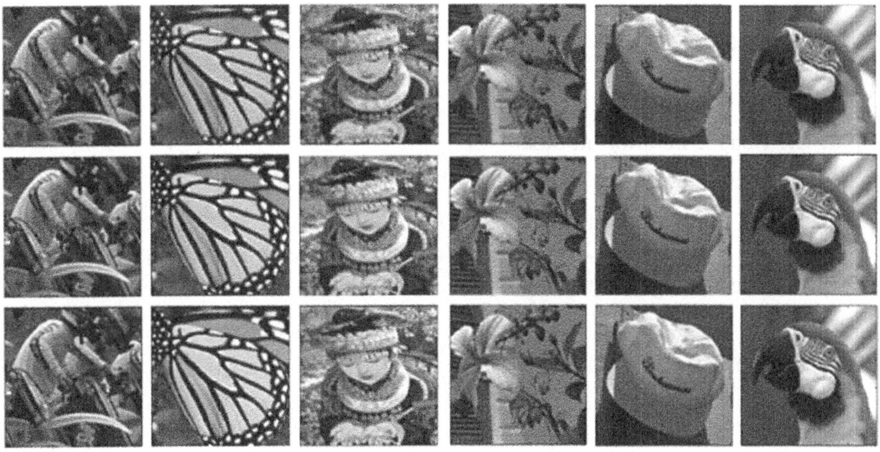

Fig. 5.15 Processed images with the decoupled gradient algorithm. The scale factors are: 4× for the top row of images, 3× for the second row of images, and 2× for the row of images at the bottom

alternative proposed with independence of the two gradient fields. Also, the test image was used to observe the sharpening effect around contours, and the results are shown in Fig. 5.13. Figure 5.14 shows the plot of the row 60, taken from the test image of Fig. 5.13, to illustrate the edge transitions for the HR recovered image.

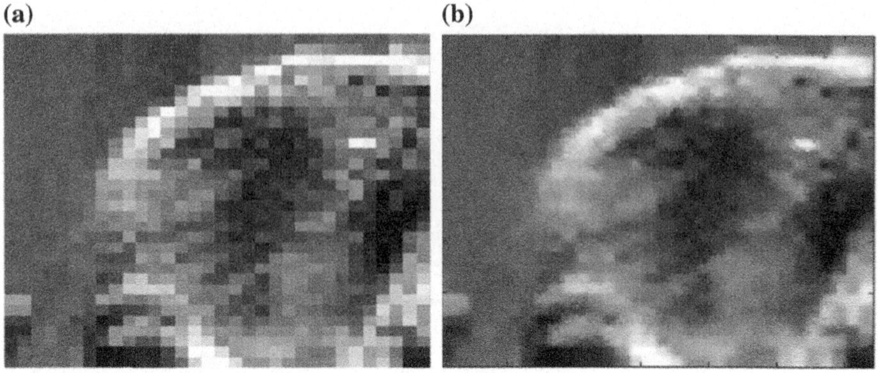

Fig. 5.16 Application of SR using the hybrid BTV and gradient management strategy with a scale of amplification of q = 4, **a** low-resolution image and **b** application of ODGM-BTV

5.3.2 *Hybrid BTV and Gradient Management*

Figure 5.16 shows the result of the proposed method ODGM-BTV using a scale of amplification of 4.

Algorithm:

Input: LR image, iteration number
For i = 1: iteration number

1. Apply the BTV algorithm to the LR input image.
2. Apply the bidirectional orthogonal gradient management.
3. Update the LR input image with the HR output image.

end
Output HR image

5.4 Metrics

The PSNR in dBs of Eq. (5.38) and SSIM of Eq. (5.39) are the metrics most used to evaluate SR algorithms.

$$\mathrm{PSNR} = 10\log_{10}\left(\frac{v_{\max}^2}{\mathrm{MSE}(x,y)}\right), \tag{5.38}$$

where x and y are the two signals to compare, $\mathrm{MSE}(x,y)$ is the mean square error, and $v_{\max}$ is the maximum possible value in the range of the signals. The SSIM factor (Wang et al. 2004) is calculated as,

$$\mathrm{SSIM}(x,y) = \frac{(2\,\mu_x\,\mu_y + c_1)(2\sigma_{xy} + c_2)}{\left(\mu_x^2 + \mu_y^2 + c_1\right)\left(\sigma_x^2 + \sigma_y^2 + c_2\right)}, \tag{5.39}$$

where μ_x and μ_y are the mean value of x and y, σ_x^2, σ_y^2, and σ_{xy} are the variance and covariance of x and y; c_1 and c_2 are constants terms. Another metric derived from the SSIM is the mean SSIM (MSSIM) of Eq. (5.40)

$$\mathrm{MSSIM} = \frac{1}{M}\sum_{j=1}^{M} \mathrm{SSIM}(x_j, y_j), \tag{5.40}$$

where M is the number of the areas being compared.

5.4.1 Discussion of the Results

Tables 5.1, 5.2, and 5.3 show an enhancement of the quality parameters SSIM and PSNR of our proposed method over the management of a single gradient. Also, the scales of amplification are greater than 3 with major increments of the quality factors for high scale factors. Our procedure employs a natural following of the gradients, and let to give a more precise dimension of the slopes, it is an important contribution to the state of the art of the algorithms of gradient management. Although the goal of our chapter is an overview of complements for super-resolution and not contributions of a novel algorithms or improvement of the results in the state of the art. The overview shows that SR is a very rich field of investigation. In each step, we can find a possibility of application of some method using the strongest principle of functioning. An example is the combination of the BTV and ODGM, the visual effect is very interesting in Fig. 5.16, and the major resolution by area can be observed. The contribution in this case avoids artifacts from gradient management, and at the same time, a less blurred image is obtained in comparison with BTV method due to the sharping procedure over the edges.

The review of the literature brings some conclusions. The investigation in this topic is extended, and the contributions for the state of the art are in the most of the cases little changes over well-known procedures. Unfortunately, the goal is based on a quality measurement and the benchmark for guide of the results is based on different configurations of the Eqs. (5.38), (5.39), and (5.40). The consequence is that the comparison between many reported algorithms and models is difficult and not always possible. In this point, the borders between classifications of the methods are diffused by this reason the comparison between methods in an overview more than attempts of classification and the explanation of the classification is not useful. Nevertheless, the great creativity exhibited in the different methods and the totally different mathematical solutions make it difficult to establish mathematical comparisons and objective conclusions without considering only empirical results based on measurement metrics.

5.5 Conclusions

SR is an exciting and diverse subject in the digital processing area and can take all possible forms. Each algorithm has a place in this area of research and is extremely complex and comprehensive. The study of these techniques should be oriented from the beginning because the development of each of them is broad and difficult to reproduce. Sometimes, a small advance can be made in one of them. Also, the initial condition is different in each case and some bases of comparison are required. In the literature, some standard measurements are proposed but the application conditions are diverse. A useful strategy to approach SR research is the knowledge of the cause of preexisting algorithms. Advantages and disadvantages are important factors to consider in order to combine characteristics that produce more convincing effects and better qualities of the output image in a system. The proposed example makes edge sharpening and average for estimation; the first method produces artifacts, but the second fails to produce clear edges. A case was proposed in which these two characteristics can be positively complemented. For future work, we continue the study of multiple possibilities in the field of SR estimation using the transformation of the image and learning from different characterizations as wavelets fluctuations with dictionary learning. Other interesting field is the minimization procedures for multiple residuals priors in the estimations as was made in works as (Ren et al. 2017).

References

Becker, S., Candès, E., & Grant, M. (2011). Templates for convex cone problems with applications to sparse signal recovery. *Mathematical Programming Computation, 3,* 165–218.

Belekos, S., Galatsanos, N., & Katsaggelos, A. (2010). Maximum a posteriori video super-resolution using a new multichannel image prior. *IEEE Transactions on Image Processing, 19*(6), 1451–1464.

Capel, D., & Zisserman, A. (1998). Automated mosaicing with super-resolution zoom. In *Proceedings of IEEE Computer Society Conference on Computer Vision and Pattern Recognition (CVPR),* Santa Barbara, California, USA, 1, 885–891.

Cheeseman, P., Kanefsky, B., Kraft, R., Stutz, J., & Hanson, R. (1994). *Super-resolved surface reconstruction from multiple images* (1st ed.). London, United Kingdom: Springer Science + Business Media.

Dai, S., Han, M., Xu, W., Wu, Y., & Gong, Y. (2007). Soft edge smooth-ness prior for alpha channel super resolution. In *Proceedings of IEEE Conference on Computer Vision and Pattern Recognition (CVPR),* Minneapolis, Minnesota, USA, *1,* 1–8.

Dai, S., Han, M., Xu, W., Wu, Y., Gong, Y., & Katsaggelos, A. (2009). SoftCuts: a soft edge smoothness prior for color image super-resolution. *IEEE Transactions on Image Processing, 18*(5), 969–981.

Debes, C., Wedi, T., Brown, C., & Zoubir, A. (2007). Motion estimation using a joint optimisation of the motion vector field and a super-resolution reference image. In *Proceedings of IEEE International Conference on Image Processing (ICIP),* San Antonio, Texas, USA, 2, 479–500.

Dong, W., Zhang, L., Shi, G., & Wu, X. (2011). Image deblurring and super-resolution by adaptive sparse domain selection and adaptive regularization. *IEEE Transactions on Image Processing, 20*(7), 1838–1856.

Elad, M., & Hel-Or, Y. (2001). A fast super-resolution reconstruction algorithm for pure translational motion and common space-invariant blur. *IEEE Transactions on Image Processing, 10*(8), 1187–1193.

Farsiu, S., Robinson, D., Elad, M., & Milanfar, P. (2003). Robust shift and add approach to super-resolution. In *Proceedings of SPIE Conference on Applications of Digital Signal and Image Processing*, San Diego, California, USA, *1*, 121–130.

Farsiu, S., Robinson, D., Elad, M., & Milanfar, P. (2004). Fast and robust multi-frame super-resolution. *IEEE Transactions on Image Processing, 13*(10), 327–1344.

Farsiu, S., Elad, M., & Milanfar, P. (2006). A practical approach to super-resolution. In *Proceedings of SPIE Conference on Visual Communications and Image Processing*, San Jose, California, USA, *6077*, 1–15.

Huang, K., Hu, R., Han, Z., Lu, T., Jiang, J., & Wang, F. (2011). A face super-resolution method based on illumination invariant feature. In *Proceedings of IEEE International Conference on Multimedia Technology (ICMT)*, Hangzhou, China, *1*, 5215–5218.

Hung, K., & Siu, W. (2012). Robust soft-decision interpolation using weighted least squares. *IEEE Transactions on Image Processing, 21*(3), 1061–1069.

Irani, M., & Peleg, S. (1990). Super-resolution from image sequences. In *Proceedings of 10th IEEE International Conference on Pattern Recognition*, Atlantic City, New Jersey, USA, 1,115–120.

Irani, M., & Peleg, S. (1991). Improving resolution by image registration. *CVGIP Graphical Models and Image Processing, 53*(3), 231–239.

Irani. M., & Peleg, S. (1992). Image sequence enhancement using multiple motions analysis. In *Proceedings of IEEE Computer Society Conference on Computer Vision and Pattern Recognition (CVPR)*, Champaign, Illinois, USA, *1*, 216–222.

Irani, M., & Peleg, S. (1993). Motion analysis for image enhancement: Resolution, occlusion, and transparency. *Journal of Visual Communication and Image Representation, 4*(4), 324–335.

Jung, C., Jiao, L., Liu, B., & Gong, M. (2011a). Position-patch based face hallucination using convex optimization. *IEEE Signal Processing Letters, 18*(6), 367–370.

Jung, M., Bresson, X., Chan, T., & Vese, L. (2011b). Nonlocal Mumford-Shah regularizers for color image restoration. *IEEE Transactions on Image Processing, 20*(6), 1583–1598.

Keren, D., Peleg, S., & Brada, R. (1998). Image sequence enhancement using subpixel displacements. In *Proceedings of IEEE Computer Society Conference on Computer Vision and Pattern Recognition (CVPR)*, Ann Arbor, Michigan, USA, *1*, 742–746.

Kim, K., & Kwon, Y. (2008). Example-based learning for single-image super-resolution. *Pattern Recognition, LNCS, 5096*, 456–465.

Kim, K., & Kwon, Y. (2010). Single-image super-resolution using sparse regression and natural image prior. *IEEE Transactions on Pattern Analysis and Machine Intelligence, 32*(6), 1127–1133.

Kong, D., Han, M., Xu, W., Tao, H., & Gong, Y. (2006). A conditional random field model for video super-resolution. In *Proceedings of 18th International Conference on Pattern Recognition (ICPR)*, Hong Kong, China, *1*, 619–622.

Kumar, S., & Nguyen, T. (2010). Total subset variation prior. In *Proceedings of IEEE International Conference on Image Processing (ICIP)*, Hong Kong, China, *1*, 77–80.

Li, X., & Orchard, M. (2001). New edge-directed interpolation. *IEEE Transactions on Image Processing, 10*(10), 1521–1527.

Li, F., Jia, X., & Fraser, D. (2008). Universal HMT based super resolution for remote sensing images. In *Proceedings of 15th IEEE International Conference on Image Processing (ICIP)*, San Diego, California, USA, 1, 333–336.

Li, X., Lam, K., Qiu, G., Shen, L., & Wang, S. (2009). Example-based image super-resolution with class-specific predictors. *Journal of Visual Communication and Image Representation, 20* (5), 312–322.

Li, X., Hu, Y., Gao, X., Tao, D., & Ning, B. (2010). A multi-frame image super-resolution method. *Signal Processing, 90*(2), 405–414.

Liu, C., & Sun, D. (2011). A Bayesian approach to adaptive video super resolution. In *Proceedings of IEEE Conference on Computer Vision and Pattern Recognition (CVPR)*, Colorado Springs, Colorado, USA, *1*, 209–216.

Mallat, S., & Yu, G. (2010). Super-resolution with sparse mixing estimators. *IEEE Transactions on Image Processing, 19*(11), 2889–2900.

Martins, A., Homem, M., & Mascarenhas, N. (2007). Super-resolution image reconstruction using the ICM algorithm. In *Proceedings of IEEE International Conference on Image Processing (ICIP)*, San Antonio, Texas, USA, *4*, 205–208.

Mochizuki, Y., Kameda, Y., Imiya, A., Sakai, T., & Imaizumi, T. (2011). Variational method for super-resolution optical flow. *Signal Processing, 91*(7), 1535–1567.

Morera, D. (2015). Determining parameters for images amplification by pulses interpolation. *Ingeniería Investigación y Tecnología, 16*(1), 71–82.

Morera, D. (2014). Amplification by pulses interpolation with high frequency restrictions for conservation of the structural similitude of the image. *International Journal of Signal Processing, Image Processing and Pattern Recognition, 7*(4), 195–202.

Omer, O., & Tanaka, T. (2010). Image superresolution based on locally adaptive mixed-norm. *Journal of Electrical and Computer Engineering, 2010*, 1–4.

Perona, P., & Malik, J. (1990). Scale-space and edge detection using anisotropic diffusion. *IEEE Transactions on Pattern Analysis and Machine Intelligence, 12*(7), 629–639.

Peyre, G., Bougleux, S., & Cohen, L. (2008). Non-local regularization of inverse problems. In *Proceedings of European Conference on Computer Vision*, Marseille, France, 5304, 57–68.

Pickup, L., Capel, D., & Roberts, S. (2006). Bayesian image super-resolution, continued. *Neural Information Processing Systems, 19*, 1089–1096.

Pickup, L. (2007). Machine learning in multi-frame image super-resolution. Ph.D. thesis, University of Oxford.

Prendergast, R., & Nguyen, T. (2008). A block-based super-resolution for video sequences. In *Proceedings of 15th IEEE International Conference on Image Processing (ICIP)*, San Diego, California, USA, *1*, 1240–1243.

Purkait, P., & Chanda, B. (2012). Super resolution image reconstruction through Bregman iteration using morphologic regularization. *IEEE Transactions on Image Processing, 21*(9), 4029–4039.

Ren, C., He, X., Teng, Q., Wu, Y., & Nguyen, T. (2006). Single image super-resolution using local geometric duality and non-local similarity. *IEEE Transactions on Image Processing, 25* (5), 2168–2183.

Ren, C., He, X., & Nguyen, T. (2017). Single image super-resolution via adaptive high-dimensional non-local total variation and adaptive geometric feature. *IEEE Transactions on Image Processing, 26*(1), 90–106.

Schultz, R., & Stevenson, R. (1994). A Bayesian approach to image expansion for improved definition. *IEEE Transactions on Image Processing, 3*(3), 233–242.

Shao, W., & Wei, Z. (2008). Edge-and-corner preserving regularization for image interpolation and reconstruction. *Image and Vision Computing, 26*(12), 1591–1606.

Song, H., Zhang, L., Wang, P., Zhang, K., & Li, X. (2010). An adaptive L1–L2 hybrid error model to super-resolution. In: *Proceedings of 17th IEEE International Conference on Image Processing (ICIP)*, Hong Kong, China, *1*, 2821–2824.

Sun, J., Xu, Z., & Shum, H. (2008). Image super-resolution using gradient profile prior. In *Proceedings of IEEE Conference on Computer Vision and Pattern Recognition (CVPR)*, Anchorage, Alaska, *1*, 1–8.

Tappen, M., Russell, B., & Freeman, W. (2003). Exploiting the sparse derivative prior for super-resolution and image demosaicing. In *Proceedings of IEEE 3rd International Workshop on Statistical and Computational Theories of Vision (SCTV)*, Nice, France, *1*, 1–24.

Villena, S., Abad, J., Molina, R., & Katsaggelos, A. (2004). Estimation of high resolution images and registration parameters from low resolution observations. *Progress in Pattern Recognition, Image Analysis and Applications, LNCS, 3287,* 509–516.

Wang, Z., Bovik, A., Sheikh, H., & Simoncelli, E. (2004). Image quality assessment: from error visibility to structural similarity. *IEEE Transactions on Image Processing, 13*(4), 600–612.

Wang, Y., Fevig, R., & Schultz, R. (2008). Super-resolution mosaicking of UAV surveillance video. In *Proceedings of 15th IEEE International Conference on Image Processing (ICIP)*, San Diego, California, USA, 1, 345–348.

Wang, L., Xiang, S., Meng, G., Wu, H., & Pan, C. (2013). Edge-directed single-image super-resolution via adaptive gradient magnitude self-interpolation. *IEEE Transactions on Circuits and Systems for Video Technology, 23*(8), 1289–1299.

Woods, N., & Galatsanos, N. (2005). Non-stationary approximate Bayesian super-resolution using a hierarchical prior model. In *Proceedings of IEEE International Conference on Image Processing (ICIP)*, Genova, Italy, 1, 37–40.

Zhang, X., & Wu, X. (2008). Image interpolation by adaptive 2-D autoregressive modeling and soft-decision estimation. *IEEE Transactions on Image Processing, 17*(6), 887–896.

Part II
Control

Chapter 6
Learning in Biologically Inspired Neural Networks for Robot Control

Diana Valenzo, Dadai Astorga, Alejandra Ciria and Bruno Lara

Abstract Cognitive robotics has focused its attention on the design and construction of artificial agents that are able to perform some cognitive task autonomously through the interaction of the agent with its environment. A central issue in these fields is the process of learning. In its attempt to imitate cognition in artificial agents, cognitive robotics has implemented models of cognitive processes proposed in areas such as biology, psychology, and neurosciences. A novel methodology for the control of autonomous artificial agents is the paradigm that has been called neuro-robotics or embedded neural cultures, which aims to embody cultures of biological neurons in artificial agents. The present work is framed in this paradigm. In this chapter, simulations of an autonomous learning process of an artificial agent controlled by artificial action potential neural networks during an obstacle avoidance task were carried out. The implemented neural model was introduced by Izhikevich (2003); this model is capable of reproducing abrupt changes in the membrane potential of biological neurons, known as action potentials. The learning strategy is based on a multimodal association process where the synaptic weights of the networks are modified using a Hebbian rule. Despite the growing interest generated by artificial action potential neural networks, there is little research that implements these models for learning and the control of autonomous agents. The present work aims to fill this gap in the literature and at the same time, serve as a guideline for the design of further experiments for in vitro experiments where neural cultures are used for robot control.

Keywords Learning · Biologically inspired neural networks · Robot control
Artificial intelligence

D. Valenzo · D. Astorga · B. Lara (✉)
Cognitive Robotics Laboratory, Centro de Investigación en Ciencias,
Universidad Autónoma del Estado de Morelos, Cuernavaca, Mexico
e-mail: bruno.lara@uaem.mx

A. Ciria
Pshycology Department, Universidad Nacional Autónoma de México (UNAM),
CDMX, Mexico

O. O. Vergara Villegas et al. (eds.), *Advanced Topics on Computer Vision, Control and Robotics in Mechatronics*,
https://doi.org/10.1007/978-3-319-77770-2_6

6.1 Introduction

Artificial neural networks have been widely used to control artificial agents (Pfeifer and Scheier 1999; Gaona et al. 2015; He et al. 2016). However, a new paradigm has emerged by fusing neuroscience and cognitive robotics. This methodology attempts to study cognitive processes, such as learning and memory in vitro. The aim is to use embed living neurons in artificial agents (DeMarse et al. 2001). By doing this, a new possibility emerges in the study of the cellular mechanisms underlying cognition.

This work attempts to give some hints and directions in this field by using simulated artificial action potential neural networks to control an artificial agent. The models used have a high biological plausibility and can serve as a guideline in the design of experiments that use in vitro neural cultures.

The chapter is divided as follows: the remainder of this section gives a short introduction to the changes of paradigm in the study of cognition in artificial intelligence. Section 6.2 presents the theoretical framework for artificial neural networks, focusing on models of action potential neurons. Section 6.3 presents the materials, methods, and results of two different experiments. Finally, in Sect. 6.4, the conclusions are presented.

6.1.1 Artificial Intelligence

Artificial intelligence, now known as classic or traditional artificial intelligence, emerges as a discipline aiming to reproduce certain aspects of human intelligence in machines. Its hypothesis was that if all aspects of learning or any other intelligence capability can be described with precision, then it could be reproduced in machines (McCarthy et al. 2006). Classical artificial intelligence (AI) was strongly influenced by the hypothesis held by the school of cognitivism, paradigm that affirms that human cognition is based on the manipulation of symbolic representations through a set of rules (Newell and Simon 1976). However, the idea that intelligent behavior can be captured by using only symbolic representations has been widely criticized. As the field has experienced, even though AI has managed significant success in specific domains such as the generation of logic algorithms capable of solving abstract problems, it has not been able to produce artificial systems capable of imitating more basic human abilities such as sensorimotor processes or perception (Mingers 2001). This is one of the reasons why critics of classic AI propose the need to design artificial agents situated in the world, putting the emphasis on the fundamental role the body plays in producing and learning intelligent behavior (Dreyfus 1967, 1972). Following these lines of thought, the new paradigm in research is born, namely new AI.

New AI is based on the embodied and embedded paradigms of cognition which highlights the importance of the interaction of the agent with its environment for the development of cognitive processes (Pfeifer and Scheier 1999). The most natural platforms to perform this interaction are artificial agents (Moravec 1984; Brooks 1991). From this perspective, Brooks argues that perception and sensorimotor abilities are the really hard problems to solve by artificial agents. Once an agent has the basic sensing and moving abilities to achieve survival and reproduction, higher-order abilities should come easier to implement. Such high-order abilities include problem solving, language expert knowledge among other things (Brooks 1991).

6.1.2 Cognitive Robotics

Following the ideas and principles of the new AI, cognitive robotics is an emerging research area postulating that the best way to imitate and study cognition is by building artificial autonomous agents. An artificial agent is defined as that machine that is capable of learning an ability by means of the interaction with its environment to successfully perform some specific cognitive task (Pfeifer and Scheier 1999). Artificial agents are then robots that have a body, sensors, and a motor system that allow them to perceive their environment and interact with it. Moreover, artificial agents can be real physical robots or computer-simulated robots that live and interact in a simulated environment.

Studies in this field focus on basic cognitive abilities, such as moving successfully in an environment. The central hypothesis is that complex behaviors emerge from the interaction of simpler behaviors, such as obstacle avoidance (Copeland 2015).

This field aims at simulating cognitive processes in robots through the implementation of models coming from other branches of cognitive sciences, such as psychology, neuroscience, and philosophy (Pfeifer and Scheier 1999). A recent area of research for control of autonomous agents arises from the fusion of robotics and neuroscience. This paradigms, neuro-robotics or embedded neural cultures, attempt to embed and embody biological neural cultures by using in vitro neurons to control artificial agents. At the same time, these agents are in direct contact with their environment and from this interaction changes in patterns and strength of connections in the embedded neurons take place.

6.2 Neurons and Neural Networks

6.2.1 Embedded Neural Cultures

Research on embedded neural cultures emerges as a field aiming at filling the gap when having, on the one hand, studies on learning and memory and, on the other, in vitro studies of the cellular mechanisms of synaptic plasticity involved in these cognitive processes (Potter and DeMarse, 2001). This methodology is an attempt to integrate top-down (cognitive, behavioral, ethological) and bottom-up (cellular, molecular) approaches in the study of learning and memory (Potter 2001; Manson 2004).

This methodological paradigm is interested in the study of hybrid neural systems that are in constant interaction with an environment. The proposed strategy is to embody biological neural cultures through simulated or real artificial agents. The general idea is that a neural culture is used to control the behavior of an artificial agent.

The hypothesis underlying this proposal is that the capabilities of an agent for learning and adaptation in a dynamic environment are the result of a continuous interaction of the nervous system with this environment mediated by the body of the agent (Novellino et al. 2007). Due to the fact that neural systems have evolved to serve a body that constantly interacts with its environment, they should be studied as embodied and embedded neural systems (Potter 2001). Given that the classical studies of in vitro cultures do not include the interaction with the environment in the study of the neural mechanisms participating in learning, it appears to be a highly limited approach. It is, following this logic that the embedding of cultures becomes an appealing paradigm (Potter 2001).

6.2.2 Simulation of Neural Cultures

A body of research in computational neuroscience has focused on the simulation of neural cultures. Useful tools in these studies are the different existing models for artificial action potential neural networks. The simulations can be of isolated as well as embodied neural cultures as this can be coupled or not with an artificial agent.

The studies with isolated cultures aim at exploring the self-organization capabilities of cell groups in complex networks of neurons. In his seminal work, Izhikevich (2004) reports the results of a simulated culture formed by 100,000 interconnected neurons. The architecture is inspired by the anatomy of the cerebral cortex. However, the aim of this work was the search for activation patterns in an isolated culture.

On the other hand, there are implementations of artificial action potential neural networks for the control of artificial agents. In these, the networks are used for controlling an artificial agent, while it navigates an environment; such is the case of Mokhtar et al. (2007) which focuses in the design of a neurocontroller based on the pyramidal neurons of the hippocampus.

In another study, Trhan (2012) reports experiments to control a Mindstorms NXT artificial agent with an artificial action potential neural network. The implemented architecture is formed by two local neural networks that process in parallel sensory information coming from the agent. One network controls the right side of the agent, while the other, the left. In this work, the network learns to control the movements of the agent based on the interaction with its environment. The system adapts to the environment through the evolutionary development of a population of individuals. The implemented evolutionary mechanism allows the adaptation of the neural network in a short period of time and the network becomes capable of controlling the agent so that it navigates safely in the environment without colliding with the walls.

In this section, the authors will include a critical analysis of the sources review related to the particular topic adressed. This part will be used for comparison and evaluation purposes. The authors should include a table in which a summary of the main features of the papers reviewed can be depicted. Interaction with its environment.

6.2.3 Bio-inspired Artificial Action Potential Neural Networks

Artificial neural networks (ANNs) are mathematical models inspired by the structure and functioning of the nervous system. These models are formed by single elements called units or neurons. Each unit is a processing element that receives some input from the environment or other units. All input signals are processed, and the unit outputs a single signal based on some mathematical function. The information is propagated through the network depending on the global architecture of the system.

ANN have been classified in three main generations according to the mathematical model, that the neurons in them use, to transform the incoming information. The first generation is based on the McCulloch-Pitts model (McCulloch and Pitts 1943). The main characteristic of this model is that outputs are binary. In the second generation, the output from the units is a continuous value (between 0 and 1 or −1 and 1) typically the result of a sigmoidal activation function. In contrast, the third generation of ANN uses action potential neurons. These models are thought to better simulate the output of biological neurons as they try to capture the nature of electrical impulses of these cells. One of the advantages of these models is that they use time as a resource as they process information due to the output of neurons is the change in their membrane potential with respect to time (Maass 1997).

There exist some of models of action potential neurons, Izhikevich (2004) provides a review of some of the most used ones. It is important to highlight that, in using these types of models, a compromise must be made between two important but seemingly mutually exclusive characteristics. On the one hand, the model must be computationally simple so as to be feasible; on the other, it must reproduce the firing patterns of biological networks (Izhikevich 2004). The most bio-physically precise model, such as the one proposed by Hodking-Huxley, have a very high computational cost given the amount of floating point number operations they perform. For this reason, the number of neurons that can be modeled in real time is limited. On the other hand, models such as integrate-and-fire are very efficient computationally, but they reproduce poorly the dynamics registered experimentally in biological networks (Izhikevich 2003). Considering all these, the model proposed by Izhikevich (2003) presents a reasonable compromise between computational efficiency and biological plausibility.

The Izhikevich model is capable of reproducing action potentials of a number of different biological cortical neurons by using only four parameters in a two-dimensional system of differential equations of the form:

$$v' = 0.04v2 + 5v + 140 - u + I$$
$$u' = a(bv - u)$$

with an adjustment condition:

$$\text{if } v \geq 30, \quad \text{then} \begin{cases} v \leftarrow c \\ u \leftarrow u + d \end{cases}$$

where v represents the membrane action potential of the neuron and u is a variable modeling the recovery of the membrane which gives negative feedback to v. Variable I is the electrical input to the neuron. The parameters a, b, c, and d of the model allow the reproduction of different neural behaviors. The effect of these parameters in the dynamics of the model are (Izhikevich 2003):

- The parameter a describes the time scale of the recovery variable u. Smaller values result in slower recovery. A typical value is $a = 0.02$.
- The parameter b describes the sensitivity of the recovery variable to the subthreshold fluctuations of the membrane potential v. Greater values couple v and u more strongly resulting in possible subthreshold oscillations and low-threshold spiking dynamics. A typical value is $b = 0.2$. The case $b < a(b > a)$ corresponds to saddle-node (Andronov–Hopf) bifurcation of the resting state (Izhikevich 2000).
- The parameter c describes the after-spike reset value of the membrane potential v caused by the fast high-threshold k^+ conductances. A typical value is $c = -65$ mV.
- The parameter d describes after-spike reset of the recovery variable caused by slow high-threshold Na^+ and k^+ conductances. A typical value is $d = 2$.

It is worth noting that, even though the Izhikevich model is a plausible biological model with low computational cost, there are few implementations of this model for the control of artificial agents in cognitive robotics. Even more, there is very few research and implemetations of learning algorithms for networks using this type of models.

6.3 Materials and Methods

Using the bio-inspired computational tools described in the previous section, it is intended that an artificial agent coupled to an artificial action potential neural network develops different obstacle avoidance behaviors. In the first two experiments, reactive agents were designed. These are used as a first approach to the simple (Izhikevich 2003) model of spiking neurons; here, the systems do not have a learning process as the purpose was to explore the parameters that control the behavior of the neurons. In the second set of experiments, the same parameters were used; however, here, an autonomous learning algorithm was implemented.

For all the experiments, a simulated artificial agent Pioneer 3DX was used. The agent has a ring of eight ultrasonic sensors (sonars) and two independent motors (with wheels), one on each side (Fig. 6.1). The Pioneer platform includes the C++ Advanced Robotics Interface for Application (ARIA) library to control the speed and movements of the agent.

The agent and its environment were simulated in MobileSim, an open-source software used to simulate mobile robot platforms and their environments. MobileSim is capable of simulating walls and other obstacles in an artificial environment. The environment designed for the learning experiments presents square obstacles with a separation of 4000 mm between them (Fig. 6.2).

The values assigned to the parameters of the neuron model during the simulations were those reported by Izhikevich (2003) to reproduce the dynamics of excitatory neurons called regular spiking (Table 6.1), because these neurons are the most abundant in the cerebral cortex of mammals. The dynamics of this cells are characterized by exhibiting a phenomenon of frequency adaptation of firing when

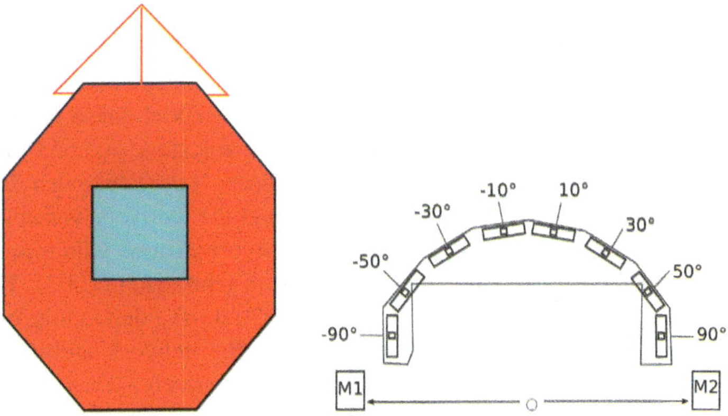

(a) Simulated robot (b) Sonars & motors

Fig. 6.1 Artificial agent Pioneer 3DX

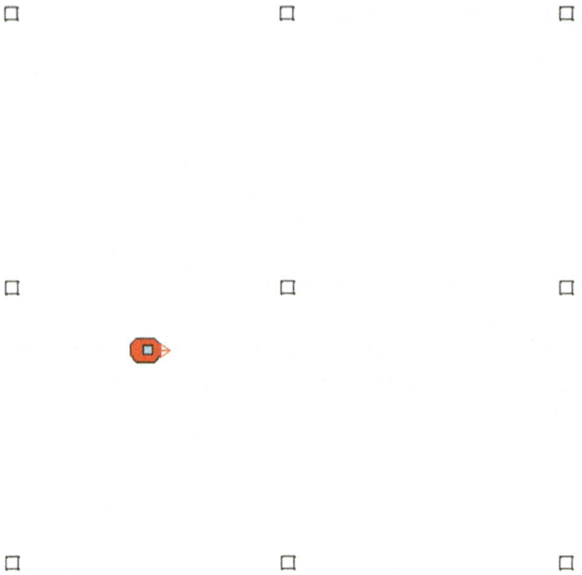

Fig. 6.2 Simulated environment during the learning experiments

Table 6.1 Values of the model to simulate excitatory neurons

Parameter	*a*	*b*	*c*	*d*
Value	0.02	0.2	−65.0	8.0

they are stimulated in a prolonged way. This means that the frequency of action potentials of these neurons decreases over time despite the fact that the stimulus persists. Another characteristic of this type of neurons is that their firing frequency increases when the current they receive increases, although they never fire too fast due to the long hyperpolarization phase they present. In the model, these parameters correspond to a more negative value of the readjustment voltage ($c = -65.0$) and a high value of the readjustment of the recovery variable $u(d = 8.0)$ (Izhikevich 2003).

The initial values for the membrane potential and the recovery variable for each neuron were $v_o = -65.0$ and $u_o = -13.0$, respectively. These values were established taking as reference the experiment reported in Izhikevich (2003). The resting membrane potential was established at −65.0 to −70.0 mV, depending on the type of neuron (sensory, interneuron, or motor) within the network architecture. This corresponds to an input stream I_{base} of 0 or 3.5, respectively.

Structure of Processing and Transmission of Information

The processing and transmission of the information used for the implementation of the systems can be divided into the following phases:

- ***Normalization of sonar values***: Originally, each of the eight sonars of the agent can register obstacles that are within a range of distance between 0 (near) and 5000 mm (far). These values were normalized to a range of 0–1, such that the maximum value indicates maximum proximity to an obstacle.

$$S_n = 1 - \frac{s}{5000}$$

  where s represents the original value of the sonar and S_n is the normalized value.
- ***Mapping of sonar information to sensory neurons***: The sonar information was mapped to an input stream of current for the sensory neurons of the networks. This input stream is proportional to the degree of activation of the sensors, in such a way that the firing frequency of sensory neurons is higher in the presence of obstacles.
- ***Propagation of information***: This process depends on the pattern of connections of each of the networks, as well as the value of the connection force or synaptic weight between the neurons. If the strength of connection is high, the current with which the presynaptic neuron contributes to the postsynaptic neuron will be enough to trigger an action potential in it. Otherwise, the contribution current would not trigger an action potential in the postsynaptic neuron.
- **Mapping of motor neuron activity to motors speeds**: Motor speed was assigned depending on the rate of action potentials of motor neurons recorded in a certain time window. Tests were performed with time windows of different duration to choose the appropriate one. A time window (W_t) of 400 ms was chosen because the shorter time windows could record very few action potentials, while very long time windows required longer simulations or higher learning rates.

$$V_{\mathrm{motor}} = 350 * \frac{\mathrm{AP}}{\mathrm{AP}_{\max}} + 150$$

  where V_{motor} refers to the speed assigned to the motor, AP is the number of action potentials within the set time window and $\mathrm{AP}_{\max}$ the maximum number of action potentials that can be generated in the time window. This speed will be assigned to the left motor if the action potentials correspond to the left motor neuron, and to the right motor if the right motor neuron is the one that triggers. If AP is zero, then a base speed is assigned to the motors, which in the equation corresponds to 150 mm/s.

6.3.1 Obstacles Avoidance with Hard-Wired Connection Weights

In these experiments, the agent does not learn, rather, the weights connecting the neurons are fixed. Two architectures are tested.

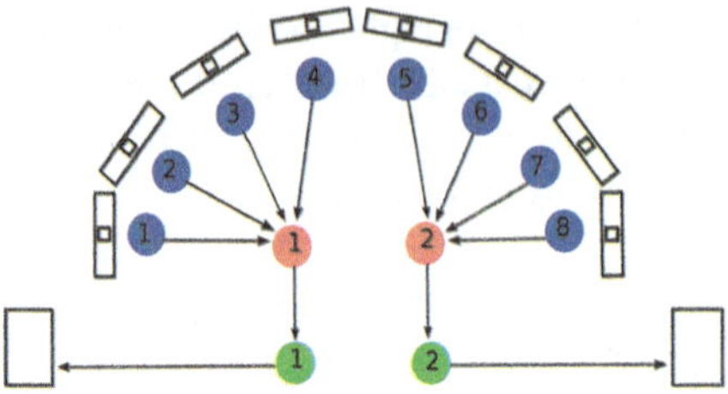

(a) Architecture 1

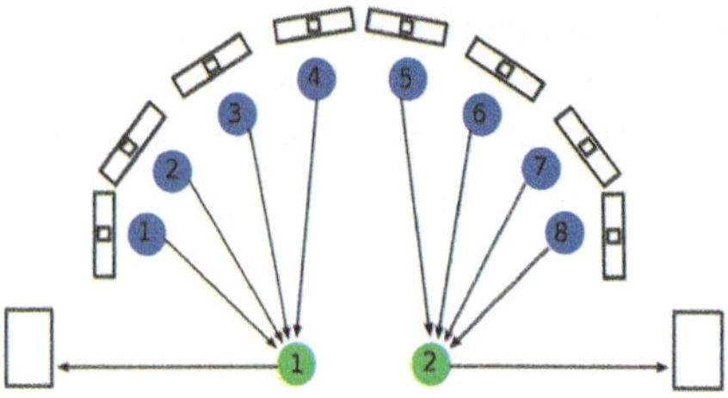

(b) Architecture 2

Fig. 6.3 Architectures with fixed weights

Neuronal Architectures

The artificial systems were modeled with an artificial action potential neural network coupled to the simulated agent (Fig. 6.3).

The two systems present a neural network architecture with eight sensory neurons and two motor neurons, which are illustrated with blue and green circles, respectively. However, one of them has two additional interneurons, shown in pink (Fig. 6.3a). In both systems, the sensory neurons are associated to a sensor of the artificial agent, in such a way that the sensor 1 is associated to the sensory neuron 1, the sensor 2 to the sensory neuron 2, and so on. Finally, the sensory neurons are connected to the interneurons or to the motor neurons, depending on the system.

Specifically, in Architecture 1 (Fig. 6.3a), sensory neurons 1, 2, 3, and 4, which correspond to the left side of the agent, are connected to interneuron 1. On the other hand, neurons 5, 6, 7, and 8 are connected to the interneuron 2. In this system, each interneuron is connected to the motor neuron that is on the same side of the architecture. In the case of Architecture 2 (Fig. 6.3b), sensory neurons are directly connected to motor neurons. The sensory neurons 1, 2, 3, and 4 are connected to the motor neuron 1, while the sensory neurons on the right side are connected to the motor neuron 2. In both systems, each motor neuron is associated with a motor. The motor neuron 1 is associated with the left motor, while the motor neuron 2 is associated with the motor on the right side of the agent. The wheels of the agent are independent of each other so that if the left wheel accelerates and the right wheel maintains its constant speed, the agent will turn to the right and vice versa.

It should be mentioned that during the experiments, a base speed was established for the motors, in such a way that when there is no obstacle near the agent, both wheels maintain the same speed and the agent advances in a straight line.

6.3.1.1 Architecture with Interneurons

In this experiment, the artificial system implemented corresponds to the neural architecture consisting of eight sensory neurons, two interneurons and two motor neurons (Fig. 6.3a). In this case, a value of 0.7 was established for each of the synaptic weights between the neurons, magnitude that remains fixed throughout the

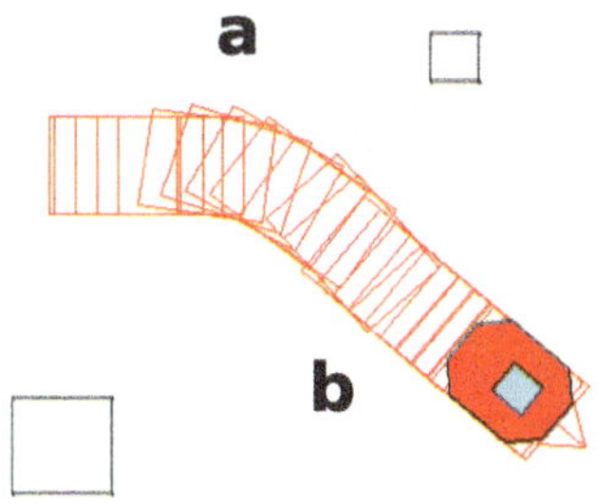

Fig. 6.4 Avoidance of two obstacles. Letter **a** indicates the moment in which the artificial agent detects one of the obstacles and turns to avoid it. When turning, the agent detects a second obstacle at time **b**

simulation. A value of 0.7 is enough to ensure that every time a neuron fires, the neurons connected to it will also fire. In this experiment, the artificial agent must evade two obstacles that are close to him; the path taken by the artificial agent during the simulation is shown in Fig. 6.4.

The agent is able to detect the obstacle in front of him at instant **a** and turns left timely. Subsequently, it detects a second obstacle at instant **b** but does not need to turn to avoid it. The recording of the sonars and the activation of the corresponding sensory neurons during the evasion of the obstacles are shown in Figs. 6.5, 6.6, 6.7, and 6.8. Figure 6.5 shows the activation of sonars 1, 2, 3, and 4, which are located on the left side of the agent. While, the recording of the activity of sensory neurons associated with each of these sonars is presented in Fig. 6.6. The graphs show that

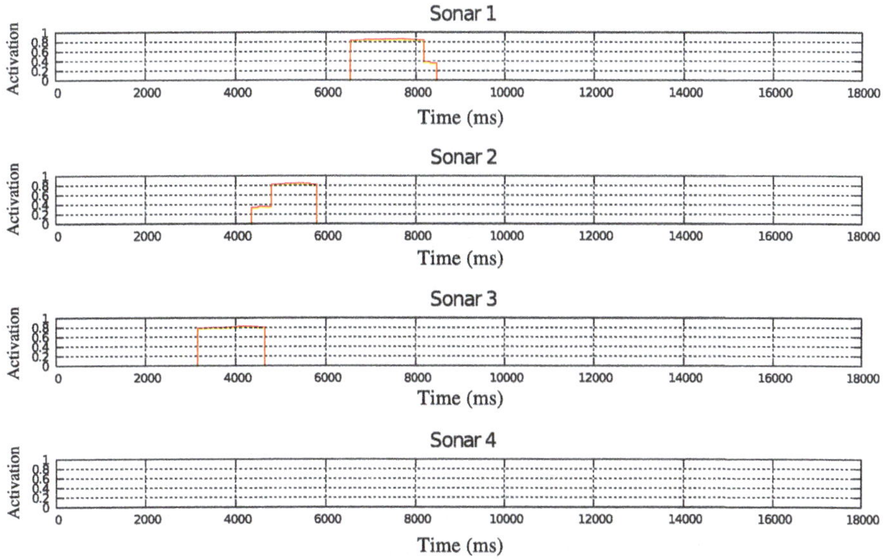

Fig. 6.5 Activation of sonars during the navigation in the environment shown in Fig. 6.4

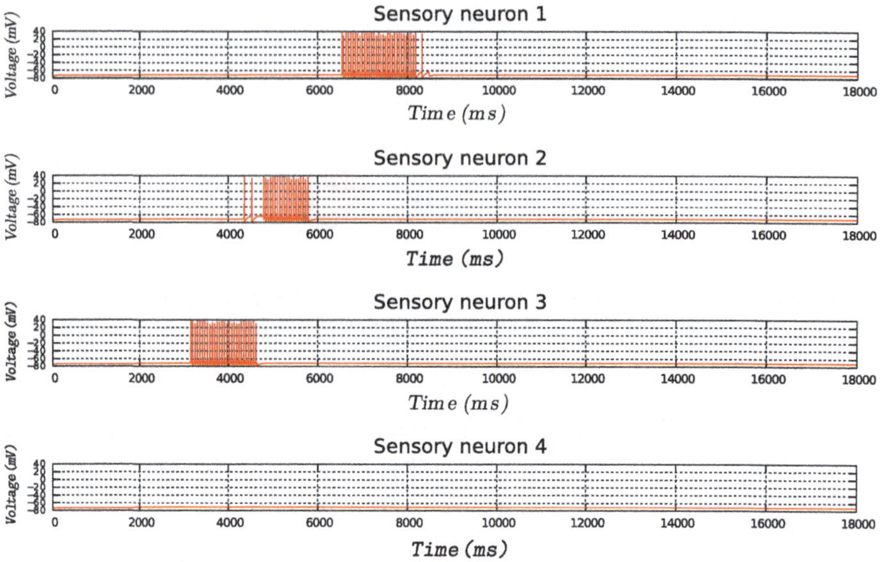

Fig. 6.6 Activation of sensory neurons while navigating environment shown in Fig. 6.4

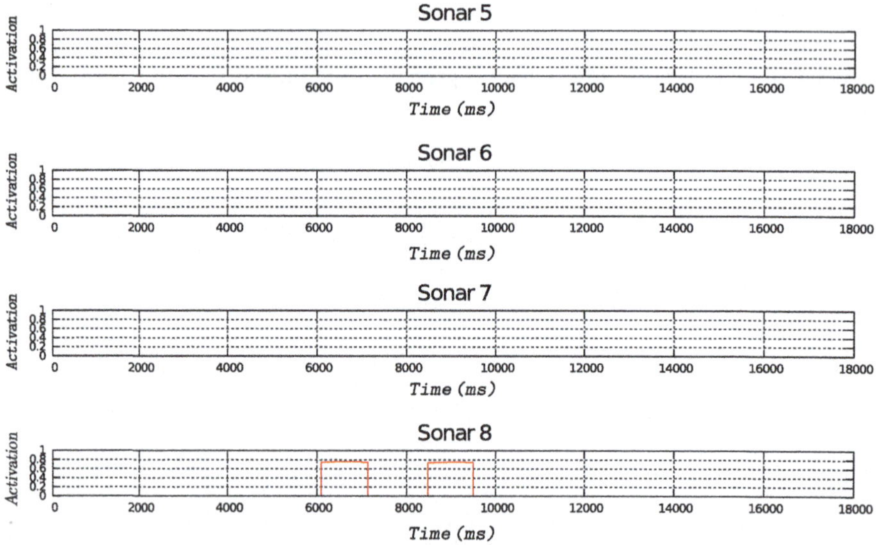

Fig. 6.7 Activation of sonars during navigation of the environment shown in Fig. 6.4

the first sensor that is activated during instant **a** is sensor 3, around 3000 ms. Its activation triggers a series of action potentials in the sensory neuron number 3, information that is transmitted to interneuron 1 and, subsequently, to the motor

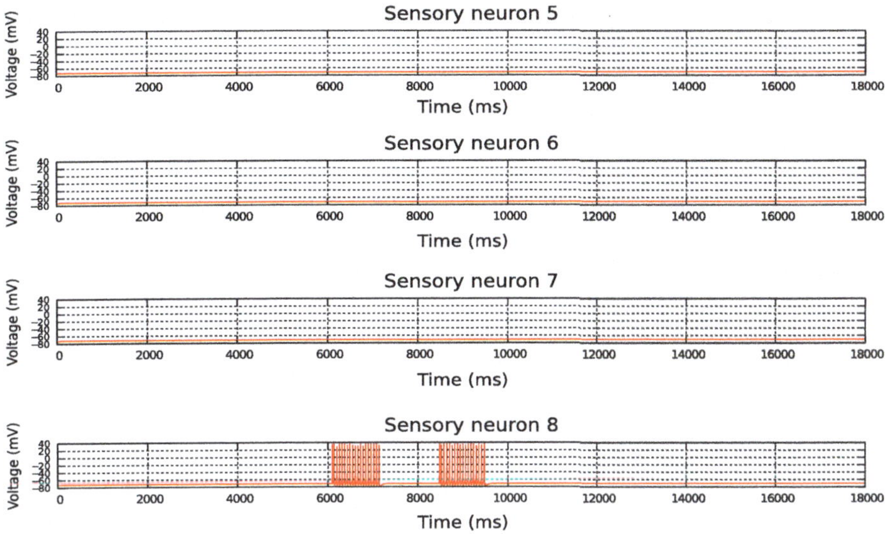

Fig. 6.8 Activation of sensory neurons during navigation of the environment shown in Fig. 6.4

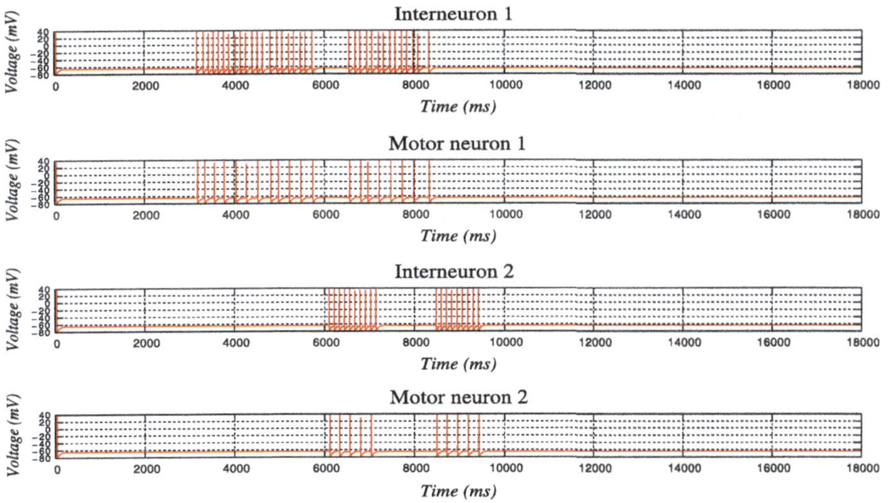

Fig. 6.9 Activation of inter- and motor neurons during navigation of the environment in Fig. 6.4

neuron 1 (Fig. 6.9). The action potentials generated by the motor neuron 1 cause the motor on the left side to increase its speed, so the agent turns to the right to avoid the obstacle. While turning, the same obstacle is detected by sensors 2 and 1 (shortly after the 4000 and 6000 ms, respectively), which generates action potentials in the sensory neurons associated with these sensors. This information is also transmitted to the motor neurons and influences the speed of the agent's turn.

Then at instant **b**, the sensor 8 on the right side of the architecture is activated at around 6000 ms due to the second obstacle (Fig. 6.7), triggering the activation of the sensory neuron 8 (Fig. 6.8), as well as interneuron 2 and motor neuron 2 (Fig. 6.9). This results in an increase in the speed of the right motor, which readjusts the direction of the agent. Changes in agent speeds, associated with the transmission of information from sonars to motor neurons, result in the successful evasion of both obstacles. Finally, the sensors do not detect any nearby obstacle, so the neurons do not fire and the speeds of both motors return to their base speed, causing the agent to move forward again.

6.3.1.2 Architecture with Direct Connections

In this experiment, the neural architecture composed of only eight sensory neurons and two motor neurons was used (Fig. 6.3b). The synaptic weight established between the connections of the network was set at a value of 0.7, as in the experiment previously described. The environment and behavior of the artificial agent during the simulation is shown in Fig. 6.10.

The artificial agent detects a barrier of obstacles with the sensors and neurons that are on the left side of its architecture. Figure 6.11 shows that 1, 2, 3 and 4 maintain a high activation during the first part of the simulation time. Therefore, the neurons associated with these sensors generate action potentials, decreasing their firing rate over time (Fig. 6.12). In turn, the result of these activations is reflected in the activity of the motor neuron 1 (Fig. 6.15), generating an increase in the speed of the left wheel. After turning, the activity of the sonars and neurons on the right side of the architecture increases due to the detection of an obstacle that is located on the right side of the environment. This happens shortly after

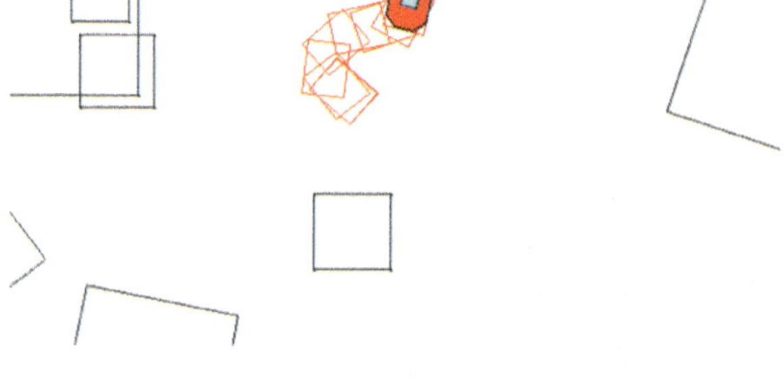

Fig. 6.10 Artificial agent detects a barrier of obstacles and turns to avoid them

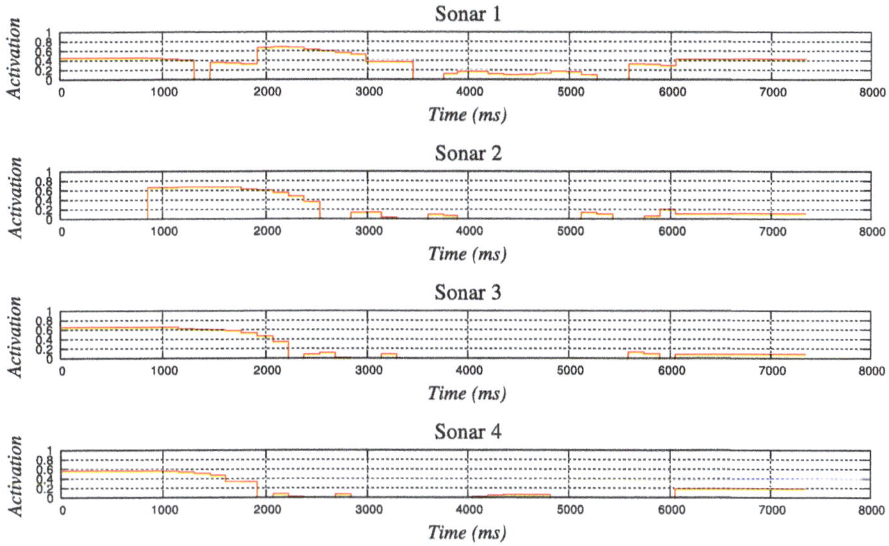

Fig. 6.11 Activation of sonars during navigation of the environment shown in Fig. 6.10

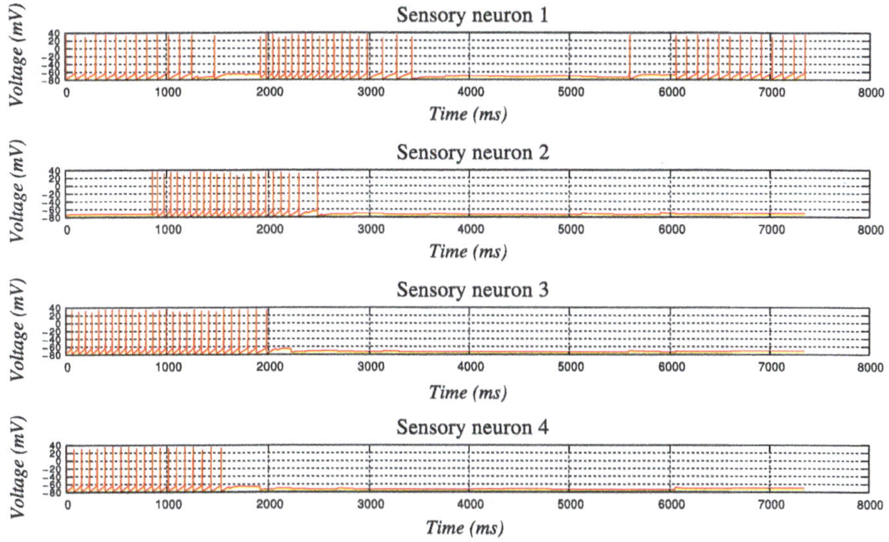

Fig. 6.12 Activation of sensory neurons during navigation of the environment shown in Fig. 6.10

3000 ms, as illustrated in Figs. 6.13 and 6.14. Finally, the sensory neuron 1 and the motor neuron 1 are activated again at the 6000 ms, generating a change in the speeds of the agent.

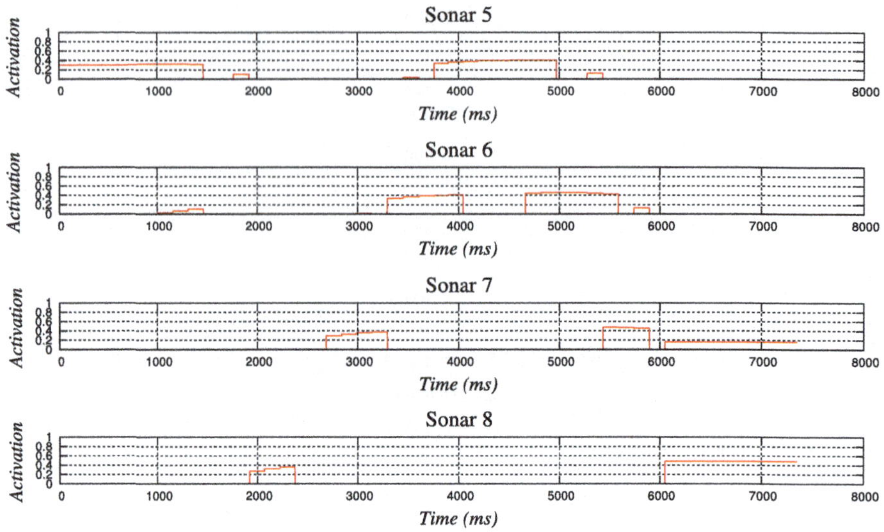

Fig. 6.13 Activation of sonars during navigation of the environment shown in Fig. 6.10

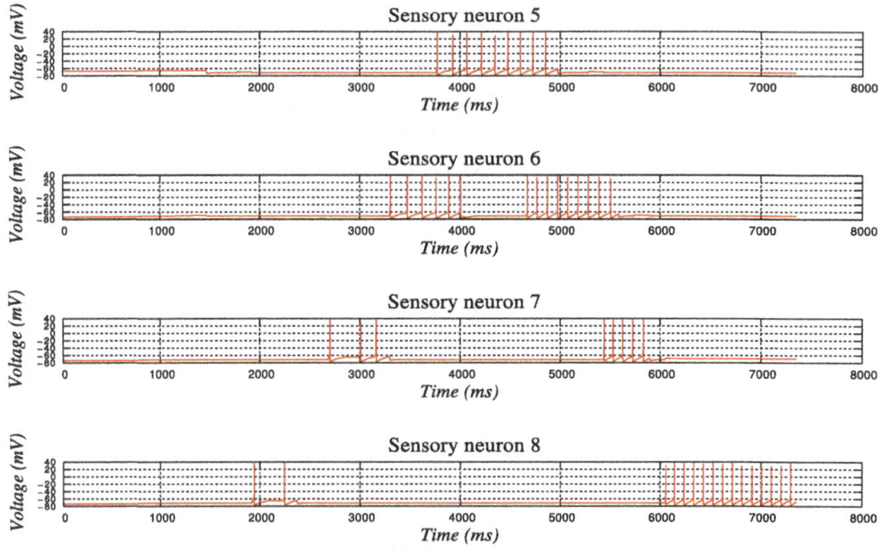

Fig. 6.14 Activation of sensory neurons during navigation of the environment shown in Fig. 6.10

6.3.2 *Learning the Weights of Connections*

In this section, the systems are expected to learn the appropriate connections between the neurons. The two architectures used are shown in Fig. 6.16.

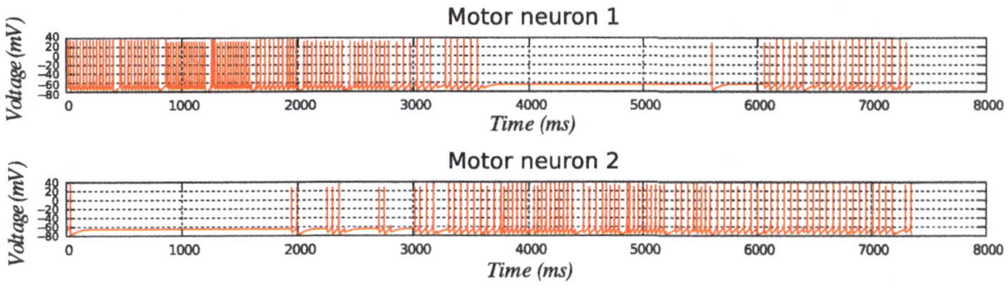

Fig. 6.15 Activation of motor neurons during navigation of the environment shown in Fig. 6.10

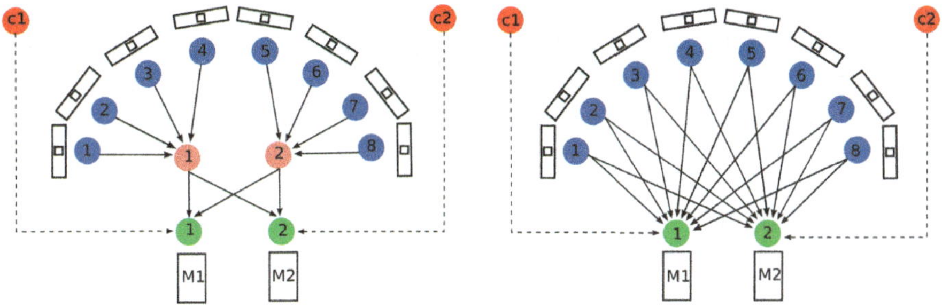

(a) Architecture 1b (b) Architecture 2b

Fig. 6.16 Architectures used in the learning experiments

These systems have a greater number of connections. In the case of Architecture 1b (Fig. 6.16a), additional connections were established only between interneurons and motor neurons (shown in pink and green, respectively). By contrast, in architecture 2b each sensory neuron is connected to the two motor neurons (Fig. 6.16b). The additional connections of these architectures have the purpose of integrating sensory information from the left side of the system with the information recorded on the right side. Also, each system has two collision sensors, one on the left side (labeled **c1** in the figure) and the other on the right side (labeled **c2**).

Both sensors have a binary activation threshold: the sensor **c1** is activated only when at least one of the sonars on the left side (sonars 1, 2, 3, or 4) has an activation equal to or greater than 0.8, which equals a distance of 1000 mm. On the other hand, sensor **c2** is activated when at least one of the sonars on the right side (sonars 5, 6, 7, or 8) has a value equal to or greater than 0.8.

The synaptic weights that participated in the self-learning were randomized and modified in each window of time by the following Hebbian learning rule:

$$\Delta w = \eta \mathrm{act}(n_{\mathrm{pre}}) \times \mathrm{act}(n_{\mathrm{post}}) - \epsilon$$

where η is the learning rate and ϵ the forgetting rate, while act(n_{pre}) and act(n_{post}) correspond to the activity of the presynaptic and postsynaptic neurons, respectively.

During the experiments, a learning rate of 0.08 and a forgetting rate of 0.000015 were used, both values determined experimentally. The value of the forgetting rate was established considering that high values caused the synaptic weights, which were reinforced during a collision, to decay quickly given the time it took the agent to meet the next obstacle. The degree of activation for the neurons was normalized in a range of 0–1, depending on the number of action potentials registered in each time window.

The systems were designed so that each time the agent hits an obstacle the activation of the corresponding collision sensor generates that the motor neuron associated with it fires. Simultaneously, the sensory neurons and/or corresponding interneurons will be activated by the proximity to the obstacle. The connections, between the sensory neurons and/or interneurons and the corresponding motor neurons, were reinforced by the Hebbian learning rule. The initial synaptic weight between the neurons involved in the learning process was established in such a way that, at the beginning of the simulations, the motor neurons are activated only by the corresponding collision sensor. However, once the system has learned, a multi-modal association occurs and the motor neurons will be activated by sensory neurons or interneurons, as the case may be, and not by the collision sensors. The functioning of the collision sensors was inspired by the experiment reported by Scheier et al. (1998).

6.3.2.1 Learning with Interneurons

The experiment was carried out with the system whose neuronal architecture presents eight sensory neurons, two interneurons and two motor neurons. The system and the values established for the simulation are specified in Fig. 6.17.

The initial synaptic weights are shown in Table 6.2. The sensory neurons are indicated with an *s*, the interneurons with an *i*, and the motor neurons with the letter *m*. The subscripts indicate the position of the neuron within the network. Each of the synaptic weights established between the sensory neurons and the interneurons has a value of 0.7, which ensures the firing of the interneurons each time one of the

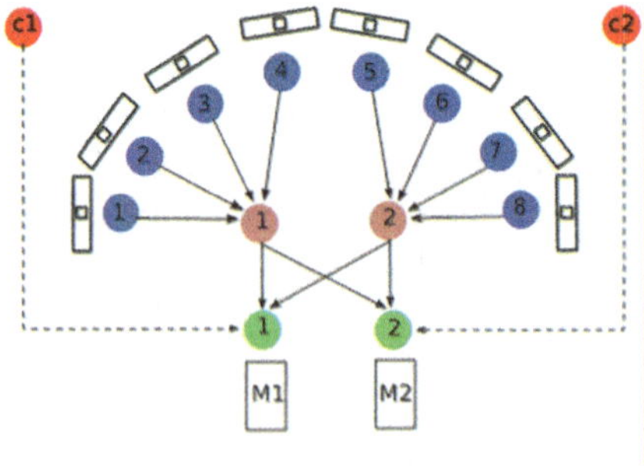

Experimental conditions	
α	0.08
β	0.000015
v_0	-65.0 mV
u_0	-13.0 mV
$u_{base\text{-}sensory}$	-70.0 mV
$u_{base\text{-}inter}$	-65.0 mV
$u_{base\text{-}motor}$	-65.0 mV
W_t	400 ms

Fig. 6.17 Network architecture and experimental conditions

Table 6.2 Initial synaptic weights for the experiment with interneurons

Neurons										
–	s_1	s_2	s_3	s_4	s_5	s_6	s_7	s_8	$\mathbf{m}_1$	$\mathbf{m}_2$
i_1	0.7	0.7	0.7	0.7	–	–	–	–	0.49	0.29
i_2	–	–	–	–	0.7	0.7	0.7	0.7	0.13	0.36

sensory neurons to which they are connected fires. The hyphens (-) indicate that there is no connection between neurons. The connection value between the sensory neurons and the interneurons remains fixed during the simulation. The synaptic weights that are modulated during the experiment are shown in boldface.

The navigation of the agent in the environment during the simulation is shown in Fig. 6.18. There were six collisions during the agent's path. The place of the collisions is indicated by an arrow and a number, marking the order in which they

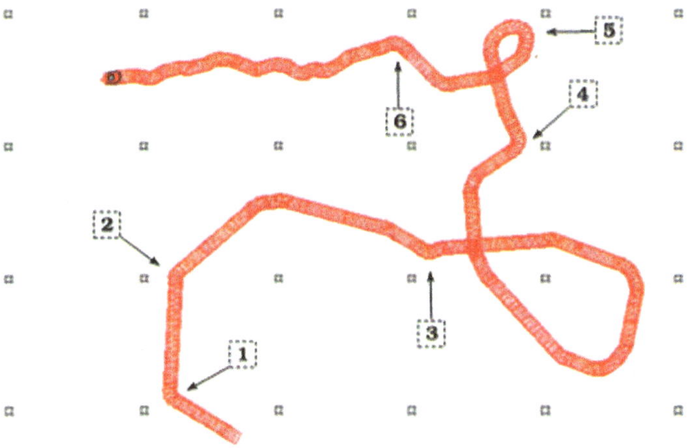

Collision							
Sonar	1	2	3	4	5	6	C1
1	Activated	Activated	-	-	Activated	-	
2	Activated	-	-	-	Activated	-	
3	-	-	-	-	Activated	-	
4	-	-	-	-	Activated	-	
5	-	-	-	-	-	-	C2
6	-	-	-	Activated	-	-	
7	-	-	-	Activated	-	Activated	
8	-	-	Activated	Activated	-	Activated	

Fig. 6.18 Path of the artificial agent during the experiment and table showing the activated sonars on each collision

happened. The information from the sensors obtained during each collision also indicates which neurons, both sensory and interneurons are activated.

The sensors that registered each of the collisions showed an activation equal to or greater than 0.8 (Fig. 6.18). The information in Table in Fig. 6.18 shows that collisions 1, 2, and 5 were registered by sensors located on the left side of the architecture, activating in these three cases the collision sensor **c1** and, therefore, neuron $\mathbf{m}_1$ was activated. In contrast, collisions 3, 4, and 6 activated collision sensor **c2**.

Figures 6.19 and 6.20 show the pattern of activation for the interneurons and motor neurons registered during the different situations the agent encounter in the environment. The change for the synaptic weights over time is also presented. These records are shown with the purpose of graphically illustrating the relationship of the activation patterns obtained with the change in the synaptic weights registered in the different situations.

One of the activation patterns registered during the experiment is shown in Fig. 6.19 and corresponds to the neuronal activity registered during collision 1. As shown, only interneuron 1 and motor neuron 1 generate action potentials. This pattern corresponds to an increase in the synaptic weight between these neurons (Fig. 6.22), while the other weights decrease due to the fact that the other two neurons do not fire during this collision. A similar activation pattern is observed

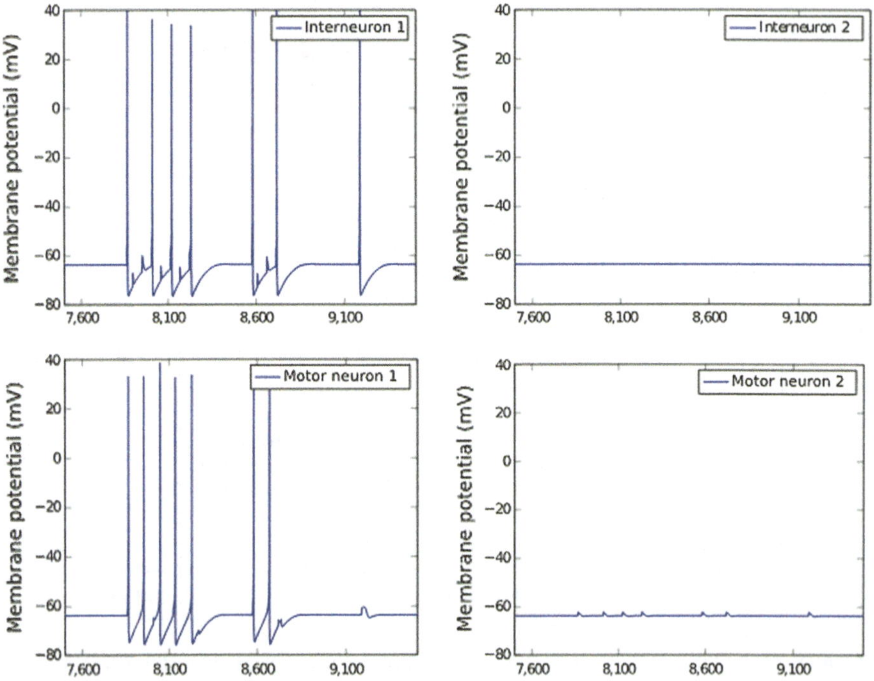

Fig. 6.19 Neural activity registered during Collision 1

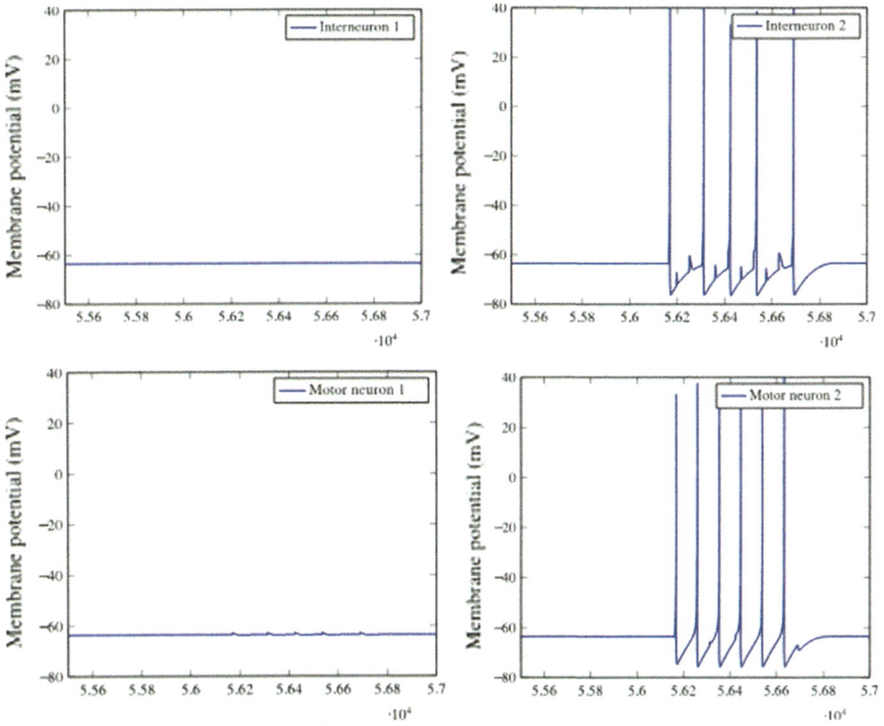

Fig. 6.20 Neural activity registered during Collision 3

during collisions 2 and 5, since in these cases the same interneurons and motor neurons are activated. The increases in synaptic weights during these collisions are also indicated in Fig. 6.22. It is important to remember that the initial synaptic weights are not high enough for the interneurons to trigger the motor neurons. Therefore, in this case, the activation of the motor neuron 1 is due to the activation of the collision sensor **c1** and not to the activation of the interneuron.

Figure 6.20 shows the activation pattern during a collision in which interneuron 2 and motor neuron 2 are activated. This is reflected in an increase in the synaptic weight of these neurons (Fig. 6.22). The other three synaptic weights do not increase. This case is similar to what happens in collisions 4 and 6, because the same interneurons and motor neurons are activated. In the same way as in Fig. 6.19, the activation of the motor neuron in Fig. 6.20 is due to the activation of the collision sensor and not to the activity of the interneuron.

Figure 6.21 illustrates an activation pattern where the four neurons are activated. In this situation, it was found that even when the four synaptic weights increase, they do it in a different proportion depending on the degree of activity between the neurons. The activation pattern shown in Fig. 6.21 is more frequent once the synaptic weights between the neurons are high enough for the interneurons to trigger the motor neurons. In this example, the activity of the motor neuron 1 is a product of the activity of the

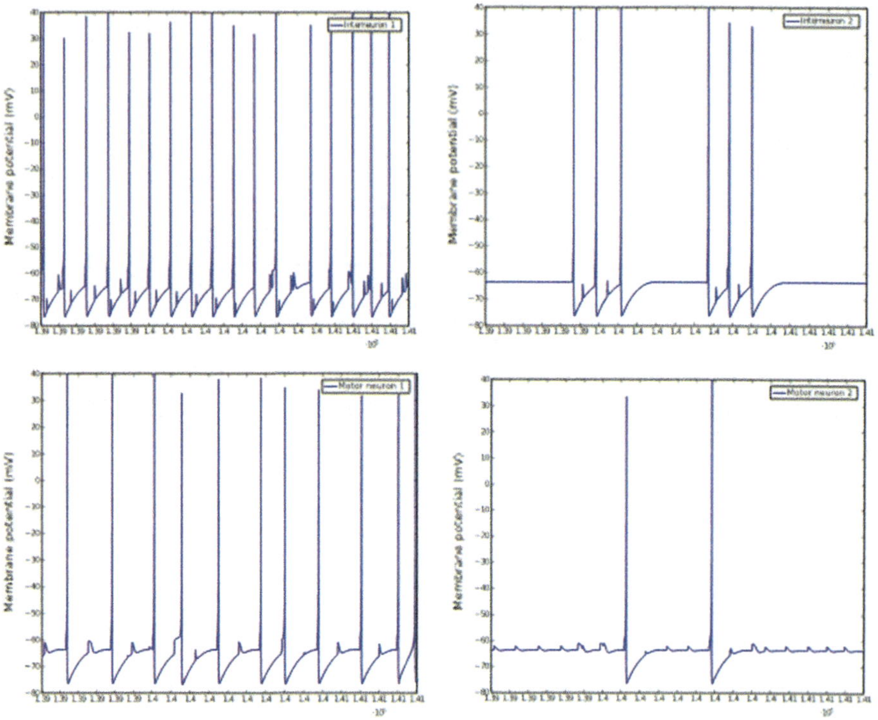

Fig. 6.21 Pattern of simultaneous activation of the four neurons

Fig. 6.22 Modulation of the synaptic weights during the experiment. The numbers in the box indicate the moment of the collisions

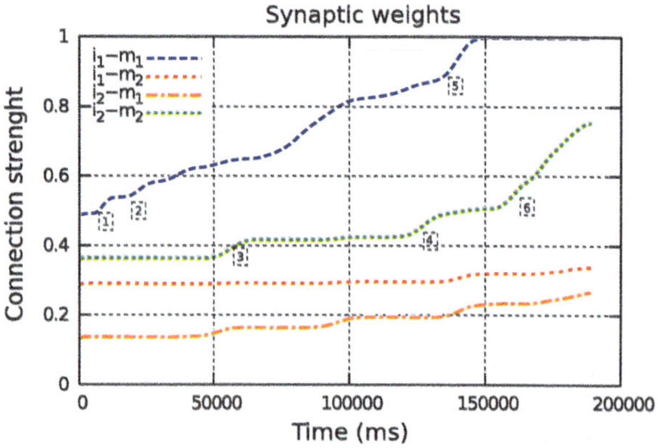

interneuron 1 and not of the collision sensor. In contrast, motor neuron 2 is still activated by collision sensor **c2**. It is possible to know this by looking at the values of the synaptic weights (Fig. 6.22).

Although there are other activation patterns, only those that generate an increase in the synaptic weight of at least some neural connection are mentioned. That is to say, activation patterns where at least one interneuron and one motor neuron fire. Patterns that do not comply with this activation condition would generate forgetfulness. For example, when only one of the four neurons is activated, the connections are not reinforced or in the case where the two interneurons are activated but not the motor neurons, the connection is not reinforced either.

The change of the synaptic weights during the experiment is shown in Fig. 6.22. The six collisions registered during the simulation are indicated by numbered boxes. The box of each of the collisions appears near the line of the synaptic weight that most varied due to this collision. As can be seen, changes in synaptic weights were recorded apart from those recorded during collisions. These changes occur because some connections between the interneurons and the motor neurons are high enough to trigger the motor neurons without the need for a collision. The values of the synaptic weights obtained at the beginning and end of the simulation are shown in Table 6.3.

The highest synaptic weights at the end of the simulation correspond to the weights between interneuron 1 and motor neuron 1, as well as the connection between interneuron 2 and motor neuron 2. Both connections are greater than 0.55, so only in these cases the activity of each motor neuron will be triggered by the activity of the corresponding interneuron and not by the collision sensor. The difference between the activation of a motor neuron due to the collision sonar and due to the interneuron associated with it is illustrated in Fig. 6.23. This difference is reflected between an action potential of interneuron 1 and motor neuron 1 at the beginning and end of the simulation.

Table 6.3 Initial and final synaptic weights

	Initial weights		Final weights	
–	m_1	m_2	m_1	m_2
i_1	0.49	0.29	**0.99**	0.34
i_2	0.13	0.36	0.26	**0.75**

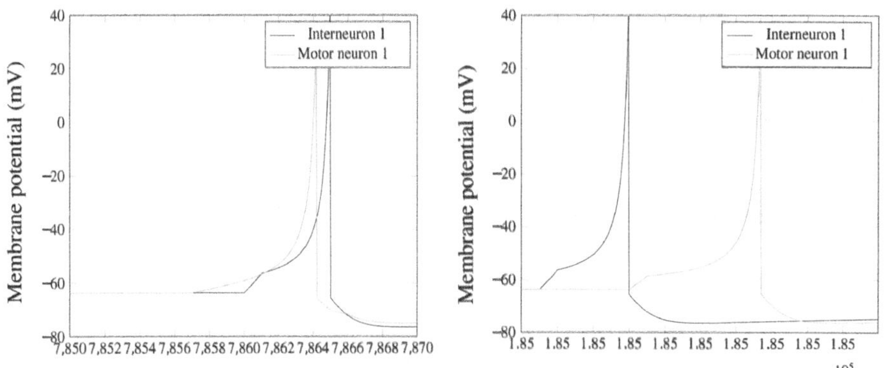

(a) Without learning (b) After learnig

Fig. 6.23 Activation when a collision occurs before learning (**a**) and after learning (**b**)

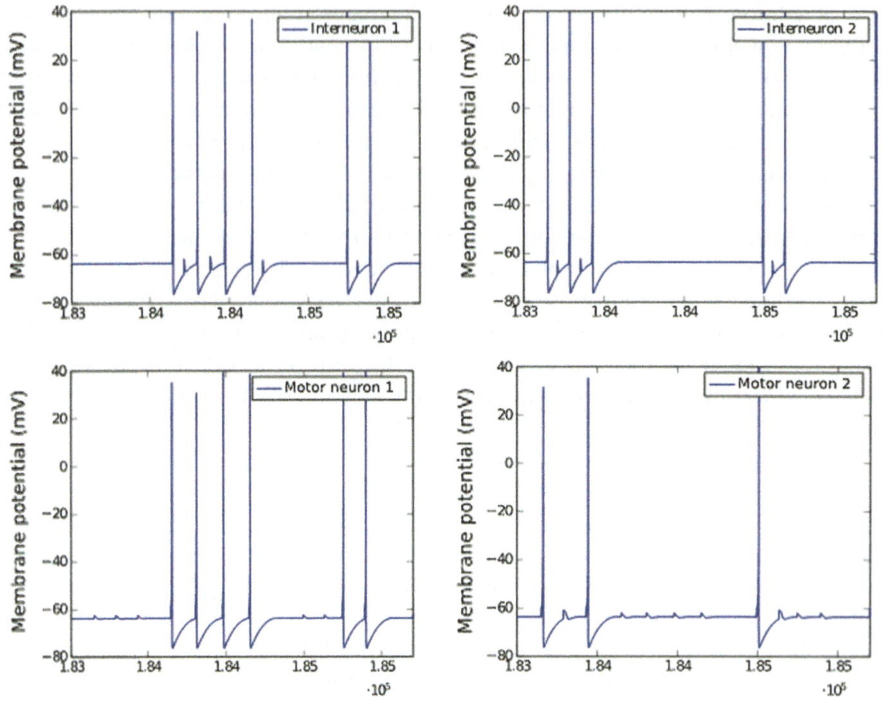

Fig. 6.24 Neuronal activity registered after the six collisions

Figure 6.24 shows the activity of the interneurons and motor neurons with the final synaptic weights. In these graphs, it can be seen that interneuron 1 and motor neuron 1 present the same number of action potentials. This is because the motor neuron fires when the interneuron fires. However, in the case of the other two neurons, motor neuron 2 does not fire even when there is activation from interneuron 2. This is because the synaptic weight is only high enough to reproduce some of the action potentials generated in the interneuron but not all of them.

The navigation of the artificial agent with the final synaptic weights is shown in Fig. 6.25. In contrast to Fig. 6.18, the agent does not come close obstacles and follows a straight path.

6.3.2.2 Learning with Direct Connections

This experiment was carried out with the neural network architecture that presents eight sensory neurons and two interconnected motor neurons. The system and the values established for the simulation are specified in Fig. 6.26.

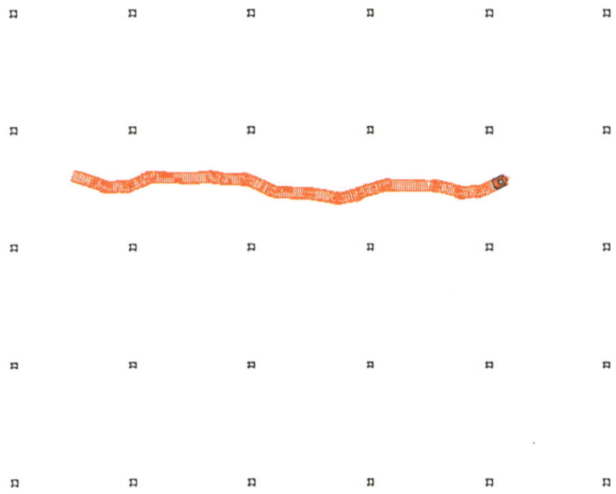

Fig. 6.25 Navigation of the agent in a cluttered environment after learning

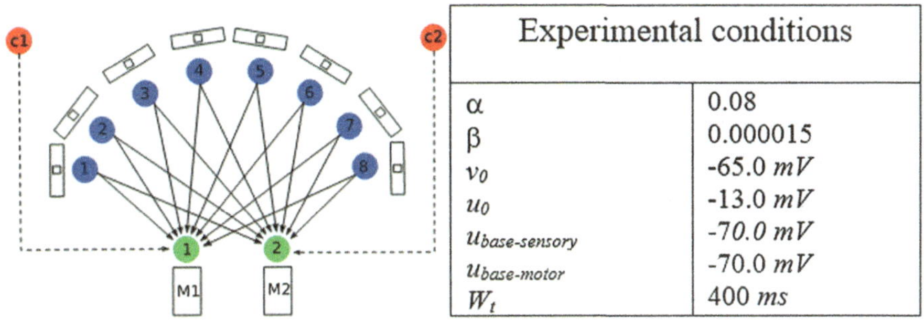

Fig. 6.26 Network architecture and experimental conditions

Table 6.4 Initial synaptic weights for the architecture with direct connections

Neurons								
–	s_1	s_2	s_3	s_4	s_5	s_6	s_7	s_8
m_1	0.39	0.12	0.17	0.29	0.42	0.17	0.21	0.43
m_2	0.28	0.16	0.42	0.43	0.33	0.35	0.41	0.14

The initial values for the synaptic weights are shown in Table 6.4. In contrast to the previous experiment, in this case, all the synaptic weights of the neural network are learned during the simulation.

The interaction of the agent with the environment during the simulation is shown in Fig. 6.27. In this Figure, the six collisions registered during the path followed by the agent are indicated by an arrow and a number.

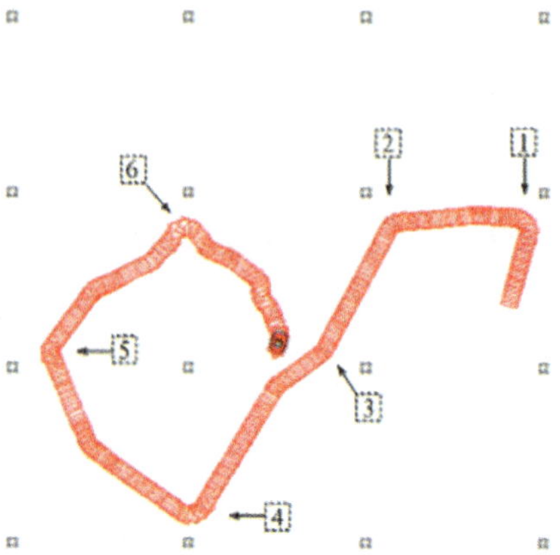

Collision

Sonar	1	2	3	4	5	6	C1
1	-	-	Activated	Activated	Activated	Activated	
2	-	-	-	Activated	-	-	
3	-	-	-	Activated	-	-	
4	-	-	-	Activated	-	Activated	C2
5	Activated	-	-	-	-	-	
6	Activated	-	-	-	-	-	
7	Activated	Activated	-	-	-	-	
8	Activated	Activated	-	-	-	-	

Fig. 6.27 Path of the artificial agent during the experiment and table showing the activated sonars on collisions

The graphs in Figs. 6.28 and 6.29 show the neuronal activity during collision 1. Figure 6.28 shows that no action potentials were registered in the neurons on the left side of the architecture. In contrast, Fig. 6.29 shows the activation of sensory neurons 5, 6, 7, and 8, which correspond to the right side of the architecture. Likewise, the same figure shows that the only active motor neuron during collision 1 was motor neuron 2. This activation pattern results in an increase in the synaptic weight connecting sensory neurons on the right side of the architecture with the motor neuron 2, which is also on the right side. This increase in synaptic weights is shown in Fig. 6.35. The synaptic weights of neurons that were not activated during this collision do not vary. The final configuration of the weights for this architecture is summarized in Table 6.5.

Figure 6.30 and 6.31 show the activation pattern corresponding to collision 4. In these figures, it is shown that sensory neurons on the left side are those that are now

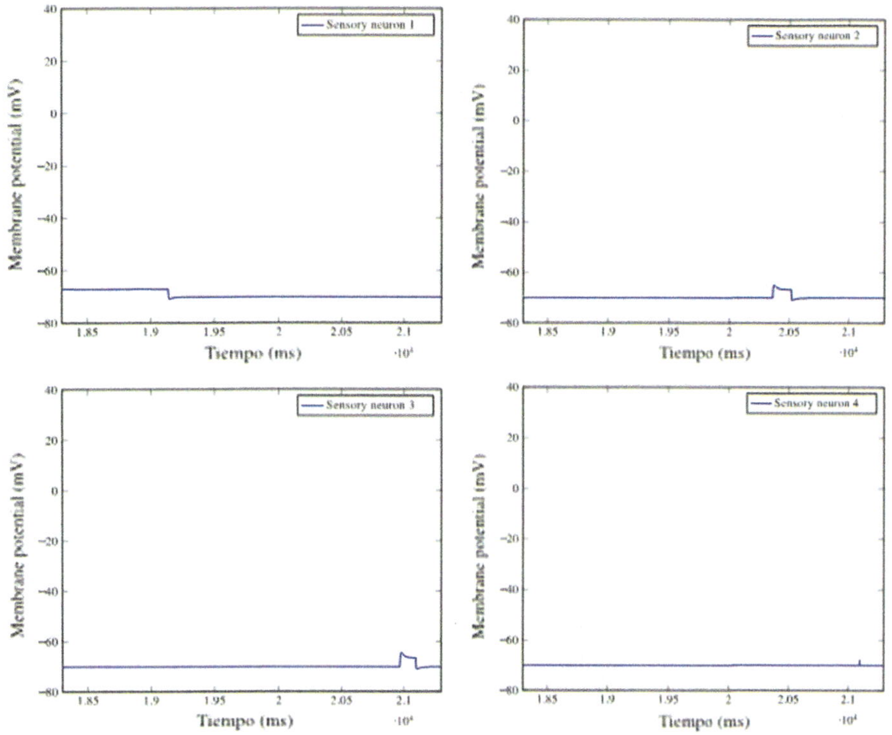

Fig. 6.28 Neural activity on the left side during Collision 1

activated, while sensory neurons 5, 6, 7, and 8 present null or very low activation in the case of neuron 8.

In contrast to collision 1, during collision 4 the synaptic weights of sensory neurons on the left side of the architecture connected to the motor neuron 1 are increased. Although, in this case the motor neuron 2 is also activated, increasing the synaptic weights between these sensory neurons and the motor neuron 2, but this increase in activation is lower than among the sensory neurons related to the motor neuron 1, because the motor neuron 2 produced a lower number of action potentials than the motor neuron 1.

The changes in synaptic weights registered during this experiment are presented in Figs. 6.32, 6.33, 6.34, and 6.35. In these figures, we can see that the highest synaptic weights correspond to the weights between motor neuron 1 and the sensory neurons 1, 3, 4, and 8 and the synaptic weights of motor neuron 2 with the sensory neurons 1, 5, 6, and 7.

The path of the artificial agent with the synaptic weights obtained at the end of the simulation is shown in Fig. 6.36. In this case, the agent continues to approach

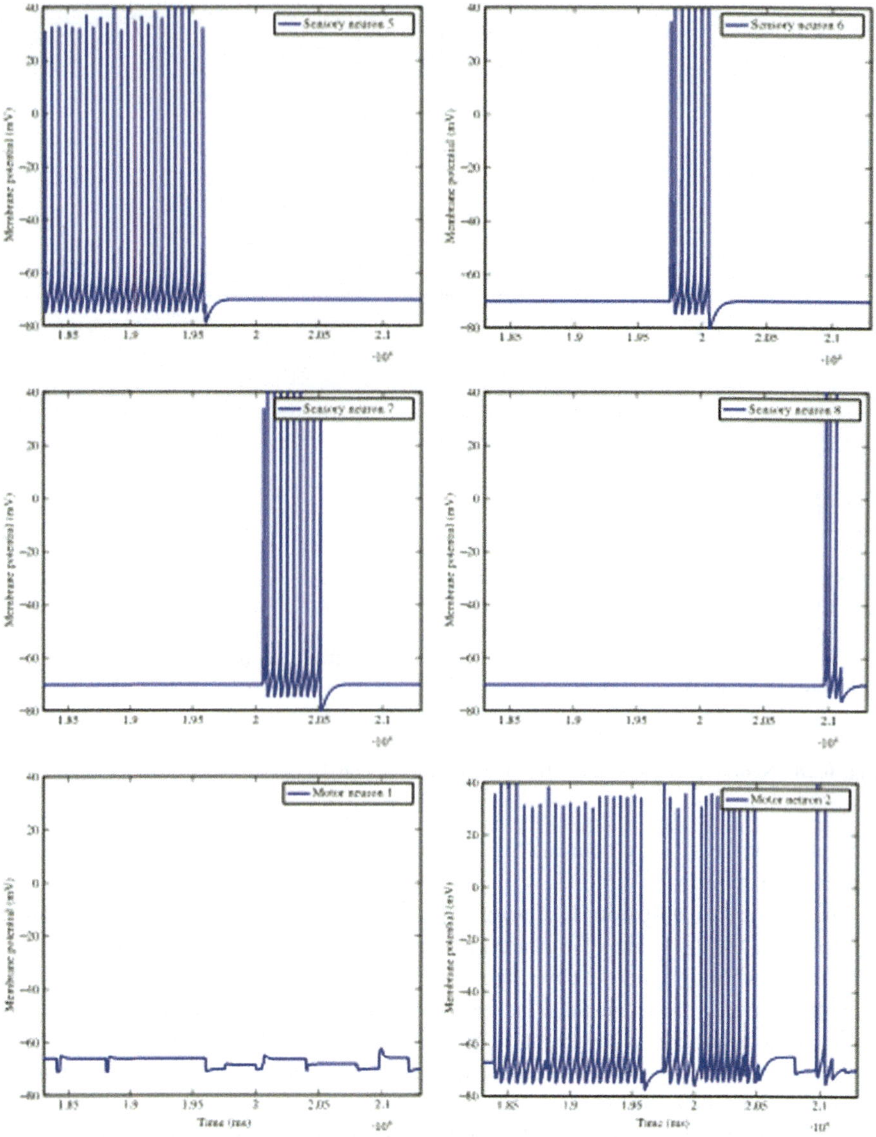

Fig. 6.29 Neural activity on the right side during Collision 1

the obstacles because some synaptic weights that transmit information on both sides of the architecture increased during the learning process. However, once the obstacles are detected by a neuron that has a strong synaptic weight with the motor neuron that is on the same side, the agent avoids the obstacle.

Table 6.5 Final synaptic weights for the experiment with direct connections

Neurons								
–	s_1	s_2	s_3	s_4	s_5	s_6	s_7	s_8
m_1	1.0	0.26	0.60	0.80	0.42	0.16	0.21	0.89
m_2	0.32	0.20	0.99	0.47	0.59	0.40	0.87	0.45

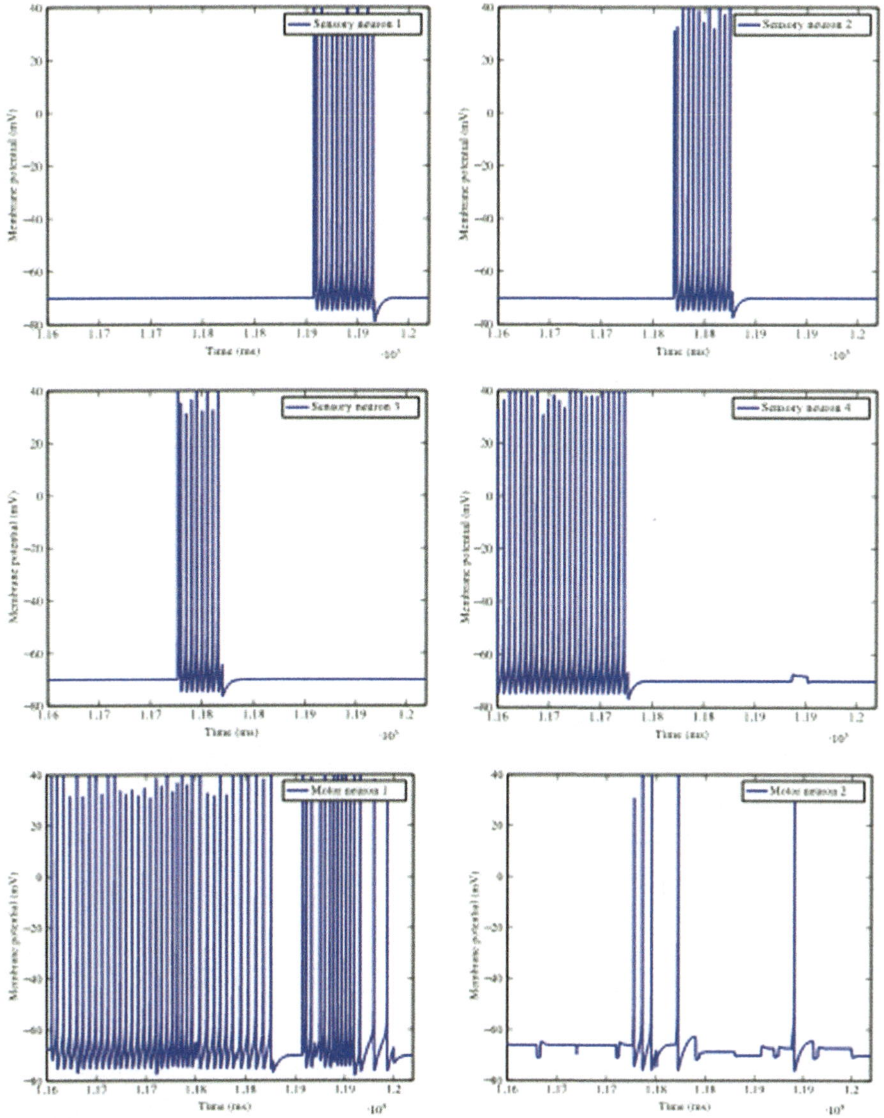

Fig. 6.30 Neural activity on the left side during Collision 4

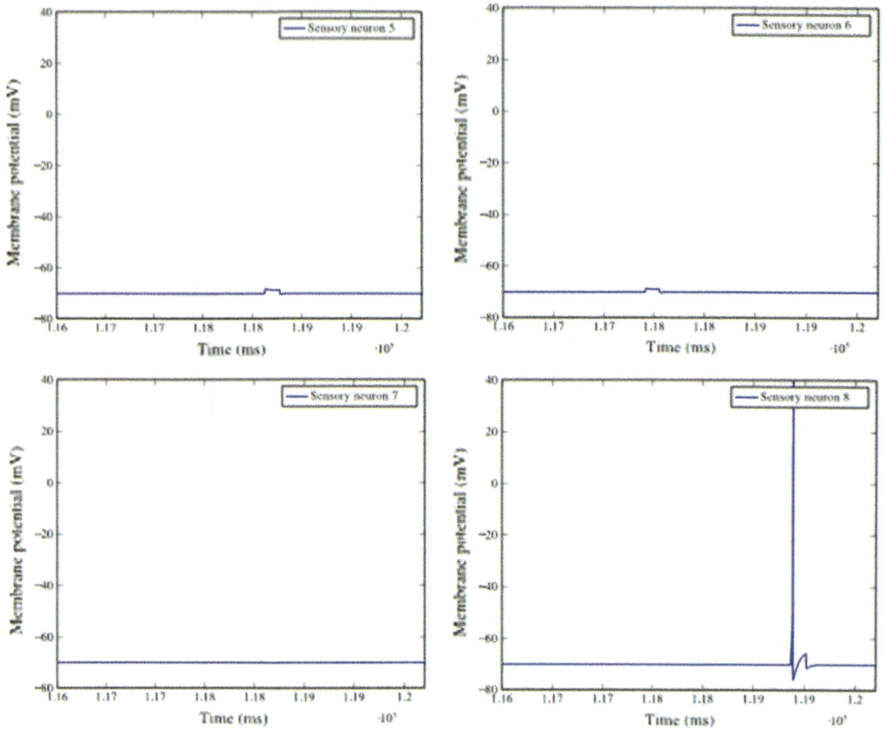

Fig. 6.31 Neural activity on the right side during Collision 4

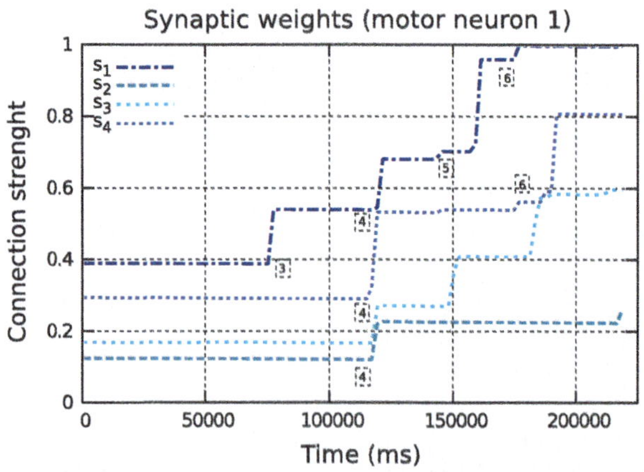

Fig. 6.32 Modulation of the synaptic weights between sensory neurons on the left side of the architecture (1, 2, 3, and 4) and motor neuron 1

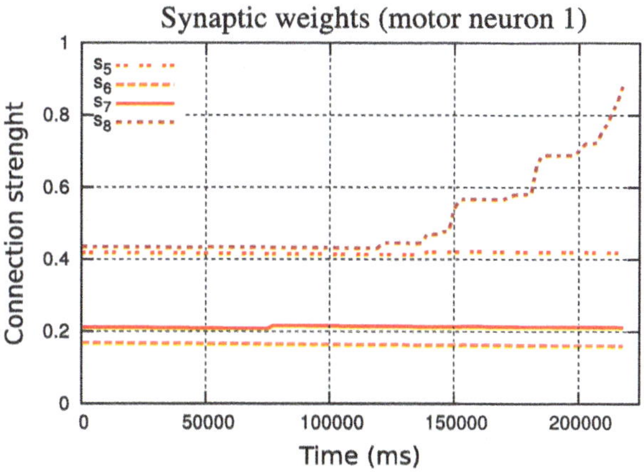

Fig. 6.33 Modulation of the synaptic weights between the sensory neurons on the right side of the architecture (5, 6, 7, and 8) and motor neuron 1

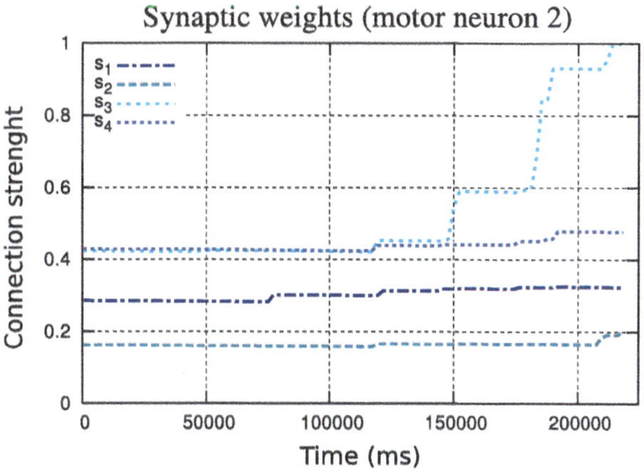

Fig. 6.34 Modulation of synaptic weights between the sensory neurons on the left side of the architecture (1, 2, 3, and 4) and motor neuron 2

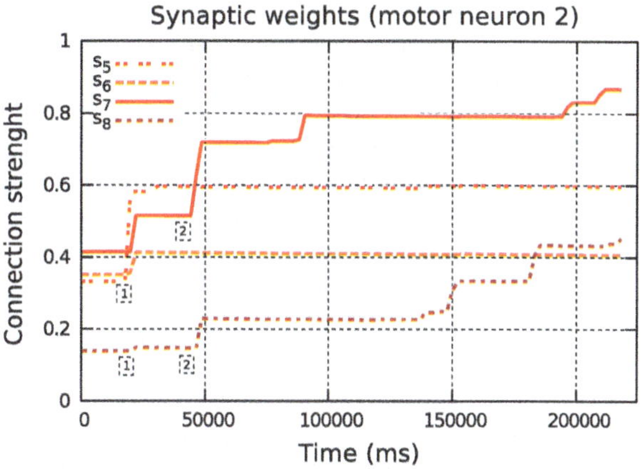

Fig. 6.35 Modulation of synaptic weights between the sensory neurons on the right side of the architecture (5, 6, 7, and 8) and motor neuron 2

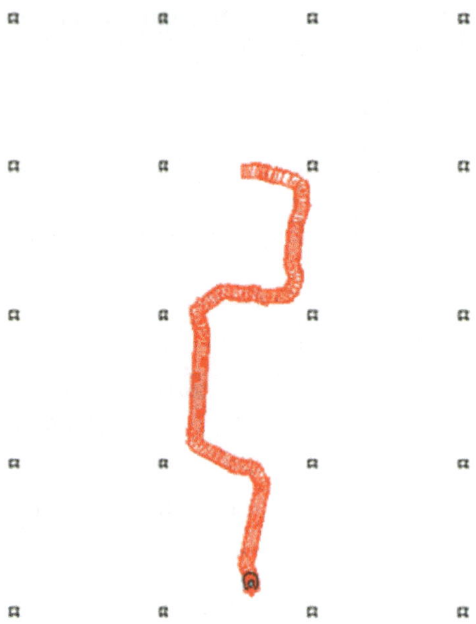

Fig. 6.36 Path of the agent with the final synaptic weights

6.4 Conclusions

The first experiments, reported in Sect. 6.3.1, show the plausibility of the action potential model proposed by Izhikevich (2003). In these experiments, the different parameters of the model were tried to obtain a coherent behavior in an artificial agent. The different network architectures studied show the potential of the models. Furthermore, in light of the results reported in Sect. 6.3.2, it is concluded that artificial action potential neural networks are useful models for the study of cognitive processes such as learning. The results of the simulations show that it is possible to observe a change in the behavior of the artificial agent before and after the Hebbian learning process for both architectures.

As already mentioned, there is scarce literature about autonomous learning algorithms for embedded artificial action potential neural networks. We consider that the main contribution of this work is the simulation of a multimodal process in an artificial action potential neural network. The learning strategy consisted on modifying the synaptic weights between the neurons of the networks by means of an implementation of Hebb's rule during a multimodal association process. This architecture was embedded in the artificial agent with eight proximity sensors and two collision sensors. The proximity sensors had a continuous activation function and were associated with sensory neurons. On the other hand, the collision sensors had a threshold of binary activation and were associated with motor neurons.

The implemented learning algorithm shows a dependency on the collision sensors at the beginning of the simulations to achieve the behavior of the obstacle avoidance. However, each time a collision occurs, the system associates the information coming from the proximity sensors with the information from the collision sensors. Therefore, the synaptic weight between the sensory neurons (or interneurons) and the motor neurons that fired simultaneously is increased. After a number of collisions, the system avoids the obstacles by learning a multimodal association using the sensory inputs to drive the motor responses.

It is worth mentioning that the behavior change exhibited by each of the architectures at the end of the simulation was different. In the case of the system with interneurons, it is observed that the agent goes closer to the obstacles at the beginning of the simulation. However, after learning, when the pattern of connections has been established, the agent does not need to detect the obstacles in its proximity in order to avoid them.

The second architecture, with direct connections, exhibits the same behavior at the beginning of the simulation. However, after the learning process, the agent acquires what we can call a more exploratory behavior. In this case, the agent tends get closer to the obstacles before avoiding them. This is due to the fact that only some of the synaptic weights increased in the way that was expected. The learning process resulted in synaptic connections between sensory input from one of the sides of the agent and the motor on the opposite side.

An explanation of the results obtained is that the number of collisions recorded during the simulations affects differently the learning in each of the architectures. The value of the connections in the interneuron system changes rapidly because each interneuron is associated with four sensory neurons, so it does not matter which sensory neuron is activated; the synaptic weight between the corresponding interneuron and motor neuron is increased more frequently. On the other hand, for the value of the connections in the second system to be the ideal, each sensory neuron must register at least one collision, depending on the initial synaptic weight.

Finally, based on the results reported here, we believe that the simulation of cognitive processes in embedded artificial neural networks can be of great interest for the development of systems that control artificial agents through cultures of biological neurons in vitro.

References

Brooks, R. A. (1991). Intelligence without representation. *Artificial Intelligence, 47*(1–3), 139–159.

Copeland, J. (2015). *Artificial intelligence: A philosophical introduction*. New York: Wiley.

DeMarse, T., Wagenaar, D., Blau, A., & Potter, S. (2001). The neurally controlled animat: Biological brains acting with simulated bodies. *Autonomous Robots, 11*(3), 305–310.

Dreyfus, H. (1967). Why computers must have bodies in order to be intelligent. *The Review of Metaphysics, 21*(1), 13–32.

Dreyfus, H. (1972). *What computers can't do*. New York: Harper & Row.

Gaona, W., Escobar, E., Hermosillo, J., & Lara, B. (2015). Anticipation by multi-modal association through an artificial mental imagery process. *Connection Science, 27*(1), 68–88.

He, W., Chen, Y., & Yin, Z. (2016). Adaptive neural network control of an uncertain robot with full-state constraints. *IEEE Transactions on Cybernetics, 46*(3), 620–629.

Izhikevich, E. (2000). Neural excitability, spiking and bursting. *International Journal of Bifurcation and Chaos, 10*(06), 1171–1266.

Izhikevich, E. (2003). Simple model of spiking neurons. *IEEE Transactions on Neural Networks, 14*(6), 1569–1572.

Izhikevich, E. (2004). Which model to use for cortical spiking neurons? *IEEE Transactions on Neural Networks, 15*(5), 1063–1070.

Maass, W. (1997). Networks of spiking neurons: The third generation of neural network models. *Neural networks, 10*(9), 1659–1671.

Manson, N. (2004). Brains, vats, and neurally-controlled animats. *Studies in History and Philosophy of Science Part C: Studies in History and Philosophy of Biological and Biomedical Sciences, 35*(2), 249–268.

McCarthy, J., Minsky, M., Rochester, N., & Shannon, C. (2006). A proposal for the dartmouth summer research project on artificial intelligence, August 31, 1955. *AI Magazine, 27*(4), 12.

McCulloch, W., & Pitts, W. (1943). A logical calculus of the ideas immanent in nervous activity. *The bulletin of mathematical biophysics, 5*(4), 115–133.

Mingers, J. (2001). Embodying information systems: The contribution of phenomenology. *Information and Organization, 11*(2), 103–128.

Mokhtar, M., Halliday, D., & Tyrrell, A. (2007, August). Autonomous navigational controller inspired by the hippocampus. In *IEEE International Joint Conference on Neural Networks* (pp. 813–818).

Moravec, H. (1984). Locomotion, vision and intelligence. In: M. Brady, & R. Paul (Eds.), *Robotics research* (pp. 215–224). Cambridge, MA: MIT Press.

Newell, A., & Simon, H. (1976). Computer science as empirical inquiry: Symbols and search. *Communications of the ACM, 19*(3), 113–126.

Novellino, A., D'Angelo, P., Cozzi, L., Chiappalone, M., Sanguineti, V., & Martinoia, S. (2007). Connecting neurons to a mobile robot: An in vitro bidirectional neural interface. *Computational Intelligence and Neuroscience, 2007.*

Pfeifer, R., & Scheier, C. (1999). *Understanding intelligence*. MIT Press.

Potter, S. (2001). Distributed processing in cultured neuronal networks. *Progress in Brain Research, 130,* 49–62.

Potter, S., & DeMarse, T. (2001). A new approach to neural cell culture for long-term studies. *Journal of Neuroscience Methods, 110*(1), 17–24.

Scheier, C., Pfeifer, R., & Kunyioshi, Y. (1998). Embedded neural networks: Exploiting constraints. *Neural Networks, 11*(7–8), 1551–1569.

Trhan, P. (2012). The application of spiking neural networks in autonomous robot control. *Computing and Informatics, 29*(5), 823–847.

Chapter 7
Force and Position Fuzzy Control: A Case Study in a Mitsubishi PA10-7CE Robot Arm

Miguel A. Llama, Wismark Z. Castañon and Ramon Garcia-Hernandez

Abstract Too many research works have focused on the problem of control of robot manipulators while executing tasks that do not involve the contact forces of the end-effector with the environment. However, many tasks require an interaction of the manipulator with the objects around it. For the correct performance of these tasks, the use of a force controller is essential. Generally, the control objective during the contact is to regulate the force and torque of the manipulator's end-effector over the environment, while simultaneously regulating the position and orientation (i.e., the pose) free coordinates of the manipulator's end-effector. Many works have been presented on this topic, in which various control strategies are presented; one of the most relevant methods is the so-called hybrid force/position control; this scheme has the advantage of being able to independently control the force in constrained directions by the environment and the pose along unconstrained directions. This work analyzes and implements the hybrid force/position control using a fuzzy logic control method, since the fuzzy control provides a solution for nonlinearities, high coupling, and variations or perturbations. The system employed is the Mitsubishi PA10-7CE robot manipulator, which is a robot of 7 degrees of freedom (DOF), but in this work, it is only used as a 6-DOF manipulator, equipped with a 6-DOF force/torque sensor in the end-effector.

7.1 Introduction

Actually, the ability to handle and manipulate physical contact between a robot and the environment that surrounds it is a demand to perform more advanced manipulation tasks. This capacity is known as the interaction of the manipulator with the physical environment in which it works.

M. A. Llama (✉) · W. Z. Castañon · R. Garcia-Hernandez
Instituto Tecnológico de La Laguna, Blvd. Revolución y Av.
Instituto Tecnológico de La Laguna S/N, Torreón, Coahuila, Mexico
e-mail: mllama@correo.itlalaguna.edu.mx

O. O. Vergara Villegas et al. (eds.), *Advanced Topics on Computer Vision, Control and Robotics in Mechatronics*,
https://doi.org/10.1007/978-3-319-77770-2_7

During the interaction, a situation called constrained movement is presented, which is a set of constraints caused by the environment during the tracking of the geometric trajectories that the robot manipulator's end-effector can follow. In this case, the force feedback becomes mandatory to achieve a robust behavior, and a safe and reliable operation, things that are very difficult to achieve with a pure movement control.

The interaction state of the manipulator with its surrounding can be described by means of estimations obtained from the presented forces in the joints of the manipulator, or more directly, from the contact force. This contact force is provided by a force/torque sensor mounted between the manipulator and the end-effector tool.

Up to date, a large number of studies have been carried out on the control of robot manipulators in interaction with the environment. Such controllers have been classified into different types according to their architecture. In the following paragraphs, the classification of the interaction controllers and their categories are described.

7.1.1 Interaction Control

The nature of the interaction between the manipulator and its environment allows classifying robotic applications in two classes. Those tasks that are involved without contact, that is, movements without restrictions in the free space, and complex robotic applications that require that the manipulator be mechanically coupled to other objects. Two categories can be distinguished within this last type of tasks. The first category is dedicated to essential tasks of force in which the end-effector is required to stabilize the physical contact with the environment and executes a specific force process. In the second category, the emphasis falls on the movement of the end-effector, which is performed on restricted surfaces (compliant motion).

7.1.1.1 Classification of Control Schemes with Restricted Movement

According to Vukobratovic et al. (2009), the type of compliance schemes can be classified as:

- Passive compliance: In which the position of the end-effector is accommodated by the contact forces themselves due to the inherent compliance in the structure of the manipulator, the servos, and the end-effector with special devices of high compliance.
- Active compliance: Where the adjustment is made through a force feedback loop in order to achieve a reaction of the robot, either through force interaction control or by generating a task with a very specific compliance.

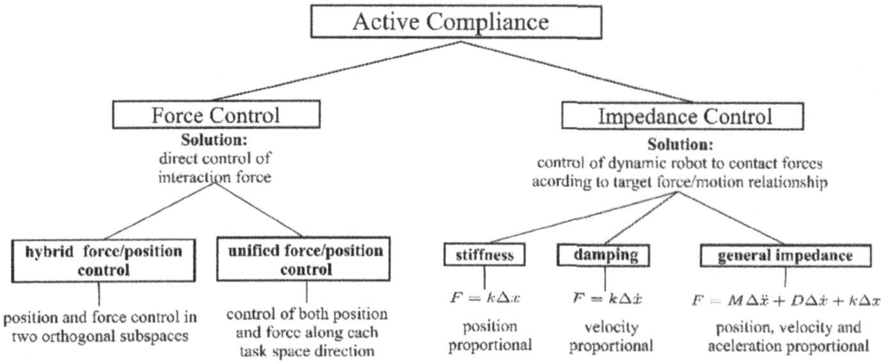

Fig. 7.1 Classification of active compliance controllers

We are interested only in the scheme of active compliance, and more specific, the hybrid control of force/position. Figure 7.1 (Vukobratovic et al. (2009) shows a control scheme that involve active compliance.

7.1.1.2 Hybrid Force/Position Control

This control methodology is based on the force and position control theory proposed by Mason (1981), depending on the mechanical and geometrical characteristics of the contact problem. This control methodology distinguishes two sets of constraints between the movement of the robot and the contact forces. The first set contains the so-called natural constraints, which arise due to the geometry of the task. The other set of constraints, called artificial constraints, is given by the characteristics associated with the execution of the specified task, i.e., the constraints are specified with respect to a framework, called a constraint framework. For instance, in a contact task where a slide is performed on a surface, it is common to adopt the Cartesian restriction framework in the way shown in Fig. 7.2 which is given in Vukobratovic et al. (2009). Assuming an ideally rigid and frictionless contact between the end-effector and the surface, it is obvious that natural constraints limit the movement of the end-effector in the direction of the z-axis, as well as rotations around the x- and y-axes.

The artificial constraints, imposed by the controller, are introduced to specify the task that will be performed by the robot with respect to the restriction framework. These restrictions divide the possible degrees of freedom (DOF) of the Cartesian movement into those that must be controlled in position and in those that must be controlled in force, in order to carry out the requested task.

In the implementation of a hybrid force/position control, it is essential to introduce two Boolean matrices S and $\overline{S}$ in the feedback loops in order to filter the forces and displacements sensed in the end-effector, which are inconsistent with the contact model of the task. The first is called the compliance selection matrix, and according

Fig. 7.2 Specification of the natural and artificial constraints for the task of sliding on a surface by means of the hybrid force/position controller

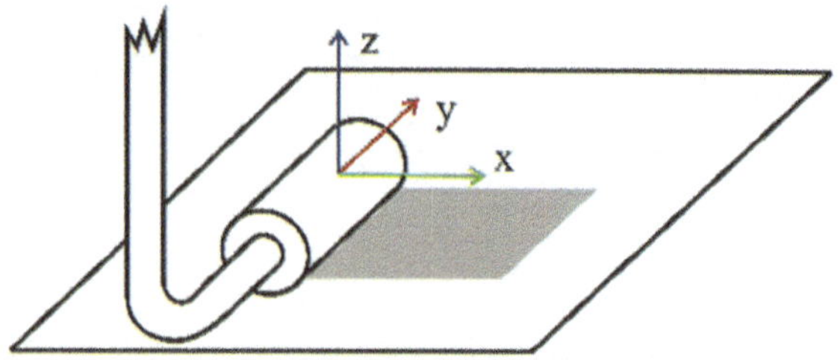

Natural constraints		Artificial constraints	
	$F_x = 0$		$v_x = V$
	$F_y = 0$		$v_y = 0$
$v_z = 0$		$F_z = 0$	
$\omega_x = 0$		$M_x = 0$	
$\omega_y = 0$		$M_y = 0$	
	$M_z = 0$		$\omega_z = 0$

to the artificial constraints specified, the *ith* diagonal element of this matrix has the value of 1 if the *i*th DOF with respect to the frame of the task has to be controlled in force, and the value of 0 if it is controlled in position. The second matrix is the selection matrix for the DOF that is controlled in position; the *ith* diagonal element of this matrix has the value of 1 if the *ith* DOF with respect to the frame of the task has to be controlled in position, and the value of 0 if controlled in force.

To specify a hybrid contact task, the following sets of information have to be defined:

- Position and orientation of the frame of the task.
- The controlled directions in position and strength with respect to the frame of the task (selection matrix).
- The desired position and strength with respect to the frame of the task.

Once the contact task is specified, the next step is to select the appropriate control algorithm.

7.1.2 Fuzzy Logic

The concept of fuzzy logic was introduced for the first time in 1965 by Professor Zadeh (1965) as an alternative to describe sets in which there is vagueness or uncertainty and, consequently, cannot be easily defined.

Fuzzy logic or fuzzy set theory is a mathematical tool based on degrees of membership that allows modeling information which contains ambiguity, imprecision, and uncertainty, by measuring the degree to which an event occurs, using for this a knowledge base or human reasoning.

7.1.2.1 Fuzzy Sets

A fuzzy set $A \in U$ can be represented as a set of ordered pairs of a generic element x and its membership value $\mu_A(x)$ that represents the degree to which the element x belongs to the fuzzy set A, that is,

$$A = \{(x, \mu_A(x)) | x \in U\}$$

A fuzzy set can also be expressed as a membership function $\mu_A(x)$, defined as $\mu_A(x)$: $A \in U \rightarrow [0, 1]$. This membership function assigns a membership value in A for each element x in the interval [1, 0]. This value is known as degree of membership.

7.1.2.2 Membership Functions

A membership function $\mu_A(x)$ can take different forms according to the system you want to describe. Among the most common forms are those described by impulsive, triangular, pseudo-trapezoidal, and Gaussian membership functions (Nguyen et al. 2003). The following describes the membership functions used later in this work.

Singleton membership function

A singleton membership function is shown in Fig. 7.3 and is defined by the following expression:

$$\delta(x; \bar{x}) = \begin{cases} 1, & \text{if} \quad x = \bar{x} \\ 0, & \text{if} \quad x \neq \bar{x} \end{cases}$$

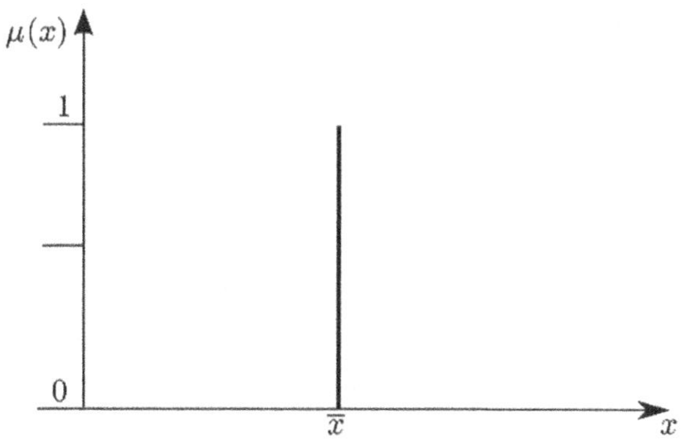

Fig. 7.3 Singleton membership function

Gaussian membership function

The Gaussian membership function G: $U \rightarrow [0, 1]$, shown in Fig. 7.4, has two parameters ρ and σ and is given by the expression

$$G(x; \rho, \sigma) = \mathrm{e}^{-\left(\frac{x-\rho}{\sigma}\right)^2}$$

Two-sided Gaussian membership function

The two-sided Gaussian membership function Υ: $U \rightarrow [0, 1]$ can be open to the left or open to the right and is created from the Gaussian function. For example, the left sigmoid membership function, which is shown in Fig. 7.5, is described by

$$\Upsilon(x; \rho, \sigma) = \begin{cases} 1, & \text{if} \quad x < \rho \\ \mathrm{e}^{-\left(\frac{x-\rho}{\sigma}\right)^2}, & \text{if} \quad x \geq \rho \end{cases}$$

Fig. 7.4 Gaussian membership function

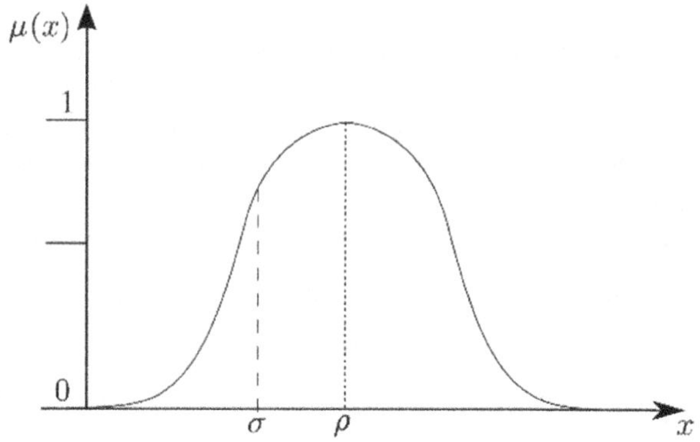

Fig. 7.5 Open to the left Gaussian membership function

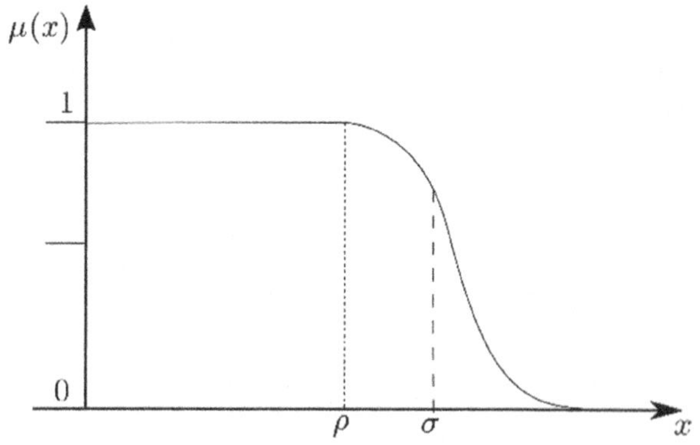

7.1.3 Control Systems Based on Fuzzy Logic

Fuzzy controllers are constructed from a set of fuzzy rules based on the knowledge of the control system and the experience of operators and experts. A fuzzy rule is expressed by

$$\text{IF } x \text{ is } A \text{ THEN } y \text{ is } B$$

where A and B are fuzzy sets.

7.1.3.1 Fuzzy Controller Structure

The structure of a fuzzy controller is composed of three main modules, shown in Fig. 7.6: fuzzification module, fuzzy rule base and inference engine module, and defuzzification module.

Fuzzification

During the transition of the fuzzification stage, the n input variables x_i ($i = 1, 2, 3, \ldots n$) are received, and for each of the N_i fuzzy sets $A_i^{l_i}$, a degree of membership $\mu_{A_i^{l_i}}(x_i^*)$ is assigned to the actual input values x_i^*, where l_i is used to identify each fuzzy set corresponding to each input i.

An important aspect when using fuzzy systems in real-time applications is to take into account the computing time to perform operations. Taking into account this, one method that reduces the computation time is the singleton fuzzification shown in Fig. 7.7, and it is defined by

$$\mu_{A_i^{l_i}}(x_i^*) = \begin{cases} \mu_{A_i^{l_i}}(x_i^*), & \text{if} \quad x_i = x_i^* \\ 0, & \text{if} \quad x_i \neq x_i^* \end{cases}$$

Fuzzy rule base and inference engine

The fuzzy rule base stores the knowledge and previous experience of the plant in which the fuzzy system will be based on. The fuzzy rule base is constructed by a group of fuzzy rules of the type IF–THEN denoted by $R_e^{l_1 \cdots l_n}$ and described as follows

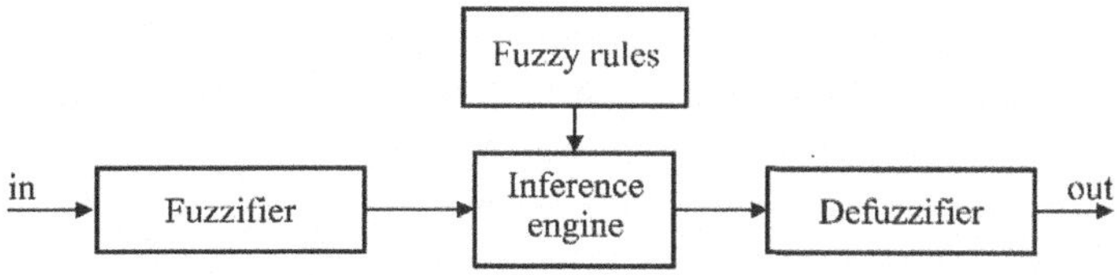

Fig. 7.6 Fuzzy controller general structure

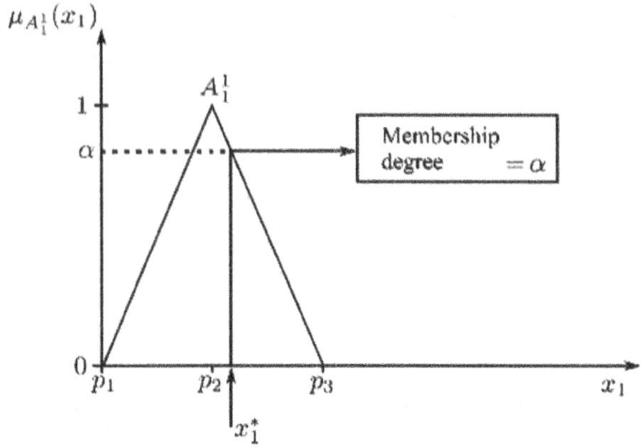

Fig. 7.7 Singleton fuzzification

Table 7.1 Fuzzy Associative Memory (FAM)

$x_2 \backslash x_1$	$l_1 = 1, A_1^1$	$l_1 = 2, A_1^2$	$l_1 = 3, A_1^3$
$l_2 = 1, A_2^1$	B^{11}	B^{21}	B^{31}
$l_2 = 2, A_2^2$	B^{12}	B^{22}	B^{32}
$l_2 = 3, A_2^3$	B^{13}	B^{23}	B^{33}

$$R_e^{l_1 \cdots l_n} : \text{ IF } x_i \text{ is } A_i^{l_i} \text{ AND} \cdots \text{AND } x_n \text{ is } A_n^{l_n} \text{ THEN } y \; is \; B^{l_1 \cdots l_n} \tag{7.1}$$

where x_i (i = 1, 2, …, n) are the input variables, and y is the output variable. The total number of rules is expressed by $M = \prod_{i=1}^{n} N_i$.

The storage of the fuzzy rule base is done through a table called Fuzzy Associative Memory (FAM) or Lookup table, such as shown in Table 7.1, which stores a fuzzy rule base with two entries (x_1 and x_2) and three membership functions for each entry.

On the other hand, we call inference to the process in which, according to the rule base, the membership values for each output membership function are computed. Inference can be seen as the process: Given a fuzzy relationship, established in advance, between an input fuzzy set A and an output fuzzy set B, i.e., IF A THEN B, we could conclude how an output B' would look like given a new input A'; i.e., IF A' THEN B'.

There exist several methods to perform the inference process; these different methods are known as inference engines. Among the most common inference engines are the max–min inference engine which uses the *min* operator between the antecedents of the rules, and the max-prod inference engine which uses the product operator between the antecedents of the rules; the latest is faster and that is the reason it is the most used in real-time applications.

max-prod inference

Taking into account the set of rules of the form Eq. (7.1), with input membership functions $\mu_{A_i^{l_i}}(x_i)$ and output $\mu_{B^{l_1 \cdots l_n}}(y)$ for all $\boldsymbol{x} = (x_1, x_2, \ldots, x_n)^T \in U \subset \mathbb{R}^n$ and $y \in V \subset \mathbb{R}$, the product inference engine is given as

$$\mu_{B'_{l_1 l_2}}(y, \boldsymbol{x}^*) = \left(\mu_{A_1^{l_1}}(x_1^*) \cdot \mu_{A_2^{l_2}}(x_2^*) \cdot \mu_{B^{l_1 l_2}}(y)\right) \tag{7.2}$$

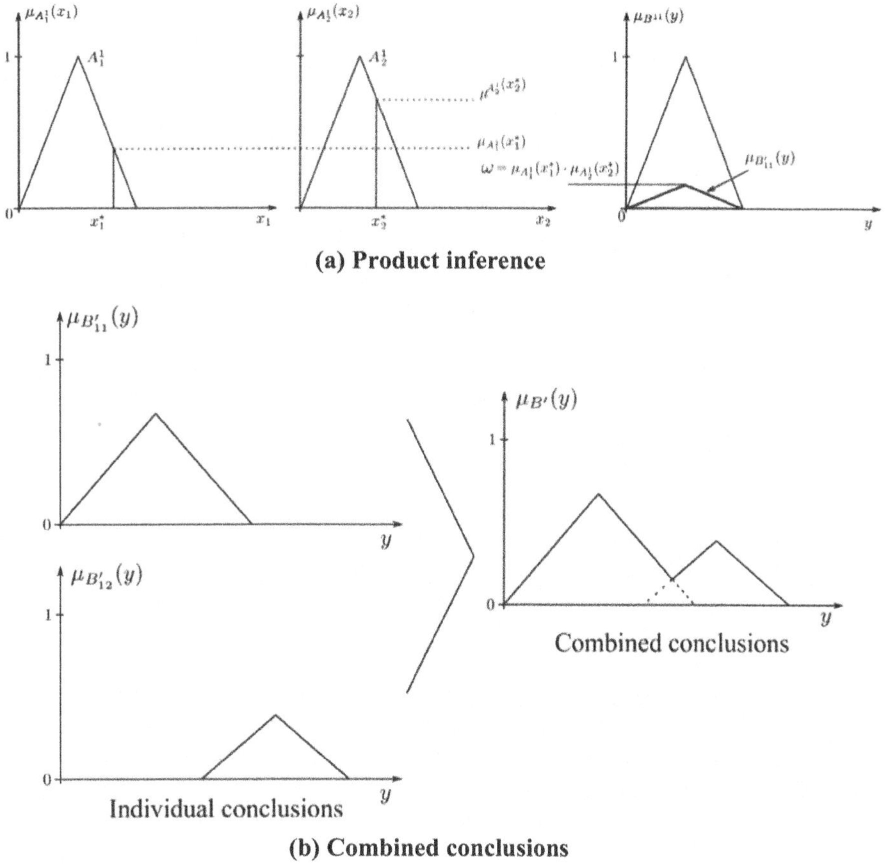

Fig. 7.8 Max-prod inference engine

Figure 7.8a describes the process of the max-product inference engine, while Fig. 7.8b describes the process of combining, by union operation, the output of several rule conclusions.

Defuzzification

In the defuzzification stage, a scalar value y^* is generated from the output $\mu_{B'}(y)$ that generates the inference engine. This value y^* is the output of the fuzzy controller that will be applied to the system to be controlled.

There are several ways to compute the output of the fuzzy controller; the most common is the center of average defuzzification which is given as

$$y^*(\boldsymbol{x}^*) = \frac{\sum_{l_1=1}^{N_1} \cdots \sum_{l_n=1}^{N_n} \bar{y}^{l_1 \cdots l_n} \omega_{l_1 \cdots l_n}(\boldsymbol{x}^*)}{\sum_{l_1=1}^{N_1} \cdots \sum_{l_n=1}^{N_n} \omega_{l_1 \cdots l_n}(\boldsymbol{x}^*)} \tag{7.3}$$

where $\bar{y}^{l_1 \cdots l_n}$ is the center of the $l_1 \ldots l_n$ output fuzzy sets, while $\omega_{l_1 \cdots l_n}$ is the height of the input membership functions, and $\boldsymbol{x}^*$ is the set of input real values.

7.2 Mitsubishi PA10-7CE Robot Arm

The Mitsubishi industrial robot manipulator PA10-7CE is one of the versions of the "Portable General-Purpose Intelligent Arm" of open architecture, developed by Mitsubishi Heavy Industries (MHI). This manipulator is composed of seven joints connected through links as shown in Fig. 7.9. The servomotors of the PA10 are three-phase brushless type and are coupled to the links by means of harmonic drives and electromagnetic brakes. In this work, the Mitsubishi is used as a 6-DOF robot arm; i.e, one of the joints is blocked, in this case the joint 3 represented by S3 in Fig. 7.9.

7.2.1 Robot Dynamics

The dynamic equation of motion for a manipulator of n DOF in interaction with the environment is expressed by Vukobratovic et al. (2009)

$$M(\boldsymbol{q})\ddot{\boldsymbol{q}} + C(\boldsymbol{q}, \dot{\boldsymbol{q}})\dot{\boldsymbol{q}} + \boldsymbol{g}(\boldsymbol{q}) = \boldsymbol{\tau} + J^T(\boldsymbol{q})\boldsymbol{f}_s \tag{7.4}$$

where $\boldsymbol{q}$ is an $n \times 1$ vector of joint displacements, $\dot{\boldsymbol{q}}$ is an $n \times 1$ vector of joint velocities, $\boldsymbol{\tau}$ is an $n \times 1$ vector of actuators applied torques, $M(\boldsymbol{q})$ is an $n \times n$ symmetric positive definite manipulator inertia matrix, $C(\boldsymbol{q}, \dot{\boldsymbol{q}})$ is an $n \times n$ matrix of centripetal and Coriolis torques, and $\boldsymbol{g}(\boldsymbol{q})$ is an $n \times 1$ vector of gravitational torques obtained as the gradient of the potential energy $U(\boldsymbol{q})$ due to gravity, $J(\boldsymbol{q}) \in \mathbb{R}^{(d+m) \times n}$ is the geometric Jacobian matrix that relates velocities in joint space with velocities in operational space; $\boldsymbol{f}_s \in \mathbb{R}^{(d+m)}$ is the vector of forces

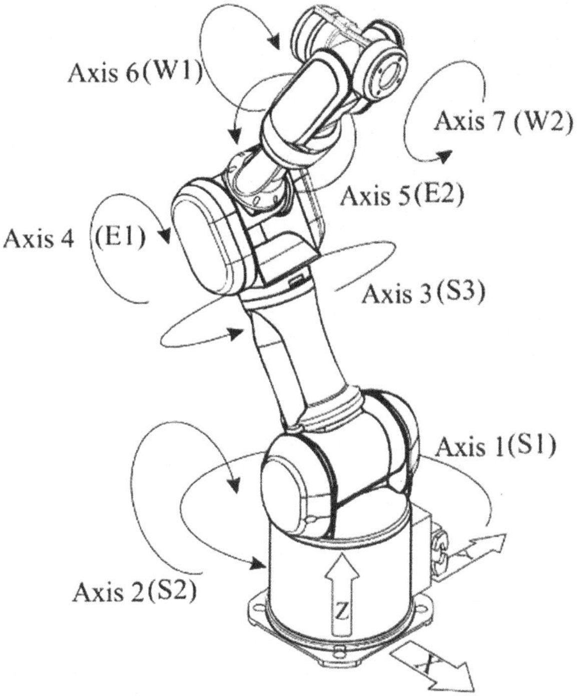

Fig. 7.9 Mechanical structure of the Mitsubishi PA10-7CE robot manipulator

and applied torques in end-effector of the manipulator, where d is the dimension of the geometric space in which the robot moves and m is the dimension of the vector space which defines the orientation. In order to avoid the resolution problem of the manipulator PA10-7CE redundancy, let the number of degrees of freedom be $n = 6$.

The matrices $M(\boldsymbol{q})$, $C(\boldsymbol{q}, \dot{\boldsymbol{q}})$ and vector $\boldsymbol{g}(\boldsymbol{q})$ were obtained by using the software HEMERO (MATLAB–Simulink toolbox for the study of manipulators and mobile robots) (Maza and Ollero 2001) in a numerical and semi-symbolic way. Some of the instructions that this package offers, and those that were used here, are

- *inertia* (dyn, $\boldsymbol{q}$) to obtain the inertia matrix $M(\boldsymbol{q})$
- *coriolis* (dyn, $\boldsymbol{q}$, $\dot{\boldsymbol{q}}$) to obtain the Coriolis matrix $C(\boldsymbol{q}, \dot{\boldsymbol{q}})$
- *gravity* (dyn, $\boldsymbol{q}$) to obtain the gravitational torques vector $\boldsymbol{g}(\boldsymbol{q})$

where dyn is a matrix containing the kinematic and dynamic parameters of the manipulator. The results obtained were compared with the results in Salinas (2011), and it was verified that the previously obtained model is correct. For this reason, the dynamic model obtained is not included in this work but can be seen in Salinas (2011).

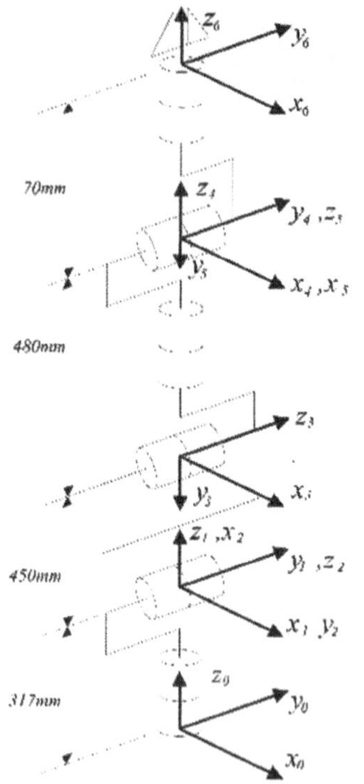

Fig. 7.10 Kinematic scheme for D-H parameters of 6-DOF reduced PA10-7CE robot manipulator

Table 7.2 D-H parameters of 6-DOF reduced PA10-7CE robot manipulator

Link	a_{i-1}[m]	α_{i-1}[rad]	d_i[m]	θ_i[rad]
1	0	0	0.317	q_1
2	0	$-\frac{\pi}{2}$	0	$q_2 - \frac{\pi}{2}$
3	0.450	0	0	$q_2 + \frac{\pi}{2}$
4	0	$\frac{\pi}{2}$	0.480	q_4
5	0	$-\frac{\pi}{2}$	0	q_5
6	0	$\frac{\pi}{2}$	0.070	q_6

7.2.2 *Robot Kinematics*

To obtain the manipulator's kinematics, the Denavit–Hartenberg convention described in Craig (2006) was used. The Denavit–Hartenberg parameters for the Mitsubishi PA10 robot were obtained with the frames assigned in Fig. 7.10 and are shown in Table 7.2.

7.2.2.1 Forward Kinematics

The position direct kinematic model of a robot manipulator is the relation that allows determining the vector $\boldsymbol{x} \in \mathbb{R}^{d+m}$ of operational coordinates according to its articular configuration $\boldsymbol{q}$:

$$\boldsymbol{x} = \boldsymbol{h}(\boldsymbol{q}) \tag{7.5}$$

where $\boldsymbol{h}$ is a vector function. Equation (7.5) is known as the direct kinematic equation of the robot manipulator (Sciavicco and Siciliano 1996). The components of the function $\boldsymbol{h}$ are determined implicitly by the product of matrices of homogeneous transformation of the manipulator.

$${}^{0}_{n}T = {}^{0}_{1}T\,{}^{1}_{2}T\,{}^{2}_{3}T\,{}^{3}_{4}T \ldots {}^{n-1}_{\;\;\;n}T \tag{7.6}$$

In general, the homogenous transformation matrix ${}^{a}_{b}T \in SE(3) \subset \mathbb{R}^{4\times 4}$ that describes the relative position of the frame $\sum_b$ with respect to the frame $\sum_a$ is given by

$${}^{a}_{b}T = \begin{bmatrix} {}^{a}_{b}R & {}^{a}_{b}p \\ 0^T & 1 \end{bmatrix} \in \mathrm{SE}(3) \subset \mathbb{R}^{4\times 4} \tag{7.7}$$

where ${}^{a}_{b}\boldsymbol{p} \in \mathbb{R}^3$ describes the position and $R \in SO(3) \subset \mathbb{R}^{3\times 3}$ is a matrix that describes the orientation.

The homogenous transformation matrix ${}^{0}_{n}T$ (with $n = 6$) for the Mitsubishi PA10-7CE robot was carried out using the HEMERO tool. The function used for that tool was

$$\mathrm{fkine}(dh, \boldsymbol{q})$$

where dh is a matrix with the Denavit–Hartenberg parameters with the following format

$$\begin{bmatrix} \cdots & \cdots & \cdots & \cdots & \cdots \\ a_{i-1} & \alpha_{i-1} & d_i & \theta_i & \sigma_i \\ \cdots & \cdots & \cdots & \cdots & \cdots \end{bmatrix} \in \mathbb{R}^{n\times 5}$$

where

- $a_{i-1}, \alpha_{i-1}, d_i, \theta_i$ are the Denavit–Hartenberg parameters according to Craig (2006).
- σ_i indicates the type of joint (it takes the value of 0 for a rotational joint and 1 if it is a prismatic one).

The elements of the homogeneous transformation matrix ${}^{0}_{n}T$ are shown in Castañon (2017).

Taking the time deriving of Eq. (7.5), we obtain

$$\dot{\boldsymbol{x}} = J_A(\boldsymbol{q})\dot{\boldsymbol{q}} \tag{7.8}$$

where $J_A(\boldsymbol{q}) \in \mathbb{R}^{(d+m)\times n}$ is the analytic Jacobian matrix of the robot. This matrix can be found in Salinas (2011). The geometric Jacobian is obtained using HEMERO tool using the following instruction

$$\text{jacob0}(dh, \boldsymbol{q})$$

The elements of this matrix can be found in Castañon (2017).

7.2.2.2 Inverse Kinematics

The position inverse kinematic model is the inverse function $\boldsymbol{h}^{-1}$ that if it exists for a given robot, it allows obtaining the necessary configuration to locate its end-effector in a given position $\boldsymbol{x}$:

$$\boldsymbol{q} = \boldsymbol{h}^{-1}(\boldsymbol{x}) \tag{7.9}$$

The expressions of the $\boldsymbol{h}^{-1}$ function of the position inverse kinematic model were calculated with the help of SYMORO+ robotics software (Khalil et al. 2014), and the results are shown in Castañon (2017).

Finally, from Eq. (7.8) the expression that characterizes the velocity inverse kinematic model is given by

$$\dot{q} = J_A(q)^{-1}\dot{\boldsymbol{x}} \tag{7.10}$$

Fig. 7.11 ATI Delta force/torque sensor

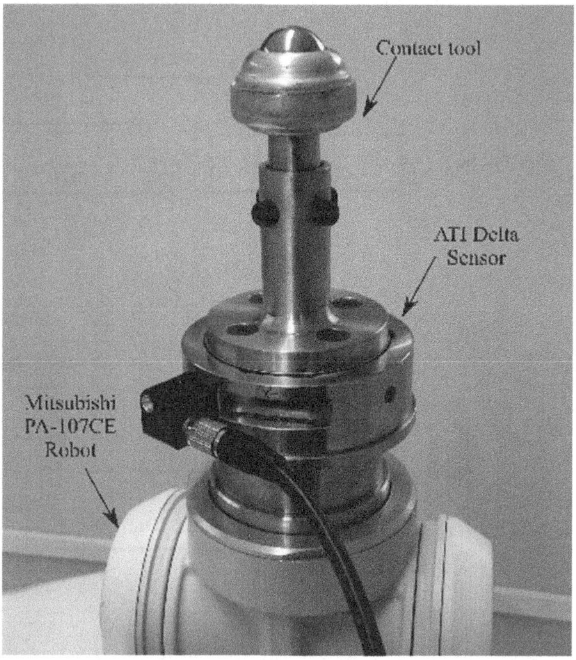

Fig. 7.12 Coupling between PA10-7CE robot, ATI Delta sensor, and contact tool

7.2.3 *Force/Torque Sensor*

The hybrid force/position controller requires the feedback of the forces and torques present in the robot's end-effector or in the contact tool used; to achieve this, the robot was fitted with a Delta model ATI force sensor shown in Fig. 7.11. This is a sensor of 6 degrees of freedom; this means that it is able to acquire the forces and torques in each of the Cartesian axes (F_x, F_y, F_z, T_x, T_y, T_z).

The main characteristics of the ATI Delta sensor are shown in Castañon (2017). For more technical information, consult (ATI Industrial Automation 2018a, b).

The ATI sensor was paired with a NI PCI-6220 DAQ card fitted in the control computer. Once the voltages signals read by the DAQ are in the MATLAB/ Simulink environment, these are converted to force/torque values. This conversion is given by the expression

$$\boldsymbol{f_s} = M_T \boldsymbol{v_c} + \boldsymbol{c_o} \tag{7.11}$$

where M_T is a 6 × 6 transformation matrix provided in the calibration file of the sensor, $\boldsymbol{v_c} \in \mathbb{R}^6$ is the vector containing the voltages sent by each gauge on the sensor, and $\boldsymbol{c_o} \in \mathbb{R}^6$ is a compensation vector that is also provided in the calibration file of the sensor found in Castañon (2017).

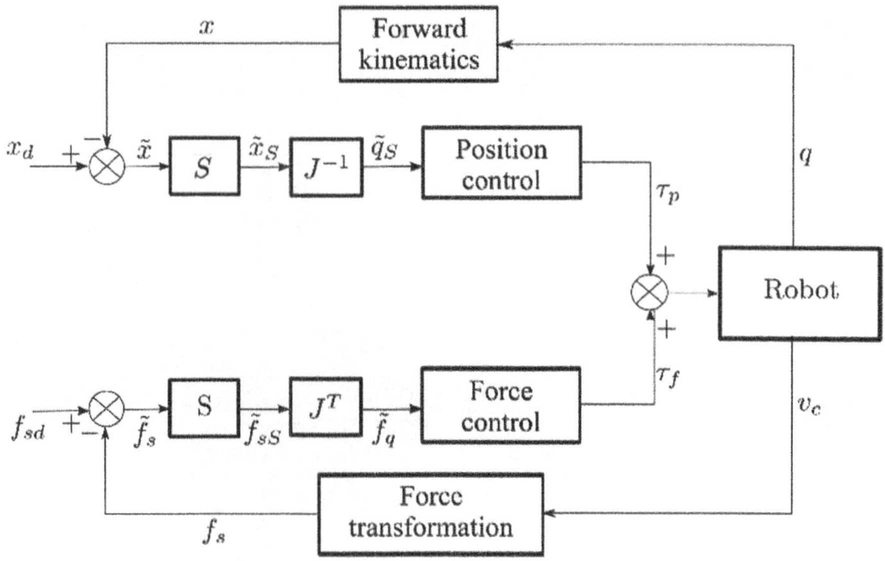

Fig. 7.13 Hybrid controller original structure

The sensor ATI was mounted between the last link of the robot and a special contact tool designed in Salinas (2011) to reduce, as much as possible, the friction between this tool and the contact surface; this is illustrated in Fig. 7.12.

7.3 Hybrid Force/Position Control with Fixed Gains

In the literature, there is a wide collection of works about different and very varied algorithms of hybrid force/position control; however, one of the most important approaches is the one proposed by Craig and Raibert (1979), which is shown in Fig. 7.13.

It contains two control loops in parallel with independent control and feedback laws for each one. The first loop is the position control which makes use of the information acquired by the position sensors in each robot joint. The second loop is the force control. This loop uses the information collected by the force sensor mounted on the end-effector. The matrix S is used to select which link will be controlled in either position or force.

In the direction controlled in force, the position errors are set to zero when multiplied by the orthogonal complement of the selection matrix (position selection matrix) defined as $\overline{S} = I - S$. This means that the part of the position control loop does not interfere with the force control loop; however, this is not the real case because there is still some coupling between both control loops.

A PD-type position control law is used with gain matrices $k_p \in \mathbb{R}^{(d+m)\times(d+m)}$ and $k_v \in \mathbb{R}^{(d+m)\times(d+m)}$, while the force control law consists of a proportional and integral action PI with their respective matrices' gains $k_{pf} \in \mathbb{R}^{(d+m)\times(d+m)}$ and $k_{if} \in \mathbb{R}^{(d+m)\times(d+m)}$, as well as a part of the feedback of the desired force in the force loop, then the control law can be written in operational space as

$$\tau_x = \tau_{fx} + \tau_{px} \tag{7.12}$$

where the control torque of the position loop is given by

$$\tau_{px} = k_p\overline{S}\tilde{x} + k_v\overline{S}\dot{\tilde{x}} \tag{7.13}$$

and for the force loop as

$$\tau_{fx} = k_{pf}\overline{S}\tilde{f}_s + k_{if}S\int_0^t \tilde{f}_s \mathrm{d}t + f_{sd} \tag{7.14}$$

where $\tau_x \in \mathbb{R}^6$ is the vector of control torques; k_p, k_v, k_{pf} and k_{if} are the 6×6 control gain diagonal matrices; $\tilde{x}$ is the vector resulting from the difference between the desired operational posture vector $x_d \in \mathbb{R}^{d+m}$and the actual operational posture vector $x \in \mathbb{R}^{d+m}$; $\dot{\tilde{x}} \in \mathbb{R}^{d+m}$ is the vector of speed errors in operational space; and $\tilde{f}_s \in \mathbb{R}^{d+m}$ is the vector obtained by the difference between the desired contact forces vector $f_{sd} \in \mathbb{R}^{d+m}$ and the instant force vector $f_{sd} \in \mathbb{R}^{d+m}$.

A problem that arises in this formulation is a dynamic instability in the force control part due to high gain effects of the feedback from the force sensor signal that is caused when a high rigidity is present in the environment, unmodeled dynamics effects caused by the arm and the elasticity of the sensor. To solve this problem, the dynamic model of the manipulator was introduced into the control law. In Shin and Lee (1985), a hybrid control of force/position is formulated in which the dynamic model of the robot is used in the control law; the expression is given by

$$\tau_x = M_x(x)\ddot{x}^* + C_x(x, \dot{x}) + g_x(x) + Sf^* \tag{7.15}$$

where $\ddot{x}^*$ is the equivalent acceleration control,

$$\ddot{x}^* = \ddot{x}_d + k_v\dot{\tilde{x}} + k_p\tilde{x} \tag{7.16}$$

and $f^* \in \mathbb{R}^6$ is the vector generated by the control law selected for the part of the force loop.

To avoid rebounding and minimize overshoots during the transition, an active damping term is added in the force control part (Khatib 1987).

$$\tau_{fx} = Sf^* - M_x(\boldsymbol{x})Sk_{vf}\dot{\boldsymbol{x}} \tag{7.17}$$

where the term k_{vf} is a diagonal matrix with Cartesian damping gains. In Bona and Indri (1992), it is proposed to modify the control law for position as

$$\tau_{px} = M_x(\boldsymbol{x})\overline{S}\left[\ddot{\boldsymbol{x}}^* - M_x^{-1}(\boldsymbol{x})(Sf^* - \boldsymbol{f}_s)\right] + C_x(\boldsymbol{x},\dot{\boldsymbol{x}}) + \boldsymbol{g}_x(\boldsymbol{x}) \tag{7.18}$$

being $M_x^{-1}(\boldsymbol{x})(Sf^* - \boldsymbol{f}_s)$ an added term to compensate the coupling between the force and position control loops, as well as the disturbances in the position controller due to the reaction force.

So far, the control laws have been handled in operational space; however in Zhang and Paul (1985), a transformation of the Cartesian space to joint space is proposed by transforming the selection matrices S and $\overline{S}$ given in Cartesian space to joint space as

$$S_q = J^{-1}SJ \tag{7.19}$$

and

$$\overline{S}_q = J^{-1}\overline{S}J \tag{7.20}$$

where J is the geometric Jacobian. With these transformations, the equivalent control law in joint space is obtained as

$$\tau = \tau_f + \tau_p \tag{7.21}$$

being

$$\tau_f = S_q\boldsymbol{f}_c^* - M(\boldsymbol{q})S_qK_{vf}\dot{\boldsymbol{q}} \tag{7.22}$$

$$\tau_p = M(\boldsymbol{q})\overline{S}_q\left[\ddot{\boldsymbol{q}}^* - M(\boldsymbol{q})^{-1}(S_q\boldsymbol{f}_c^* - J^T\boldsymbol{f}_s)\right] + C(\boldsymbol{q},\dot{\boldsymbol{q}})\dot{\boldsymbol{q}} + \boldsymbol{g}(\boldsymbol{q}) \tag{7.23}$$

where

$$\ddot{\boldsymbol{q}}^* = \ddot{\boldsymbol{q}}_d + K_v\dot{\tilde{\boldsymbol{q}}} + K_p\tilde{\boldsymbol{q}} \tag{7.24}$$

and

$$\boldsymbol{f}_c^* = K_{pf}J^T\tilde{\boldsymbol{f}}_s + K_{if}J^T\int_0^t \tilde{\boldsymbol{f}}_s \mathrm{d}t \tag{7.25}$$

$M(\boldsymbol{q}) \in \mathbb{R}^{n\times n}$, $C(\boldsymbol{q},\dot{\boldsymbol{q}}) \in \mathbb{R}^{n\times n}$ and $\boldsymbol{g}(\boldsymbol{q}) \in \mathbb{R}^n$ are the dynamic joint components of the manipulator; $\tilde{\boldsymbol{q}} \in \mathbb{R}^n$ is the vector of differences between the vector of desired

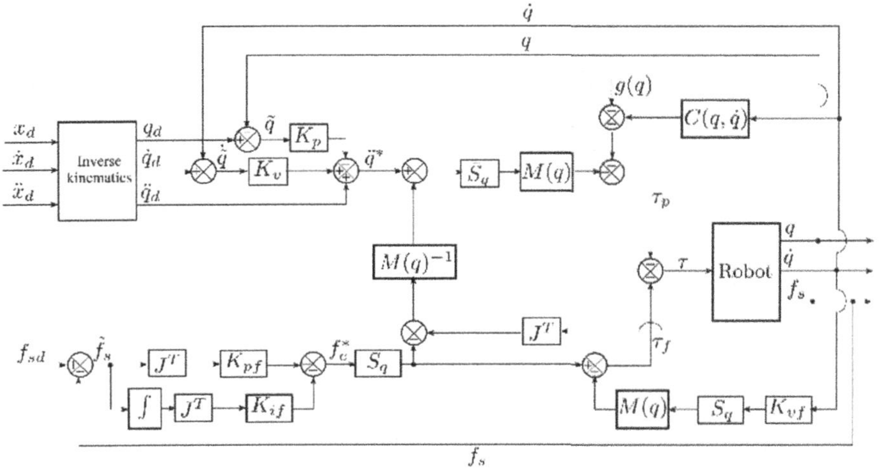

Fig. 7.14 Block diagram for hybrid controller with fixed gains in joint space

joint positions $\boldsymbol{q}_d \in \mathbb{R}^n$ and the vector of instantaneous joint positions $\boldsymbol{q} \in \mathbb{R}^n$; $K_v \in \mathbb{R}^{n\times n}$, $K_p \in \mathbb{R}^{n\times n}$ are gain diagonal matrices for the position control loop, and $K_{pf} \in \mathbb{R}^{n\times n}$, $K_{if} \in \mathbb{R}^{n\times n}$, and $K_{vf} \in \mathbb{R}^{n\times n}$ are gain diagonal matrices for the loop of force control in joint space.

The block diagram of the hybrid controller with fixed gains in joint space is shown in Fig. 7.14.

7.4 Hybrid Force/Position Control with Fuzzy Gains

Based on experimental results with the control law expressed by Eqs. (7.21)–(7.25) performed in the Mitsubishi PA10 robot arm, it was observed that the control performance changes depending on the rigidity of the contact environment; hence, to improve the performance from one surface to another, it was necessary to retune the control gains. A similar approach was proposed by Shih-Tin and Ang-Kiong (1998) but in a hierarchical way by tuning the scaling factor of the fuzzy logic controller.

Our experimental results showed that K_{pf} was the most sensible gain compared to K_{vf} and K_{if} gains. For this reason, we proposed that only the gain K_{pf} be supervised in a fuzzy manner and that K_{vf} and K_{if} were configured with constant values.

The proposed fuzzy control design is based on the control laws of Eqs. (7.21)–(7.25) with the difference that a supervisory fuzzy system is used to tune in the control gain $\widehat{K}_{pf}$ in the force control loop. Since now the gain is given by the function $\widehat{K}_{pf}(x)$, the Eq. (7.25) in the force control loop is given by

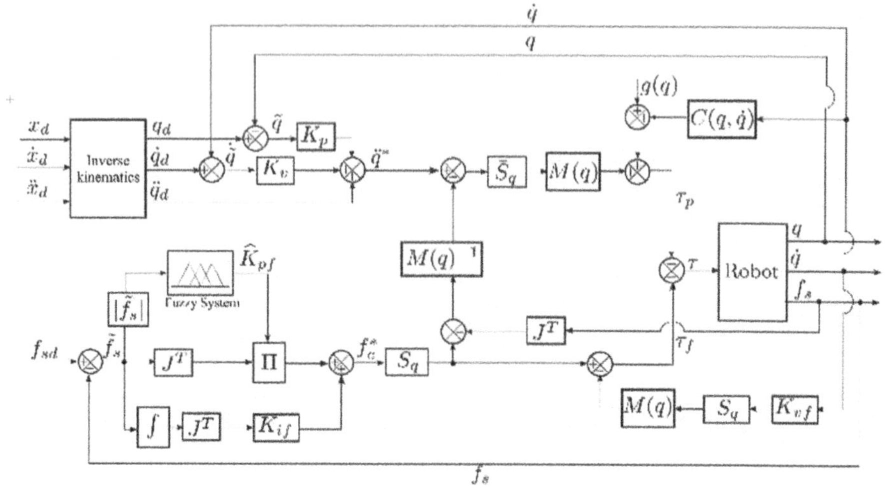

Fig. 7.15 Block diagram for hybrid controller with fuzzy gains in joint space

$$f_c^* = \widehat{K}_{pf}(x)J^T\tilde{f}_s + K_{if}J^T\int_0^t \tilde{f}_s dt \tag{7.26}$$

A fuzzy system $\widehat{K}_{pf}(x)$ like that represented in (7.3) with one input $x_1 = \tilde{f}_{sz}$ and one output $\bar{y}_1^{l_1}$ is designed. We define N_1 fuzzy sets $A_1^{l_1}(l_1 = 1, 2, \ldots, N_1)$ for input x_1, each of them described by a Gaussian membership function $\mu_{A_1^{l_1}}(x_1)$. For the output, singleton membership functions are selected.

The fuzzy system can be built from the set of N_1 fuzzy IF–THEN rules of the form

$$\text{IF } x_1 \text{ is } A_1^{l_1} \text{ THEN } \widehat{K}_{pf}(x) \text{ is } \bar{y}_1^{l1}$$

If singleton fuzzification, product inference engine, and center average defuzzification are used, the system can be described by

$$\widehat{K}_{pf}(x) = \frac{\sum_{l_1=1}^{N_1} \bar{y}_1^{l_1}(\mu_{A_1^{l_1}}(x_1))}{\sum_{l_1=1}^{N_1}(\mu_{A_1^{l_1}}(x_1))} \tag{7.27}$$

The block diagram of the hybrid controller with fuzzy gains in joint space is shown in Fig. 7.15.

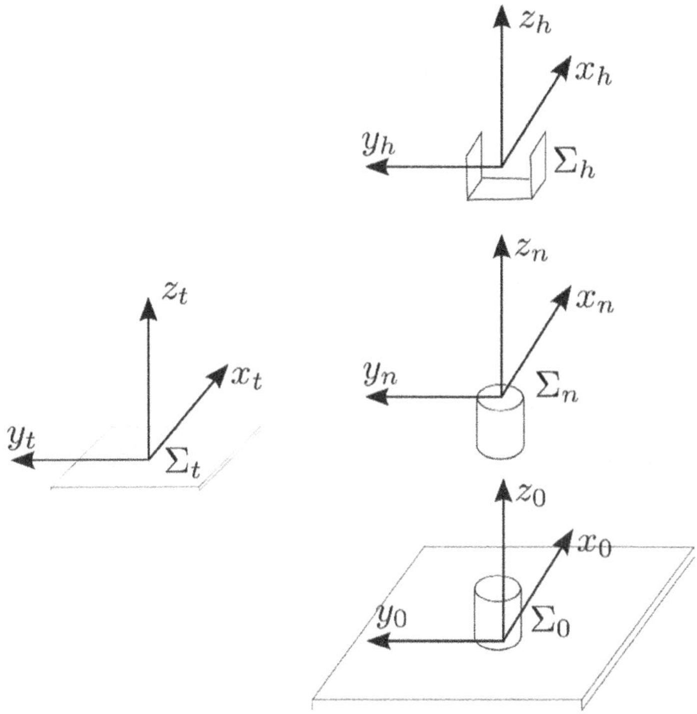

Fig. 7.16 Situation of the task frame referred to the frame of the base of the robot

7.5 Experimental Results

The controller was programmed in the MATLAB/Simulink environment. Also, the QUARC-BLOCKSET-PA10-Real-Time Control Software for Windows was used, which provides a simple way to make a communication between the internal control system of the PA10-7CE robot arm and the Simulink environment, thus obtaining easily the signals of the joint positions and joint speeds of the robot and send the control signals in torque mode back to the manipulator. In addition, the voltage signals provided by the Delta ATI force sensor could be also read and used within Simulink environment. The sampling time used for the whole system was 5 ms.

The task to be performed is to apply a desired force f_{szd} only on the z-axis of the frame of the task Σ_t (frame associated with where the contact is made) and which coincides in orientation with the base frame of the robot Σ_0 (frame of reference of the robot), with the frame of the last link Σ_n and with the frame of the end-effector Σ_h (see Fig. 7.16), while a tracking task of a straight line is carried out on the x-axis of the frame of the task.

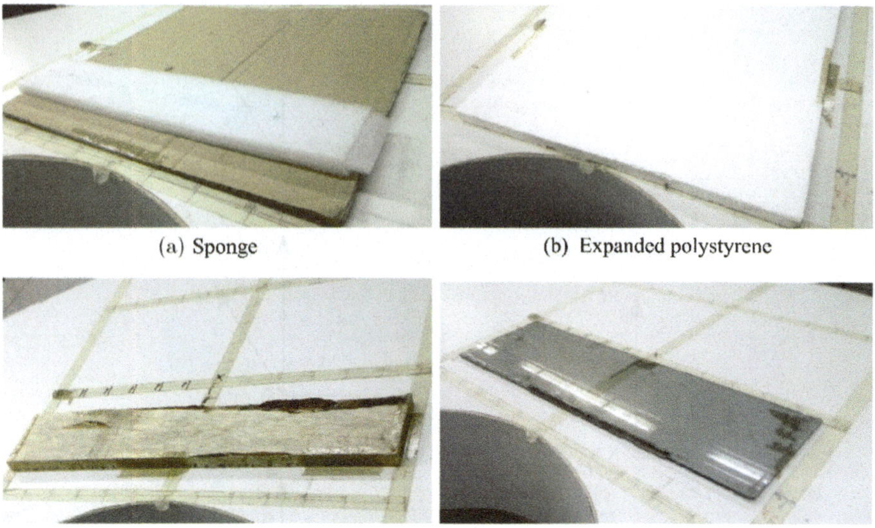

(a) Sponge (b) Expanded polystyrene (c) Wood (d) Glass

Fig. 7.17 Materials used as contact surface

Different materials were selected, with different degrees of stiffness, to place them as contact surfaces. The selected materials with a large stiffness coefficient K_s, which have greater resistance to deformation, were a wood board with a thickness of ¾ in., and a glass with a thickness of ¼ in. Materials with a low K_s stiffness coefficient, which tend to deform more easily, were a block of expanded polystyrene and a sponge. The selected materials are shown in Fig. 7.17.

7.5.1 Hybrid Force/Position Control with Fuzzy Gains

The controller was implemented by using Eqs. (7.21)–(7.24) and (7.26). The selection matrices for position S and for force $\overline{S}$ were selected as Eqs. (7.28) and (7.29), respectively. On the other hand, the values of the diagonal gain matrix for the position control loop are given in Table 7.3 and the fixed gains for the force control loop are shown in Table 7.4. K_{pf} was selected as the variable gain, and a fuzzy logic tuner was implemented for tuning such a gain.

$$S = \begin{bmatrix} 0 & 0 & 0 & 0 & 0 & 0 \\ 0 & 0 & 0 & 0 & 0 & 0 \\ 0 & 0 & 1 & 0 & 0 & 0 \\ 0 & 0 & 0 & 0 & 0 & 0 \\ 0 & 0 & 0 & 0 & 0 & 0 \\ 0 & 0 & 0 & 0 & 0 & 0 \end{bmatrix} \tag{7.28}$$

Table 7.3 Gains of the position control loop τ_p in the hybrid position/force controller

Joint	K_{pp} [1/s^2]	K_{vp} [1/s]
1	1250	100
2	1750	15
3	2750	12.5
4	500	10
5	500	200
6	2500	9000

Table 7.4 Fixed parameters of the force control part τ_f in the hybrid force/position controller with fuzzy gains

	K_{vf} [1/rad]	K_{if}
Value	0.1	3

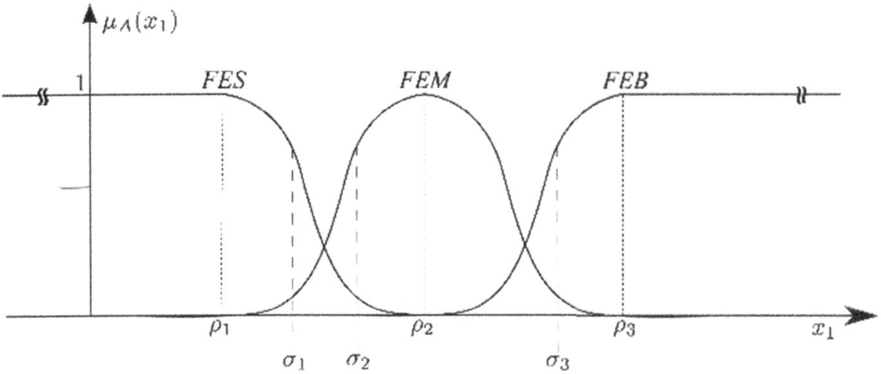

Fig. 7.18 Input membership functions

$$\overline{S} = \begin{bmatrix} 1 & 0 & 0 & 0 & 0 & 0 \\ 0 & 1 & 0 & 0 & 0 & 0 \\ 0 & 0 & 0 & 0 & 0 & 0 \\ 0 & 0 & 0 & 1 & 0 & 0 \\ 0 & 0 & 0 & 0 & 1 & 0 \\ 0 & 0 & 0 & 0 & 0 & 1 \end{bmatrix} \tag{7.29}$$

To approximate the gain through the fuzzy system $\widehat{K}_{pf}(x)$, it receives an input $x_1 = |\tilde{f}_{sz}|$ with a universe of discourse partitioned into $N_1 = 3$ fuzzy sets: $A_1^1 = \text{FES}$ (Force Error Small), $A_1^2 = \text{FEM}$ (Force Error Medium), $A_1^3 = \text{FEB}$ (Force Error Big). To build the fuzzy system, we propose to use an open to the left Gaussian

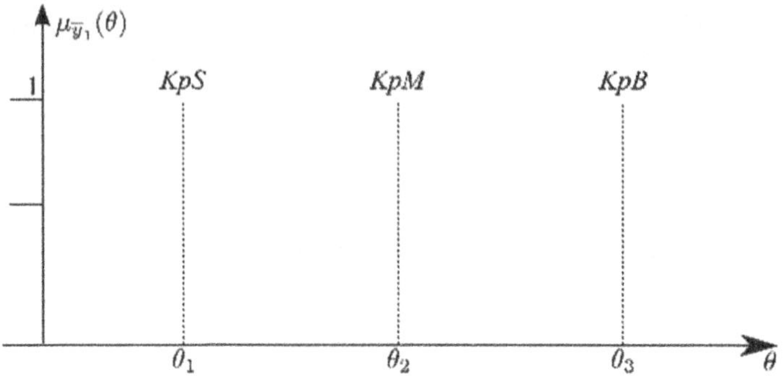

Fig. 7.19 Output membership functions

function, a Gaussian function, and an open to the right Gaussian function as shown in Fig. 7.18.

The partitions of the universe of discourse, using the notation $\rho_{A_1} = \{\rho_1, \rho_2, \rho_3\}$, were selected as

$$\rho_{|\tilde{f}_{sz}|} = \{0.5, 2, 4\}[\mathbf{N}]$$

and the standard deviations as

$$\sigma_{|\tilde{f}_{sz}|} = \{-6, 0.5, 3\}[\mathbf{N}]$$

As already mentioned, the fuzzy system consists of singleton functions for the output variable. The universe of discourse of the output is also partitioned into 3 impulsive functions: *KpS* (Small K_{pf} Gain), *KpM* (Medium K_{pf} Gain), and *KpB* (Big K_{pf} Gain); this is shown in Fig. 7.19, where each parameter θ corresponds to the position of the impulse functions. Taking the notation $\theta_{\bar{y}_1} = \{\theta_1, \theta_2, \theta_3\}$, the partitions of the universe of discourse for the output variable were selected like

$$\theta_{K_{pf}} = \{0.1, 0.15, 0.2\}$$

Fuzzy rules are selected as:

$$\text{IF } |\tilde{f}_{sz}| \text{ is FES THEN } \widehat{K}_{pf}(x) \text{ is } KpB$$

$$\text{IF } |\tilde{f}_{sz}| \text{ is FEM THEN } \widehat{K}_{pf}(x) \text{ is } KpM$$

$$\text{IF } |\tilde{f}_{sz}| \text{ is FEB THEN } \widehat{K}_{pf}(x) \text{ is } KpS$$

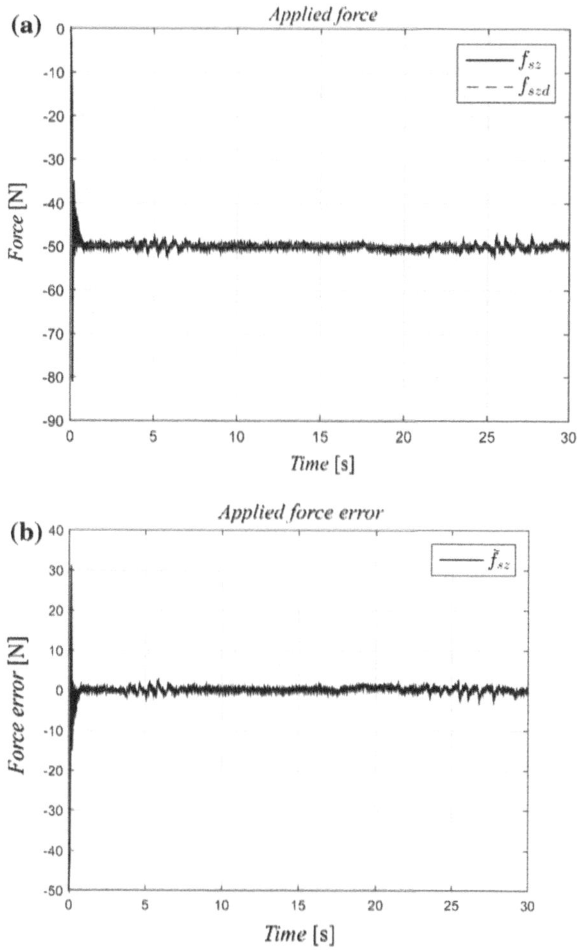

Fig. 7.20 **a** Applied force and **b** force error with the fuzzy gain hybrid controller for a desired force reference $f_{szd} = -50$ N on sponge

7.5.2 *Force Regulation and Position Tracking on a Sponge*

The experiments carried out on the Mitsubishi PA10 robot arm with the hybrid controller with fuzzy gains were made on different materials (shown in Fig. 7.17) and a desired force reference $f_{szd} = -50$ N (force applied down on the z-axis). Figure 7.20 shows the response f_{sz} to the applied force reference f_{szd} and the force error $\tilde{f}_{sz}$ on the z-axis applied on a sponge surface like the one in Fig. 7.17a.

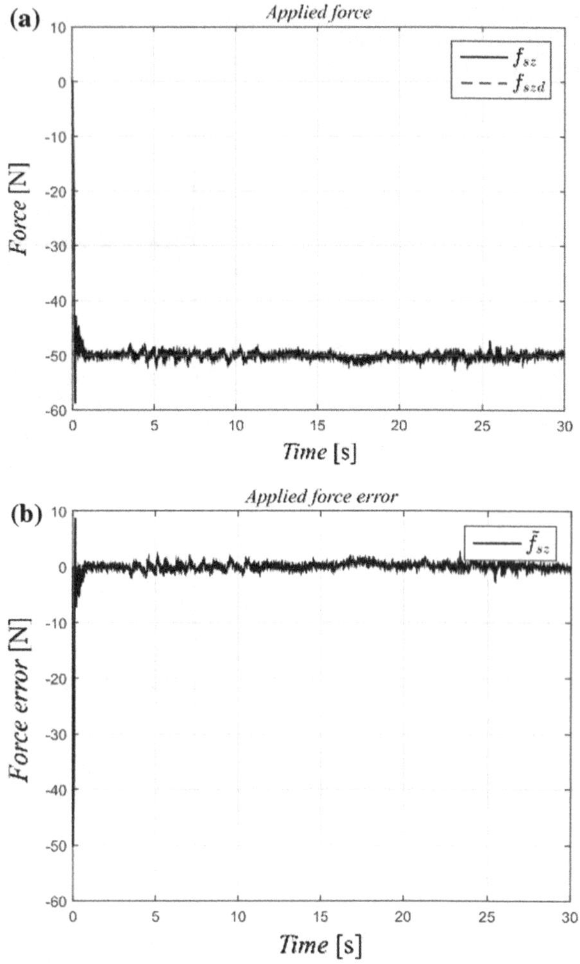

Fig. 7.21 **a** Applied force and **b** force error with the fuzzy gain hybrid controller for a desired force reference $f_{szd} = -50$ N on expanded polystyrene

7.5.3 Force Regulation and Position Tracking on Expanded Polystyrene

The results of the hybrid controller with fuzzy gains applied to the PA10 robot manipulator in interaction with an expanded polystyrene surface are shown in Fig. 7.21. This figure shows the response f_{sz} to the applied force reference $f_{szd} = -50$ N and the force error $\tilde{f}_{sz}$ on the z-axis applied on an expanded polystyrene surface like the one in Fig. 7.17b.

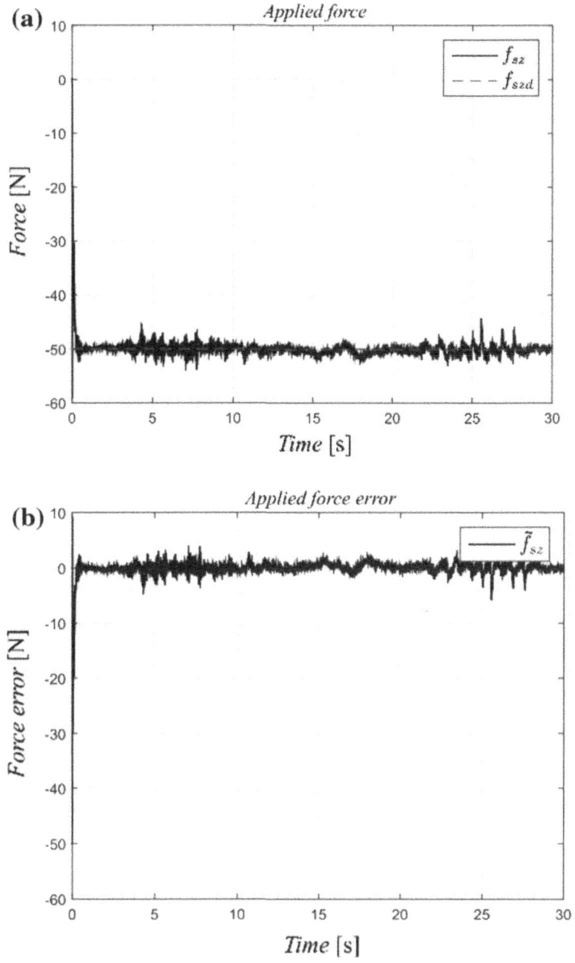

Fig. 7.22 **a** Applied force and **b** force error with the fuzzy gain hybrid controller for a desired force reference $f_{szd} = -50$ N on a wood board

7.5.4 *Force Regulation and Position Tracking on a Wood Board*

This section shows the results of the hybrid controller with fuzzy gains applied to the PA10 robot manipulator in interaction with a wood board surface like the one in Fig. 7.17c. Figure 7.22 shows the response f_{sz} to the applied force reference $f_{szd} = -50$ N and the force error $\tilde{f}_{sz}$ on the z-axis.

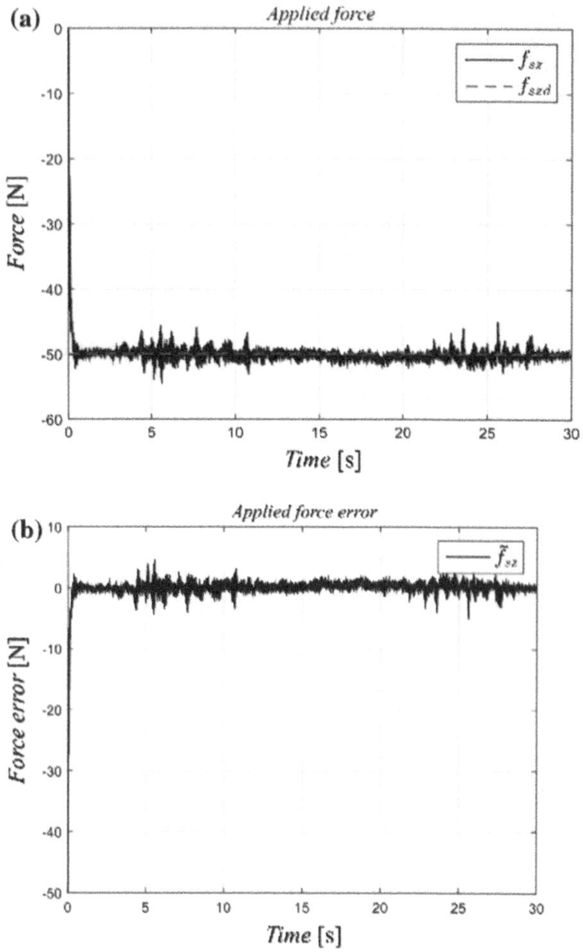

Fig. 7.23 **a** Applied force and **b** force error with the fuzzy gain hybrid controller for a desired force reference $f_{szd} = -50$ N on glass

7.5.5 *Force Regulation and Position Tracking on Glass*

The following figures show the results of the hybrid controller with fuzzy gains applied to the PA10 robot manipulator in interaction with a glass surface like the one in Fig. 7.17d. Figure 7.23 shows the response f_{sz} to the applied force reference $f_{szd} = -50$ N and the force error $\tilde{f}_{sz}$ on the z-axis.

The position and orientation errors are all very small, and they are reported in Castañon (2017).

7.6 Conclusions

The proposed hybrid force/position controller with fuzzy gains has the great advantage over its corresponding fixed gain controller that it does not require to retune the gains to exert a desired force in different types of materials with good performance. Conversely, the hybrid force/position controller with fixed gains requires the retuning of its gains for each material; in other words, for the fixed gain controller, the best gains obtained for soft materials cannot be used in hard materials because the system becomes unstable and very violent vibrations occur. This problem is not present in the proposed fuzzy version.

References

ATI Industrial Automation, I. (2018a). ATI F/T Catalogs and Manuals. Recuperado el 16 de enero de 2018, a partir de http://www.ati-ia.com/products/ft/ft_literature.aspx.

ATI Industrial Automation, I. (2018b). ATI Industrial Automation: Multi-Axis Force / Torque Sensors. Recuperado el 16 de enero de 2018, a partir de http://www.ati-ia.com/products/ft/sensors.aspx.

Bona, B., & Indri, M. (1992). Exact decoupling of the force-position control using the operational space formulation. In Proceedings of IEEE International Conference on Robotics and Automation, Nice, France, May.

Castañon, W. Z. (2017). Control difuso de fuerza para el robot manipulador Mitsubishi PA10-7CE, Master dissertation, Instituto Tecnológico de la Laguna, Torreón, Coahuila, México.

Craig, J. J. (2006). Robotica. Upper Saddle River: Prentice Hall.

Craig, J. J., & Raibert, M. H. (1979). A systematic method of hybrid position/force control of a manipulator. In Proceedings of the IEEE Computer Software and Applications Conference, Chicago, IL, USA.

Khalil, W., Vijayalingam, A., Khomutenko B., Mukhanov I., Lemoine P., & Ecorchard, G. (2014). OpenSYMORO: An open-source software package for Symbolic Modelling of Robots. IEEE/ASME International Conference on Advanced Intelligent Mechatronics, Besancon, France. pp. 1206–1211.

Khatib, O. (1987). A unified approach for motion and force control of robot manipulators: The operational space formulation. *IEEE Journal on Robotics and Automation, 3*(1), 43–53.

Mason, M. T. (1981). Compliance and force control for computer controlled manipulators. *IEEE Transactions on Systems, Man and Cybernetics, 11*(6), 418–432.

Maza, J. I., & Ollero, A. (2001). HEMERO: Herramienta MATLAB/Simulink para el estudio de manipuladores y robots moviles, Marcombo-Boixareu.

Nguyen, H., Prasad, R., & Walker, C. (2003). *A first course in fuzzy and neural control*. USA: Chapman & Hall/CRC.

Salinas, A. (2011). *Análisis e implementación de esquemas de control de interacción activa para robots manipuladores: Aplicación al robot Mitsubishi PA10, Master dissertation, Instituto Tecnológico de la Laguna, Torreón*. Diciembre: Coah.

Sciavicco, L., & Siciliano B., (1996). Modelling and control of robot manipulators. Berlin: Springer.

Shih-Tin, L., & Ang-Kiong, H. (1998, August). Hierarchical fuzzy force control for industrial robots. *IEEE Transactions on Industrial Electronics, 45*(4).

Shin, K. G., & Lee, C. P. (1985). Compliant control of robotic manipulators with resolved acceleration. In Proceedings of 24th IEEE Conference on Decision and Control, Ft. Lauderdale, FL, USA, December.

Vukobratovic, M., Surdilovic, D., Ekalo, Y., & Katic, D. (2009). Dynamics and robust control of robot-environment interaction. Singapore: World Scientific.

Zadeh, L. A. (1965). Fuzzy sets, *Information and Control, 8,* 338–353.

Zhang, H., & Paul, R. (1985). Hybrid control of robot manipulator. In Proceedings of the IEEE International Conference on Robotics and Automation.

Chapter 8
Modeling and Motion Control of the 6-3-PUS-Type Hexapod Parallel Mechanism

Ricardo Campa, Jaqueline Bernal and Israel Soto

Abstract This chapter reports the kinematics and dynamics models of the parallel mechanism known as Hexapod, which has a structure of the type known as 6-3-PUS. For computing the dynamics model, we start considering a non-minimal set of generalized coordinates and employ the Euler–Lagrange formulation; after that, we apply the so-called projection method to get a minimal model. It is worth noticing that the modeling approach presented here can be used for similar robotic structures, and the resulting models are suitable for automatic control applications. The computed analytical kinematics and dynamics models are validated by comparing their results with numerical simulations carried out using the SolidWorks Motion platform. In addition, this chapter describes the implementation of two motion tracking controllers in a real Hexapod robot. The tested controllers are one with a two-loop structure (a kinematic controller in the outer loop and a PI velocity controller in the inner loop) and other with an inverse dynamics structure. The experimental results of both controllers show a good performance.

Keywords Modeling · Motion control · 6-3-PUS-type Hexapod

8.1 Introduction

First robot manipulators were inspired in the human arm; that is the reason why they had open kinematic chains and were later known as serial manipulators. However, with the passage of time, it was necessary to use a different type of

---

R. Campa · J. Bernal
Tecnológico Nacional de México, Instituto Tecnológico de la laguna,
Torreón, Coahuila, Mexico

I. Soto (✉)
Universidad Autónoma de Ciudad Juárez, Instituto de Ingeniería y Tecnología,
Ciudad Juárez, Chihuahua, Mexico
e-mail: angel.soto@uacj.mx

O. O. Vergara Villegas et al. (eds.), *Advanced Topics on Computer Vision, Control and Robotics in Mechatronics*,
https://doi.org/10.1007/978-3-319-77770-2_8

mechanical structures: those containing closed kinematic chains, of which parallel robot manipulators are a particular class.

It is well known that serial manipulators have large workspace and dexterous maneuverability. But even though possessing these features, serial manipulators do not have a good precision positioning capability and their payload capacity is poor. Additionally, the links of serial manipulators tend to bend under heavy loads and vibrate at high speeds. Hence, in applications where high payload capacity, good dynamic performance, and precise positioning are of paramount importance, parallel manipulators are essential (Dasgupta and Mruthyunjaya 2000).

Most of parallel robot manipulators are constituted by one mobile platform connected through some kinematic chains (aka legs) to a fixed base. Because of the closed kinematic chains, in parallel robots some joints are passive (not actuated), but the number of actuators is usually selected to be equal to the number of degrees of freedom (DOF) of the mechanism. Moreover, each leg of a parallel robot can be chosen to have different number and type of joints, and it is this choice which determines the workspace, mobility, and actuator requirements for a given robot.

The first platform-based parallel robot was designed by E. Gough in the early 1950s. Curiously, that mechanism is best known as the Stewart platform, due to another structure, similar in functionality but different in design, that was published in 1965 by D. Stewart (see Dasgupta and Mruthyunjaya 2000). Since then, there have been several variants of the now sometimes known as Gough–Stewart platform.

Employing the current way of naming parallel robots, the original Gough–Stewart platform would be of the type known as 6-U$\underline{\text{P}}$S, where the preceding number is the number of legs, and the final alphabetic string indicates the type of joints that constitute the kinematic chain of each leg from the base to the platform, highlighting the actuated joint by the underscore; thus, in this case we have a universal (U) joint, an actuated prismatic (P) joint, and a spherical (S) joint. Other common variants of this original architecture are the 6-$\underline{\text{P}}$US and the equivalent 6-$\underline{\text{P}}$SU type (see Hopkins and Williams II 2002).

The parallel robot studied in this document is known as Hexapod, and it is produced by Quanser Inc. (see Fig. 8.1). This mechanism has six prismatic actuated joints embedded in the base which imprint linear motion (along the straight rails forming an equilateral triangle in the base) to six universal joints. The legs are constituted by two rigid links connecting the universal joints in the same side of the base triangle to one of the spherical joints in the vertices of the mobile platform, which also constitutes an equilateral triangle.

The Hexapod is a parallel robot of the 6-3-$\underline{\text{P}}$US type, where now the prefix *a*-*b* is used as in Liu et al. (2012) to denote parallel robots where the number of joints in the base (*a*) is not equal to the number of joints in the platform (*b*).

Several textbooks deal with the modeling of parallel mechanisms in general (see, e.g., Tsai 1999; Merlet 2006). For the specific case of parallel robots with a mobile platform, in this paper the following definitions are considered:

Fig. 8.1 Quanser's Hexapod parallel robot

- The *pose kinematics model* gives the relation between the generalized coordinates employed to describe the robot's configuration, and those used for describing the platform's position and orientation (i.e., its pose) in space.
- The *velocity kinematics model* gives the relation between the first derivatives of the generalized coordinates (or generalized velocities) and the linear and angular velocity vectors of the platform.
- The *dynamics model* establishes the relation between the generalized coordinates, their first and second derivatives (generalized velocities and accelerations), and the generalized forces applied to the robot in order to produce its motion.
- The *statics model* is a particular case of the dynamics model, when no motion occurs; in other words, it gives the relation between the generalized coordinates and the generalized forces.

It is a well-known fact that in the case of the kinematics of platform-type parallel robots, a major difficulty arises when computing the platform's pose from a given set of generalized coordinates; that is called the forward pose kinematics model (or FPK model, for simplicity). There exist several methods (both analytical and numerical) to deal with this problem, but it can be shown that it always has multiple solutions. On the other hand, the velocity kinematics model is useful for the analysis of singularities in the robot's workspace.

In recent years, many research works have been conducted on the dynamics modeling of parallel manipulators. Several methods or formulations have been proposed to find the equations of motion governing the dynamics of such mechanisms, being two of the most important the Newton–Euler formulation and the Euler–Lagrange formulation.

Despite its widespread use, the Newton–Euler formulation requires the computation of all constraint forces and moments between the links, but these are usually not necessary for simulation and control purposes. On the other hand, the Euler–Lagrange formulation has several advantages, such as (a) the possibility of using generalized coordinates, (b) the use of energy (rather than forces) which is a scalar quantity, and (c) the possibility of excluding from the analysis the constraint forces that do not directly produce the motion of the robot.

But independently of the formulation employed to compute the dynamics equations, it is now a common practice to employ a non-minimal set of generalized coordinates, and then to apply a method for reducing those equations and getting the minimal dynamics model. Such a method is in general known as the projection method (see, e.g., Arczewski and Blajer 1996; Blajer 1997; Ghorbel et al. 2000; Betsch 2005).

It is worth mentioning here that although the dynamics of the original Gough–Stewart platform (UPS type) has been subject of numerous studies (see Geng et al. 1992; Liu et al. 2000), little has been reported about the dynamics of platform mechanisms with different kinematic chains. Narayanan et al. (2010) and Carbonari et al. (2011) deal with the kinematics of a 6-3-$\underline{P}$US platform such as the one studied in this paper, but, as far as the authors' knowledge, there is no previous study about the dynamics of such mechanism.

The literature regarding the control of parallel mechanisms is not as vast as for serial manipulators. Nevertheless, since the work of Murray and Lovell (1989), it has become apparent that the possibility of getting a minimal model for a closed-chained mechanism allows to apply to this kind of systems the same type of controllers as for serial robots. As pointed out by Ghorbel et al. (2000), the main issue to take into account when proceeding this way is that the (Lyapunov) stability conclusions will at best be local due to the structural singularities of parallel mechanisms.

The aim of this paper is threefold. First, we recall the basics on kinematics and dynamics modeling of parallel robots; in the case of the dynamics model, we focus on the Euler–Lagrange formulation and explain the generalities of the projection method in order to show its application for computing the minimal dynamics model. Secondly, after describing the Quanser's Hexapod robot, we compute both its kinematics and dynamics models, and they are validated by comparing the results generated numerically by SolidWorks Motion. Moreover, we show the experimental results of the application of two model-based motion controllers to the Hexapod robot.

The chapter is organized as follows. Section 8.2 recalls the generalities of the kinematics and dynamics modeling of parallel robots. Section 8.3 introduces the Quanser's Hexapod robot, while Sects. 8.4 and 8.5 describe the derivation of the kinematics and dynamics models of such mechanism, respectively. The validation of such models is provided in Sect. 8.6, and the real-time experiments are described in Sect. 8.7. Finally, Sect. 8.8 gives concluding remarks.

8.2 Modeling of Parallel Robots

8.2.1 Pose Kinematics Modeling

Let $\boldsymbol{\rho} \in \mathbb{R}^m$ be the vector of generalized coordinates employed to describe the configuration of an n-DOF platform-type parallel robot (with $m \geq n$). If $m > n$ then $\boldsymbol{\rho}$ is known as the vector of *non-minimal* generalized coordinates, and there should exist $r = m - n$ independent holonomic constraints among the elements of $\boldsymbol{\rho}$. We must recall here that a holonomic constraint among the set of variables $\{\rho_1, \rho_2, \ldots \rho_m\}$ is an equation of the form $\gamma(\rho_1, \rho_2, \ldots, \rho_m) = 0$, where $\gamma : \mathbb{R}^m \to \mathbb{R}$. Thus, for a parallel robot described by non-minimal generalized coordinates, the r holonomic constraints $\gamma_i(\boldsymbol{\rho})$, with $i = 1, 2, \ldots, r$, can be grouped in a vector given by

$$\boldsymbol{\gamma}(\boldsymbol{\rho}) = \mathbf{0} \in \mathbb{R}^r. \tag{8.1}$$

Now consider the particular case where we choose a set of $m = n$ generalized coordinates to describe the system; in order to highlight its relevance, in such a case we will use $\mathbf{q}$ instead of $\boldsymbol{\rho}$, so that $\mathbf{q} \in \mathbb{R}^n$ will be the vector of minimal generalized coordinates. And, as it is common that those minimal coordinates correspond to the variables associated with each of the actuated (or active) joints in a fully actuated robot, then $\mathbf{q}$ is also known as the vector of *active joint coordinates*. Furthermore, it is important to notice that in this case there are no holonomic constraints.

For the forthcoming analysis, let us assume that the relation between $\mathbf{q}$ and $\boldsymbol{\rho}$ is given by

$$\mathbf{q} = \boldsymbol{\alpha}(\boldsymbol{\rho}) \tag{8.2}$$

where $\boldsymbol{\alpha} : \mathbb{R}^m \to \mathbb{R}^n$ is a given smooth function.

According to Ghorbel et al. (2000), as the non-minimal generalized coordinates are suitably chosen to be readily visual, then function $\boldsymbol{\alpha}$ in Eq. (8.2) is usually easy to obtain explicitly. And following the analysis in Ghorbel et al. (2000), we can write:

$$\boldsymbol{\psi}(\boldsymbol{\rho}) = \begin{bmatrix} \boldsymbol{\alpha}(\boldsymbol{\rho}) \\ \boldsymbol{\gamma}(\boldsymbol{\rho}) \end{bmatrix} \in \mathbb{R}^m$$

and

$$J_{\psi}(\boldsymbol{\rho}) = \frac{\partial \boldsymbol{\psi}(\boldsymbol{\rho})}{\partial \boldsymbol{\rho}} = \begin{bmatrix} \frac{\partial \alpha(\boldsymbol{\rho})}{\partial \boldsymbol{\rho}} \\ \frac{\partial \gamma(\boldsymbol{\rho})}{\partial \boldsymbol{\rho}} \end{bmatrix} = \begin{bmatrix} J_{\alpha}(\boldsymbol{\rho}) \\ J_{\gamma}(\boldsymbol{\rho}) \end{bmatrix} \in \mathbb{R}^{m \times m}. \tag{8.3}$$

Let us now define the set

$$\Omega_{\rho} = \left\{\boldsymbol{\rho} \in \mathbb{R}^m : \boldsymbol{\gamma}(\boldsymbol{\rho}) = \mathbf{0}, \det(\mathbf{J}_{\psi}(\boldsymbol{\rho})) \neq 0\right\}.$$

It follows that Ω_ρ is the workspace region in the $\boldsymbol{\rho}$ coordinates where the mechanical system satisfies the holonomic constraints and, in addition, the columns of $\mathbf{J}_\alpha(\boldsymbol{\rho})^{\mathrm{T}}$ (indicating the direction of maximum growth of the n minimal generalized coordinates) and those of $\mathbf{J}_\gamma(\boldsymbol{\rho})^{\mathrm{T}}$ (which indicate the direction of the r constraint forces) are linearly independent.

Moreover, from Eq. (8.2), if we define

$$\bar{\boldsymbol{\psi}}(\mathbf{q},\boldsymbol{\rho}) = \boldsymbol{\psi}(\boldsymbol{\rho}) - \begin{bmatrix} \mathbf{q} \\ \mathbf{0} \end{bmatrix} = \begin{bmatrix} \boldsymbol{\alpha}(\boldsymbol{\rho}) - \mathbf{q} \\ \boldsymbol{\gamma}(\boldsymbol{\rho}) \end{bmatrix} = \mathbf{0} \in \mathbb{R}^m \tag{8.4}$$

and notice that

$$\frac{\partial \bar{\boldsymbol{\psi}}(\mathbf{q},\boldsymbol{\rho})}{\partial \boldsymbol{\rho}} = \mathbf{J}_\psi(\boldsymbol{\rho}), \tag{8.5}$$

then we can apply the implicit function theorem (see Dontchev and Rockafellar 2014) to show that, for any $\boldsymbol{\rho}_0 \in \Omega_\rho$, there is a neighborhood N_ρ of $\boldsymbol{\rho}_0$, and a neighborhood N_q of $\mathbf{q}_0 = \boldsymbol{\alpha}(\boldsymbol{\rho}_0)$ such that, for any $\mathbf{q} \in N_q$, there exist a unique $\boldsymbol{\rho} \in N_\rho$ and a continuously differentiable function $\boldsymbol{\sigma} : N_q \to N_\rho$ such that

$$\boldsymbol{\rho} = \boldsymbol{\sigma}(\mathbf{q}) \tag{8.6}$$

with Jacobian satisfying

$$\mathbf{A}(\mathbf{q}) = \frac{\partial \boldsymbol{\sigma}(\mathbf{q})}{\partial \mathbf{q}} = -\left(\frac{\partial \bar{\boldsymbol{\psi}}(\mathbf{q},\boldsymbol{\rho})}{\partial \boldsymbol{\rho}}\right)^{-1} \frac{\partial \bar{\boldsymbol{\psi}}(\mathbf{q},\boldsymbol{\rho})}{\partial \mathbf{q}}\bigg|_{\boldsymbol{\rho}=\boldsymbol{\sigma}(\mathbf{q})}$$

so that, considering Eqs. (8.4) and (8.5):

$$\mathbf{A}(\mathbf{q}) = \mathbf{J}_\psi(\boldsymbol{\sigma}(\mathbf{q}))^{-1} \begin{bmatrix} \mathbf{I} \\ \mathbf{O} \end{bmatrix} \in \mathbb{R}^{m \times n}. \tag{8.7}$$

where $\mathbf{I} \in \mathbb{R}^{n \times n}$ and $\mathbf{O} \in \mathbb{R}^{r \times n}$ are the identity and null matrices, respectively.

Let $\Omega_\rho^* \subset \Omega_\rho$ denote the largest subset of Ω_ρ containing $\boldsymbol{\rho}_0$ for which the unique parameterization Eq. (8.6) holds, and let Ω_q be the corresponding domain of $\boldsymbol{\sigma}$. Then we have a diffeomorphism from Ω_ρ^* to Ω_q as follows:

$$\Omega_\rho^* \xrightarrow{\alpha} \Omega_q \xrightarrow{\sigma} \Omega_\rho^*. \tag{8.8}$$

Notice that unlike $\boldsymbol{\alpha}$, which can be easily found, $\boldsymbol{\sigma}$ cannot in general be expressed explicitly in an analytical form (sometimes it can only be computed iteratively by numerical methods), but the previous analysis shows that whenever $\mathbf{q} \in \Omega_q$, there is always a unique solution $\boldsymbol{\rho} = \boldsymbol{\sigma}(\mathbf{q}) \in \Omega_\rho^*$ for which $\mathbf{q} = \boldsymbol{\alpha}(\boldsymbol{\rho}) \in \Omega_q$ holds (Ghorbel et al. 2000). An estimate of the domain Ω_q is also proposed in Ghorbel et al. (2000).

It is worth noting also that for a given $\mathbf{q} \in \Omega_q$, we can also find other solutions for the mapping $\Omega_q \to \Omega_\rho$ different from $\boldsymbol{\sigma}$. Let us denote $\boldsymbol{\sigma}'$ any of those solutions, then

$$\boldsymbol{\rho}' = \boldsymbol{\sigma}'(\mathbf{q})$$

is an element of Ω_ρ but not of Ω_ρ^*.

On the other hand, let $\boldsymbol{\xi} \in \mathbb{R}^\eta$ be the vector of coordinates that describe the pose (i.e., the position and orientation) of the robot's platform. For computing such pose coordinates, it is common to attach a frame Σ_F to the platform, at a particular point F; the pose of the platform can then be described by means of a position vector $\mathbf{r}_F \in \mathbb{R}^3$ and a rotation matrix ${}^0\mathbf{R}_F \in SO(3)$, which respectively give the position of point F and the orientation of frame Σ_F with respect to a fixed frame Σ_0, usually placed at the robot base; the pose coordinates usually employed in robotics are the Cartesian coordinates (x, y, z) taken from the position vector $\mathbf{r}_F \in \mathbb{R}^3$, and the Euler angles (λ, μ, ν), which can be extracted from ${}^0\mathbf{R}_F$ using standard formulas (see, e.g., Siciliano et al. 2009).

In sum, we usually have $\eta = 6$ and

$$\boldsymbol{\xi} = \begin{bmatrix} x & y & z & \lambda & \mu & \nu \end{bmatrix}^T \in \mathbb{R}^6,$$

although it is also possible to employ any other parameterization of the orientation (see Campa and de la Torre 2009).

For the purpose of this paper, the problem of forward pose kinematics consists on determining the vector of pose coordinates of the platform $\boldsymbol{\xi}$ as a function of the vector of minimal generalized coordinates $\mathbf{q}$; that is to say, we need to find the FPK model:

$$\boldsymbol{\xi} = \mathbf{h}(\boldsymbol{q}), \tag{8.9}$$

where $\mathbf{h} : \Omega_q \to \Omega_\xi$ (with Ω_ξ being the set of all admissible poses of the platform) is known as the FPK function of the robot. But it is a well-known fact that in the case of parallel robots, the FPK model has multiple solutions, in the sense that a single set of active joints can produce different poses of the platform.

Now, let us define a function $\boldsymbol{\chi} : \Omega_\rho \to \Omega_\xi$. For $\boldsymbol{\rho} \in \Omega_\rho^* \subset \Omega_\rho$, we can write $\boldsymbol{\xi} = \boldsymbol{\chi}(\boldsymbol{\rho}) \in \Omega_\xi^* \subset \Omega_\xi$, and using Eq. (8.6) we get

$$\boldsymbol{\xi} = \boldsymbol{\chi}(\boldsymbol{\sigma}(\mathbf{q})), \tag{8.10}$$

while for those $\boldsymbol{\rho}' \in \Omega_\rho$ and $\boldsymbol{\rho}' \notin \Omega_\rho^*$, we have

$$\boldsymbol{\xi}' = \boldsymbol{\chi}(\boldsymbol{\rho}') = \boldsymbol{\chi}(\boldsymbol{\sigma}'(\boldsymbol{q})), \tag{8.11}$$

comparing Eqs. (8.10) and (8.11) with Eq. (8.9) we conclude that the FPK function $\mathbf{h}$ can be either $\mathbf{h} = \boldsymbol{\chi} \circ \boldsymbol{\sigma}$ or $\mathbf{h} = \boldsymbol{\chi} \circ \boldsymbol{\sigma}'$ with $\circ$ the standard symbol for function composition.

Figure 8.2 shows the diagram of sets Ω_q, Ω_ρ, Ω_ξ and the functions among them. The relevance of sets Ω_ρ^* and Ω_q^* lies in the fact that they can correspond to actual (or real) configurations of the robot. Indeed, if $\boldsymbol{\rho}_0 \in \Omega_\rho$ is chosen to be a known configuration of the real robot (e.g., its home configuration), then the definition of a smooth function $\boldsymbol{\alpha}$ and the implicit function theorem guarantee the existence of sets Ω_ρ^*, Ω_ξ^* and functions $\boldsymbol{\sigma}$ and $\boldsymbol{\chi}$.

In general, the computation of the FPK model becomes a major problem due to the complexity of the equations involved and the difficulty to find a closed set of solutions. The methods for solving the FPK model can be classified into analytical and numerical methods. The analytical methods allow to get all the possible solutions of the FPK model (even those that are not physically realizable, due to mechanical constraints); however, we are often interested in knowing only the solution that describes the actual pose of the platform (corresponding to $\boldsymbol{\xi}^*$), so iterative numerical methods are sufficient. Several analytical methods can be employed for solving the FPK model of a parallel robot (see, e.g., Merlet 2006); among them, the so-called elimination methods are of particular interest (Kapur 1995).

The main idea of an elimination method is to manipulate the equations of the FPK in order to reduce the problem to the solution of a univariate polynomial whose real roots enable to determine all the possible poses of the platform. A drawback of this procedure is that it can be performed in several different ways, not all of them leading to the same degree of the resulting polynomial (Merlet 1999). Therefore, it is necessary to find the univariate polynomial with the least degree. Such degree can be obtained, for example, using the Bezout's theorem (Merlet 2006). But once the roots of the polynomial are computed, it is necessary to determine which one gives the actual configuration of the robot. We will consider that such configuration (given by $\boldsymbol{\rho} \in \Omega_\rho^*$) and the corresponding pose of the platform ($\boldsymbol{\xi} \in \Omega_\xi^*$) can be determined by considering the diffeomorphism Eq. (8.8).

It is worth mentioning here that the FPK of a large number of mechanisms can be determined by studying equivalent mechanisms for which the univariate polynomial can be easily extracted (Merlet 2006). For example, in the case of the mechanism under study, its 6-3-PUS structure can be analyzed as a 3-PRPS type (Carbonari et al. 2011).

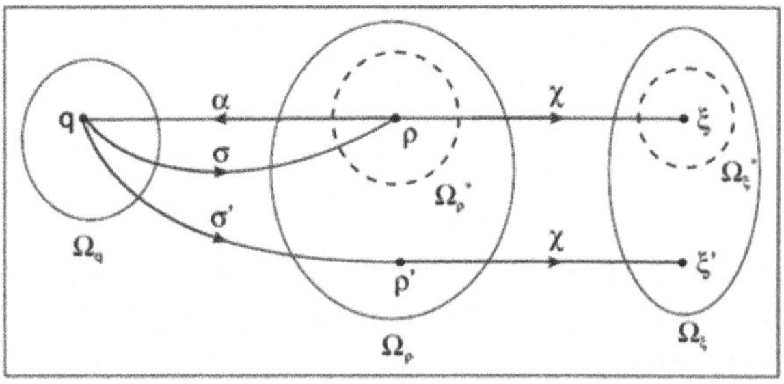

Fig. 8.2 Relations among sets Ω_q, Ω_ρ and Ω_ξ

8.2.2 Velocity Kinematics Modeling

Now, let $\boldsymbol{\upsilon} \in \mathbb{R}^3$ and $\boldsymbol{\omega} \in \mathbb{R}^3$ be, respectively, the vectors of linear and angular velocities of the center of mass (com) of the platform. Then the forward velocity kinematics (FVK) model can be written as:

$$\begin{bmatrix} \boldsymbol{\upsilon} \\ \boldsymbol{\omega} \end{bmatrix} = \mathbf{J}(\mathbf{q})\dot{\mathbf{q}}$$

where $\dot{\mathbf{q}} = \frac{\mathrm{d}}{\mathrm{dt}}\mathbf{q} \in \mathbb{R}^n$, and $\mathbf{J}(\mathbf{q}) \in \mathbb{R}^{6\times n}$ is known as the *geometric Jacobian* matrix of the robot.

Taking the time derivative of Eq. (8.1), we get

$$\mathbf{J}_\gamma(\boldsymbol{\rho})\dot{\boldsymbol{\rho}} = \mathbf{0} \in \mathbb{R}^r \tag{8.12}$$

where

$$\mathbf{J}_\gamma(\boldsymbol{\rho}) = \frac{\partial \boldsymbol{\gamma}(\boldsymbol{\rho})}{\partial \boldsymbol{\rho}} \in \mathbb{R}^{r\times m} \tag{8.13}$$

denoted here as the *constraint Jacobian*, was already used in Eq. (8.3). Moreover, taking the time derivative of Eq. (8.6) we get the relation between the vectors of minimal and non-minimal generalized velocities, i.e.,

$$\dot{\boldsymbol{\rho}} = \mathbf{A}(\mathbf{q})\dot{\mathbf{q}} \in \mathbb{R}^m \tag{8.14}$$

where $\mathbf{A}(\mathbf{q})$ is the transformation Jacobian matrix given by:

$$\mathbf{A}(\mathbf{q}) = \frac{\partial \boldsymbol{\sigma}(\mathbf{q})}{\partial \mathbf{q}} \in \mathbb{R}^{m\times n} \tag{8.15}$$

which can be also computed in terms of $\boldsymbol{\gamma}$ and $\boldsymbol{\alpha}$ as in Eq. (8.7).

Combining Eqs. (8.12) and (8.14), we have

$$\mathbf{J}_\gamma(\boldsymbol{\rho})\mathbf{A}(\mathbf{q})\dot{\mathbf{q}} = \mathbf{0} \in \mathbb{R}^r \tag{8.16}$$

and it can be shown (see, e.g., Blajer 1997) that Eq. (8.16) implies:

$$\mathbf{J}_\gamma(\boldsymbol{\rho})\mathbf{A}(\mathbf{q}) = \mathbf{O} \in \mathbb{R}^{r\times n}$$

or equivalently

$$\mathbf{A}(\mathbf{q})^{\mathrm{T}}\mathbf{J}_\gamma(\boldsymbol{\rho})^{\mathrm{T}} = \mathbf{O} \in \mathbb{R}^{n\times r} \tag{8.17}$$

which is a very useful property, as will be shown in the following subsection.

Now let us assume that the pose of the platform is known, and given in terms of the position vector $\mathbf{r}_F$ and the rotation matrix ${}^0\mathbf{R}_F$, as functions of $\boldsymbol{\rho}$; in other words, we know $\mathbf{r}_F(\boldsymbol{\rho})$, and ${}^0\mathbf{R}_F(\boldsymbol{\rho})$ (which is a parameterization of $\boldsymbol{\chi}(\boldsymbol{\rho})$). Then the linear velocity vector $\boldsymbol{\upsilon}$ is simply computed as

$$\boldsymbol{\upsilon} = \dot{\mathbf{r}}_F(\boldsymbol{\rho}) = \frac{\partial \mathbf{r}_F(\boldsymbol{\rho})}{\partial \boldsymbol{\rho}}\dot{\boldsymbol{\rho}} = \frac{\partial \mathbf{r}_F(\boldsymbol{\rho})}{\partial \boldsymbol{\rho}}\mathbf{A}(\mathbf{q})\dot{\mathbf{q}}, \tag{8.18}$$

where we have employed the chain rule and Eq. (8.14). And if the columns of matrix ${}^0\mathbf{R}_F(\boldsymbol{\rho})$ are the orthonormal vectors $\hat{\boldsymbol{x}}_F(\boldsymbol{\rho})$, $\hat{\boldsymbol{y}}_F(\boldsymbol{\rho})$ and $\hat{\boldsymbol{z}}_F(\boldsymbol{\rho})$, i.e.,

$${}^0\mathbf{R}_F(\boldsymbol{\rho}) = [\hat{\boldsymbol{x}}_F(\boldsymbol{\rho}) \quad \hat{\boldsymbol{y}}_F(\boldsymbol{\rho}) \quad \hat{\boldsymbol{z}}_F(\boldsymbol{\rho})] \in SO(3),$$

then it is possible to show (see Campa and de la Torre 2009) that the angular velocity vector $\boldsymbol{\omega}$ can be obtained using the following:

$$\boldsymbol{\omega} = \frac{1}{2}\left[\mathbf{S}(\hat{\boldsymbol{x}}_F(\boldsymbol{\rho}))\dot{\hat{\boldsymbol{x}}}_F(\boldsymbol{\rho}) + \mathbf{S}(\hat{\boldsymbol{y}}_F(\boldsymbol{\rho}))\dot{\hat{\boldsymbol{y}}}_F(\boldsymbol{\rho}) + \mathbf{S}(\hat{\boldsymbol{z}}_F(\boldsymbol{\rho}))\dot{\hat{\boldsymbol{z}}}_F(\boldsymbol{\rho})\right] \tag{8.19}$$

where for a vector $\mathbf{v} = \begin{bmatrix} v_1 & v_2 & v_3 \end{bmatrix}^{\mathrm{T}} \in \mathbb{R}^3$, the skew-symmetric matrix operator $\mathbf{S}(\mathbf{v})$ is given by

$$\mathbf{S}(\mathbf{v}) = \begin{bmatrix} 0 & -v_3 & v_2 \\ v_3 & 0 & -v_1 \\ -v_2 & v_1 & 0 \end{bmatrix}.$$

Again, using the chain rule and Eq. (8.14), we can rewrite Eq. (8.19) as

$$\boldsymbol{\omega} = \frac{1}{2}\left[\mathbf{S}(\hat{\boldsymbol{x}}_F(\boldsymbol{\rho}))\frac{\partial \hat{\boldsymbol{x}}_F(\boldsymbol{\rho})}{\partial \boldsymbol{\rho}} + \mathbf{S}(\hat{\boldsymbol{y}}_F(\boldsymbol{\rho}))\frac{\partial \hat{\boldsymbol{y}}_F(\boldsymbol{\rho})}{\partial \boldsymbol{\rho}} + \mathbf{S}(\hat{\boldsymbol{z}}_F(\boldsymbol{\rho}))\cdot\frac{\partial \hat{\boldsymbol{z}}_F(\boldsymbol{\rho})}{\partial \boldsymbol{\rho}}\right]\mathbf{A}(\mathbf{q})\dot{\mathbf{q}} \tag{8.20}$$

To complete this section, let us consider the time derivative of the FPK model Eq. (8.9), i.e.,

$$\dot{\boldsymbol{\xi}} = \frac{\partial \mathbf{h}(\mathbf{q})}{\partial \mathbf{q}}\dot{\mathbf{q}} = \mathbf{J}_A(\mathbf{q})\dot{\mathbf{q}}$$

where $J_A(\mathbf{q})$ is known as the analytical *Jacobian*.

8.2.3 Dynamics Modeling

The minimal dynamics model of a robot manipulator is the mapping between the generalized forces exerted on the links by the active joints (named here as $\boldsymbol{\tau}_q$), and the minimal generalized coordinates, velocities, and accelerations (i.e., $\mathbf{q}$, $\dot{\mathbf{q}}$, and $\ddot{\mathbf{q}}$, respectively).

According to Siciliano et al. (2009), the *direct dynamics* problem consists of determining, for $t \geq t_o$, the vectors $\mathbf{q}(t)$, $\dot{\mathbf{q}}(t)$, and $\ddot{\mathbf{q}}(t)$, resulting from a given $\tau_q(t)$, once the initial state of the system, given by $\mathbf{q}(t_o)$ and $\dot{\mathbf{q}}(t_o)$, is known. On the other hand, the *inverse dynamics* problem consists of determining the $\tau_q(t)$ needed to generate the motion specified by $\mathbf{q}(t)$, $\dot{\mathbf{q}}(t)$, and $\ddot{\mathbf{q}}(t)$. We are interested in the inverse dynamics in this document.

There are several methods (aka formulations) for computing the inverse dynamics model of a robot, being the most common the Newton–Euler and the Euler–Lagrange formulations. In this paper, we use the latter for computing the inverse dynamics model of the 6-3-PUS parallel robot.

Let us start by considering a general platform-type n-DOF parallel manipulator described either by a minimal ($\mathbf{q} \in \mathbb{R}^n$) or by a non-minimal ($\boldsymbol{\rho} \in \mathbb{R}^m$) set of coordinates. Let $\mathcal{K}$ be the total kinetic energy, and $\mathcal{U}$ be the total potential energy of such a system (without arguments for the sake of generalization). If the robot is composed of b rigid bodies (links), then the total kinetic and potential energies are simply:

$$\mathcal{K} = \sum_{l=1}^{b} \mathcal{K}_l, \quad \mathcal{U} = \sum_{l=1}^{b} \mathcal{U}_l \tag{8.21}$$

where $\mathcal{K}_l$ and $\mathcal{U}_l$ are, respectively, the kinetic and potential energies of the l-th rigid body.

In order to compute $\mathcal{K}_l$ and $\mathcal{U}_l$, it is customary to consider again the fixed (inertial) coordinate frame Σ_o and a coordinate frame Σ_l attached to the l-th body, usually in accordance with the Denavit–Hartenberg convention (see, e.g., Siciliano et al. 2009). We now can write:

$$\mathcal{K}_l = \frac{1}{2}\left[m_l \boldsymbol{\upsilon}_l^{\mathrm{T}} \boldsymbol{\upsilon}_l + {}^{l}\boldsymbol{\omega}_l^{\mathrm{T}} I_l^{l} \boldsymbol{\omega}_l\right] \tag{8.22}$$

$$\mathcal{U}_l = -m_l \boldsymbol{p}_l^{\mathrm{T}} \mathbf{g}_0 \tag{8.23}$$

where

- m_l mass of the l-th body.
- I_l inertia tensor of the l-th body with respect to a frame with origin at its com, but oriented as Σ_l.
- $\mathbf{p}_l$ position vector of the com of the l-th body, with respect to frame Σ_o.
- $\boldsymbol{\upsilon}_l$ linear velocity vector of the com of the l-th body, with respect to frame $\Sigma_o(\boldsymbol{\upsilon}_l = \dot{\boldsymbol{p}}_l)$.
- ${}^{l}\boldsymbol{\omega}_l$ angular velocity vector of the l-th body, with respect to frame Σ_o, expressed in the coordinates of Σ_l.
- $\mathbf{g}_o$ constant vector of gravitational acceleration, with respect to frame Σ_o.

Consider the case where the system is described by the minimal set of generalized coordinates $\mathbf{q} \in \mathbb{R}^n$. If the pose of the l-th rigid body is given by the position vector $\mathbf{p}_l(\mathbf{q}) \in \mathbb{R}^3$, and the rotation matrix

$$
{}^0\mathbf{R}_l(\mathbf{q}) = [\hat{x}_l(q) \quad \hat{y}_l(q) \quad \hat{z}_l(q)] \in \mathrm{SO}(3),
$$

then the linear velocity vector is simply $\mathbf{\upsilon}_l = \dot{\mathbf{p}}_l$, and the angular velocity vector ${}^l\boldsymbol{\omega}_l$ can be computed using an expression like Eq. (8.19), i.e.,

$$
\frac{1}{2} {}^0\mathbf{R}_F^{\mathrm{T}} \left[\mathbf{S}(\hat{x}_l(q))\dot{\hat{x}}_l(q) + \mathbf{S}(\hat{y}_l(q))\dot{\hat{y}}_l(q) + \mathbf{S}(\hat{z}_l(q))\dot{\hat{z}}_l(q) \right]
$$

where the left-multiplying rotation matrix allows to express the angular velocity vector $\boldsymbol{\omega}_l$ (computed with respect to Σ_o) in the coordinates of Σ_l.

By using Eqs. (8.21)–(8.23), we can now compute the total kinetic and potential energies, $\mathcal{K}(\mathbf{q}, \dot{\mathbf{q}})$ and $\mathcal{U}(\mathbf{q})$.

Then the *Lagrangian function* of the robot is defined as

$$
\mathcal{L}(\mathbf{q}, \dot{\mathbf{q}}) = \mathcal{K}(\mathbf{q}, \dot{\mathbf{q}}) - \mathcal{U}(\mathbf{q})
$$

and it can be shown that the inverse dynamics model of such a system is given by the so-called Euler–Lagrange equations of motion:

$$
\frac{\mathrm{d}}{\mathrm{d}t}\left(\frac{\partial \mathcal{L}(\mathbf{q}, \dot{\mathbf{q}})}{\partial \dot{\mathbf{q}}} \right) - \frac{\partial \mathcal{L}(\mathbf{q}, \dot{\mathbf{q}})}{\partial \mathbf{q}} = \boldsymbol{\tau}_q. \tag{8.24}
$$

By expanding and collecting terms, Eq. (8.24) can be rewritten as:

$$
\mathbf{M}_q(\mathbf{q})\ddot{\mathbf{q}} + \mathbf{C}_q(\mathbf{q}, \dot{\mathbf{q}})\dot{\mathbf{q}} + \mathbf{g}_q(\mathbf{q}) = \boldsymbol{\tau}_q,
$$

where $\mathbf{M}_q(\mathbf{q}) \in \mathbb{R}^{n\times n}$ is known as the robot inertia matrix, $\mathbf{C}_q(\mathbf{q}, \dot{\mathbf{q}}) \in \mathbb{R}^{n\times n}$ is the matrix of terms arising from the centrifugal and Coriolis forces, and $\mathbf{g}_q(\mathbf{q}) \in \mathbb{R}^n$ represents the vector of forces due to gravity.

But if we choose a non-minimal set of generalized coordinates, given by $\boldsymbol{\rho} \in \mathbb{R}^m$ to describe the same system, then the dynamics must also include the holonomic constraints given by Eq. (8.1). The total kinetic and potential energies should now be $\mathcal{K}(\boldsymbol{\rho}, \dot{\rho})$ and $\mathcal{U}(\boldsymbol{\rho})$, respectively, and for their computation we can also use Eqs. (8.21)–(8.23).

If the pose of the l-th rigid body is given by $\mathbf{p}_l(\boldsymbol{\rho}) \in \mathbb{R}^3$ and ${}^0\mathbf{R}_l(\mathbf{q}) = [\hat{x}_l(\boldsymbol{\rho}) \quad \hat{y}_l(\boldsymbol{\rho}) \quad \hat{z}_l(\boldsymbol{\rho})] \in \mathrm{SO}(3)$, then $\mathbf{\upsilon}_l$ and ${}^l\boldsymbol{\omega}_l$ can be computed using the following expressions

$$
\mathbf{\upsilon}_l = \frac{\partial \mathbf{p}_l(\boldsymbol{\rho})}{\partial \boldsymbol{\rho}} \mathbf{A}(\mathbf{q})\dot{\mathbf{q}} \tag{8.25}
$$

$$
{}^l\boldsymbol{\omega}_l = \frac{1}{2} {}^0\mathbf{R}_l(\boldsymbol{\rho})^{\mathrm{T}} \left[\mathbf{S}(\hat{x}_l(\boldsymbol{\rho})) \frac{\partial \hat{x}_l(\boldsymbol{\rho})}{\partial \boldsymbol{\rho}} + \mathbf{S}(\hat{y}_l(\boldsymbol{\rho})) \frac{\partial \hat{y}_l(\boldsymbol{\rho})}{\partial \boldsymbol{\rho}} + \mathbf{S}(\hat{z}_l(\boldsymbol{\rho})) \frac{\partial \hat{z}_l(\boldsymbol{\rho})}{\partial \boldsymbol{\rho}} \right] \mathbf{A}(\mathbf{q})\dot{\mathbf{q}} \tag{8.26}
$$

where the time derivative of $\boldsymbol{\rho}$ is not explicitly required (and, as will be shown in Sect. 8.5, that fact is useful when computing the dynamics model of parallel robots).

The Lagrangian function would become

$$\mathcal{L}(\boldsymbol{\rho},\dot{\boldsymbol{\rho}}) = \mathcal{K}(\boldsymbol{\rho},\dot{\boldsymbol{\rho}}) - \mathcal{U}(\boldsymbol{\rho}),$$

and the expansion of the Lagrange equations of motion in this case leads to the following expression:

$$\mathbf{M}_\rho(\boldsymbol{\rho})\ddot{\boldsymbol{\rho}} + \mathbf{C}_\rho(\boldsymbol{\rho},\dot{\boldsymbol{\rho}})\dot{\boldsymbol{\rho}} + \mathbf{g}_\rho(\boldsymbol{\rho}) = \boldsymbol{\tau}_\rho + \mathbf{J}_\gamma^{\mathrm{T}}(\boldsymbol{\rho})\boldsymbol{\lambda} \tag{8.27}$$

where now $\mathbf{M}_\rho(\boldsymbol{\rho}) \in \mathbb{R}^{m\times m}$ represents the inertia matrix, $\mathbf{C}_\rho(\boldsymbol{\rho},\dot{\boldsymbol{\rho}}) \in \mathbb{R}^{m\times m}$ the matrix of centrifugal and Coriolis forces, and $\mathbf{g}_\rho(\boldsymbol{\rho}) \in \mathbb{R}^m$ the vector of gravitational forces; $\mathbf{J}_\gamma(\boldsymbol{\rho})$ is defined in Eq. (8.13), and $\boldsymbol{\lambda}$ is the vector of Lagrange multipliers, which ensures that the constraints in Eq. (8.1) are fulfilled.

It is worth mentioning that an alternative for computing matrices $\mathbf{M}_\rho(\boldsymbol{\rho})$, $\mathbf{C}_\rho(\boldsymbol{\rho},\dot{\boldsymbol{\rho}})$, and $\mathbf{g}(\boldsymbol{\rho})$, without explicitly expanding the Lagrange's equations of motion, is employing the following properties (Kelly et al. 2005):

$$\mathcal{K}(\boldsymbol{\rho},\dot{\boldsymbol{\rho}}) = \frac{1}{2}\dot{\boldsymbol{\rho}}^{\mathrm{T}}\mathbf{M}_\rho(\boldsymbol{\rho})\dot{\boldsymbol{\rho}} \tag{8.28}$$

$$\frac{\partial\mathcal{K}(\boldsymbol{\rho},\dot{\boldsymbol{\rho}})}{\partial\boldsymbol{\rho}} = \left[\dot{\mathbf{M}}_\rho(\boldsymbol{\rho}) - \mathbf{C}_\rho(\boldsymbol{\rho},\dot{\boldsymbol{\rho}})\right]\dot{\boldsymbol{\rho}} \tag{8.29}$$

$$\frac{\partial\mathcal{U}(\boldsymbol{\rho})}{\partial\boldsymbol{\rho}} = \mathbf{g}_\rho(\boldsymbol{\rho}) \tag{8.30}$$

Notice that given the total kinetic $\mathcal{K}(\boldsymbol{\rho},\dot{\boldsymbol{\rho}})$ and potential $\mathcal{U}(\boldsymbol{\rho})$ energies, we can use Eqs. (8.28), (8.29), and (8.30) to, respectively, get $\mathbf{M}_\rho(\boldsymbol{\rho})$, $\mathbf{C}_\rho(\boldsymbol{\rho},\dot{\boldsymbol{\rho}})$, and $\mathbf{g}_\rho(\boldsymbol{\rho})$.

Now if we left-multiply Eq. (8.27) by $\mathbf{A}(\mathbf{q})^{\mathrm{T}}$, and employ Eq. (8.17), together with Eq. (8.14) and its time derivative, we get the following minimal dynamics model:

$$\overline{\mathbf{M}}_q(\boldsymbol{\rho})\ddot{\mathbf{q}} + \overline{\mathbf{C}}_q(\boldsymbol{\rho},\dot{\boldsymbol{\rho}})\dot{\mathbf{q}} + \bar{\mathbf{g}}_q(\boldsymbol{\rho}) = \boldsymbol{\tau}_q. \tag{8.31}$$

where

$$\overline{\mathbf{M}}_q(\boldsymbol{\rho}) = \mathbf{A}(\mathbf{q})^{\mathrm{T}}\mathbf{M}_\rho(\boldsymbol{\rho})\mathbf{A}(\mathbf{q}), \tag{8.32}$$

$$\overline{\mathbf{C}}_q(\boldsymbol{\rho},\dot{\boldsymbol{\rho}}) = \mathbf{A}(\mathbf{q})^{\mathrm{T}}\mathbf{C}_\rho(\boldsymbol{\rho},\dot{\boldsymbol{\rho}})\mathbf{A}(\mathbf{q}) + \mathbf{A}(\mathbf{q})^{\mathrm{T}}\mathbf{M}_\rho(\boldsymbol{\rho})\dot{\mathbf{A}}(\mathbf{q},\mathbf{q}) \tag{8.33}$$

$$\bar{\mathbf{g}}_q = \mathbf{A}(\mathbf{q})^{\mathrm{T}} \mathbf{g}_\rho(\boldsymbol{\rho}), \tag{8.34}$$

$$\boldsymbol{\tau}_q = \mathbf{A}(\mathbf{q})^{\mathrm{T}} \boldsymbol{\tau}_\rho. \tag{8.35}$$

This last step is known in general as the projection method and has been employed by different authors (see Arczewski and Blajer 1996; Blajer 1997; Ghorbel et al. 2000; Betsch 2005) to reduce the dynamics of constrained mechanical systems. It should be clear that Eq. (8.31) represents the minimal inverse dynamics model of the robot.

As mentioned in Ghorbel et al. (2000), however, the reduced dynamics model described above in Eq. (8.31) has two special characteristics which make it different from regular dynamics models of open-chain mechanical systems. First, the above reduced model is valid only (locally) for $\mathbf{q}$ in the compact set Ω_q. Second, since the parameterization $\boldsymbol{\rho} = \boldsymbol{\sigma}(\mathbf{q})$ is implicit, it is an implicit model.

Now, after mentioning the above concepts, we can list the necessary steps to get the dynamics model of a system subject to holonomic constraints using the Euler–Lagrange formulation and the projection method:

1. Define the sets of minimal and non-minimal coordinates so that $\mathbf{q} \in \Omega_q \subset \mathbb{R}^n$ and $\boldsymbol{\rho} \in \boldsymbol{\Omega}_\rho^* \subset \mathbb{R}^m$.
2. Determine functions $\gamma(\boldsymbol{\rho})$, $\boldsymbol{\alpha}(\boldsymbol{\rho})$, and $\boldsymbol{\sigma}(\mathbf{q})$ (if possible), so that Eqs. (8.1), (8.2), and (8.6) are met.
3. Compute $\mathbf{J}_\psi(\boldsymbol{\rho})$ and $\mathbf{A}(\mathbf{q})$ using Eqs. (8.3) and (8.7) (or Eq. (8.15), if possible).
4. Express the pose of each of the b rigid bodies in the robot in terms of $\boldsymbol{\rho}$, i.e., find $\mathbf{p}_l(\boldsymbol{\rho})$ and ${}^0\mathbf{R}_l(\boldsymbol{\rho})$.
5. Compute the vectors of linear velocity $\boldsymbol{\upsilon}_l$ and angular velocity ${}^l\boldsymbol{\omega}_l$, using Eqs. (8.25) to (8.26).
6. Compute the total kinetic $\mathcal{K}(\boldsymbol{\rho}, \dot{\boldsymbol{\rho}})$ and potential $\mathcal{U}(\boldsymbol{\rho})$ energies using Eqs. (8.21)–(8.23).
7. Find the matrices of the non-minimal model Eq. (8.27) either expanding Eq. (8.24) or using Eqs. (8.28)–(8.30).
8. Left-multiply model Eq. (8.27) by $\mathbf{A}(\mathbf{q})^{\mathrm{T}}$ to get the minimal dynamics model Eq. (8.31).

8.3 Hexapod's Geometric Description

Figure 8.3 is again a picture of the Hexapod mechanism, but now with some marks which will be useful for modeling the robot kinematics. These marks will be described in the following paragraphs.

Points T_1, T_2, and T_3 define the vertices of an equilateral triangle which is fixed to the base and has a side length L_B. For simplicity, it is assumed that the centers of the six universal joints (labeled $D_0, D_1, \ldots, D_5$ in Fig. 8.3) lie in one of the sides of the $T_1T_2T_3$ triangle. Attached to the center of this base triangle is the reference

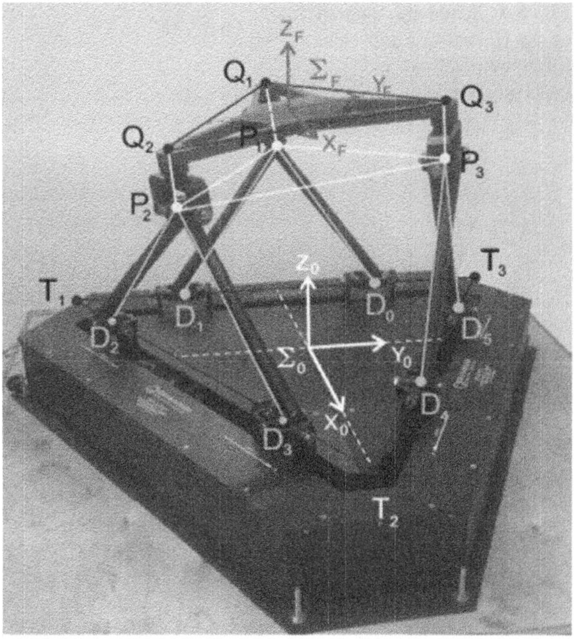

Fig. 8.3 Hexapod robot with describing marks

frame $\Sigma_0(X_0, Y_0, Z_0)$, with an orientation such that the X_0 axis points toward the vertex T_2, the Y_0 axis is parallel to the side T_1T_3, and the Z_0 axis points upward.

The points denoted as Q_1, Q_2, and Q_3 are placed where the last axis of each spherical joint meets the surface of the mobile platform, and they form a rigid equilateral triangle. The coordinate frame $\Sigma_F(X_F, Y_F, Z_F)$ is attached to the mobile platform, with its origin placed at the geometric center of triangle $Q_1Q_2Q_3$, and its orientation is such that the X_F axis points to the center of the Q_2Q_3 side, and the Z_F axis is normal to the $Q_1Q_2Q_3$ triangle, and points upward. Due to the mechanical design of the spherical joints, the triangle defined by the points P_1, P_2, and P_3 (points located at the center of the spherical joints) is rigid and equilateral (see Fig. 8.3); moreover, the $P_1P_2P_3$ triangle is always parallel to the $Q_1Q_2Q_3$ triangle and has its same dimensions, so that they together constitute a rigid right triangular prism whose length of each side is $L_P = L_B/2$, and its height is H_{PQ}.

The intrinsic symmetry of this mechanism greatly simplifies its kinematic analysis. Assuming hereinafter that the triad (i, j, k) is an element of the set $\{(1, 2, 3), (2, 3, 1), (3, 1, 2)\} \equiv S_3$ of cyclic permutations, it is possible to obtain expressions for one side of the base equilateral triangle which are similar to those of the other two sides, simply changing the indexes in the corresponding expressions.

In order to simplify the forthcoming analysis, a frame $\Sigma_{T_i}(X_{T_i}, Y_{T_i}, Z_{T_i})$ is assigned to each vertex T_i of the base triangle. And, as it is shown in the schematic diagram of Fig. 8.4a, this frame is such that the axis X_{Ti} has the direction of the vector $\mathbf{r}_{T_kT_i}$, which goes from the point T_k to the point T_i (hereinafter, unless otherwise indicated, this vector notation will be used), and the axis Z_{Ti} is always

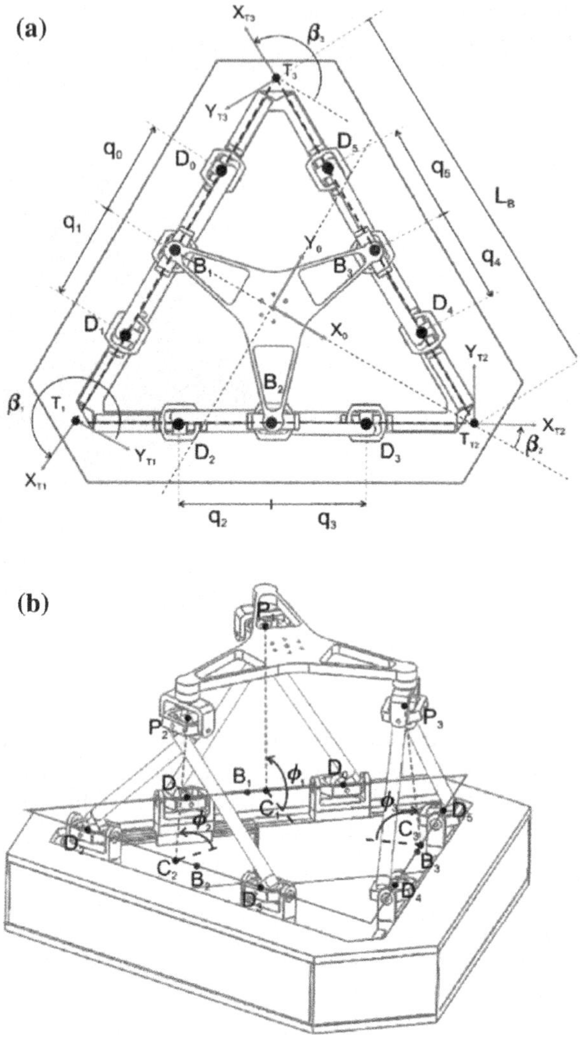

Fig. 8.4 Schematic diagram of the Hexapod: **a** top view (in home position) and **b** perspective view

perpendicular to the base, and points upward. The matrix relating the orientation of frames Σ_{T_i} and Σ_0 is:

$$ {}^{0}\mathbf{R}_{T_i}(\beta_i) = \begin{bmatrix} \cos(\beta_i) & -\sin(\beta_i) & 0 \\ \sin(\beta_i) & \cos(\beta_i) & 0 \\ 0 & 0 & 1 \end{bmatrix}, \tag{8.36} $$

where β_i is the angle from X_0 to X_{Ti}, around the Z_0 axis, so that $\beta_1 = 270°$, $\beta_2 = 30°$, and $\beta_3 = 150°$.

Besides the centers of the six universal joints, denoted by $D_0, D_1, \ldots, D_5$, Fig. 8.4 also shows the midpoints of the segments T_kT_i, indicated by B_i. The active joint variables are named $q_0, q_1, \ldots, q_5$ (see Fig. 8.4a). Note that q_{2i-2} and q_{2i-1} $(i = 1, 2, 3)$ are the distances from the point B_i to points D_{2i-2} and D_{2i-1}, respectively, which are on the same side of the $T_1T_2T_3$ triangle.

The active joint variables can be grouped into the following vector of active joint coordinates:

$$\mathbf{q} = [q_0 \quad q_1 \quad q_2 \quad q_3 \quad q_4 \quad q_5]^{\mathrm{T}} \in \mathbb{R}^6.$$

As mentioned before, each universal joint is connected to a spherical joint through a rigid link with length L. In this way, as can be seen in Fig. 8.4b, the points D_{2i-2} and D_{2i-1} (along to the X_{Ti} axis) form an isosceles triangle together with the vertex P_i of the mobile platform triangle. The segment $D_{2i-2}D_{2i-1}$ has a length equal to $q_{2i-2} + q_{2i-1}$, and its midpoint is denoted by C_i. Moreover, it should be noted that vector $\mathbf{r}_{C_iP_i}$, which is the height of the triangle $D_{2i-2}P_iD_{2i-1}$, is always perpendicular to the X_{Ti} axis and forms an angle ϕ_i with the Y_{Ti} axis.

In addition, on each side i of the robot base the following coordinate frames were assigned (see Fig. 8.5):

- The frame $\Sigma_{C_i}(X_{C_i}, Y_{C_i}, Z_{C_i})$ gives the inclination of the triangle $D_{2i-2}P_iD_{2i-1}$, and its orientation corresponds to that of frame Σ_{T_i} but after a rotation of an angle ϕ_i around the X_{Ti} axis;
- The frames $\Sigma_{D_{2i-2}}(X_{D_{2i-2}}, Y_{D_{2i-2}}, Z_{D_{2i-2}})$ and $\Sigma_{D_{2i-1}}(X_{D_{2i-1}}, Y_{D_{2i-1}}, Z_{D_{2i-1}})$ give the inclination of the segments $D_{2i-2} - P_i$ and $D_{2i-1} - P_i$, respectively; their orientation corresponds to that of the frame Σ_{C_i} but after a rotation of the corresponding angle, $-\alpha_i$ or α_i around the $Z_{D_{2i-2}}$ or $Z_{D_{2i-1}}$ axis, respectively (see Fig. 8.5).

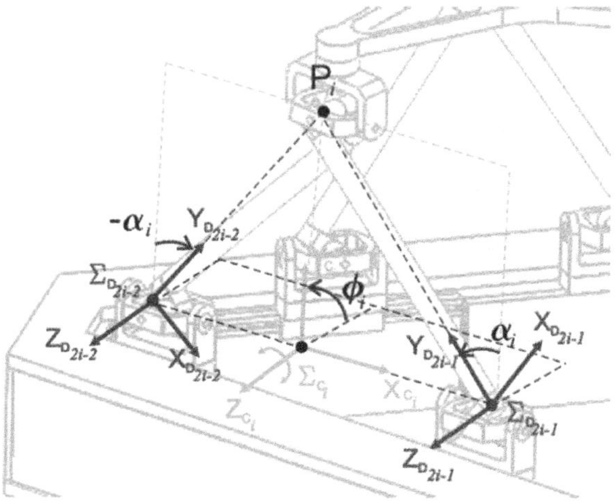

Fig. 8.5 Frames used for the orientation of the robot legs

8.4 Kinematics

In this section, we describe the computation of the pose and velocity kinematics models of the Hexapod parallel robot. It is worth mentioning that the kinematics analysis of this mechanism has been previously reported in Campa et al. (2016).

8.4.1 *Forward Pose Kinematics*

As mentioned in Sect. 8.2.1, to get the FPK model of a parallel robot we require to calculate the position and orientation of the mobile platform, given respectively by $\mathbf{r}_F(\mathbf{q})$ and ${}^0\mathbf{R}_F(\mathbf{q})$. In order to get these expressions, we will first compute the position vector with respect to frame Σ_0 of each point P_i, i.e., $\mathbf{r}_{P_i}$, as a function of $\mathbf{q}$.

From Fig. 8.4b, we can verify that the vector $\mathbf{r}_{P_i}$ is given by:

$$\mathbf{r}_{P_i} = \mathbf{r}_{B_i} + \mathbf{r}_{B_iC_i} + \mathbf{r}_{C_iP_i} \tag{8.37}$$

In the following, we explain how to obtain $\mathbf{r}_{B_i}$, $\mathbf{r}_{B_iC_i}$, and $\mathbf{r}_{C_iP_i}$.

From Fig. 8.4, considering Eq. (8.36) and recalling that the distance of the center of an equilateral triangle of side l_0 to each of its vertices is $l_0/\sqrt{3}$, it is possible to verify that:

$$\mathbf{r}_{B_i} = {}^0\mathbf{R}_{T_i}(\beta_i)^i\mathbf{r}_{B_i} = -\frac{L_B}{2\sqrt{3}}\,{}^0\mathbf{R}_{T_i}(\beta_i)\begin{bmatrix}0\\1\\0\end{bmatrix}. \tag{8.38}$$

Vectors $\mathbf{r}_{B_iC_i}$ and $\mathbf{r}_{C_iP_i}$ can be expressed in terms of the joint variables of the corresponding side (i.e., q_{2i-2} and q_{2i-1}) as:

$$\mathbf{r}_{B_iC_i} = {}^0\mathbf{R}_{T_i}(\beta_i)^i\mathbf{r}_{B_iC_i} = \left(\frac{q_{2i-1} - q_{2i-2}}{2}\right){}^0\mathbf{R}_{T_i}(\beta_i)\begin{bmatrix}1\\0\\0\end{bmatrix} \tag{8.39}$$

while $\mathbf{r}_{C_iP_i}$ is given by:

$$\mathbf{r}_{C_iP_i} = \left(\frac{\sqrt{4L^2 - (q_{2i-1} + q_{2i-2})^2}}{2}\right){}^0\mathbf{R}_{T_i}(\beta_i)\begin{bmatrix}0\\ \cos(\phi_i)\\ \sin(\phi_i)\end{bmatrix}, \tag{8.40}$$

where ϕ_i (the angle between Y_{Ci} and Y_{Ti} around X_{Ti}) is in general a function of $\mathbf{q}$.

Substituting Eqs. (8.38), (8.39), and (8.40) in Eq. (8.37), we get:

$$\mathbf{r}_{P_i} = {}^0\mathbf{R}_{T_i}(\beta_i)\begin{bmatrix} r_{B_iC_i} \\ -\frac{L_B}{2\sqrt{3}} + r_{C_iP_i}\cos(\phi_i) \\ r_{C_iP_i}\sin(\phi_i) \end{bmatrix} \tag{8.41}$$

where

$$r_{B_iC_i} = \frac{q_{2i-1} - q_{2i-2}}{2}, \tag{8.42}$$

$$r_{C_iP_i} = \frac{\sqrt{4L^2 - \bar{q}_i^2}}{2}, \tag{8.43}$$

with

$$\bar{q}_i = q_{2i-1} + q_{2i-2} \tag{8.44}$$

It is worth noticing here that Eqs. (8.42) and (8.43) can be considered to perform a change of coordinates from (q_{2i-1}, q_{2i-2}) to $(\mathrm{r}_{BiCi},\ \mathrm{r}_{CiPi})$ and, as shown in Eq. (8.41), r_{BiCi} and r_{CiPi} together with ϕ_i can be used to describe the position of point P_i. The same rationale is applied in Carbonari et al. (2011) to transform the 2-1-PUS mechanisms in each side of the Hexapod into the equivalent PRPS structure.

Further, it should be noted that given the vectors $\mathbf{r}_{Pi}$, it is possible to compute:

$$\mathbf{r}_{ij} = \mathbf{r}_{P_j} - \mathbf{r}_{P_i} \tag{8.45}$$

whose magnitude is constant and equals to L_p:

$$\left\|\mathbf{r}_{ij}\right\| = \left\|\mathbf{r}_{P_j} - \mathbf{r}_{P_i}\right\| = L_p, \forall(i,j,k) \in S_3. \tag{8.46}$$

On the other hand, from the geometry of the robot, it is possible to verify that the rotation matrix ${}^0\mathbf{R}_F$ (whose columns 1, 2, and 3 are unit vectors in the direction of the axes X_F, Y_F, and Z_F, respectively, with respect to Σ_0) is given by:

$${}^0\mathbf{R}_F = \begin{bmatrix} \frac{2}{\sqrt{3}L_P^3}\mathbf{r}_{23} \times (\mathbf{r}_{23} \times \mathbf{r}_{21}) & \frac{1}{L_P}\mathbf{r}_{23} & \frac{2}{\sqrt{3}L_P^2}\mathbf{r}_{23} \times \mathbf{r}_{21} \end{bmatrix}; \tag{8.47}$$

or, by using the cross-product identity $\mathbf{a} \times (\mathbf{b} \times \mathbf{c}) = \mathbf{b}(\mathbf{a}^\mathrm{T}\mathbf{c}) - \mathbf{c}(\mathbf{a}^\mathrm{T}\mathbf{b})$, with $\mathbf{a}, \mathbf{b}, \mathbf{c} \in \mathbb{R}^3$, we get:

$${}^0\mathbf{R}_F = [\hat{\boldsymbol{x}}_F(\boldsymbol{\rho}) \quad \hat{\boldsymbol{y}}_F(\boldsymbol{\rho}) \quad \hat{\boldsymbol{z}}_F(\boldsymbol{\rho})] \tag{8.48}$$

where

$$\hat{\boldsymbol{x}}_F(\boldsymbol{\rho}) = \frac{2}{\sqrt{3}L_P}\left(\frac{1}{2}\mathbf{r}_{23} - \mathbf{r}_{21}\right),\ \hat{\boldsymbol{y}}_F(\boldsymbol{\rho}) = \frac{1}{L_P}\mathbf{r}_{23},\ \hat{\boldsymbol{z}}_F(\boldsymbol{\rho}) = \frac{2}{\sqrt{3}L_P^2}\mathbf{r}_{23} \times \mathbf{r}_{21} \tag{8.49}$$

Besides, the position vector $\mathbf{r}_F$ is given by

$$\mathbf{r}_F = \mathbf{r}_{P_1} + \mathbf{r}_{P_1F} = \mathbf{r}_{P_1} + {}^0R_F\, {}^F\mathbf{r}_{P_1F}, \tag{8.50}$$

where ${}^F\mathbf{r}_{P_1F}$ is the vector from P_1 to F with respect to frame Σ_F; but it should be noted that $\mathbf{r}_{P_1F}$ has components only in the direction of the X_F and Z_F axes, i.e., ${}^F\mathbf{r}_{P_1F} = \begin{bmatrix} L_p/\sqrt{3} & 0 & H_{PQ} \end{bmatrix}^{\mathrm{T}}$, so that substituting Eq. (8.48) in Eq. (8.50) we get:

$$\mathbf{r}_F = \mathbf{r}_{P_1} + \frac{1}{3}\mathbf{r}_{23} - \frac{2}{3}\mathbf{r}_{21} + \frac{2H_{PQ}}{\sqrt{3}L_P^2}(\mathbf{r}_{23} \times \mathbf{r}_{21}),$$

or considering Eq. (8.45) and the definition of $\hat{\mathbf{z}}_F$ in Eq. (8.48):

$$\mathbf{r}_F = \frac{1}{3}(\mathbf{r}_{P_1} + \mathbf{r}_{P_2} + \mathbf{r}_{P_3}) + H_{PQ}\hat{\mathbf{z}}_F. \tag{8.51}$$

From Eqs. (8.45), (8.47), and (8.51), we have that in order to get the FPK model of the Hexapod it is sufficient to know the vectors $\mathbf{r}_{P_1}$, $\mathbf{r}_{P_2}$, and $\mathbf{r}_{P_3}$. However, to obtain $\mathbf{r}_{P_i}$ as functions of $\mathbf{q}$, from Eq. (8.41) we have that we need to compute ϕ_1, ϕ_2, and ϕ_3 as functions of $\mathbf{q}$, and that is precisely the most difficult task when obtaining the FPK model in this robot.

In order to do so, we start by noticing that substituting Eq. (8.41) in Eq. (8.46), we can get three expressions of the form

$$a_i\mathrm{c}(\phi_i) + b_i\mathrm{c}(\phi_j) + c_i\mathrm{c}(\phi_i)\mathrm{c}(\phi_j) + d_i s(\phi_i)s(\phi_j) + e_i = 0 \tag{8.52}$$

where $(i,j,k) \in S_3$ $\mathrm{c}(\cdot) = \cos(\cdot)$, $s(\cdot) = \sin(\cdot)$, and the coefficients:

$$a_i = -\sqrt{3}r_{C_iP_i}\left(r_{B_jC_j} + \frac{1}{2}L_B\right) \tag{8.53}$$

$$b_i = \sqrt{3}r_{C_jP_j}\left(r_{B_iC_i} - \frac{1}{2}L_B\right), \tag{8.54}$$

$$c_i = r_{C_iP_i}\, r_{C_jP_j}, \tag{8.55}$$

$$d_i = -2r_{C_iP_i}\, r_{C_jP_j} \tag{8.56}$$

$$e_i = \frac{1}{4}L_B^2 + \frac{1}{2}L_B(r_{B_jC_j} - r_{B_iC_i}) + r_{B_iC_i}\, r_{B_jC_j} + r_{B_iC_i}^2 + r_{C_iP_i}^2 + r_{B_jC_j}^2 + r_{C_jP_j}^2 - L_P^2 \tag{8.57}$$

are in general functions of $\mathbf{q}$, and we have employed the fact that $\beta_j - \beta_i = 120^{\circ}$, for all $(i,j,k) \in S_3$.

It is worth noticing here that if we choose the vector of non-minimal generalized coordinates to be

$$\boldsymbol{\rho} = [q_0 \quad q_1 \quad q_2 \quad q_3 \quad q_4 \quad q_5 \quad \phi_1 \quad \phi_2 \quad \phi_3]^{\mathrm{T}} \in \mathbb{R}^9 \tag{8.58}$$

or, in a compact form

$$\boldsymbol{\rho} = \begin{bmatrix} \mathbf{q}^{\mathrm{T}} & \boldsymbol{\phi}^{\mathrm{T}} \end{bmatrix}^{\mathrm{T}}$$

with $\boldsymbol{\phi} = [\phi_1 \quad \phi_2 \quad \phi_3]^{\mathrm{T}} \in \mathbb{R}^3$, ϕ_1, ϕ_2, and ϕ_3 being the actual inclination angles of the isosceles triangles formed by the legs (see Fig. 8.4b), then the three expressions in Eq. (8.52) can be taken as the holonomic constraints to be fulfilled. That is to say that the constraint vector $\boldsymbol{\gamma}(\boldsymbol{\rho}) = \mathbf{0}$ would be:

$$\begin{aligned}\boldsymbol{\gamma}(\boldsymbol{\rho}) &= \begin{bmatrix} a_1(\mathbf{q})\mathrm{c}(\phi_1) + b_1(\mathbf{q})\mathrm{c}(\phi_2) + c_1(\mathbf{q})\mathrm{c}(\phi_1)\mathrm{c}(\phi_2) + d_1(\mathbf{q})\mathrm{s}(\phi_1)\mathrm{s}(\phi_2) + e_1(\mathbf{q}) \\ a_2(\mathbf{q})\mathrm{c}(\phi_2) + b_2(\mathbf{q})\mathrm{c}(\phi_3) + c_2(\mathbf{q})\mathrm{c}(\phi_2)\mathrm{c}(\phi_3) + d_2(\mathbf{q})\mathrm{s}(\phi_2)\mathrm{s}(\phi_3) + e_2(\mathbf{q}) \\ a_3(\mathbf{q})\mathrm{c}(\phi_3) + b_3(\mathbf{q})\mathrm{c}(\phi_1) + c_3(\mathbf{q})\mathrm{c}(\phi_3)\mathrm{c}(\phi_1) + d_3(\mathbf{q})\mathrm{s}(\phi_3)\mathrm{s}(\phi_1) + e_3(\mathbf{q}) \end{bmatrix} \\ &= \mathbf{0},\end{aligned} \tag{8.59}$$

Moreover, notice that for the given selection of $\boldsymbol{\rho}$, we can choose

$$\mathbf{q} = \boldsymbol{\alpha}(\boldsymbol{\rho}) = [\mathbf{I} \quad \mathbf{O}]\boldsymbol{\rho} = [\rho_1 \quad \rho_2 \quad \rho_3 \quad \rho_4 \quad \rho_5 \quad \rho_6]^{\mathrm{T}},$$

which clearly is a smooth (or continuously differentiable) function; then the application of the implicit function theorem ensures the existence of function $\boldsymbol{\rho} = \boldsymbol{\sigma}(\mathbf{q})$.

In the following, we explain how to solve the system of equations given by Eq. (8.52) using an analytic method. To start, all the sine and cosine trigonometric functions in Eq. (8.52) are replaced according to:

$$\mathrm{c}(\phi_i) = \frac{1 - x_i^2}{1 + x_i^2} \quad \text{and} \quad \mathrm{s}(\phi_i) = \frac{2x_i}{1 + x_i^2}$$

where $x_i = \tan(\phi_i/2)$. Rearranging terms in $\gamma_1(\boldsymbol{\rho})$, $\gamma_2(\boldsymbol{\rho})$, and $\gamma_3(\boldsymbol{\rho})$, we obtain the equations

$$x_2^2(f_1x_1^2 + g_1) + h_1x_1^2 + 4d_1x_1x_2 + k_1 = 0 \tag{8.60}$$

$$x_3^2(f_2x_2^2 + g_2) + h_2x_2^2 + 4d_2x_2x_3 + k_2 = 0 \tag{8.61}$$

$$x_1^2(f_3x_3^2 + g_3) + h_3x_3^2 + 4d_3x_3x_1 + k_3 = 0 \tag{8.62}$$

where $f_i = -a_i - b_i + c_i + e_i$, $g_i = a_i - b_i - c_i + e_i$, $h_i = a_i + b_i - c_i + e_i$, and $k_i = a_i + b_i + c_i + e_i$, with a_i, b_i, c_i, d_i, and e_i are given by Eqs. (8.53)–(8.57).

In order to obtain a polynomial that depends only on the variable x_1, first, we will eliminate x_3 from Eqs. (8.61) and (8.62), obtaining an equation that contains only x_1 and x_2, using the Bezout's method (Nanua et al. 1990). Finally, employing Bezout's method again, from the equation obtained in the previous step and Eq. (8.60), it is possible to get an equation that depends only on x_1.

For the sake of simplicity, we rewrite Eqs. (8.61) and (8.62) as:

$$A_1 x_3^2 + B_1 x_3 + C_1 = 0 \tag{8.63}$$

$$A_2 x_3^2 + B_2 x_3 + C_2 = 0 \tag{8.64}$$

where $A_1 = f_2 x_2^2 + g_2$, $B_1 = 4d_2 x_2$, $C_1 = h_2 x_2^2 + k_2$, $A_2 = f_3 x_1^2 + h_3$, $B_2 = 4d_3 x_1$, and $C_2 = g_3 x_1^2 + k_3$.

Now, following the procedure described in Nanua et al. (1990), from Eqs. (8.63) and (8.64) we can get the equation:

$$\det\begin{bmatrix} \det\begin{bmatrix} A_1 & C_1 \\ A_2 & C_2 \end{bmatrix} & \det\begin{bmatrix} B_2 & B_1 \\ A_2 & A_1 \end{bmatrix} \\ \det\begin{bmatrix} B_1 & C_1 \\ B_2 & C_2 \end{bmatrix} & \det\begin{bmatrix} A_1 & C_1 \\ A_2 & C_2 \end{bmatrix} \end{bmatrix} = Dx_2^4 + Ex_2^3 + Fx_2^2 + Gx_2 + H = 0 \tag{8.65}$$

where $\det[\cdot]$ denotes the determinant of the matrix in its argument, $D = l_1 x_1^4 + n_1 x_1^2 + o_1$, $E = m_1 x_1^3 + p_1 x_1$, $F = l_2 x_1^4\ n_2 x_1^2 + o_2$, $G = m_2 x_1^3 + p_2 x_1$, and $H = l_3 x_1^4 + n_3 x_1^2 + o_3$, with

$$\begin{aligned}
l_1 &= f_2^2 g_3^2 - 2f_2 f_3 g_3 h_2 + f_3^2 h_2^2 \\
m_1 &= -16 d_2 d_3 f_2 g_3 - 16 d_2 d_3 f_3 h_2 \\
n_1 &= 16 d_3^2 f_2 h_2 + 2 f_2^2 g_3 k_3 - 2 f_2 f_3 h_2 k_3 - 2 f_2 g_3 h_2 h_3 + 2 f_3 h_2^2 h_3 \\
o_1 &= f_2^2 k_3^2 - 2 f_2 h_2 h_3 k_3 + h_2^2 h_3^2 \\
p_1 &= -16 d_2 d_3 f_2 k_3 - 16 d_2 d_3 h_2 h_3 \\
l_2 &= 16 d_2^2 f_3 g_3 - 2 f_2 f_3 g_3 k_2 + 2 f_2 g_2 g_3^2 + 2 f_3^2 h_2 k_2 - 2 f_3 g_2 g_3 h_2 m_2 - 16 d_2 d_3 f_3 k_2 - 16 d_2 d_3 g_2 g_3 \\
n_2 &= 16 d_2^2 f_3 k_3 + 16 d_2^2 g_3 h_3 + 16 d_3^2 f_2 k_2 + 16 d_3^2 g_2 h_2 - 2 f_2 f_3 k_2 k_3 + 4 f_2 g_2 g_3 k_3 \\
&\quad - 2 f_2 g_3 h_3 k_2 - 2 f_3 g_2 h_2 k_3 + 4 f_3 h_2 h_3 k_2 - 2 g_2 g_3 h_2 h_3 \\
o_2 &= 16 d_2^2 h_3 k_3 + 2 f_2 g_2 k_3^2 - 2 f_2 h_3 k_2 k_3 \\
p_2 &= -16 d_2 d_3 g_2 k_3 - 16 d_2 d_3 h_3 k_2 \\
l_3 &= f_3^2 k_2^2 - 2 f_3 g_2 g_3 k_2 + g_2^2 g_3^2 \\
n_3 &= 16 d_3^2 g_2 k_2 - 2 f_3 g_2 k_2 k_3 + 2 f_3 h_3 k_2^2 + 2 g_2^2 g_3 k_3 - 2 g_2 g_3 h_3 k_2 - 2 g_2 g_3 h_3 k_2 \\
o_3 &= g_2^2 k_3^2 - 2 g_2 h_3 k_2 k_3 + h_3^2 k_2^2
\end{aligned}$$

After following a similar method to the one given in Nanua et al. (1990), we can eliminate x_2 from Eqs. (8.60) and (8.65) and get the following equation:

$$r_1x_1^{16} + r_2x_1^{14} + r_3x_1^{12} + r_4x_1^{10} + r_5x_1^8 + r_6x_1^6 + r_7x_1^4 + r_8x_1^2 + r_9 = 0 \tag{8.66}$$

where the coefficients can be computed through the expression:

$$\det\begin{bmatrix} EA_3 - DB_3 & FA_3 - DC_3 & GA_3 & HA_3 \\ FA_3 - DC_3 & FB_3 - EC_3 + GA_3 & GB_3 + HA_3 & HB_3 \\ A_3 & B_3 & C_3 & 0 \\ 0 & A_3 & B_3 & C_3 \end{bmatrix} = 0$$

where $A_3 = f_1x_1^2 + g_1$, $B_3 = 4d_1x_1$, and $C_3 = h_1x_1^2 + k_1$.

Given the kinematics parameters of the robot, it is possible to compute the 16 roots of Eq. (8.66) through software. Note that each root of Eq. (8.66) is equal to $\tan(\phi_i/2)$, so it is possible to solve ϕ_1,ϕ_2, and ϕ_3. It is worth noticing that the degree of the polynomial in Eq. (8.66) can be reduced using a new variable $\varsigma = x_1^2$, and solving for ς we finally get $x_1 = \pm\sqrt{\varsigma}$. That means that the 16 solutions of Eq. (8.66) come in pairs (i.e., if $\bar{x}_1$ is a solution of Eq. (8.66), then also $-\bar{x}_1$ is).

It is worth mentioning that only one of these 16 solutions correspond to the angles of the real configuration of the robot (i.e., only one is part of $\mathbf{\rho}^* \in \Omega_\rho^*$). To determine which is the real solution, we can evaluate the FPK model for each of the solutions and compare with the actual configuration of the robot. A complete study of the singularities of the Hexapod, including those configurations for which $\mathbf{J}_\psi(\mathbf{\rho})$ is singular, is to be performed, although intuition seems to indicate that due to the symmetry of the robot and the physical limits of each active joint variable, the whole real workspace of the Hexapod is a subset of Ω_ρ^*.

Finally, notice that $\mathbf{J}_\psi(\mathbf{\rho})$ for the Hexapod robot becomes:

$$\mathbf{J}_\psi(\mathbf{\rho}) = \begin{bmatrix} \frac{\partial\alpha(\mathbf{\rho})}{\partial\mathbf{\rho}} \\ \frac{\partial\gamma(\mathbf{\rho})}{\partial\mathbf{\rho}} \end{bmatrix} = \begin{bmatrix} \frac{\partial\alpha(\mathbf{\rho})(\mathbf{q},\phi)}{\partial\mathbf{q}} & \frac{\partial\alpha(\mathbf{\rho})(\mathbf{q},\phi)}{\partial\phi} \\ \frac{\partial\gamma(\mathbf{\rho})(\mathbf{q},\phi)}{\partial\mathbf{q}} & \frac{\partial\gamma(\mathbf{\rho})(\mathbf{q},\phi)}{\partial\phi} \end{bmatrix} = \begin{bmatrix} \mathbf{I} & \mathbf{O} \\ \frac{\partial\gamma(\mathbf{\rho})(\mathbf{q},\phi)}{\partial\mathbf{q}} & \frac{\partial\gamma(\mathbf{\rho})(\mathbf{q},\phi)}{\partial\phi} \end{bmatrix} \in \mathbb{R}^{9\times 9}$$

so that $\mathbf{J}_\psi(\mathbf{\rho})$ is invertible if and only if $\frac{\partial\gamma(\mathbf{q},\phi)}{\partial\phi}$ is invertible, that is to say

$$\det[\mathbf{J}_\psi(\mathbf{\rho})] \neq 0 \Leftrightarrow \det\left[\frac{\partial\gamma(\mathbf{q},\phi)}{\partial\phi}\right] \neq 0. \tag{8.67}$$

8.4.2 *Forward Velocity Kinematics*

The vectors of linear velocity $\boldsymbol{\upsilon}$ and angular velocity $\boldsymbol{\omega}$ of the Hexapod's platform can be computed using Eqs. (8.18) and (8.20).

Thus, considering Eqs. (8.48) and (8.51) we have

$$\boldsymbol{\upsilon} = \dot{\mathbf{r}}_F = \left[\frac{1}{3}(\mathbf{J}_1 + \mathbf{J}_2 + \mathbf{J}_3) + H_{PQ}\mathbf{J}_z\right]\mathbf{A}(\mathbf{q})\dot{\mathbf{q}}$$

$$\boldsymbol{\omega} = \frac{1}{2}\left[\mathbf{S}(\hat{\boldsymbol{x}}_F)\mathbf{J}_x + \mathbf{S}(\hat{\boldsymbol{y}}_F)\mathbf{J}_y + \mathbf{S}(\hat{\boldsymbol{z}}_F)\mathbf{J}_z\right]\mathbf{A}(\mathbf{q})\dot{\mathbf{q}}$$

where

$$\mathbf{J}_i = \frac{\partial \mathbf{r}_{P_i}}{\partial \boldsymbol{\rho}} \qquad (i = 1, 2, 3) \tag{8.68}$$

and

$$\mathbf{J}_x = \frac{\partial \hat{\boldsymbol{x}}_F}{\partial \boldsymbol{\rho}}, \quad \mathbf{J}_y = \frac{\partial \hat{\boldsymbol{y}}_F}{\partial \boldsymbol{\rho}}, \quad \mathbf{J}_z = \frac{\partial \hat{\boldsymbol{z}}_F}{\partial \boldsymbol{\rho}} \tag{8.69}$$

For the computation of matrix $\mathbf{A}(\mathbf{q})$, let us consider the following analysis, which is a contribution of this work.

The constraint vector for the Hexapod robot is given by Eq. (8.59), and taking its time derivative, we get

$$\frac{\partial \boldsymbol{\gamma}(\mathbf{q}, \boldsymbol{\phi})}{\partial \mathbf{q}}\dot{\mathbf{q}} + \frac{\partial \boldsymbol{\gamma}(\mathbf{q}, \boldsymbol{\phi})}{\partial \boldsymbol{\phi}}\dot{\boldsymbol{\phi}} = \mathbf{0}; \tag{8.70}$$

assuming that $\boldsymbol{\phi}$ is chosen to satisfy $\boldsymbol{\rho} = [\mathbf{q} \quad \boldsymbol{\phi}]^{\mathrm{T}} \subset \Omega_{\rho}^{*}$, then $\mathbf{J}_{\psi}(\boldsymbol{\rho}) \neq \mathbf{0}$ and we can use Eq. (8.67) to solve Eq. (8.70) for $\dot{\boldsymbol{\phi}}$ to get

$$\dot{\boldsymbol{\phi}} = \frac{\partial \boldsymbol{\phi}(\mathbf{q})}{\partial \mathbf{q}}\dot{\mathbf{q}} = -\left(\frac{\partial \boldsymbol{\gamma}(\mathbf{q}, \boldsymbol{\phi})}{\partial \boldsymbol{\phi}}\right)^{-1}\frac{\partial \boldsymbol{\gamma}(\mathbf{q}, \boldsymbol{\phi})}{\partial \mathbf{q}}\dot{\mathbf{q}} \tag{8.71}$$

or

$$\frac{\partial \boldsymbol{\phi}(\mathbf{q})}{\partial \mathbf{q}} = -\left(\frac{\partial \boldsymbol{\gamma}(\mathbf{q}, \boldsymbol{\phi})}{\partial \boldsymbol{\phi}}\right)^{-1}\frac{\partial \boldsymbol{\gamma}(\mathbf{q}, \boldsymbol{\phi})}{\partial \mathbf{q}}.$$

Besides, as function $\boldsymbol{\sigma}(q)$ is given by

$$\boldsymbol{\sigma}(\mathbf{q}) = \begin{bmatrix} \mathbf{q} \\ \boldsymbol{\phi}(\mathbf{q}) \end{bmatrix} \tag{8.72}$$

then from Eq. (8.72) the matrix $\mathbf{A}(\mathbf{q})$ is

$$\mathbf{A}(\mathbf{q}) = \frac{\partial \boldsymbol{\sigma}(\mathbf{q})}{\partial \mathbf{q}} = \begin{bmatrix} \mathbf{I} \\ \frac{\partial \boldsymbol{\phi}(\mathbf{q})}{\partial \mathbf{q}} \end{bmatrix} = \begin{bmatrix} \mathbf{I} \\ -\left(\frac{\partial \boldsymbol{\gamma}(\mathbf{q}, \boldsymbol{\phi})}{\partial \boldsymbol{\phi}}\right)^{-1}\frac{\partial \boldsymbol{\gamma}(\mathbf{q}, \boldsymbol{\phi})}{\partial \mathbf{q}} \end{bmatrix} \in \mathbb{R}^{9\times 6}. \tag{8.73}$$

It is worth noticing that Eq. (8.73) allows to compute matrix $\mathbf{A}(\mathbf{q})$ without explicitly knowing function $\boldsymbol{\sigma}(q)$. The key for this useful result was the selection of the non-minimal coordinates given by Eq. (8.58), and this suggests that a similar procedure could be applied to other types of parallel robots.

Now, for the computation of the analytical Jacobian of the Hexapod robot (which will be required for the implementation of the two-loop controller described in Sect. 8.7.1), let us consider the fact that the pose of the platform can be written in terms of $\mathbf{q}$ and $\boldsymbol{\phi}$, i.e.,

$$\boldsymbol{\xi} = \boldsymbol{\chi}(\boldsymbol{\rho}) = \boldsymbol{\chi}(\mathbf{q}, \boldsymbol{\phi}),$$

and taking the time derivative:

$$\dot{\boldsymbol{\xi}} = \frac{\partial \boldsymbol{\chi}(\mathbf{q}, \boldsymbol{\phi})}{\partial \mathbf{q}} \dot{\mathbf{q}} + \frac{\partial \boldsymbol{\chi}(\mathbf{q}, \boldsymbol{\phi})}{\partial \boldsymbol{\phi}} \dot{\boldsymbol{\phi}} \tag{8.74}$$

Substituting $\dot{\boldsymbol{\phi}}$ in Eq. (8.74), we get

$$\dot{\boldsymbol{\xi}} = \frac{\partial \boldsymbol{\chi}(\mathbf{q}, \boldsymbol{\phi})}{\partial \mathbf{q}} \dot{\mathbf{q}} - \frac{\partial \boldsymbol{\chi}(\mathbf{q}, \boldsymbol{\phi})}{\partial \boldsymbol{\phi}} \left(\frac{\partial \boldsymbol{\gamma}(\mathbf{q}, \boldsymbol{\phi})}{\partial \boldsymbol{\phi}} \right)^{-1} \frac{\partial \boldsymbol{\gamma}(\mathbf{q}, \boldsymbol{\phi})}{\partial \mathbf{q}} \dot{\mathbf{q}}$$

and the analytical Jacobian becomes

$$\mathbf{J}_A(\mathbf{q}) = \frac{\partial \boldsymbol{\chi}(\mathbf{q}, \boldsymbol{\phi})}{\partial \mathbf{q}} - \frac{\partial \boldsymbol{\chi}(\mathbf{q}, \boldsymbol{\phi})}{\partial \boldsymbol{\phi}} \left(\frac{\partial \boldsymbol{\gamma}(\mathbf{q}, \boldsymbol{\phi})}{\partial \boldsymbol{\phi}} \right)^{-1} \frac{\partial \boldsymbol{\gamma}(\mathbf{q}, \boldsymbol{\phi})}{\partial \mathbf{q}} \tag{8.75}$$

8.5 Dynamics

In this section, we show the application of the procedure described in Sect. 8.2.3 for computing the inverse dynamics model of the Hexapod parallel robot.

The analysis considers that the Hexapod robot consists of a total of $b = 25$ mobile rigid bodies, distributed as indicated below:

- The platform: 1.
- The legs: 6.
- The links between a P joint and a U joint: 6.
- The links between the two ends of a U joint: 6.
- The two links between the two ends of an S joint: $2 \times 3 = 6$.

8.5.1 *Computation of the Linear and Angular Velocities for Each Body*

The following subsections explain how to compute $\boldsymbol{\upsilon}_l$ and ${}^l\boldsymbol{\omega}_l$ for the different rigid bodies in the Hexapod. In each subsection, we first show how to describe the pose of the rigid body l, via the position vector $\mathbf{p}_l \in \mathbb{R}^3$ and the rotation matrix ${}^0\mathbf{R}_l \in SO(3)$; after that, we compute $\boldsymbol{\upsilon}_l$ and ${}^l\boldsymbol{\omega}_l$ using Eqs. (8.25) and (8.26). But in order to simplify the subsequent analysis, we employ the Jacobians $\mathbf{J}_{G_l}$ and $\mathbf{K}_{G_l}$ satisfying

$$\begin{bmatrix} \boldsymbol{\upsilon}_l \\ {}^l\boldsymbol{\omega}_l \end{bmatrix} = \begin{bmatrix} \mathbf{J}_{G_l} \\ \mathbf{K}_{G_l} \end{bmatrix} \dot{\boldsymbol{\rho}} \tag{8.76}$$

so that

$$\mathbf{J}_{G_l} = \frac{\partial \mathbf{p}_l(\boldsymbol{\rho})}{\partial \boldsymbol{\rho}} \tag{8.77}$$

$$\mathbf{K}_{G_l} = \frac{1}{2}{}^0R_l(\boldsymbol{\rho})^{\mathrm{T}}\left[\mathbf{S}(\hat{x}_l(\boldsymbol{\rho}))\frac{\partial \hat{x}_l(\boldsymbol{\rho})}{\partial \boldsymbol{\rho}} + \mathbf{S}(\hat{y}_l(\boldsymbol{\rho}))\frac{\partial \hat{y}_l(\boldsymbol{\rho})}{\partial \boldsymbol{\rho}} + \mathbf{S}(\hat{z}_l(\boldsymbol{\rho}))\frac{\partial \hat{z}_l(\boldsymbol{\rho})}{\partial \boldsymbol{\rho}}\right] \tag{8.78}$$

8.5.1.1 Platform

Let us consider that $l = 1$ in the case of the platform. From Eq. (8.51), the position vector of its com is

$$\mathbf{p}_1 = \frac{1}{3}(\mathbf{r}_{P_1} + \mathbf{r}_{P_2} + \mathbf{r}_{P_3}) + H^*\hat{\mathbf{z}}_F$$

where H^* is the distance from the center of the triangle $P_1P_2P_3$ to the com of the platform. As we can choose frame Σ_F to compute the angular velocity of the platform, we can write

$${}^0\mathbf{R}_1 = {}^0\mathbf{R}_F = [\,\hat{x}_F(\boldsymbol{\rho}) \quad \hat{y}_F(\boldsymbol{\rho}) \quad \hat{z}_F(\boldsymbol{\rho})\,],$$

considering Eqs. (8.48) and (8.49).

Thus, the linear and angular velocity vectors can be computed via Eq. (8.76), with $l = 1$, using:

$$\mathbf{J}_{G_1} = \frac{1}{3}(\mathbf{J}_1 + \mathbf{J}_2 + \mathbf{J}_3) + H^*\mathbf{J}_z \tag{8.79}$$

and

$$\mathbf{K}_{G_1} = \frac{1}{2}{}^0\mathbf{R}_F^{\mathrm{T}}\left[\mathbf{S}(\hat{\mathbf{x}}_F)\mathbf{J}_x + \mathbf{S}(\hat{\mathbf{y}}_F)\mathbf{J}_y + \mathbf{S}(\hat{\mathbf{z}}_F)\mathbf{J}_z\right] \tag{8.80}$$

where the auxiliary Jacobians $\mathbf{J}_i$, $\mathbf{J}_x$, $\mathbf{J}_y$, and $\mathbf{J}_z$ are defined in Eqs. (8.68) and (8.69).

8.5.1.2 Legs

There are six legs in the Hexapod robot, two in each side of the base triangle. Figure 8.6 shows the side i ($i = 1,2,3$) with the legs $2i-2$ and $2i-1$ (which are coupled to the active joints with the same number). For the analysis, let us consider that the leg $2i-2$ corresponds to the body $l = 2i$ and the leg $2i-1$ to the body $l = 2i+1$. Thus, in the count of $b = 25$ rigid bodies, the legs are those with the numbers $l = 2,3,\ldots,7$.

For the sake of simplicity, let us assume that each of the legs is symmetric, so that its com, labeled either as G_{2i-2} or G_{2i-1}, is at the midpoint of the corresponding segment, either P_iD_{2i-2} or P_iD_{2i-2}.

As it can be seen in Fig. 8.6, E_{2i-2} and E_{2i-1} are the midpoints of the segments C_iD_{2i-2} and C_iD_{2i-1}, respectively, therefore $\mathbf{r}_{C_iE_{2i-1}} = -\mathbf{r}_{C_iE_{2i-2}} = (\bar{q}_i/4){}^0\mathbf{R}_{T_i}(\beta_i)[1\ \ 0\ \ 0]^{\mathrm{T}}$ and $\mathbf{r}_{E_{2i-1}G_{2i-1}} = \mathbf{r}_{E_{2i-2}G_{2i-2}} = (1/2)\mathbf{r}_{C_iP_i}$; therefore, the position vectors of G_{2i-2} and G_{2i-1}, $\mathbf{r}_{G_{2i-2}}$ and $\mathbf{r}_{G_{2i-1}}$, are given by:

$$\begin{aligned}
\mathbf{p}_{2i} &= \mathbf{r}_{G_{2i-2}} = \mathbf{r}_{C_i} + \mathbf{r}_{C_iE_{2i-2}} + \mathbf{r}_{E_{2i-2}G_{2i-2}} \\
&= {}^0\mathbf{R}_{T_i}(\beta_i)\begin{bmatrix} r_{B_iC_i} - \frac{\bar{q}_i}{4} \\ -\frac{L_B}{2\sqrt{3}} + \frac{1}{2}r_{C_iP_i}\cos(\phi_i) \\ \frac{1}{2}r_{C_iP_i}\sin(\phi_i) \end{bmatrix} = \frac{1}{2}\mathbf{r}_{P_i} + {}^0\mathbf{R}_{T_i}(\beta_i)\begin{bmatrix} -\frac{q_{2i-2}}{2} \\ -\frac{L_B}{4\sqrt{3}} \\ 0 \end{bmatrix} \\
\mathbf{p}_{2i+1} &= \mathbf{r}_{G_{2i-1}} = \mathbf{r}_{C_i} + \mathbf{r}_{C_iE_{2i-1}} + \mathbf{r}_{E_{2i-1}G_{2i-1}} \\
&= {}^0\mathbf{R}_{T_i}(\beta_i)\begin{bmatrix} r_{B_iC_i} + \frac{\bar{q}_i}{4} \\ -\frac{L_B}{2\sqrt{3}} + \frac{1}{2}r_{C_iP_i}\cos(\phi_i) \\ \frac{1}{2}r_{C_iP_i}\sin(\phi_i) \end{bmatrix} = \frac{1}{2}\mathbf{r}_{P_i} + {}^0\mathbf{R}_{T_i}(\beta_i)\begin{bmatrix} \frac{q_{2i-1}}{2} \\ -\frac{L_B}{4\sqrt{3}} \\ 0 \end{bmatrix}
\end{aligned}$$

Now, by considering frames Σ_{C_i}, $\Sigma_{D_{2i-2}}$, and $\Sigma_{D_{2i-1}}$, defined at the end of Sect. 8.3 and shown in Fig. 8.5, we have that the orientation of the legs $2i-1$ and $2i-2$, with respect to the base frame, is respectively given by

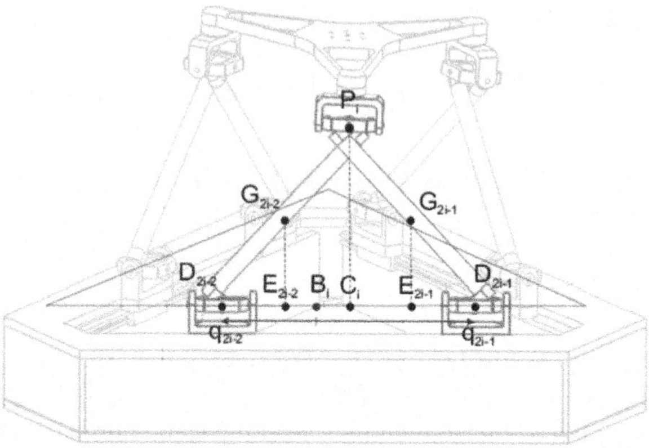

Fig. 8.6 Localization of the centers of mass of the Hexapod's legs $2i-2$ and $2i-1$

$$
\begin{aligned}
{}^0\mathbf{R}_{2i} &= {}^0\mathbf{R}_{D_{2i-1}} = {}^0\mathbf{R}_{T_i}(\beta_i)\mathbf{R}_x(\phi_i)\mathbf{R}_z(\alpha_i),\\
{}^0\mathbf{R}_{2i+1} &= {}^0\mathbf{R}_{D_{2i-2}} = {}^0\mathbf{R}_{T_i}(\beta_i)\mathbf{R}_x(\phi_i)\mathbf{R}_z(-\alpha_i),
\end{aligned}
$$

where we employ some elementary rotation matrices, defined as

$$
\begin{aligned}
\mathbf{R}_x(\cdot) &= \begin{bmatrix} 1 & 0 & 0 \\ 0 & \cos(\cdot) & -\sin(\cdot) \\ 0 & \sin(\cdot) & \cos(\cdot) \end{bmatrix}, \ \mathbf{R}_y(\cdot) = \begin{bmatrix} \cos(\cdot) & 0 & \sin(\cdot) \\ 0 & 1 & 0 \\ -\sin(\cdot) & 0 & \cos(\cdot) \end{bmatrix},\\
\mathbf{R}_z(\cdot) &= \begin{bmatrix} \cos(\cdot) & -\sin(\cdot) & 0 \\ \sin(\cdot) & \cos(\cdot) & 0 \\ 0 & 0 & 1 \end{bmatrix},
\end{aligned}
\tag{8.81}
$$

and we apply $\sin(\alpha_i) = \frac{\|\mathbf{r}_{C_iD_{2i-1}}\|}{L} = \frac{\bar{q}_i}{2L}$, $\cos(\alpha_i) = \frac{r_{C_iP_i}}{L}$, with $r_{C_iP_i}$ and $\bar{q}_i$ being defined in Eqs. (8.43) and (8.44), respectively, and L the length of the link.

Now, applying Eqs. (8.77) and (8.78) for $l = 2i$ and $l = 2i+1$, we can verify that

$$
\begin{aligned}
\mathbf{J}_{G_{2i}} &= \frac{1}{2}\mathbf{J}_i + {}^0\mathbf{R}_{T_i}(\beta_i)\frac{\partial}{\partial \boldsymbol{\rho}}\left[-\tfrac{q_{2i-2}}{2} \ \ 0 \ \ 0\right]^{\mathrm{T}},\\
\mathbf{J}_{G_{2i+1}} &= \frac{1}{2}\mathbf{J}_i + {}^0\mathbf{R}_{T_i}(\beta_i)\frac{\partial}{\partial \boldsymbol{\rho}}\left[\tfrac{q_{2i-1}}{2} \ \ 0 \ \ 0\right]^{\mathrm{T}},\\
\mathbf{K}_{G_{2i}} &= \frac{r_{CiPi}}{L}\frac{\partial}{\partial \boldsymbol{\rho}}\begin{bmatrix} \phi_i \\ 0 \\ 0 \end{bmatrix} + \frac{\bar{q}_i}{2L}\frac{\partial}{\partial \boldsymbol{\rho}}\begin{bmatrix} 0 \\ \phi_i \\ 0 \end{bmatrix} - \frac{1}{2r_{CiPi}}\frac{\partial}{\partial \boldsymbol{\rho}}\begin{bmatrix} 0 \\ 0 \\ \bar{q}_i \end{bmatrix},
\end{aligned}
\tag{8.82}
$$

and

$$\mathbf{K}_{G_{2i+1}} = \frac{r_{CiPi}}{L}\frac{\partial}{\partial \boldsymbol{\rho}}\begin{bmatrix}\phi_i\\0\\0\end{bmatrix} - \frac{\bar{q}_i}{2L}\frac{\partial}{\partial \boldsymbol{\rho}}\begin{bmatrix}0\\\phi_i\\0\end{bmatrix} + \frac{1}{2r_{CiPi}}\frac{\partial}{\partial \boldsymbol{\rho}}\begin{bmatrix}0\\0\\\bar{q}_i\end{bmatrix}, \tag{8.83}$$

where $r_{C_iP_i}$ and $\bar{q}_i$ are defined in Eqs. (8.43) and (8.44), respectively, so that

$$\begin{aligned}
\boldsymbol{\upsilon}_{2i} &= \frac{1}{2}\dot{\mathbf{r}}_{P_i} + {}^{0}\mathbf{R}_{T_i}(\beta_i)\begin{bmatrix}-\frac{\dot{q}_{2i-2}}{2} & 0 & 0\end{bmatrix}^{\mathrm{T}},\\
\boldsymbol{\upsilon}_{2i+1} &= \frac{1}{2}\dot{\mathbf{r}}_{P_i} + {}^{0}\mathbf{R}_{T_i}(\beta_i)\begin{bmatrix}\frac{\dot{q}_{2i-1}}{2} & 0 & 0\end{bmatrix}^{\mathrm{T}},\\
{}^{2i}\boldsymbol{\omega}_{2i} &= \begin{bmatrix}\frac{r_{CiPi}}{L}\dot{\phi}_i & \frac{\bar{q}_i}{2L}\dot{\phi}_i & -\frac{1}{2r_{CiPi}}\dot{\bar{q}}_i\end{bmatrix}^{\mathrm{T}}, \text{ and}\\
{}^{2i+1}\boldsymbol{\omega}_{2i+1} &= \begin{bmatrix}\frac{r_{CiPi}}{L}\dot{\phi}_i & -\frac{\bar{q}_i}{2L}\dot{\phi}_i & \frac{1}{2r_{CiPi}}\dot{\bar{q}}_i\end{bmatrix}^{\mathrm{T}}.
\end{aligned}$$

8.5.1.3 P and U Joints

Notice that the prismatic (P) and the universal (U) joints can be seen as a 2-DOF compound joint between the robot base and each of its legs. Each PU joint contains two intermediate rigid bodies (or parts) which can be appreciated in Fig. 8.7a:

- Part 1 of PU joint: This is the base of the U joint, which moves along the side of the base triangle; its configuration depends only on the active joint coordinate corresponding to the P joint.
- Part 2 of PU joint: This is the coupling link between the part 1 of a PU joint and the corresponding leg; its configuration depends on the active joint coordinate of the P joint and the corresponding ϕ_i angle, which gives a rotation around the same axis.

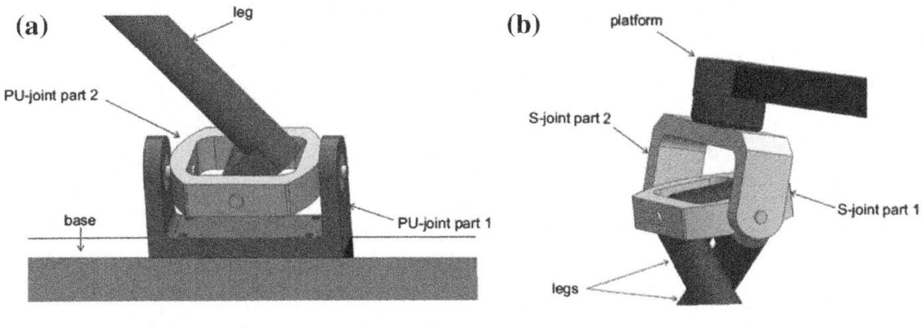

Fig. 8.7 Rigid bodies in the Hexapod's joints: **a** PU joint and **b** S joint

Part 1 of PU joints

Let us consider that the part 1 of the PU joint $2i-2$ corresponds to the body $l=2i+6$ and the same part of the PU joint $2i-1$ to the body $l=2i+7$. Thus, the first parts of the PU joints of the Hexapod are those rigid bodies with the numbers $l=8,9,\ldots,13$.

Notice that the position vector of the com for the first part of a PU joint is either

$$\mathbf{p}_{2i+6} = {}^{0}\mathbf{R}_{T_i}(\beta_i)[-q_{2i-2} \quad d \quad -l_{c_1}]^{\mathrm{T}} \text{ or}$$
$$\mathbf{p}_{2i+7} = {}^{0}\mathbf{R}_{T_i}(\beta_i)[q_{2i-1} \quad d \quad -l_{c_1}]^{\mathrm{T}}$$

where $d=-L_B/2\sqrt{3}$ and l_{c_1} is the distance from either point D_{2i-2} or D_{2i-1} (with $i=1,2,3$) to the com of the corresponding PU joint part 1. Thus

$$\begin{aligned}\mathbf{J}_{G_{2i+6}} &= {}^{0}\mathbf{R}_{T_i}(\beta_i)\frac{\partial}{\partial \boldsymbol{\rho}}[-q_{2i-2} \quad d \quad -l_{c_1}]^{\mathrm{T}} \quad \text{and}\\ \mathbf{J}_{G_{2i+7}} &= {}^{0}\mathbf{R}_{T_i}(\beta_i)\frac{\partial}{\partial \boldsymbol{\rho}}[-q_{2i-1} \quad d \quad -l_{c_1}]^{\mathrm{T}}\end{aligned} \tag{8.84}$$

And as these parts do not rotate, then

$$\mathbf{K}_{G_{2i+6}} = \mathbf{K}_{G_{2i+7}} = \mathbf{O} \in \mathbb{R}^{3\times 9}. \tag{8.85}$$

Part 2 of PU joints

For the count of rigid bodies, let us consider that the part 2 of the PU joint $2i-2$ corresponds to the body $l=2i+12$ and the same part of the PU joint $2i-1$ to the body $l=2i+13$. Thus, the second parts of the PU joints are those rigid bodies with the numbers $l=14,15,\ldots,19$.

The position vector of the second part of a PU joint is either

$$\mathbf{p}_{2i+12} = {}^{0}\mathbf{R}_{T_i}(\beta_i)[-q_{2i-2} \quad d \quad 0]^{\mathrm{T}} \text{ or } \mathbf{p}_{2i+13} = {}^{0}\mathbf{R}_{T_i}(\beta_i)[-q_{2i-1} \quad d \quad 0]^{\mathrm{T}}$$

where $d=-L_B/(2\sqrt{3})$. Thus

$$\begin{aligned}\mathbf{J}_{G_{2i+12}} &= {}^{0}\mathbf{R}_{T_i}(\beta_i)\frac{\partial}{\partial \boldsymbol{\rho}}[-q_{2i-2} \quad d \quad 0]^{\mathrm{T}} \quad \text{and}\\ \mathbf{J}_{G_{2i+13}} &= {}^{0}\mathbf{R}_{T_i}(\beta_i)\frac{\partial}{\partial \boldsymbol{\rho}}[q_{2i-1} \quad d \quad 0]^{\mathrm{T}}\end{aligned} \tag{8.86}$$

The orientation of this part for either $l=2i+12$ or $l=2i+13$ is given by

$$ {}^{0}\mathbf{R}_{2i+12} = {}^{0}\mathbf{R}_{2i+13} = {}^{0}\mathbf{R}_{T_i}(\beta_i)\mathbf{R}(\phi_i) $$

thus

$$ \mathbf{K}_{G_{2i+12}} = \mathbf{K}_{G_{2i+13}} = \frac{\partial}{\partial \boldsymbol{\rho}}[\phi_i \quad 0 \quad 0]^{\mathrm{T}}. \tag{8.87} $$

8.5.1.4 S Joints

There are only three spherical joints in the Hexapod robot. The spherical joint i is placed in the side i of the base triangle, and it joins the vertex Q_i of the platform triangle with the legs $2i-2$ and $2i-1$. Figure 8.8 shows this side of the robot in foreground.

Each spherical joint is assumed to be formed by three independent rotational joints; let θ_{1i}, θ_{2i}, and θ_{3i} be the joint coordinates of the S joint in the i side of the robot. Notice in Fig. 8.8 that these three angles can be considered as the Euler angles of the XYZ convention, which allow to express the relative orientation of frame Σ_F with respect to a frame denoted in the same figure as Σ_{N_i}. In other words, θ_{1i}, θ_{2i}, and θ_{3i} are the angles that Σ_{N_i} has to be rotated in order to have the same orientation as Σ_F.

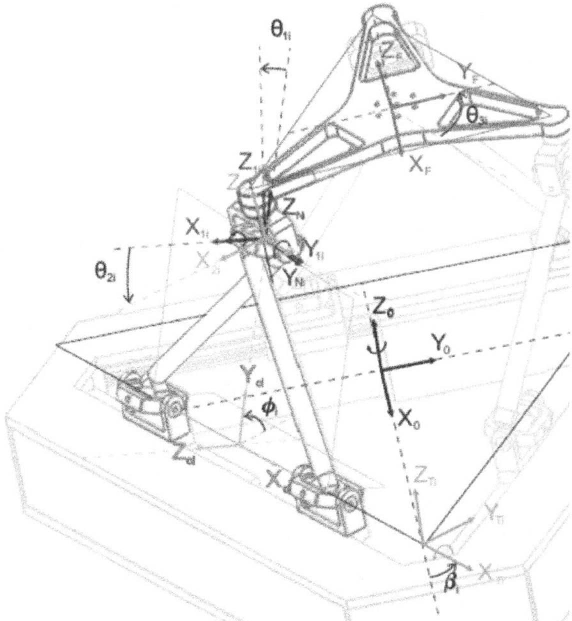

Fig. 8.8 Coordinate frames for modeling the spherical joints

Now, with respect to the pose of frame Σ_{N_i}, it is worth noticing that the origin of this frame is at the point P_i, and its orientation is given by the following composition of rotation matrices:

$$^0\mathbf{R}_{N_i} = {}^0\mathbf{R}_{T_i}(\beta_i)\mathbf{R}_x(\phi_i){}^{C_i}\mathbf{R}_{N_i}$$

where

$$^{C_i}\mathbf{R}_{N_i} = \begin{bmatrix} 0 & 1 & 0 \\ 0 & 0 & 1 \\ 1 & 0 & 0 \end{bmatrix} \in \mathrm{SO}(3)$$

is the rotation matrix giving the relative orientation of frame Σ_{N_i} with respect to Σ_{C_i}

Analyzing the coordinate frames in Fig. 8.8, it should be clear that the following expression is valid:

$$^0\mathbf{R}_F = {}^0\mathbf{R}_{T_i}(\beta_i)\mathbf{R}_x(\phi_i){}^{C_i}\mathbf{R}_{N_i}\mathbf{R}_x(\theta_{1i})\mathbf{R}_y(\theta_{2i})\mathbf{R}_z(\theta_{3i}) \tag{8.88}$$

where the elementary rotation matrices $\mathbf{R}_x(\cdot)$, $\mathbf{R}_y(\cdot)$, and $\mathbf{R}_z(\cdot)$ were defined in Eq. (8.81).

Applying the property $\mathbf{R}^{-1} = \mathbf{R}^{\mathrm{T}}$ of any rotation matrix $\mathbf{R} \in \mathrm{SO}(3)$, we can rewrite Eq. (8.88) as

$$\mathbf{R}_x(\theta_{1i})\mathbf{R}_y(\theta_{2i})\mathbf{R}_z(\theta_{3i}) = {}^{C_i}\mathbf{R}_{N_i}^{\mathrm{T}}\mathbf{R}_x(\phi_i)^{\mathrm{T}}\,{}^0\mathbf{R}_{T_i}(\beta_i)^{\mathrm{T}}\,{}^0\mathbf{R}_F = {}^{N_i}\mathbf{R}_F(\boldsymbol{\rho}) \tag{8.89}$$

The (Euler) angles θ_{1i}, θ_{2i}, and θ_{3i} can now be computed using the standard formulas for the XYZ convention (see Craig 2004), that is:

$$\theta_{1i} = \operatorname{atan2}\left(-R_{2,3}(\boldsymbol{\rho}), R_{3,3}(\boldsymbol{\rho})\right) \tag{8.90}$$

$$\theta_{2i} = \operatorname{atan2}\left(R_{1,3}(\boldsymbol{\rho}), \sqrt{R_{2,3}(\boldsymbol{\rho})^2 + R_{3,3}(\boldsymbol{\rho})^2}\right) \tag{8.91}$$

$$\theta_{3i} = \operatorname{atan2}\left(-R_{1,2}(\boldsymbol{\rho}), R_{1,1}(\boldsymbol{\rho})\right)$$

where $R_{u,v}(\boldsymbol{\rho})$ is the element (u, v) of the matrix $^{N_i}\mathbf{R}_F(\boldsymbol{\rho})$ defined in Eq. (8.89).

Moreover, there are two rigid bodies (parts) in every spherical joint (see Fig. 8.7b):

- Part 1 of S joint: This is the link between the legs and the second part of the S joint; its configuration can be computed from the corresponding leg's configuration and the angle θ_{1i}.
- Part 2 of S joint: This is the link between the part 1 of the same S joint and the platform; its configuration can be computed from the configuration of the part 1 and the angle θ_{2i}.

Part 1 of S joints

The first parts of the S joints correspond to the rigid bodies with $l = 20, 21, 22$, or, in terms of i, $l = 19 + i$. For the sake of simplicity, let us consider that the com of the S joint part 1 is at point P_i, that is to say that

$$\mathbf{p}_{19+i} = \mathbf{r}_{P_i}$$

so that

$$\mathbf{J}_{G_{19+i}} = \mathbf{J}_i, \tag{8.92}$$

with $\mathbf{J}_i = \frac{\partial \mathbf{r}_{Pi}}{\partial \boldsymbol{\rho}}$.

The frame attached to this body is labeled as Σ_{1i} in Fig. 8.8, and its orientation is given by

$${}^{0}\mathbf{R}_{19+i}(\boldsymbol{\rho}) = {}^{0}\mathbf{R}_{T_i}(\beta_i)\mathbf{R}_x(\phi_i){}^{C_i}\mathbf{R}_{N_i}\mathbf{R}_x(\theta_{1i});$$

it can be shown that

$$\mathbf{K}_{G_{19+i}} = \mathbf{R}_x(\theta_{1i})^{\mathrm{T}}\,{}^{C_i}\mathbf{R}_{N_i}^{\mathrm{T}}\frac{\partial}{\partial \boldsymbol{\rho}}[\phi_i \quad 0 \quad 0]^{\mathrm{T}} + \frac{\partial}{\partial \boldsymbol{\rho}}[\theta_{1i} \quad 0 \quad 0]^{\mathrm{T}} \tag{8.93}$$

where the term $\frac{\partial \theta_{1i}}{\partial \boldsymbol{\rho}}$ can be computed by taking the partial derivative of Eq. (8.90) and results in

$$\frac{\partial \theta_{1i}}{\partial \boldsymbol{\rho}} = \frac{R_{2,3}(\boldsymbol{\rho})\frac{\partial R_{3,3}(\boldsymbol{\rho})}{\partial \boldsymbol{\rho}} - R_{3,3}(\boldsymbol{\rho})\frac{\partial R_{2,3}(\boldsymbol{\rho})}{\partial \boldsymbol{\rho}}}{R_{2,3}(\boldsymbol{\rho})^2 + R_{3,3}(\boldsymbol{\rho})^2}.$$

The angular velocity thus becomes:

$${}^{19+i}\boldsymbol{\omega}_{19+i} = \mathbf{R}_x(\theta_{1i})^{\mathrm{T}}\,{}^{C_i}\mathbf{R}_{N_i}^{\mathrm{T}}\begin{bmatrix}\dot{\phi}_i \\ 0 \\ 0\end{bmatrix} + \begin{bmatrix}\dot{\theta}_{i1} \\ 0 \\ 0\end{bmatrix}$$

Part 2 of S joints

The second parts of the S joints correspond to the rigid bodies with $l = 23, 24, 25$, or, in terms of i, $l = 22 + i$. In this case, the com of the S joint part 2 is not at point P_i but at a distance l_c in the direction of $\hat{\mathbf{z}}_F$, that is to say that

$$\mathbf{p}_{22+i} = \mathbf{r}_{P_i} + l_c\hat{\mathbf{z}}_F$$

so that

$$\mathbf{J}_{G_{22+i}} = \mathbf{J}_i + l_c \mathbf{J}_z; \tag{8.94}$$

where $\mathbf{J}_i = \frac{\partial \mathbf{r}_{Pi}}{\partial \boldsymbol{\rho}}$ and $\mathbf{J}_z = \frac{\partial \hat{\mathbf{z}}_F}{\partial \boldsymbol{\rho}}$.

The frame attached to this body is labeled as Σ_{2i} in Fig. 8.8, and its orientation is given by

$${}^0\mathbf{R}_{22+i}(\boldsymbol{\rho}) = {}^0\mathbf{R}_{T_i}(\beta_i)\mathbf{R}_x(\phi_i)^{C_i}\mathbf{R}_{N_i}\mathbf{R}_x(\theta_{1i})\mathbf{R}_y(\theta_{2i});$$

it can be shown that

$$\mathbf{K}_{G_{22+i}} = \mathbf{R}_y(\theta_{2i})^{\mathrm{T}} \mathbf{K}_{G_{19+i}} + \frac{\partial}{\partial \boldsymbol{\rho}} [0 \quad \theta_{2i} \quad 0]^{\mathrm{T}} \tag{8.95}$$

where the term $\frac{\partial \theta_{2i}}{\partial \boldsymbol{\rho}}$ can be computed by taking the partial derivative of Eq. (8.91) and results in

$$\frac{\partial \theta_{2i}}{\partial \boldsymbol{\rho}} = \frac{1}{\sqrt{R_{2,3}(\boldsymbol{\rho})^2 + R_{3,3}(\boldsymbol{\rho})^2}} \left(\left(R_{2,3}(\boldsymbol{\rho})^2 + R_{3,3}(\boldsymbol{\rho})^2 \right) \frac{\partial R_{1,3}(\boldsymbol{\rho})}{\partial \boldsymbol{\rho}} - R_{1,3}(\boldsymbol{\rho}) \left(R_{2,3}(\boldsymbol{\rho}) \frac{\partial R_{2,3}(\boldsymbol{\rho})}{\partial \boldsymbol{\rho}} + R_{3,3}(\boldsymbol{\rho}) \frac{\partial R_{3,3}(\boldsymbol{\rho})}{\partial \boldsymbol{\rho}} \right) \right)$$

The angular velocity thus becomes:

$${}^{22+i}\boldsymbol{\omega}_{22+i} = \mathbf{R}_y(\theta_{2i})^{\mathrm{T}}\, {}^{19+i}\boldsymbol{\omega}_{19+i} + \begin{bmatrix} 0 \\ \dot{\theta}_{2i} \\ 0 \end{bmatrix}$$

Notice that the angular velocity of the part 2 of each S joint is given as a function of the angular velocity of the part 1 of the same S joint.

8.5.2 *Computation of the Inverse Dynamics Model*

According to Eqs. (8.21)–(8.23), the total kinetic and potential energies of the Hexapod robot are given by:

$$\mathcal{K}(\boldsymbol{\rho}, \dot{\boldsymbol{\rho}}) = \sum_{l=1}^{25} \mathcal{K}_l = \frac{1}{2} \sum_{l=1}^{25} \left[m_l \boldsymbol{\upsilon}_l^{\mathrm{T}} \boldsymbol{\upsilon}_l + {}^l\boldsymbol{\omega}_l \mathbf{I}_l^l \boldsymbol{\omega}_l \right] \tag{8.96}$$

and

$$\mathcal{U}(\boldsymbol{\rho}) = \sum_{l=1}^{25} \mathcal{U}_l = -\left[\sum_{l=1}^{25} m_l \mathbf{p}_l^{\mathrm{T}} \right] \mathbf{g}_o, \tag{8.97}$$

Now, replacing Eq. (8.76) in Eq. (8.96) we get

$$\mathcal{K}(\boldsymbol{\rho}, \dot{\boldsymbol{\rho}}) = \frac{1}{2}\dot{\boldsymbol{\rho}}^{\mathrm{T}} \left[\sum_{l=1}^{25} m_l \mathbf{J}_{G_l}^{\mathrm{T}} \mathbf{J}_{G_l} + \mathbf{K}_{G_l}^{\mathrm{T}} \mathbf{I}_l \mathbf{K}_{G_l} \right] \dot{\boldsymbol{\rho}}$$

and comparing with Eq. (8.28), it should be clear that

$$\mathbf{M}(\boldsymbol{\rho}) = \sum_{l=1}^{25} \left[m_l \mathbf{J}_{G_l}^{\mathrm{T}} \mathbf{J}_{G_l} + \mathbf{K}_{G_l}^{\mathrm{T}} \mathbf{I}_l \mathbf{K}_{G_l} \right] \tag{8.98}$$

where the Jacobians $\mathbf{J}_{G_l}$ and $\mathbf{K}_{G_l}$, with $l = 1, 2, \ldots, 25$, for the Hexapod robot were found in the previous subsection [Eqs. (8.79), (8.80), (8.82)–(8.87), and (8.92)–(8.95)], and the dynamics parameters m_l and $\mathbf{I}_l$ are to be determined for the Hexapod robot either by an experimental procedure or via CAD modeling.

Once $\mathbf{M}(\boldsymbol{\rho})$ is computed, the vector of centrifugal and Coriolis forces can be obtained by rewriting Eq. (8.29) as:

$$\mathbf{C}(\boldsymbol{\rho}, \dot{\boldsymbol{\rho}})\dot{\boldsymbol{\rho}} = \dot{\mathbf{M}}(\boldsymbol{\rho})\dot{\boldsymbol{\rho}} - \frac{\partial}{\partial \boldsymbol{\rho}} \left(\frac{1}{2} \dot{\boldsymbol{\rho}}^{\mathrm{T}} \mathbf{M}(\boldsymbol{\rho}) \dot{\boldsymbol{\rho}} \right)$$

and taking $\dot{\rho}$ out:

$$\mathbf{C}(\boldsymbol{\rho}, \dot{\boldsymbol{\rho}}) = \dot{\mathbf{M}}(\boldsymbol{\rho}) - \frac{1}{2} \frac{\partial}{\partial \boldsymbol{\rho}} (\dot{\boldsymbol{\rho}}^{\mathrm{T}} \mathbf{M}(\boldsymbol{\rho})) \tag{8.99}$$

Finally, the vector of gravitational forces can be obtained from (8.30) and (8.97), i.e.,

$$\mathbf{g}(\boldsymbol{\rho}) = \frac{\partial \mathcal{U}(\boldsymbol{\rho})}{\partial \boldsymbol{\rho}} = - \left[\sum_{l=1}^{25} m_l \frac{\partial \mathbf{p}_l(\boldsymbol{\rho})}{\partial \boldsymbol{\rho}}^{\mathrm{T}} \right] \mathbf{g}_o = - \left[\sum_{l=1}^{25} m_l \mathbf{J}_{G_l}^{\mathrm{T}} \right] \mathbf{g}_o \tag{8.100}$$

where $\mathbf{g}_o = [0 \quad 0 \quad g_o]^{\mathrm{T}}$, with $g_o = 9.81[\mathrm{m/s^2}]$, is the gravitational acceleration vector for the Hexapod robot.

Once the matrices of the non-minimal dynamics model given by Eqs. (8.98), (8.99), and (8.100) are computed, we can use the matrix $\mathbf{A}(\mathbf{q})$ given by Eq. (8.73), i.e.,

$$\mathbf{A}(\mathbf{q}) = \begin{bmatrix} \mathbf{I} \\ -\frac{\partial \gamma(\mathbf{q}, \phi)}{\partial \phi}^{-1} \frac{\partial \gamma(\mathbf{q}, \phi)}{\partial \mathbf{q}} \end{bmatrix} \in \mathbb{R}^{9 \times 6},$$

to find the corresponding matrices of the minimal dynamics model Eq. (8.31) using Eqs. (8.32)–(8.35).

8.6 Validation of the Models

In order to show the validity of the Hexapod's forward kinematics model and inverse dynamics model, obtained in the previous sections, we carried out some simulations in which the results produced by our analytical models for a given motion were compared with the results generated by the SolidWorks Motion software tool.

SolidWorks Motion is a module of the SolidWorks® product family which is useful for the analysis and design of mechanisms, when their SolidWorks CAD model is provided. SolidWorks Motion can compute the solution of kinematics and dynamics models numerically for a time-based motion.

For the validation of the kinematics and dynamics models, we designed a motion profile in which the active joint variables were taken as inputs and time-varying functions were applied to them. Such motion profile was then used in simulations employing the analytical expressions of the FPK and inverse dynamics models, developed in Sects. 8.4 and 8.5, respectively, and compared with the results given by SolidWorks Motion.

The trajectory for the active joints is given by the vector

$$\mathbf{q}_d(t) \; = \mathbf{q}(0) + \mathbf{u} \quad [m]$$

where $\mathbf{u} = [u_1 \quad u_2 \quad u_3 \quad u_4 \quad u_5 \quad u_6]^{\mathrm{T}} \in \mathbb{R}^6$, with components $u_i = c_i(1 - e^{-\kappa t^3}) \sin(\omega_i t)$, and $\mathbf{q}(0)$ corresponds to the vector of active joints at the home configuration (where the robot starts at $t = 0$) which, according to the specifications of the Hexapod robot, is given by $\mathbf{q}(0) = 0.1765[1 \quad 1 \quad 1 \quad 1 \quad 1 \quad 1]^{\mathrm{T}}$m, which corresponds to the home pose of the platform given by $\mathbf{r}_F(0) = [x(0) \quad y(0) \quad z(0)]^{\mathrm{T}} = [0 \quad 0 \quad 0.424]^{\mathrm{T}}$m and ${}^0\mathbf{R}_F = \mathbf{I}$ (or $[\lambda(0) \quad \mu(0) \quad v(0)]^{\mathrm{T}} = [0 \; 0 \; 0]^{\mathrm{T}}$rad).

Then, by using SolidWorks Motion, we computed: (a) the vector $\boldsymbol{\xi} \in \mathbb{R}^6$ of coordinates describing the pose of the platform (the three Cartesian coordinates and the three ZYX-convention Euler angles), which is the output of the FPK model; and (b) the vector of active joint generalized forces $\boldsymbol{\tau}_q$, which is the output of the inverse dynamics model.

The parameters of the trajectory were chosen to be $c_1 = c_4 = c_6 = 0.05$ [m], $c_2 = c_3 = c_5 = 0.08$ [m], $\kappa = 1$ [s^{-3}], and $\omega_1 = \omega_2 = 2\omega_3 = 2\omega_4 = 4\omega_5 = 4\omega_6 = 3$ [rad/s]. It is worth noticing that this trajectory starts at the home configuration with null velocity and null acceleration (i.e., $\mathbf{q}_d(t) = \dot{\mathbf{q}}_d(t) = \ddot{\mathbf{q}}_d(t) = 0$ for $t = 0$). Also notice that as $t \to \infty$, the desired trajectory reduces to simple sinusoidal functions in each axis.

Once the position vector $\mathbf{r}_F$ and the rotation matrix ${}^0\mathbf{R}_F$ were computed by following the steps at the end of Sect. 8.4.1, the ZYX Euler angles were determined using the following expressions:

$$\lambda = \mathrm{atan}\left(\frac{{}^{0}R_{F_{2,1}}}{{}^{0}R_{F_{1,1}}}\right),\ \mu = \mathrm{atan}\left(\frac{-{}^{0}R_{F_{3,1}}}{\sqrt{{}^{0}R_{F_{1,1}}^{2} + {}^{0}R_{F_{2,1}}^{2}}}\right),\ \nu = \mathrm{atan}\left(\frac{{}^{0}R_{F_{3,2}}}{{}^{0}R_{F_{3,3}}}\right)$$

where ${}^{0}\mathbf{R}_{F_{u,v}}$ is the element (u, v) of matrix ${}^{0}\mathbf{R}_{F}$, and we take $L_B = 0.866\,[\mathrm{m}]$, $L = 0.3689\,[\mathrm{m}]$, and $H_{PQ} = 0.090\,[\mathrm{m}]$ as kinematic parameters.

Figure 8.9 shows the time evolution of the six variables giving the pose of the platform obtained by SolidWorks and our analytical expressions. It should be noticed that the graphs for each coordinate are very similar. Figure 8.10 shows the norm of the vectors giving the difference between the SolidWorks and analytical models for both the position and orientation parts of the pose. If we consider the first 20 s shown in Fig. 8.9, the maximum deviation between both graphs is given by $\tilde{x} = 0.55$ [mm], $\tilde{y} = 1.28$ and $\tilde{x} = 0.235$ [mm], $\tilde{\alpha} = 0.1314^{\circ}$, $\tilde{\beta} = 0.1050^{\circ}$ and $\tilde{\gamma} = 0.1317^{\circ}$.

In the case of the dynamics model, we also employed $H^{*} = 0.0791\,[\mathrm{m}]$, $l_{c_1} = 0.0287\,[\mathrm{m}]$, $l_{c_2} = 0.03081\,[\mathrm{m}]$, and the dynamics parameters given in Table 8.1. Notice in this table that the last column gives the moment of inertia tensor with respect to the frame associated with the corresponding body (i.e., I_l), but only the terms in its diagonal are considered.

Figure 8.11 shows the time evolution of the resulting joint generalized forces obtained from SolidWorks and using the minimal dynamics model Eq. (8.32).

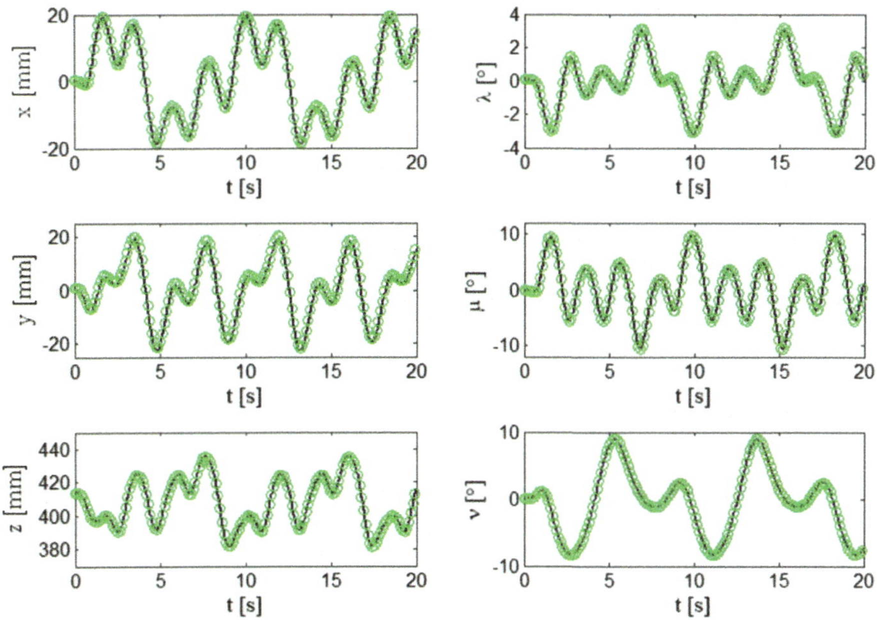

Fig. 8.9 Pose coordinates of the platform, computed by both the analytical model (black solid line) and the CAD model (green circled line)

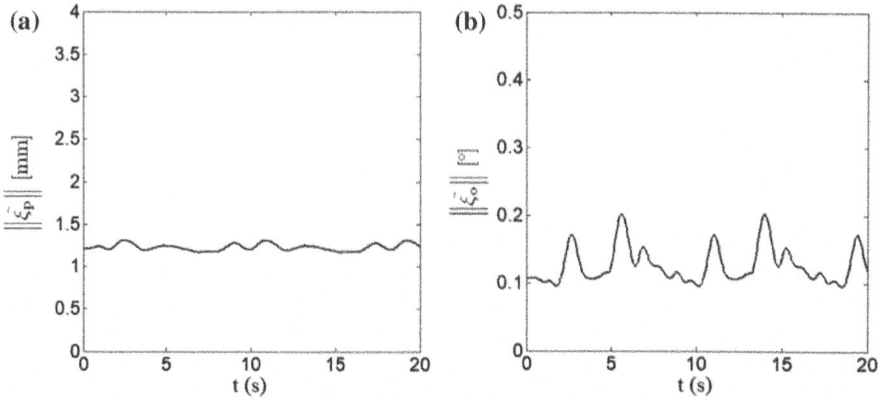

Fig. 8.10 Norm of the vector giving the difference between the analytical and the CAD model for the platform's **a** position and **b** orientation

Table 8.1 Dynamics parameters of the Hexapod

Rigid body	Mass [Kg]	(Ixx, Iyy, Izz)[Kg cm^2]
Mobile platform	2.085	(198.58, 199.84, 396.31)
Leg k	0.44917	(54.66, 0.50, 54.78)
PU joint part 1	0.3194	(4.94, 5.51, 2.51)
PU joint part 2	0.2200	(1.76, 3.36, 1.75)
S joint part 1	0.2200	(1.75, 1.76, 3.36)
S joint part 2	0.3025	(5.864, 1.22, 5.218)

Figure 8.12 shows the time evolution of the norm of the vector formed by the difference between the generalized forces computed by those models for all the six joints. The maximum deviation from both graphs is given by: $\tilde{\tau}_0 = 0.179$ [N], $\tilde{\tau}_1 = 0.239$ [N], $\tilde{\tau}_2 = 0.143$ [N], $\tilde{\tau}_3 = 0.116$ [N], $\tilde{\tau}_4 = 0.094$ [N], $\tilde{\tau}_5 = 0.096$ [N].

It is clear, by the results presented in this section, that the analytical expressions we obtained for the FPK model and inverse dynamics model of the Hexapod robot are validated by the SolidWorks Motion software.

8.7 Real-Time Experiments

In this section, we describe the implementation and the results of real-time experiments carried out in the Hexapod parallel robot of the Mechatronics and Control Laboratory of the Instituto Tecnológico de la Laguna.

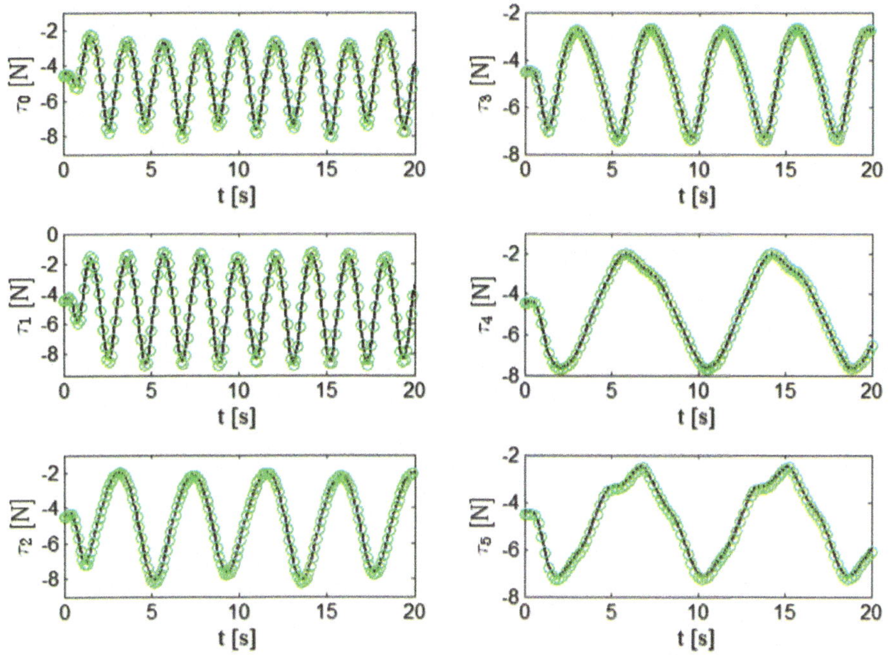

Fig. 8.11 Active joint generalized forces, computed by both the analytical model (black solid line) and the CAD model (green circled line)

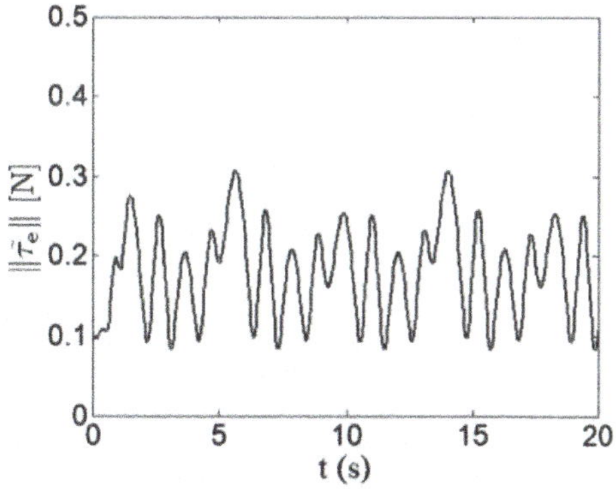

Fig. 8.12 Norm of the vector giving the difference between the active joint generalized forces computed by the analytical and CAD models

Two controllers were tested:

1. A two-loop hierarchical controller consisting of a kinematic controller (one whose output is the vector of desired joint velocities) of the type known as resolved-rate motion control (RMRC) in the outer loop, and a PI joint velocity controller in the inner loop.
2. An inverse dynamics controller, whose output is the vector of desired joint torques required to cancel the nonlinear dynamics of the robot.

It is worth mentioning that the kinematic controller requires the computation of the inverse of the analytical Jacobian while the dynamic controller uses the inverse dynamics model of the robot.

Experiments were implemented using the software provided by Quanser, via the MATLAB/Simulink platform and the QUARC toolbox for real-time control.

8.7.1 Two-Loop Controller

Figure 8.13 shows the block diagram of the two-loop controller proposed as a tracking controller for the Hexapod robot. This controller, applied to serial robot manipulators, has been studied, and its stability is analyzed by Camarillo et al. (2008).

By kinematic control, we refer to any scheme that uses an inverse Jacobian algorithm to resolve the desired joint velocities directly from the pose variables of the desired task. Thus, a kinematic controller is often employed as the outer loop of a two-loop controller such as the one in Fig. 8.13. In this paper, we use as kinematic controller the so-called resolved motion rate controller (RMRC), which was first proposed by Whitney (1969). Using this scheme, the desired joint velocity for the inner loop can be written as

$$\boldsymbol{v}_d = \mathbf{J}_A(\mathbf{q})^{-1}\left[\dot{\xi}_d + \mathbf{K}\tilde{\xi}\right], \tag{8.101}$$

where $\boldsymbol{v}_d \in \mathbb{R}^n$ is the desired joint velocity vector, $\dot{\xi}_d \in \mathbb{R}^6$ is the time derivative of the ξ_d, $\tilde{\xi} = \xi_d - \xi \in \mathbb{R}^6$ is the pose error vector, and $\mathbf{K} \in \mathbb{R}^{6\times 6}$ is a symmetric positive definite matrix of control gains.

Under the assumption of perfect velocity tracking, that is, $\dot{\mathbf{q}} \equiv \boldsymbol{v}_d$, pre-multiplying both sides of Eq. (8.101) by $\mathbf{J}_A(\mathbf{q})$ and substituting $\dot{\xi} = \mathbf{J}_A(\mathbf{q})\dot{\mathbf{q}}$, we can write

$$\dot{\tilde{\xi}} = -K\tilde{\xi}. \tag{8.102}$$

As $\boldsymbol{K}$ is a symmetric positive definite matrix, we conclude that $\xi \to \mathbf{0}$ as $t \to \infty$.

However, a real joint velocity controller does not ensure the instantaneous tracking of the desired velocity $\boldsymbol{v}_d$. In practice, we get asymptotic velocity tracking

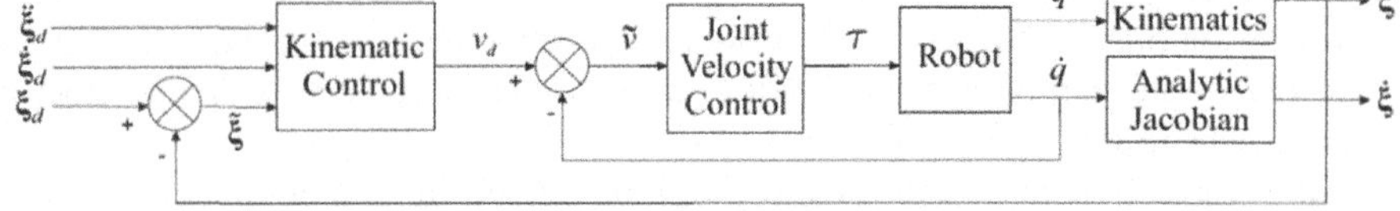

Fig. 8.13 Block diagram of the two-loop controller

instead of ideal velocity tracking. So, the implementation of the kinematic control given by Eq. (8.101) requires the design of a joint velocity controller. To this end, let us define the joint velocity error as

$$\tilde{\mathbf{v}} = \mathbf{v}_d - \dot{\mathbf{q}} \in \mathbb{R}^n. \tag{8.103}$$

From $\xi = \mathbf{J}_A(\mathbf{q})\dot{\mathbf{q}}$, Eqs. (8.101) and (8.103), we get

$$\dot{\tilde{\xi}} = -\mathbf{K}\tilde{\xi} + \mathbf{J}(\mathbf{q})\tilde{\mathbf{v}},$$

instead o Eq. (8.102).

For the inner loop in Fig. 8.13, we consider the classical joint-velocity proportional-integral (PI) controller commonly used in industrial robots, which can be written as

$$\tau = \mathbf{K}_p\tilde{\mathbf{v}} + \mathbf{K}_i \int_0^t \tilde{\mathbf{v}} dt,$$

where $\mathbf{K}_p$ and $\mathbf{K}_i \in \mathbb{R}^{6\times 6}$ are diagonal positive definite matrices.

8.7.1.1 Experiments

Experiments started at the home configuration of the Hexapod, which is given by the following vector of joint coordinates:

$$\mathbf{q}(0) = 0.1765[1 \quad 1 \quad 1 \quad 1 \quad 1 \quad 1]^{\mathrm{T}}\mathrm{m},$$

which corresponds to:

$$\begin{aligned} \mathbf{r}_F(0) &= [0 \quad 0 \quad 0.424]^{\mathrm{T}}\mathrm{m} \\ \psi(0) &= [0 \quad 0 \quad 0]^{T}\mathrm{rad} \end{aligned} \tag{8.104}$$

As desired trajectory we employed:

$$p_d = \begin{bmatrix} c_1(1 - e^{-\gamma t^3})\cos(\omega t) \\ c_2(1 - e^{-\gamma t^3})\sin(\omega t) \\ c_3(1 - e^{-\gamma t^3})(\sin(\omega t) - \frac{z_0}{c_3}) + 0.424 \end{bmatrix} \mathrm{m}$$

$$\psi_d = \begin{bmatrix} c_4(1 - e^{-\gamma t^3})\cos(\omega t) \\ c_5(1 - e^{-\gamma t^3})\sin(\omega t) \\ c_6(1 - e^{-\gamma t^3})\sin(\omega t) \end{bmatrix} \mathrm{rad}$$

with $c_1 = c_2 = c_3 = z_0 = 0.02$ [m], $\gamma = .01$ [s^{-3}], $\omega = 0.20\pi$ [rad/s], and $c_4 = c_5 = c_6 = \pi/60$ [rad]. It is worth noticing that this trajectory was chosen so that the corresponding vectors of pose coordinates $\xi_d = \left[\mathbf{p}_d^{\mathrm{T}}\ \boldsymbol{\phi}_d^{\mathrm{T}}\right]^{\mathrm{T}}$ started from the home position, while the vectors of pose velocities $\dot{\xi}_d$ and accelerations $\ddot{\xi}_d$ at $t = 0$ are zero. Also notice that as $t \to \infty$, the desired trajectory reduces to simple sinusoidal functions in each axis.

For the experiments, the gain matrix for the kinematic controller was chosen to be diagonal and equal to $\mathbf{K} = \mathrm{diag}\{15, 15, 10, 15, 10, 10\}$, and for the velocity controller the matrices were $\mathbf{K}_p = 3000\,\mathbf{I}$ [Nm s/rad], and $\mathbf{K}_i = 4750.6\,\mathbf{I}$ [Nm/rad], where $\mathbf{I} \in \mathbb{R}^{6\times6}$ is the identity matrix.

Figures 8.14 and 8.15 show, respectively, the time evolution of the norm of the position error vector (in Cartesian coordinates) and the orientation error vector (in ZYX Euler angles). Notice that the whole pose error is close to zero meaning that the end effector pose of the mobile platform follows the desired trajectory. Figure 8.16 shows the torques applied to the actuated prismatic joints.

8.7.2 *Inverse Dynamics Control*

Inverse dynamics control is a classical technique to perform tracking control of robots manipulators whose objective is to find a control law that linearizes and

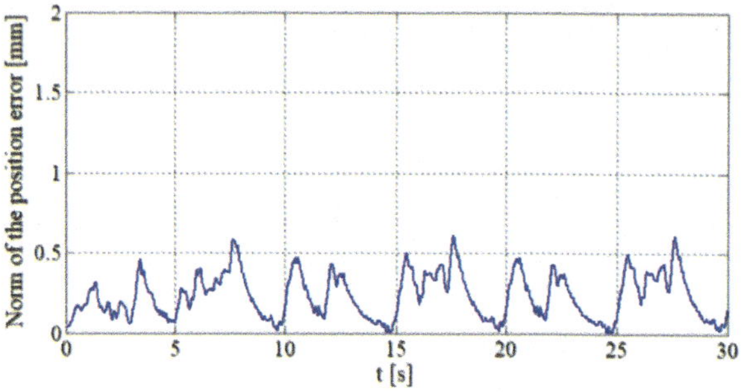

Fig. 8.14 Two-loop controller: norm of the position error vector

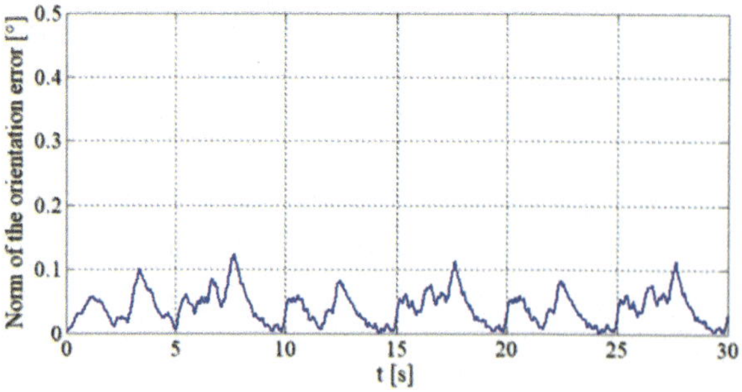

Fig. 8.15 Two-loop controller: norm of the orientation error vector

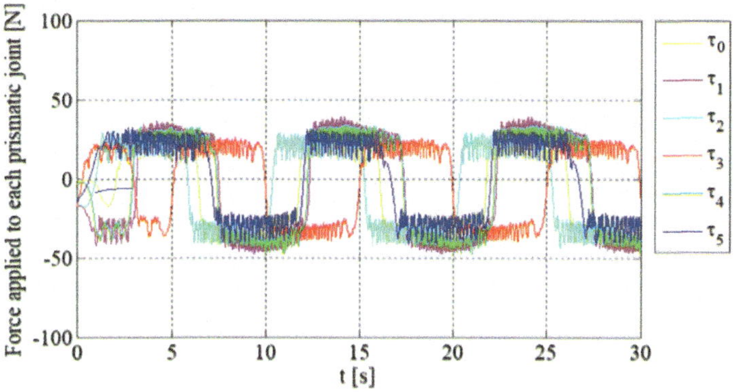

Fig. 8.16 Two-loop controller: forces applied to the prismatic joints

decouples the mechanical system, by adding the necessary nonlinear terms to the control law (Cheah and Haghighi 2014).

In this work, we employ the inverse dynamics controller in operational space proposed by Khatib (1987), which uses Euler angles to parameterize the orientation. This controller is given by:

$$\boldsymbol{\tau}_q = \bar{\mathbf{M}}_q(\boldsymbol{\rho})\mathbf{J}_A(\mathbf{q})^{-1}\left[\ddot{\boldsymbol{\xi}}_d + \mathbf{K}_V\dot{\tilde{\boldsymbol{\xi}}} + \mathbf{K}_P\tilde{\boldsymbol{\xi}} - \dot{\mathbf{J}}_A(\mathbf{q})\dot{\mathbf{q}}\right] + \bar{\mathbf{C}}_q(\boldsymbol{\rho},\dot{\boldsymbol{\rho}})\dot{\mathbf{q}} + \bar{\mathbf{g}}_q(\boldsymbol{\rho}), \quad (8.105)$$

where $\mathbf{J}_A(\mathbf{q})$ is the analytical Jacobian; $\mathbf{K}_P$ and $\mathbf{K}_V$ are diagonal matrices of control gains, and $\tilde{\boldsymbol{\xi}} = \boldsymbol{\xi}_d - \boldsymbol{\xi}$, being $\boldsymbol{\xi}_d$, $\dot{\boldsymbol{\xi}}_d$, $\ddot{\boldsymbol{\xi}}_d$ the vectors of desired pose coordinates, velocities, and accelerations, respectively.

Figure 8.17 shows the block diagram of the controller given by Eq. (8.105).

Substituting this control law in the minimal robot dynamics Eq. (8.31), and assuming that $\mathbf{J}_A(\mathbf{q})$ is invertible in the region of the workspace where the robot operates, it is possible to demonstrate, using $\dot{\boldsymbol{\xi}} = \mathbf{J}_A(\mathbf{q})\dot{\mathbf{q}}$ and its time derivative, that the closed-loop system is $\ddot{\tilde{\boldsymbol{\xi}}} + \mathbf{K}_V\dot{\tilde{\boldsymbol{\xi}}} + \mathbf{K}_P\tilde{\boldsymbol{\xi}} = \mathbf{0} \in \mathbb{R}^6$, which is a linear system whose stability is easy to demonstrate.

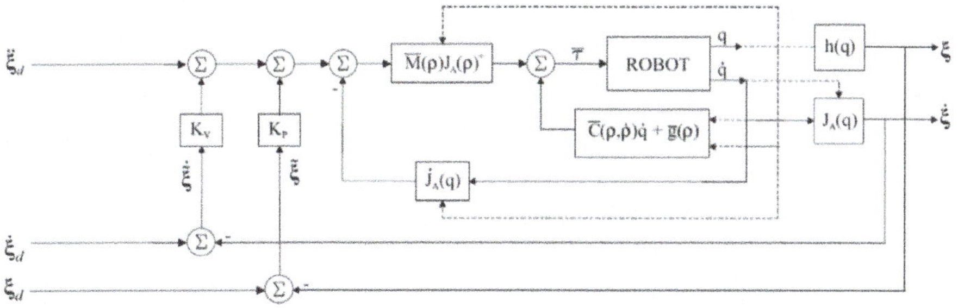

Fig. 8.17 Block diagram of the inverse dynamics controller

8.7.2.1 Experiments

For the implementation of the inverse dynamics controller, we used the same desired trajectory and initial conditions as for the two-loop controller. Moreover, the gain matrices were defined as $\mathbf{K}_P = \text{diag}\{35, 35, 50, 80, 100, 90\}[10^3/\text{s}^2]$ and $\mathbf{K}_V = \text{diag}\{190, 250, 120, 550, 500\}[1/\text{s}]$.

Figures 8.18 and 8.19 show the time evolution of the norm of the position error (in Cartesian coordinates) and the orientation error (in ZYX Euler angles) parts of $\tilde{\xi}$. Note that both norms are kept bounded, meaning that the position of the mobile platform follows the desired path with a relatively small error.

Figure 8.20 shows the generalized forces applied to the prismatic joints.

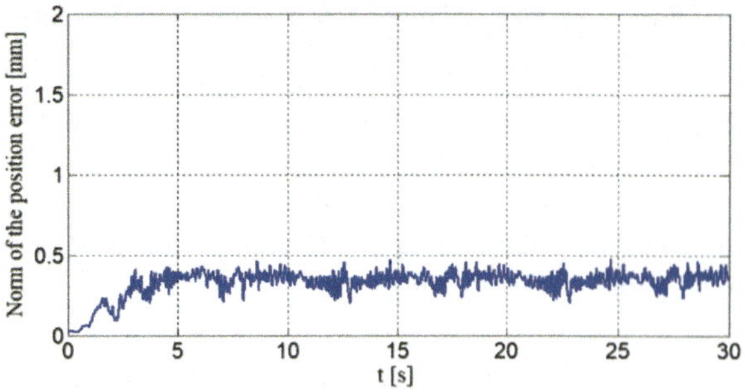

Fig. 8.18 Inverse dynamics controller: norm of the position error vector

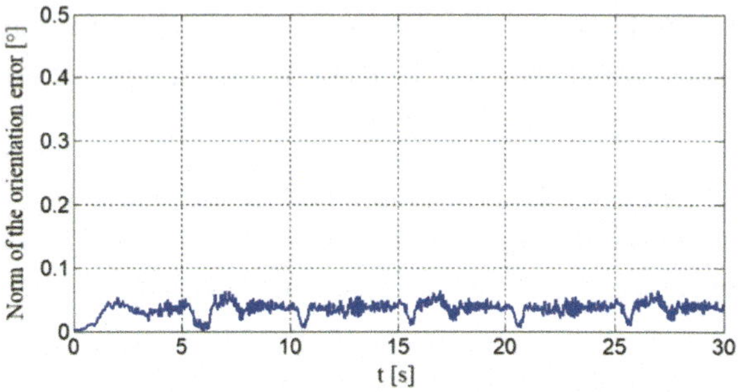

Fig. 8.19 Inverse dynamics controller: norm of the orientation error vector

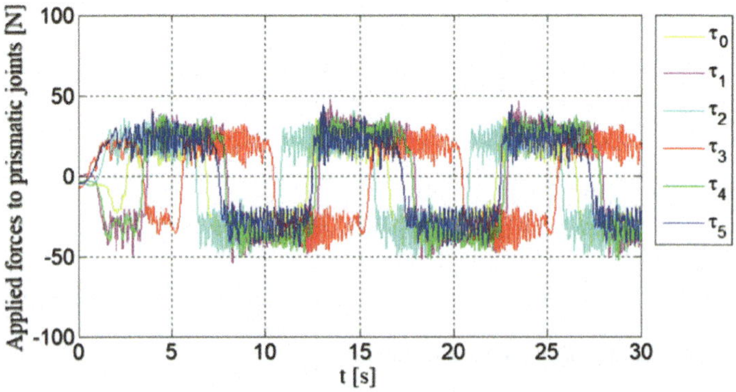

Fig. 8.20 Inverse dynamics controller: forces applied to prismatic joints

8.8 Conclusions

This work first recalls the kinematics and dynamics modeling of platform-type parallel manipulators. The Lagrangian formulation, together with the projection method, is suggested for obtaining the minimal dynamics model. The proposed methodology is then employed to model a 6-3-PUS-type parallel robot, known as Hexapod. The effect of all mechanical parts of the robot (including those of the joints) is taken into account, and the computed kinematics and dynamics models are validated by comparing them with numerical simulations using SolidWorks Motion. It is worth noticing that the proposed method can be used for similar parallel robotic structures.

Additionally, we show how to implement in the Hexapod robot two tracking controllers in operational space (i.e., employing Euler angles for describing the orientation). The first controller has a two-loop structure: a resolved motion rate controller (RMRC) in the outer loop and a joint velocity PI controller in the inner loop. The second controller is of the inverse dynamics type, and it requires the computation of the inverse dynamics model. The experimental results show a good performance for both controllers, and this also allows to conclude the validity of the kinematics and dynamics models we have obtained for the mechanism under study.

References

Arczewski, K., & Blajer, W. (1996). A unified approach to the modelling of holonomic and nonholonomic mechanical systems. *Mathematical Modelling of Systems, 2*(3), 157–174.

Betsch, P. (2005). The discrete null space method for the energy consistent integration of constrained mechanical systems: Part I: Holonomic constraints. *Computer Methods in Applied Mechanics and Engineering, 194,* 5159–5190.

Blajer, W. (1997). A geometric unification of constrained system dynamics. *Multibody System Dynamics, 1,* 3–21.

Camarillo, K., Campa, R., Santibáñez, V., & Moreno-Valenzuela, J. (2008). Stability analysis of the operational space control for industrial robots using their own joint velocity PI controllers. *Robotica, 26*(6), 729. https://doi.org/10.1017/S0263574708004335.

Campa, R., Bernal, J., & Soto, I. (2016). Kinematic modeling and control of the Hexapod parallel robot. In *Proceedings of the 2016 American Control Conference* (pp. 1203–1208). IEEE. http://doi.org/10.1109/ACC.2016.7525081.

Campa, R., & de la Torre, H. (2009). Pose control of robot manipulators using different orientation representations: A comparative review. In *Proceedings of the American Control Conference*. St. Louis, MO, USA.

Carbonari, L., Krovi, V. N., & Callegari, M. (2011). Polynomial solution to the forward kinematics problem of a 6-PUS parallel-architecture robot (in Italian). In *Proceedings of the Congresso dell'Associazione Italiana di Meccanica Teorica e Applicata*. Bologna, Italy.

Cheah, C. C., & Haghighi, R. (2014). Motion control of robot manipulators. In *Handbook of Manufacturing Engineering and Technology* (pp. 1–40). London: Springer London. http://doi.org/10.1007/978-1-4471-4976-7_93-1.

Craig, J. J. (2004). *Introduction to robotics: Mechanics and control*. Pearson.

Dasgupta, B., & Mruthyunjaya, T. S. (2000). The Stewart platform manipulator: A review. *Mechanism and Machine Theory, 35*(1), 15–40.

Dontchev, A. L., & Rockafellar, R. T. (2014). *Implicit functions and solution mappings: A view from variational analysis*. Springer.

Geng, Z., Haynes, L. S., Lee, J. D., & Carroll, R. L. (1992). On the dynamic model and kinematic analysis of a class of Stewart platforms. *Robotics and Autonomous Systems, 9*(4), 237–254.

Ghorbel, F. H., Chételat, O., Gunawardana, R., & Longchamp, R. (2000). Modeling and set point control of closed-chain mechanisms: Theory and experiment. *IEEE Transactions on Control Systems Technology, 8*(5), 801–815.

Hopkins, B. R., & Williams, R. L., II. (2002). Kinematics, design and control of the 6-PSU platform. *Industrial Robot: An International Journal, 29*(5), 443–451.

Kapur, D. (1995). Algorithmic elimination methods. In *Tutorial Notes of the International Symposium on Symbolic and Algebraic Computation*. Montreal, Canada.

Kelly, R., Santibáñez, V., & Loría, A. (2005). *Control of robot manipulators in joint space*. Springer.

Khatib, O. (1987). A unified approach for motion and force control of robot manipulators: The operational space formulation. *IEEE Journal on Robotics and Automation, 3*(1), 43–53. https://doi.org/10.1109/JRA.1987.1087068.

Liu, C. H., Huang, K. C., & Wang, Y. T. (2012). Forward position analysis of 6-3 Linapod parallel manipulators. *Meccanica, 47*(5), 1271–1282.

Liu, M. J., Li, C. X., & Li, C. N. (2000). Dynamics analysis of the Gough-Stewart platform manipulator. *IEEE Transactions on Robotics and Automation, 16*(1), 94–98.

Merlet, J.-P. (1999). Parallel robots: Open problems. In *Proceedings of the International Symposium of Robotics Research*. Snowbird, UT, USA.

Merlet, J.-P. (2006). *Parallel robots*. Springer.

Murray, J. J., & Lovell, G. H. (1989). Dynamic modeling of closed-chain robotic manipulators and implications for trajectory control. *IEEE Transactions on Robotics and Automation, 5*(4), 522–528. https://doi.org/10.1109/70.88066.

Nanua, P., Waldron, K. J., & Murthy, V. (1990). Direct kinematic solution of a Stewart platform. *IEEE Transactions on Robotics and Automation, 6*(4), 438–444.

Narayanan, M. S., Chakravarty, S., Shah, H., & Krovi, V. N. (2010). Kinematic, static and workspace analysis of a 6-PUS parallel manipulator. In *Volume 2: 34th Annual Mechanisms and Robotics Conference, Parts A and B* (pp. 1456–1456.8). Montreal, Canada: ASME. http://doi.org/10.1115/DETC2010-28978.

Siciliano, B., Sciavicco, L., Villani, L., & Oriolo, G. (2009). *Robotics*. London: Springer London. https://doi.org/10.1007/978-1-84628-642-1.

Tsai, L. W. (1999). *Robot analysis: The mechanics of serial and parallel manipulators*. Wiley.

Whitney, D. (1969). Resolved motion rate control of manipulators and human prostheses. *IEEE Transactions on Man Machine Systems, 10*(2), 47–53. https://doi.org/10.1109/TMMS.1969.299896.

Chapter 9
A Finite-Time Nonlinear PID Set-Point Controller for a Parallel Manipulator

Francisco Salas, Israel Soto, Raymundo Juarez and Israel U. Ponce

Abstract In recent years, finite-time controllers have attracted attention from some researchers in control, who have formulated applications to several processes and systems, including serial robotic manipulators. In this work, we report the application of a finite-time nonlinear PID controller to a Five-Bar Mechanism, which is a parallel manipulator, for set-point controller. The stability analysis of the closed-loop system shows global finite-time stability of the system. The dynamic model of the Five-Bar Mechanism developed in this work is a so-called reduced model, which has a structure similar to a serial robot. Moreover, the results of the numerical simulations carried out confirm the usefulness of the proposed application. The contribution of this work is to show the feasibility of the application of a finite-time nonlinear controller to a Five-Bar Mechanism and the usefulness of the proposed approach by numerical simulations.

Keywords Nonlinear controller · Finite-time PID · Parallel manipulator

9.1 Introduction

In recent years, finite-time controllers have attracted attention from some researchers in control.

As a result, the fundamental theory has been developed (Dorato 1961; Michel 1970; Weiss and Infante 1967) and enriched by many contributions (Bhat and Bernstein 1998, 2000, 2005; Polyakov and Poznyak 2009; Polyakov 2014). According to Amato et al. (2013), the finite-time stability is a property related to the quantitative behavior of the states of a system over a period of time.

F. Salas · R. Juarez
Universidad Autónoma de Coahuila, Facultad de Contaduría y Administración,
Blvd. Revolución 151 Oriente, Col. Centro, 27000 Torreón, Coahuila, CP, Mexico

I. Soto (✉) · I. U. Ponce
Universidad Autónoma de Ciudad Juárez, Instituto de Ingeniería
y Tecnología, Ciudad Juárez, Chihuahua, Mexico
e-mail: angel.soto@uacj.mx

O. O. Vergara Villegas et al. (eds.), *Advanced Topics on Computer Vision, Control and Robotics in Mechatronics*,
https://doi.org/10.1007/978-3-319-77770-2_9

A system is finite-time stable if, given a bound on an initial condition, the weighted norm of the state does not exceed a certain threshold over a specific time period. Moreover, finite-time stability and Lyapunov asymptotic stability are independent concepts, although not exclusively one from another. The existence of one does not imply the existence of the other. Some advantages of the finite-time stabilization of dynamic systems are that it can produce faster transient responses and high-precision performance, as well as convergence to the equilibrium in finite time. Some previous works in applications of finite-time controllers to robotic manipulators are Feng et al. (2002), Gruyitch and Kokosy (1999), Hong et al. (2002), Yu et al. (2005), Su and Zheng (2009, 2010), Zhao et al. (2010). In Su and Zheng (2009), a finite-time nonlinear PID-like control for regulation of robotic manipulators is presented. The authors propose a not model-based controller, to take advantage of the robustness to parametric uncertainty of the model. This work is improved in Su and Zheng (2010) by adding a nonlinear filter to estimate velocity when measurements are not available.

On the other hand, parallel robots are closed-chain mechanisms that possess some particular features such as high-speed capabilities and high stiffness that make them useful for some tasks as machining (Barnfather et al. 2017; Kelaiaia 2017), welding (Li et al. 2015; Wu et al. 2008), packaging (Pierrot et al. 1990; Xie and Liu 2016) as well as flight simulators (Huang and Cao 2005) and telescopes (Enferadi and Shahi 2016; Nan et al. 2011). Some recent approaches of control of this kind of robotic manipulators include not model-based controllers (Bourbonnais et al. 2015; Ren et al. 2007) and model-based controllers (Diaz-Rodriguez et al. 2013; Ren et al. 2007; Salinas et al. 2016). In Ren et al. (2007), a comparison of several control approaches for robot tracking of three degrees of freedom (DOF) parallel robot is presented. They compare the performance of an adaptive controller, a PI-type synchronized controller (model-based), a conventional PID controller, and an adaptive synchronized controller (not model-based). In Bourbonnais et al. (2015), a computed torque controller and a conventional PID controller are implemented for a novel Five-Bar parallel robot. In Diaz-Rodriguez et al. (2013), a reduced model-based controller of a three DOF parallel robot is proposed. The reduced model is obtained by considering a simplified model with a set of relevant parameters. In Salinas et al. (2016), a family of nonlinear PID-like controllers in which an integral action of a nonlinear function of the position error is added to the control signal.

In this work, inspired in the work of Su and Zheng (2017) on a finite-time controller for set-point controller of a serial robot, we propose the application of this controller to a parallel manipulator, in order to prove the finite-time stability of the closed-loop system by developing the stability analysis and to prove the feasibility and the usefulness of such an application. The dynamic model of the Five-Bar Mechanism constitutes a set of differential algebraic equations (DAEs). Based on Soto and Campa (2014) and Khan et al. (2005), a procedure is carried out in order to transform the set of DAEs into a set of ordinary differential equations (ODEs). By using such a model, the Lyapunov stability analysis and the finite-time stability analysis of the closed-loop system can be developed. As a result, the global finite-time stability of the closed-loop system is proven. Moreover, such a model of ODEs representing the dynamics of the Five-Bar Mechanism let us to carry out numerical simulations of the system. The results of the simulations confirm the validity and usefulness of the application.

9.1.1 Mathematical Preliminaries

In this work, vectors are denoted with italic–bold lowercase letters, e.g., $\boldsymbol{x}$ or $\boldsymbol{\omega}$. Matrices are denoted with italic capital letters, e.g., $\boldsymbol{A}$. $\|\boldsymbol{x}\| = \sqrt{\boldsymbol{x}^T\boldsymbol{x}}$ represents the Euclidean norm of vector $\boldsymbol{x}$. $\lambda_{\max}\{A\}$ and $\lambda_{\min}\{A\}$ represent the largest and the smallest eigenvalues of matrix A, respectively.

In the following, based on Su and Zheng (2017) we define some useful vectors and vector functions, as well as a definition for the control design and analysis.

$$\mathrm{Sig}^{\alpha}(\boldsymbol{x}) = [|x_1|^{\alpha}\mathrm{sign}(x_1), \ldots, |x_n|^{\alpha}\mathrm{sign}(x_n)]^T \in \Re^n \tag{9.1}$$

$$\mathrm{Sech}(\boldsymbol{x}) = \mathrm{diag}(\mathrm{sech}(x_1), \ldots, \mathrm{sech}(x_n)) \in \Re^{n\times n} \tag{9.2}$$

where α_0 and α are positive constants, and $\boldsymbol{x} \in \Re^n$. Furthermore, $0<\alpha<1$, $\mathrm{sign}(\cdot)$, and $\mathrm{sech}(\cdot)$ are the standard scalar functions signum and hyperbolic secant, respectively, and $\mathrm{diag}(\cdot)$ denotes a diagonal matrix. By defining the vector function

$$\mathrm{Tanh}(\boldsymbol{x}) = [\tanh(x_1), \ldots, \tanh(x_n)]^T \in \Re^n, \tag{9.3}$$

the validity of the following expressions

$$\boldsymbol{x}^T\mathrm{Sig}^{\alpha}(\boldsymbol{x}) = \sum_{i=1}^{n} |x_i|^{\alpha+1} \geq \mathrm{Tanh}^T(\boldsymbol{x})\mathrm{Sig}^{\alpha}(\boldsymbol{x}) \geq 0 \tag{9.4}$$

$$|x_i|^{\alpha+1} \geq \tanh^2(x_i) \tag{9.5}$$

$$(\mathrm{Sech}^2(\boldsymbol{x}))_M = 1 \tag{9.6}$$

can be proven for all $\boldsymbol{x} \neq 0 \in \Re^n$.

9.1.2 Fundamentals of Finite-Time Stability Analysis

Although finite-time stability concepts in control systems literature can be traced back to decade of the 1960s, it was until the works reported in Bhat and Bernstein (1998) and (2000) when the foundations of finite-time stability theory were rigorously established. In Bhat and Bernstein (2005) were further studied some conditions for finite-time stability, in relation to the homogeneity of a system. In the following, some definitions will be exposed in order to clarify the concepts of finite-time stability.

Definition 1 A function $V : \Re^n \to \Re$ is homogeneous of degree d with respect to the weights $\boldsymbol{p} = (p_1, \ldots, p_n) \in \Re^n$ if for any given $\delta > 0, V(\delta^{p_1} x_1, \ldots, \delta^{p_n} x_n) = \delta^d V(\boldsymbol{x}), \forall \boldsymbol{x} \in \Re^n$. A vector field $\boldsymbol{h}$ is homogeneous of degree d with respect to the weights $\boldsymbol{p} = (p_1, \ldots, p_n) \in \Re^n_+$, if for all $1 \leq i \leq n$, the ith component h_i is a homogeneous function of degree $p_i + d$.

Definition 2 Consider the system

$$\dot{\boldsymbol{x}} = \boldsymbol{h}(\boldsymbol{x}), \boldsymbol{h}(0) = 0, \boldsymbol{x} \in \Re^n \tag{9.7}$$

with $\boldsymbol{h} : U_0 \to \Re^n$ continuous on an open neighborhood U_0 of the origin. Suppose that system Eq. (9.1) possesses unique solutions in forward time for all initial condition. The equilibrium $\boldsymbol{x} = 0$ of system Eq. (9.1) is (locally) finite-time stable if it is Lyapunov stable and finite-time convergent in a neighborhood $U \subset U_0$ of the origin. The finite-time convergence means the existence of a function $T(\boldsymbol{x}_0) : U \backslash \{0\} \to (0, \infty)$, such that, $\forall \boldsymbol{x}_0 \in U \subset \Re^n$, the solution of Eq. (9.1) denoted by $s_t(\boldsymbol{x}_0)$ with $\boldsymbol{x}_0$ as the initial condition is defined and $s_t(\boldsymbol{x}_0) \in U \backslash \{0\}$ for $t \in [0, T(\boldsymbol{x}_0))$ and $\lim_{t \to T(\boldsymbol{x}_0)} s_t(\boldsymbol{x}_0) = 0$. When $U = \Re^n$, the global finite-time stability is obtained.

Remark 1 The system (7) is homogeneous if $\boldsymbol{h}(\cdot)$ is homogeneous.

The following results represent sufficient conditions for finite-time stability of the closed-loop system.

Lemma 1 (Hong et al. 2002; Huang and Cao 2005)
Consider the system

$$\dot{\boldsymbol{x}} = \boldsymbol{h}(\boldsymbol{x}) + \hat{\boldsymbol{h}}(\boldsymbol{x}), \boldsymbol{h}(0) = 0, \hat{\boldsymbol{h}}(0) = 0, \boldsymbol{x} \in \Re^n \tag{9.8}$$

where $\boldsymbol{h}(\boldsymbol{x})$ *is a continuous homogeneous vector field of degree* $d < 0$, *with respect to* $(p_1, \ldots, p_n)$. *Assume that* $\boldsymbol{x} = 0$ *is an asymptotically stable equilibrium of system* Eq. (9.7). *Then,* $\boldsymbol{x} = 0$ *is a locally finite-time stable equilibrium of system* Eq. (9.8) if

$$\lim_{\delta \to 0} \frac{\hat{f}_i(\delta^{p_1} x_1, \ldots, \delta^{p_n} x_n)}{\delta^{d + p_i}} = 0, \; i = 1, \ldots, n, \forall \boldsymbol{x} \neq 0 \tag{9.9}$$

Lemma 2 (Hong et al. 2002; Su and Zheng (2017)
Global asymptotical stability and finite-time stability imply global finite-time stability.

9.2 Dynamic Model of a Parallel Robot Manipulator

A parallel robot manipulator is a closed-chain mechanism (CCM) that consists of kinematic chains which are connected in loops. Consider a CCM with n actuated joints. Due to its closed configuration, the CCM is subject to holonomic constraints. In this work, the actuated joints are represented by the vector $\boldsymbol{q} \in \Re^n$, while the non-actuated joints are represented by the vector $\boldsymbol{\beta} \in \Re^m$ and the holonomic constraints are represented by the vector $\gamma \in \Re^r$. Let us define the vector of generalized coordinates $\boldsymbol{\rho}$ that fully explicitly represents the configuration of the CCM as

$$\boldsymbol{\rho} = \left[\boldsymbol{q}^T \boldsymbol{\beta}^T\right]^T \in \Re^s$$

with $s = n + m$. By applying the Euler–Lagrange formulation, the dynamic model of a parallel robot with viscous friction is in general formulated as

$$\begin{aligned} &M'(\boldsymbol{\rho})\ddot{\boldsymbol{\rho}} + C'(\boldsymbol{\rho}, \dot{\boldsymbol{\rho}})\dot{\boldsymbol{\rho}} + \mathbf{g}'(\boldsymbol{\rho}) + F'\dot{\boldsymbol{\rho}} = \boldsymbol{\tau}' + D^T(\boldsymbol{\rho})\boldsymbol{\lambda} \\ &\gamma(\boldsymbol{\rho}) = 0 \end{aligned} \tag{9.10}$$

where $M'(\boldsymbol{\rho}) \in \Re^{s\times s}$ represents the inertia matrix, $C'(\boldsymbol{\rho}, \dot{\boldsymbol{\rho}}) \in \Re^{s\times s}$ is the matrix of terms arising from the centripetal and Coriolis forces, $\mathbf{g}'(\boldsymbol{\rho}) \in \Re^s$ represents the vector of forces due to gravity, $F' \in \Re^{s\times s}$ represents the diagonal matrix of viscous friction coefficients, $\boldsymbol{\tau}' \in R^s$ is the vector of generalized forces associated with scalar variables of $\boldsymbol{\rho}, D(\boldsymbol{\rho}) = \frac{\partial \gamma(\boldsymbol{\rho})}{\partial \boldsymbol{\rho}} \in \Re^{r\times s}$ is the Jacobian matrix of the system holonomic constraints $\gamma(\boldsymbol{\rho}) \in \Re^r$, or the constraint Jacobian, and $\boldsymbol{\lambda} \in \Re^r$ is the vector of Lagrange multipliers.

Notice that Eq. (9.1) constitutes a set of DAEs. There are several methods to transform the DAEs into ODEs (Khan et al. 2005; Soto and Campa 2014). The purpose of such a transformation is to be able to apply standard numerical methods for solving the ODEs rather than the DAEs. One of the most important methods is the method of projection via the constraint Jacobian. According to Soto and Campa (2014, 2015), this method consists of finding a matrix $R(\boldsymbol{\rho})$ whose column space belongs to the null space of $D(\boldsymbol{\rho})$, i.e., $D(\boldsymbol{\rho})R(\boldsymbol{\rho}) = 0$. By considering $\dot{\boldsymbol{q}} = \mathrm{d}\boldsymbol{q}/\mathrm{d}t$ as the vector of independent velocities and $\dot{\boldsymbol{\rho}} = \mathrm{d}\boldsymbol{\rho}/\mathrm{d}t$ as the vector of feasible dependent velocities of a constrained body that belong to the space spanned by the columns of $R(\boldsymbol{\rho})$, we obtain the expression

$$\dot{\boldsymbol{\rho}} = R(\boldsymbol{\rho})\dot{\boldsymbol{q}} \tag{9.11}$$

with $R(\boldsymbol{\rho}) \in \Re^{s\times n}$. Notice that, given the differential kinematic model $\dot{\boldsymbol{\beta}} = J_\beta(\boldsymbol{\rho})\dot{\boldsymbol{q}}$, the matrix $R(\boldsymbol{\rho})$ can be constructed as

$$R(\boldsymbol{\rho}) = \begin{bmatrix} I_n \\ J_\beta(\boldsymbol{\rho}) \end{bmatrix} \tag{9.12}$$

where I_n is the identity matrix of dimensions $n \times n$. It can be proven that, by substituting the expression in Eq. (9.11) and its temporal derivative in model Eq. (9.10), it can be written as

$$M(\boldsymbol{\rho})\boldsymbol{q} + C(\boldsymbol{\rho}, \dot{\boldsymbol{\rho}})\dot{\boldsymbol{q}} + \boldsymbol{g}(\boldsymbol{\rho}) + F\dot{\boldsymbol{q}} = \boldsymbol{\tau} \tag{9.13}$$

where

$$M(\boldsymbol{\rho}) = R^T(\boldsymbol{\rho})M'(\boldsymbol{\rho})R(\boldsymbol{\rho}) \tag{9.14}$$

$$C(\boldsymbol{\rho}, \dot{\boldsymbol{\rho}}) = R^T(\boldsymbol{\rho})M'(\boldsymbol{\rho})\dot{R}(\boldsymbol{\rho}) + R^T(\boldsymbol{\rho})C'(\boldsymbol{\rho}, \dot{\boldsymbol{\rho}})R(\boldsymbol{\rho}) \tag{9.15}$$

$$\boldsymbol{g}(\boldsymbol{\rho}) = R^T(\boldsymbol{\rho})\boldsymbol{g}'(\boldsymbol{\rho}) \tag{9.16}$$

$$F = R^T(\boldsymbol{\rho})F'R(\boldsymbol{\rho}) \tag{9.17}$$

$$\boldsymbol{\tau} = R^T(\boldsymbol{\rho})\boldsymbol{\tau}' \tag{9.18}$$

Notice that the term of Eq. (9.10) containing the product of the constraint Jacobian by the Lagrange multipliers vanishes because it belongs to the null space of $R(\boldsymbol{\rho})$, as it was pointed above.

Ghorbel et al. (2000) proven that there exist a unique parametrization $\boldsymbol{\rho} = \boldsymbol{\eta}(\boldsymbol{q})$ of $\boldsymbol{\rho} \in N_\rho$ inside a neighborhood N_ρ, whenever the system is not in a singular configuration. Moreover, Muller (2005) established that, for a parallel machine, a subset $\boldsymbol{q}$ of $\boldsymbol{n}$ joint variables determines its configuration, in virtue of that exist a smooth mapping φ that assigns to each $\boldsymbol{\rho}$ the parallel machine configuration as $\boldsymbol{\rho} = \varphi(\boldsymbol{q})$, where the map φ^{-1} is a local parametrization of the n dimensional manifold V, such as $V = \{\boldsymbol{\rho} \in V^n; \gamma(\boldsymbol{\rho}) = 0\}$, where V represents the set of all admissible configurations of the parallel machine, and $\gamma(\boldsymbol{\rho}) = 0$ represents the holonomic constraints.

In consequence, we can write down, without loss of generality, the matrices and vectors of the dynamic model $M(\boldsymbol{\rho})$, $C(\boldsymbol{\rho}, \dot{\boldsymbol{\rho}})$ and $g(\boldsymbol{\rho})$ as $M(\boldsymbol{q})$, $C(\boldsymbol{q}, \dot{\boldsymbol{q}})$ and $g(\boldsymbol{q})$, respectively. Thus, the dynamic model Eq. (9.10) takes the form

$$M(\boldsymbol{q})\boldsymbol{q} + C(\boldsymbol{q}, \dot{\boldsymbol{q}})\dot{\boldsymbol{q}} + \boldsymbol{g}(\boldsymbol{q}) + F\dot{\boldsymbol{q}} = \boldsymbol{\tau} \tag{9.19}$$

The model Eq. (9.19) exhibits the following properties.

Property 1 Cheng et al. 2003
The inertia matrix $M(\boldsymbol{q})$ is symmetric and positive definite.

Property 2 *The inertia matrix $M(\boldsymbol{q})$ is bounded as*

$$\|M(\boldsymbol{q})\| \leq M_M$$

where M_M is a positive finite constant, whenever the robot is not in singular configuration.

Proof Since $||A^T B|| \leq ||A|| ||B||$, from Eq. (9.14) we can write

$$||R^T M' R|| \leq ||R^T|| \, ||M'|| \, ||R||$$

Norm $||M'||$ is upper bounded whenever its entries are finite. For robots with only revolute joints, this is assured because entries of matrix M' are sinusoidal functions of joint variables with constant coefficients. On the other hand, $||R||$ is upper bounded whenever its entries are finite, that is to say, matrix $||R||$ is well posed. From Eq. (9.12) it can be noticed that $||R||$ is well posed whenever there exists a continuous mapping between $\dot{\boldsymbol{p}}$ and $\dot{\boldsymbol{q}}$; i.e., the robot is not in singular configuration.

Property 3 (Ghorbel et al. 2000; Cheng et al. 2003)
The matrix $\frac{1}{2}\dot{M}(\boldsymbol{q}) - C(\boldsymbol{q}, \dot{\boldsymbol{q}})$ is skew-symmetric.

Property 4 (Khalil and Dombre 2004)
There exists a constant $k_C > 0$ such that $||C_r(\boldsymbol{q}, \dot{\boldsymbol{q}})|| \leq k_C ||\dot{\boldsymbol{q}}||$, for all $\boldsymbol{q} \in \Re^n$.

Property 5 *The friction matrix F can be bounded as*

$$f_m I \leq F \leq f_M I$$

9.3 Finite-Time Nonlinear PID Controller

The solution of the problem of global finite-time regulation of a robot manipulator implies finding input torques for the actuators of the manipulator in order to reach a desired position $\boldsymbol{q}_d$, such that for any initial state $(\boldsymbol{q}(0), \dot{\boldsymbol{q}}(0))$, $\tilde{\boldsymbol{q}} = \boldsymbol{q} - \boldsymbol{q}_d \rightarrow 0$ and $\dot{\boldsymbol{q}}(t) \rightarrow 0$ in finite time.

In this work, we propose to apply the finite-time regulation controller, inspired in Su and Zheng (2017)

$$\tau = -K_p \operatorname{Sig}^{\alpha_1}(\tilde{\boldsymbol{q}}) - K_d \operatorname{Sig}^{\alpha 2}(\boldsymbol{\eta}) - k_{p0} - K_I \int_0^t \boldsymbol{\eta}(\sigma)\, \mathrm{d}\sigma - k_{d0}\dot{\boldsymbol{q}} \tag{9.20}$$

to a CCM, with

$$\eta = \dot{\boldsymbol{q}} + \alpha_0 \tanh(\tilde{\boldsymbol{q}}) \tag{9.21}$$

and

$$\varphi = \int_0^t \eta(\sigma)\,\mathrm{d}\sigma \tag{9.22}$$

K_p, K_I and K_d are positive definite constant diagonal control gain matrices, respectively; k_{p0} and k_{d0} are positive constants and $0<\alpha_1<1$, while $\alpha_2 = 2\alpha_1/(\alpha_1+1)$. By substituting Eqs. (9.20) and (9.22) in Eq. (9.19), we obtain

$$\begin{aligned} &M(\boldsymbol{q})\ddot{\boldsymbol{q}} + C(\boldsymbol{q},\dot{\boldsymbol{q}})\dot{\boldsymbol{q}} + F\dot{\boldsymbol{q}} \\ &\quad + K_p \operatorname{Sig}^{\alpha_1}(\tilde{\boldsymbol{q}}) + K_d \operatorname{Sig}^{\alpha_2}(\boldsymbol{\eta}) + k_{p0}\tilde{\boldsymbol{q}} + k_{d0}\dot{\boldsymbol{q}} + K_I\varphi = 0 \end{aligned} \tag{9.23}$$

With Eqs. (9.22) and (9.23), and taking into account that $\dot{\tilde{\boldsymbol{q}}} = \dot{\boldsymbol{q}}$ when $\boldsymbol{q}_d = 0$, the closed-loop equation can be written as

$$\frac{\mathrm{d}}{\mathrm{d}t}\begin{bmatrix} \tilde{\boldsymbol{q}} \\ \dot{\boldsymbol{q}} \\ \varphi \end{bmatrix} = \begin{bmatrix} \dot{\boldsymbol{q}} \\ -M^{-1}(\boldsymbol{q})[C(\boldsymbol{q},\dot{\boldsymbol{q}})\dot{\boldsymbol{q}} + F\dot{\boldsymbol{q}} + K_p \operatorname{Sig}^{\alpha_1}(\tilde{\boldsymbol{q}}) + K_d \operatorname{Sig}^{\alpha_2}(\boldsymbol{\eta}) \\ +K_I\varphi + k_{p0}\tilde{\boldsymbol{q}} + k_{d0}\dot{\boldsymbol{q}}] \\ \dot{\boldsymbol{q}} + \alpha_0 \operatorname{Tanh}(\tilde{\boldsymbol{q}}) \end{bmatrix} \tag{9.24}$$

Notice that the origin of the system Eq. (9.24) is the only equilibrium of the system.

9.3.1 Stability Analysis of the Closed-Loop System

By proceeding inspired in Su and Zheng (2017), we study the global asymptotical stability of Eq. (9.24). First, we propose the Lyapunov function candidate

$$\begin{aligned} V(\tilde{\boldsymbol{q}},\dot{\boldsymbol{q}},\varphi) = &\frac{1}{2}\dot{\boldsymbol{q}}^T M(\boldsymbol{q})\dot{\boldsymbol{q}} + \alpha_0 \operatorname{Tanh}^T(\tilde{\boldsymbol{q}})M(\boldsymbol{q})\dot{\boldsymbol{q}} + \frac{1}{2}k_{p0}\tilde{\boldsymbol{q}}^T\tilde{\boldsymbol{q}} \\ &+ \frac{1}{\alpha_1+1}\sum_{i=1}^{n} k_{pi}|\tilde{q}_i|^{\alpha_1+1} + \alpha_0\sum_{i=1}^{n}(f_i + k_{d0})\ln(\cosh(\tilde{q}_i)) + \frac{1}{2}\varphi^T K_I \varphi \end{aligned} \tag{9.25}$$

where f_i is the ith entry of friction matrix F. In order to investigate positive definiteness of Eq. (9.25), notice that in virtue of

$$\begin{aligned} &\frac{1}{4}\dot{\boldsymbol{q}}^T M(\boldsymbol{q})\dot{\boldsymbol{q}} + \alpha_0 \operatorname{Tanh}^T(\tilde{\boldsymbol{q}})M(\boldsymbol{q})\dot{\boldsymbol{q}} + \frac{1}{2(\alpha_1+1)}\sum_{i=1}^{n} k_{pi}|\tilde{q}_i|^{\alpha_1+1} \\ &\geq \frac{1}{2(\alpha_1+1)}\sum_{i=1}^{n}\left[k_{pi} - 2(\alpha_1+1)\alpha_0^2 M_M\right]\tanh^2(\tilde{q}_i) \end{aligned}$$

where we have used Property 2 and Eq. (9.5), we can lower bound Eq. (9.25) as

$$V \geq \frac{1}{2(\alpha_1+1)} \sum_{i=1}^{n} \left[k_{pi} - 2(\alpha_1+1)\alpha_0^2 M_M\right] \tanh^2(\tilde{q}_i) + \frac{1}{2} k_{p0} \tilde{\boldsymbol{q}}^T \tilde{\boldsymbol{q}} \\ + \frac{1}{2} \varphi^T K_I \varphi + \frac{1}{4} \dot{\boldsymbol{q}}^T M(\boldsymbol{q}) \dot{\boldsymbol{q}} + \alpha_0 \sum_{i=1}^{n} (f_i + k_{d0}) \ln(\cosh(\tilde{q}_i)) \tag{9.26}$$

The three last terms of the right side of inequality Eq. (9.26) can be lower bounded as

$$\frac{1}{2} \varphi^T K_I \varphi \geq \frac{1}{2} \lambda_{\min}\{K_I\} \|\varphi\|^2 > 0, \forall \varphi \neq 0 \in \Re^n$$
$$\frac{1}{4} \dot{\boldsymbol{q}}^T M(\boldsymbol{q}) \dot{\boldsymbol{q}} \geq \frac{1}{4} \lambda_{\min}\{M(\boldsymbol{q})\} \|\dot{\boldsymbol{q}}\|^2 > 0, \forall \dot{\boldsymbol{q}} \neq 0 \in \Re^n$$
$$\alpha_0 \sum_{i=1}^{n} (f_i + k_{d0}) \ln(\cosh(\tilde{q}_i)) \geq \alpha_0 \sum_{i=1}^{n} (f_i + k_{d0}) \mathrm{e} > 0$$

The second term of the right side of inequality Eq. (9.26) is positive definite since $\frac{1}{2} k_{p0} \tilde{\boldsymbol{q}}^T \tilde{\boldsymbol{q}} = \frac{1}{2} k_{p0} \|\tilde{\boldsymbol{q}}\|^2$. Notice that the first term or the right side of Eq. (9.26) is positive as long as $k_{pi} - 2(\alpha_1+1)\alpha_0^2 M_M$ is positive, i.e.,

$$k_{pi} > 2(\alpha_1+1)\alpha_0^2 M_M \tag{9.27}$$

Therefore, since the fourth last terms of the right side of Eq. (9.26) are positive definite for all $\tilde{\boldsymbol{q}}, \dot{\boldsymbol{q}}, \varphi \neq 0 \in \Re^n$, the Lyapunov candidate function Eq. (9.25) is positive definite while Eq. (9.27) is satisfied.

The temporal derivative of the Lyapunov function candidate Eq. (9.25) is as follows:

$$\dot{V}(\tilde{\boldsymbol{q}}, \dot{\boldsymbol{q}}, \varphi) = \frac{1}{2} \dot{\boldsymbol{q}}^T \dot{M}(\boldsymbol{q}) \dot{\boldsymbol{q}} + \dot{\boldsymbol{q}}^T M(\boldsymbol{q}) \ddot{\boldsymbol{q}} + \alpha_0 (\mathrm{Sech}^2(\tilde{\boldsymbol{q}}) \dot{\tilde{\boldsymbol{q}}})^T M(\boldsymbol{q}) \dot{\boldsymbol{q}} \\ + \alpha_0 \mathrm{Tanh}^T(\tilde{\boldsymbol{q}}) \dot{M}(\boldsymbol{q}) \dot{\boldsymbol{q}} + \alpha_0 \mathrm{Tanh}^T(\tilde{\boldsymbol{q}}) M(\boldsymbol{q}) \ddot{\boldsymbol{q}} + k_{p0} \dot{\tilde{\boldsymbol{q}}}^T \tilde{\boldsymbol{q}} \\ + \dot{\tilde{\boldsymbol{q}}}^T K_p \,\mathrm{Sig}^{\alpha_1}(\tilde{\boldsymbol{q}}) + \alpha_0 \mathrm{Tanh}^T(\tilde{\boldsymbol{q}})(F + k_{d0} I) \dot{\tilde{\boldsymbol{q}}} + \dot{\varphi}^T K_I \varphi \tag{9.28}$$

Along the trajectories of the closed-loop system in Eq. (9.24), we obtain

$$\dot{V} = -\dot{\boldsymbol{q}}^T F \dot{\boldsymbol{q}} - k_{d0} \dot{\boldsymbol{q}}^T \dot{\boldsymbol{q}} - \alpha_0 \mathrm{Tanh}(\tilde{\boldsymbol{q}}) K_p \,\mathrm{Sig}^{\alpha_1}(\tilde{\boldsymbol{q}}) - \boldsymbol{\eta}^T K_d \,\mathrm{Sig}^{\alpha_2}(\boldsymbol{\eta}) \\ + \alpha_0 [\mathrm{Tanh}(\tilde{\boldsymbol{q}}) C(\boldsymbol{q}, \dot{\boldsymbol{q}}) \dot{\boldsymbol{q}} + (\mathrm{Sech}^2(\tilde{\boldsymbol{q}}) \dot{\boldsymbol{q}})^T M(\boldsymbol{q}) \dot{\boldsymbol{q}}] - \alpha_0 \mathrm{k}_{p0} \mathrm{Tanh}(\tilde{\boldsymbol{q}}) \tilde{\boldsymbol{q}} \tag{9.29}$$

where we have used the Property 3 (skew symmetry). Here, we neglect the gravitational forces vector from Eq. (9.24) since the CCM is a horizontal Five-Bar

Mechanism, in which motion of interest is not subject to gravitational forces. The two parts of the fifth term of the right side of Eq. (9.29) can be upper bounded as

$$\begin{aligned}
\mathrm{Tanh}(\tilde{\boldsymbol{q}})C(\boldsymbol{q},\dot{\boldsymbol{q}})\dot{\boldsymbol{q}} &\le \|\mathrm{Tanh}(\tilde{\boldsymbol{q}})C(\boldsymbol{q},\dot{\boldsymbol{q}})\dot{\boldsymbol{q}}\| \\
&\le \|\mathrm{Tanh}(\tilde{\boldsymbol{q}})\|\|C(\boldsymbol{q},\dot{\boldsymbol{q}})\|\|\dot{\boldsymbol{q}}\| \\
&\le \sqrt{n}k_C\|\dot{\boldsymbol{q}}\|^2 \\
(\mathrm{Sech}^2(\tilde{\boldsymbol{q}})\dot{\boldsymbol{q}})^T M(\boldsymbol{q})\dot{\boldsymbol{q}} &\le \|(\mathrm{Sech}^2(\tilde{\boldsymbol{q}})\dot{\boldsymbol{q}})^T M(\boldsymbol{q})\dot{\boldsymbol{q}}\| \\
&\le \|(\mathrm{Sech}^2(\tilde{\boldsymbol{q}})\dot{\boldsymbol{q}})^T\|\|M(\boldsymbol{q})\|\|\dot{\boldsymbol{q}}\| \\
&\le M_M\|\dot{\boldsymbol{q}}\|^2
\end{aligned}$$

where we have used Eq. (9.2), Property 2, and Property 4. Thus, the fifth term of the right side of Eq. (9.29) can be upper bounded as

$$\begin{aligned}
&\alpha_0[\mathrm{Tanh}(\tilde{\boldsymbol{q}})C(\boldsymbol{q},\dot{\boldsymbol{q}})\dot{\boldsymbol{q}} + (\mathrm{Sech}^2(\tilde{\boldsymbol{q}})\dot{\boldsymbol{q}})^T M(\boldsymbol{q})\dot{\boldsymbol{q}}] \\
&\le \alpha_0(\sqrt{n}k_C + M_M)\|\dot{\boldsymbol{q}}\|^2
\end{aligned} \tag{9.30}$$

In addition, by using Property 5 the first term of the right side of Eq. (9.29) can be upper bounded as

$$-\dot{\boldsymbol{q}}^T F\dot{\boldsymbol{q}} \le -f_m\|\dot{\boldsymbol{q}}\|^2 \tag{9.31}$$

After substituting Eqs. (9.30) and (9.31) in Eq. (9.29) and rearranging terms, we can upper bound Eq. (9.29) as

$$\begin{aligned}
\dot{V} \le &- [f_m + k_{d0} - \alpha_0(\sqrt{n}k_C + M_M)]\|\dot{\boldsymbol{q}}\|^2 - \alpha_0\mathrm{Tanh}(\tilde{\boldsymbol{q}})K_p\,\mathrm{Sig}^{\alpha_1}(\tilde{\boldsymbol{q}}) \\
&- \boldsymbol{\eta}^T K_d\,\mathrm{Sig}^{\alpha_2}(\boldsymbol{\eta}) - \alpha_0 \mathrm{k}_{\mathrm{p0}}\mathrm{Tanh}(\tilde{\boldsymbol{q}})\tilde{\boldsymbol{q}}
\end{aligned}$$

In virtue of that, $\tanh(\boldsymbol{x})$ and $\boldsymbol{x}$ have the same sign, and then $\mathrm{Tanh}(\tilde{\boldsymbol{q}})\tilde{\boldsymbol{q}} > 0, \forall\tilde{\boldsymbol{q}} \neq 0$. Therefore, we can write

$$\begin{aligned}
\dot{V} \le &- [f_m + k_{d0} - \alpha_0(\sqrt{n}k_C + M_M)]\|\dot{\boldsymbol{q}}\|^2 - \alpha_0\mathrm{Tanh}(\tilde{\boldsymbol{q}})K_p\,\mathrm{Sig}^{\alpha_1}(\tilde{\boldsymbol{q}}) \\
&- \boldsymbol{\eta}^T K_d\,\mathrm{Sig}^{\alpha_2}(\boldsymbol{\eta})
\end{aligned} \tag{9.32}$$

After using the expression in Eq. (9.4), Eq. (9.32) can be rewritten as

$$\begin{aligned}
\dot{V} \le &- [f_m + k_{d0} - \alpha_0(\sqrt{n}k_C + M_M)]\|\dot{\boldsymbol{q}}\|^2 - \alpha_0\sum_{i=1}^{n} k_{pi}|\tanh(\tilde{q}_i)||\tilde{q}_i|^{\alpha_1} \\
&- \sum_{i=1}^{n} k_{di}|\eta_i|^{\alpha_2+1}
\end{aligned} \tag{9.33}$$

where k_{pi} and k_{di} represent the ith diagonal elements of matrices K_p and K_d, respectively. Therefore, we can conclude that $\dot{V} \leq 0$ as long as

$$k_{d0} > \alpha_0(\sqrt{n}k_C + M_M) - f_m \tag{9.34}$$

is satisfied. In order to conclude the global asymptotical stability of the closed-loop system Eq. (9.24), by LaSalle's theorem (Kelly et al. 2005), we have that $\tilde{\boldsymbol{q}}(t) \to 0$, $\dot{\boldsymbol{q}}(t) \to 0$ and $\varphi(t) \to 0$ when $t \to \infty$ for any initial state. Thus, we conclude the global asymptotical stability of origin of the closed-loop system Eq. (9.24).

9.3.2 Finite-Time Stability

In this section, we will apply the concepts of finite-time stability in order to establish, in a similar way to Su and Zheng (2017), the stability of the closed-loop system in finite time. We first define the state vector $\boldsymbol{y} = \begin{bmatrix} \boldsymbol{y}_1^T & \boldsymbol{y}_2^T & \boldsymbol{y}_3^T \end{bmatrix}^T$ where $\boldsymbol{y}_1 = \tilde{\boldsymbol{q}}$, $\boldsymbol{y}_2 = \boldsymbol{\eta}$ and $\boldsymbol{y}_3 = \varphi$. The closed-loop system of state vector $\boldsymbol{y}$ can be written as

$$\frac{\mathrm{d}}{\mathrm{d}t}\begin{bmatrix} \boldsymbol{y}_1 \\ \boldsymbol{y}_2 \\ \boldsymbol{y}_3 \end{bmatrix} = \begin{bmatrix} \boldsymbol{y}_2 - \alpha_0 \mathrm{Tanh}(\boldsymbol{y}_1) \\ -\mathrm{M}^{-1}(\boldsymbol{y}_1 + \boldsymbol{q}_d)[(C(\boldsymbol{y}_1 + \boldsymbol{q}_d, \boldsymbol{y}_2 - \alpha_0\,\mathrm{Tanh}(\boldsymbol{y}_1)) \\ + F + k_{d0}I)(\boldsymbol{y}_2 - \alpha_0\,\mathrm{Tanh}(\boldsymbol{y}_1)) + K_p\,\mathrm{Sig}^{\alpha_1}(\boldsymbol{y}_1) \\ + K_d \mathrm{Sig}^{\alpha_2}(\boldsymbol{y}_2) + k_{p0}\boldsymbol{y}_1 + K_I \boldsymbol{y}_3] \\ + \alpha_0(\mathrm{Sech}^2(\boldsymbol{y}_1)(\boldsymbol{y}_2 - \alpha_0\,\mathrm{Tanh}(\boldsymbol{y}_1))) \\ \boldsymbol{y}_2 \end{bmatrix} \tag{9.35}$$

Notice that the origin $\boldsymbol{y} = 0 \in \Re^{3n}$ is the equilibrium of Eq. (9.35). Equation (9.35) can be rewritten as

$$\frac{\mathrm{d}}{\mathrm{d}t}\begin{bmatrix} \boldsymbol{y}_1 \\ \boldsymbol{y}_2 \\ \boldsymbol{y}_3 \end{bmatrix} = \begin{bmatrix} \boldsymbol{y}_2 + \hat{\boldsymbol{h}}_1(y) \\ -\mathrm{M}^{-1}(\boldsymbol{q}_d)\left[K_p\,\mathrm{Sig}^{\alpha_1}(\boldsymbol{y}_1) + K_d \mathrm{Sig}^{\alpha_2}(\boldsymbol{y}_2)\right] + \hat{\boldsymbol{h}}_2(y) \\ \boldsymbol{y}_2 \end{bmatrix} \tag{9.36}$$

where

$$\hat{\boldsymbol{h}}_1 = -\alpha_0 \mathrm{Tanh}(\boldsymbol{y}_1) \tag{9.37}$$

$$\begin{aligned} \hat{\boldsymbol{h}}_2 = &-M^{-1}(\boldsymbol{y}_1 + \boldsymbol{q}_d)[(C(\boldsymbol{y}_1 + \boldsymbol{q}_d, \boldsymbol{y}_2 - \alpha_0 \mathrm{Tanh}(\boldsymbol{y}_1)) \\ &+ F + k_{d0}I)(\boldsymbol{y}_2 - \alpha_0 \mathrm{Tanh}(\boldsymbol{y}_1)) + k_{p0}\boldsymbol{y}_1 + K_I(\boldsymbol{y}_3)\boldsymbol{y}_3] \\ &- \widetilde{M}(\boldsymbol{y}_1, \boldsymbol{q}_d)[K_p\,\mathrm{Sig}^{\alpha_1}(\boldsymbol{y}_1) + K_d\,\mathrm{Sig}^{\alpha_2}(\boldsymbol{y}_2)] \\ &+ \alpha_0(\mathrm{Sech}^2(\boldsymbol{y}_1)(\boldsymbol{y}_2 - \alpha_0\,\mathrm{Tanh}\,(\boldsymbol{y}_1))) \end{aligned} \tag{9.38}$$

$$\widetilde{M}(y_1,q_d) = -M^{-1}(y_1+q_d) - M^{-1}(y_1+q_d) \tag{9.39}$$

Now consider the closed-loop system

$$\frac{\mathrm{d}}{\mathrm{d}t}\begin{bmatrix} y_1 \\ y_2 \\ y_3 \end{bmatrix} = \begin{bmatrix} y_2 \\ -\mathrm{M}^{-1}(q_d)\left[K_p \operatorname{Sig}^{\alpha_1}(y_1) + K_d \operatorname{Sig}^{\alpha_2}(y_2)\right] \\ y_2 \end{bmatrix} \tag{9.40}$$

which, according to Definition 1, is homogeneous if the following expressions are satisfied:

$$\begin{aligned} p_2 &= d + p_1 \\ \alpha_1 p_1 &= \alpha_2 p_2 = d + p_2 \\ p_2 &= d + p_3 \end{aligned} \tag{9.41}$$

It can be verified that with the values $p_1 = 2$, $p_2 = \alpha_1 + 1$, $p_3 = 2, \alpha_2 = 2\alpha_1/(\alpha_1+1)$, and $d = \alpha_1 - 1$, Eq. (9.41) is satisfied. Moreover, selecting α_1 such that $0 < \alpha_1 < 1$ results in $d = \alpha_1 - 1 < 0$. Then, it can be concluded that Eq. (9.40) is homogeneous of degree $d = \alpha_1 - 1 < 0$. Notice that $\boldsymbol{h}(0) = 0$, from Eq. (9.40), and $\hat{\boldsymbol{h}}(0) = 0$ from Eqs. (9.37) to (9.38).

In order to prove the asymptotical stability of the equilibrium $\boldsymbol{y} = 0$ of the system Eq. (9.40), we propose the positive definite Lyapunov candidate function

$$V_2 = \frac{1}{\alpha_1+1}\sum_{i=1}^{n} k_{pi}|y_{1i}|^{\alpha_1+1} + \frac{1}{2}y_2^T M(q_d)y_2 + \frac{1}{2}(y_1-y_3)^T(y_1-y_3) \tag{9.42}$$

where y_{1i} denotes the ith component of vector y_1. The temporal derivative of Eq. (9.42) is

$$\dot{V}_2 = \dot{\boldsymbol{y}}_1^T K_p \operatorname{Sig}^{\alpha_1}(\boldsymbol{y}_1) + \boldsymbol{y}_2^T M(\boldsymbol{q}_d)\dot{y}_2 + (\dot{y}_1 - \dot{y}_3)^T(\boldsymbol{y}_1 - y_3) \tag{9.43}$$

where it was taken into account the fact that $\dot{M}(\boldsymbol{q}_d) = 0$ when $\boldsymbol{q}_d$ is constant. After substituting Eq. (9.40) in Eq. (9.43), we have

$$\dot{V}_2 = \boldsymbol{y}_2^T K_d \operatorname{Sig}^{\alpha_2}(\boldsymbol{y}_2) \tag{9.44}$$

By using Eq. (9.4) in Eq. (9.44), it can be concluded that $\dot{V}_2 \le 0$, which implies that the origin is a stable equilibrium. By using the LaSalle invariance theorem (Kelly et al. 2005), the global asymptotical stability of the origin can be concluded.

Now consider the hyperbolic tangent function:

$$\mathrm{Tanh}(\varepsilon^{p_1}\mathbf{y}_1) = o(\varepsilon^{p_1}\mathbf{y}_1) \tag{9.45}$$

where $o(\varepsilon^{p_1}\mathbf{y}_1)$ means to be of order $\varepsilon^{p_1}\mathbf{y}_1$ as $\varepsilon^{p_1}\mathbf{y}_1 \to 0$. Therefore, for any fixed $\mathbf{y} = (\mathbf{y}_1^T\, \mathbf{y}_2^T\, \mathbf{y}_3^T)^T \in \Re^{3n}$, we have

$$\begin{aligned}\lim_{\varepsilon\to 0}\frac{\hat{\mathbf{h}}_1(\varepsilon^{p_1}\mathbf{y}_1, \varepsilon^{p_2}\mathbf{y}_2, \varepsilon^{p_3}\mathbf{y}_3)}{\varepsilon^{d+p_1}} &= -\alpha_0 \lim_{\varepsilon\to 0}\frac{\mathrm{Tanh}(\varepsilon^{p_1}\mathbf{y}_1)}{\varepsilon^{d+p_1}} \\ &= -\alpha_0 \lim_{\varepsilon\to 0} o(\varepsilon^{-d}\mathbf{y}_1) = 0\end{aligned} \tag{9.46}$$

Since $M^{-1}(\mathbf{y}_1 + \mathbf{q}_d)$ and $C(\mathbf{y}_1 + \mathbf{q}_d, \mathbf{y}_2)$ are smooth [see Hong et al. (2002); Su and Zheng (2009)], we obtain

$$\begin{aligned}&\lim_{\varepsilon\to 0} -\frac{M^{-1}(\varepsilon^{p_1}\mathbf{y}_1 + \mathbf{q}_d)}{\varepsilon^{d+p_2}}[(C(\varepsilon^{p_1}\mathbf{y}_1 + \mathbf{q}_d, \varepsilon^{p_2}\mathbf{y}_2 - \alpha_0\,\mathrm{Tanh}\,(\varepsilon^{p_1}\mathbf{y}_1)) \\ &\quad + F + k_{d0}I)(\varepsilon^{p_2}\mathbf{y}_2 - \alpha_0\,\mathrm{Tanh}(\varepsilon^{p_1}\mathbf{y}_1))] \\ &= -M^{-1}(\mathbf{q}_d)[(C(\mathbf{q}_d, 0) + F + k_{d0}I) \\ &\quad \times (\mathbf{y}_2 \lim_{\varepsilon\to 0}\varepsilon^{-d} - \alpha_0 \lim_{\varepsilon\to 0} o(\varepsilon^{p_1-d-p_2}\mathbf{y}_1))] = 0\end{aligned} \tag{9.47}$$

and

$$\begin{aligned}&\lim_{\varepsilon\to 0} -\frac{M^{-1}(\varepsilon^{p_1}\mathbf{y}_1 + \mathbf{q}_d)}{\varepsilon^{d+p_2}}[k_{p0}\varepsilon^{p_1}\mathbf{y}_1 + K_I(\varepsilon^{p_3}\mathbf{y}_3)] \\ &= -M^{-1}(\mathbf{q}_d)\left[k_{p0}\mathbf{y}_1 \lim_{\varepsilon\to 0}\varepsilon^{p_1-d-p_2} - K_I\mathbf{y}_3 \lim_{\varepsilon\to 0}\varepsilon^{p_3-d-p_2}\right] = 0\end{aligned}$$

After applying the mean value theorem to each entry of $\widetilde{M}(y_1, q_d)$ yields

$$\widetilde{M}(\varepsilon^{p_1}\mathbf{y}_1, \mathbf{q}_d) = M^{-1}(\varepsilon^{p_1}\mathbf{y}_1 + \mathbf{q}_d) - M^{-1}(\mathbf{q}_d) = o(\varepsilon^{p_1}) \tag{9.48}$$

which results in

$$\begin{aligned}&\lim_{\varepsilon\to 0} -\frac{\widetilde{M}(\varepsilon^{p_1}\mathbf{y}_1, \mathbf{q}_d)\lfloor K_p\,\mathrm{Sig}^{\alpha_1}(\varepsilon^{p_1}\mathbf{y}_1) + K_d\,\mathrm{Sig}^{\alpha_2}(\varepsilon^{p_2}\mathbf{y}_2)\rfloor}{\varepsilon^{d+p_2}} \\ &= \lim_{\varepsilon\to 0} o(\varepsilon^{p_1-d-p_2}) = 0\end{aligned} \tag{9.49}$$

Moreover, in virtue of a property of the ordinary hyperbolic secant function applied to Eq. (9.2), $\mathrm{Sech}^2(0) = I$. Then,

$$\lim_{\varepsilon\to 0} -\frac{\alpha_0(\mathrm{Sech}^2(\varepsilon^{p_1}\boldsymbol{y}_1)(\varepsilon^{p_2}\boldsymbol{y}_2 - \alpha_0\mathrm{Tanh}(\boldsymbol{y}_1)))}{\varepsilon^{d+p_2}} \\ = \alpha_0\boldsymbol{y}_2 \lim_{\varepsilon\to 0}\varepsilon^{-d} - \alpha_0^2 \lim_{\varepsilon\to 0} o(\varepsilon^{p_1-d-p_2}\boldsymbol{y}_1) = 0 \tag{9.50}$$

where $p_3 - d - p_2 = p_1 - d - p_2 = 2(1-\alpha_1) > 0$ and $-d = 1 - \alpha_1 > 0$ for $0 < \alpha_1 < 1$. Then, for any fixed $\boldsymbol{y} = (\boldsymbol{y}_1^T\ \boldsymbol{y}_2^T\ \boldsymbol{y}_3^T)^T \in \Re^{3n}$, we have

$$\lim_{\varepsilon\to 0}\frac{\hat{\boldsymbol{h}}_2(\varepsilon^{p_1}\boldsymbol{y}_1, \varepsilon^{p_2}\boldsymbol{y}_2, \varepsilon^{p_3}\boldsymbol{y}_3)}{\varepsilon^{d+p_2}} = 0 \tag{9.51}$$

Thus, according to Lemma 1, the finite-time stability of the system Eq. (9.35) is proven. Moreover, by invoking Lemma 2, the global finite-time stability of the system Eq. (9.35) is proven.

9.4 Simulations

In order to show the feasibility of the proposed application of the finite-time regulation controller for a parallel manipulator, we carried out numerical simulations. Simulations of the finite-time nonlinear PID controller applied to the model of a real horizontal Five-Bar Mechanism were carried out.

9.4.1 Model of the Five-Bar Mechanism

A Five-Bar Mechanism is a planar parallel manipulator of two degrees of freedom. A scheme of the Five-Bar Mechanism is shown in Fig. 9.1. Notice that the structure of the mechanism is shown as an open structure. However, the extreme ends of the links 3 and 4 are joined. In the current section, the matrices $M'(\boldsymbol{\rho})$, $C'(\boldsymbol{\rho}, \dot{\boldsymbol{\rho}})$ of the model Eq. (9.10) and the matrices $M(\boldsymbol{q})$ Eq. (9.14), $C(\boldsymbol{q}, \dot{\boldsymbol{q}})$ Eq. (9.15) and $F(q)$ Eq. (9.17) of the model Eq. (9.19) (Soto and Campa 2014), including the transformation matrix $R(\boldsymbol{\rho})$ Eq. (9.12) and its temporal derivative, are shown. Note that since the Five-Bar Mechanism is horizontal, the gravitational forces vector is zero. The matrices of the model Eq. (9.10) are

$$M'(\boldsymbol{\rho}) = \begin{bmatrix} m'_{11} & 0 & m'_{13} & 0 \\ 0 & m'_{22} & 0 & m'_{24} \\ m'_{31} & 0 & m'_{33} & 0 \\ 0 & m'_{42} & 0 & m'_{44} \end{bmatrix}, C'(\boldsymbol{\rho}, \dot{\boldsymbol{\rho}}) = \begin{bmatrix} c'_{11} & 0 & c'_{13} & 0 \\ 0 & c'_{22} & 0 & c'_{24} \\ c'_{31} & 0 & 0 & 0 \\ 0 & c'_{42} & 0 & 0 \end{bmatrix}$$

Fig. 9.1 Five-Bar Mechanism

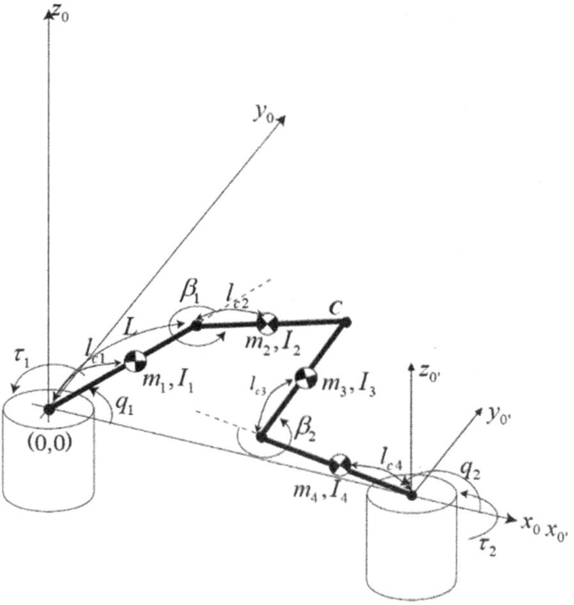

where

$$
\begin{aligned}
m'_{11} &= m_1 l_{c1}^2 + m_3(L_1^2 + l_{c3}^2 + 2L_1 l_{c3}\cos(\beta_1)) + I_1 + I_3 \\
m'_{13} &= m_3(l_{c3}^2 + L_1 l_{c3}\cos(\beta_1)) + I_3 \\
m'_{31} &= m'_{13} \\
m'_{22} &= m_2 l_{c2}^2 + m_4(L_2^2 + l_{c4}^2 + 2L_2 l_{c4}\cos(\beta_2)) + I_2 + I_4 \\
m'_{24} &= m_4(l_{c4}^2 + L_2 l_{c4}\cos(\beta_2)) + I_4 \\
m'_{42} &= m'_{24} \\
m'_{33} &= m_3 l_{c3}^2 + I_3 \\
m'_{44} &= m_4 l_{c4}^2 + I_4
\end{aligned}
$$

and

$$
\begin{aligned}
c'_{11} &= -m_3 L_1 l_{c3}\sin(\beta_1)\dot{\beta}_1 \\
c'_{13} &= -m_3 L_1 l_{c3}\sin(\beta_1)(\dot{q}_1 + \dot{\beta}_1) \\
c'_{22} &= -m_4 L_2 l_{c4}\sin(\beta_2)\dot{\beta}_2 \\
c'_{24} &= -m_4 L_2 l_{c4}\sin(\beta_2)(\dot{q}_2 + \dot{\beta}_2) \\
c'_{31} &= m_3 L_1 l_{c3}\sin(\beta_1)\dot{q}_1 \\
c'_{42} &= m_4 L_2 l_{c4}\sin(\beta_2)\dot{\beta}_2
\end{aligned}
$$

The transformation matrix $R(\rho)$ and its temporal derivative $\dot{R}(\rho)$ are

$$R(\rho)=\begin{bmatrix}1 & 0\\ 0 & 1\\ r_{11} & r_{12}\\ r_{21} & r_{22}\end{bmatrix}, \dot{R}(\rho)=\begin{bmatrix}0 & 0\\ 0 & 0\\ \dot{r}_{11} & \dot{r}_{12}\\ \dot{r}_{21} & \dot{r}_{22}\end{bmatrix}$$

where

$$\begin{aligned}
r_{11} &= -\frac{\sin(q_1-q_2-\beta_2)}{\sin(q_1-q_2+\beta_1-\beta_2)}-1\\
r_{12} &= -\frac{\sin(\beta_2)}{\sin(q_1-q_2+\beta_1-\beta_2)}\\
r_{21} &= \frac{\sin(\beta_1)}{\sin(q_1-q_2+\beta_1-\beta_2)}\\
r_{22} &= -\frac{\sin(q_1-q_2-\beta_1)}{\sin(q_1-q_2+\beta_1-\beta_2)}-1
\end{aligned}$$

The matrices of the model Eq. (9.19) are

$$M(\boldsymbol{q})=\begin{bmatrix}m_{11} & m_{12}\\ m_{21} & m_{22}\end{bmatrix} \quad C(\boldsymbol{q},\dot{\boldsymbol{q}})=\begin{bmatrix}c_{11} & c_{12}\\ c_{21} & c_{22}\end{bmatrix},$$

where

$$\begin{aligned}
m_{11} &= m'_{44}r_{22}^2+m'_{11}+m'_{13}r_{11}+r_{11}(m'_{13}+m'_{33}r_{11})\\
m_{12} &= m'_{24}r_{21}+r_{12}(m'_{13}+m'_{33}r_{11})+m'_{44}r_{21}r_{22}\\
m_{21} &= m'_{24}r_{21}+r_{12}(m'_{13}+m_{33}r_{11})+m'_{44}r_{21}r_{22}\\
m_{22} &= m'_{33}r_{12}^2+m'_{22}+m'_{24}r_{22}+r_{22}(m'_{24}+m'_{44}r_{22})
\end{aligned}$$

and

$$\begin{aligned}
c_{11} &= c'_{11}+c'_{13}r_{11}+c'_{31}r_{11}+\dot{r}_{11}(m'_{13}+m'_{33}r_{11})+m'_{44}r_{21}\dot{r}_{21}\\
c_{12} &= c'_{13}r_{12}+c'_{42}r_{21}+\dot{r}_{12}(m'_{13}+m'_{33}r_{11})+m'_{44}r_{21}\dot{r}_{22}\\
c_{21} &= c'_{31}r_{12}+c'_{24}r_{21}+\dot{r}_{21}(m'_{24}+m'_{44}r_{22})+m'_{33}r_{12}\dot{r}_{11}\\
c_{22} &= c'_{22}+c'_{24}r_{22}+c'_{42}r_{22}+\dot{r}_{22}(m'_{24}+m'_{44}r_{22})+m'_{33}r_{12}\dot{r}_{12}
\end{aligned}$$

The friction coefficients matrices of the model Eq. (9.10) and of the model Eq. (9.19) are

$$F'=\begin{bmatrix}f'_{11} & 0 & 0 & 0\\ 0 & f'_{22} & 0 & 0\\ 0 & 0 & f'_{33} & 0\\ 0 & 0 & 0 & f'_{44}\end{bmatrix}, F=\begin{bmatrix}f_{11} & f_{12}\\ f_{21} & f_{22}\end{bmatrix}$$

where

$$
\begin{aligned}
f_{11} &= f'_{11} + r_{11}^2 f'_{33} + r_{21}^2 f'_{44} \\
f_{12} &= r_{11} r_{12} f'_{33} + r_{21} r_{22} f'_{44} \\
f_{21} &= r_{11} r_{12} f'_{33} + r_{21} r_{22} f'_{44} \\
f_{22} &= f'_{22} + r_{12}^2 f'_{33} + r_{22}^2 f'_{44}
\end{aligned}
$$

The parameters of the dynamic model are shown in Table 9.1.

The elements of the matrix F of friction coefficients of the model Eq. (9.10) are shown in Table 9.2.

The desired values of the joint variables were computed using the inverse kinematic model [see Soto and Campa (2015)] based on the desired position of the end effector to reach a point $P_1 = (x_h - \Delta, y_h + \Delta)$ from the initial or home position $P_h = (x_h, y_h)$. Notice that $\Delta = 0.02$ (m). The coordinates of the initial position are $x_h = L_1,\ y_h = L_2$, with respect to the origin located at the rotation axis of the joint q_1 (see Fig. 9.1). The values of the joint variables that correspond to the home position are $q_1 = 0$ (rad) and $q_2 = 1.5708$ (rad), while the values that correspond to the point P_1 are $q_1 = 0.1686$ (rad) and $q_2 = 1.7441$ (rad). The gains and parameters of the finite-time nonlinear PID controller used in the simulations are shown in Table 9.3. These gains were selected by try and error procedure in order to achieve the best performance in terms of small position errors and at the same time, avoiding to exceed a maximum value of torque of 0.2 Nm. This maximum torque value is similar to the maximum values of electric motors that usually drive a small Five-Bar Mechanism for academic purposes.

Table 9.1 Parameters of the dynamic model of the Five-Bar Mechanism

Parameter	Value (units)	Parameter	Value (units)
L_1, L_2, L_3, L_4	0.127 (m)	m_2	0.121 (kg)
l_{c1}	0.047 (m)	m_3	0.085 (kg)
l_{c2}	0.045 (m)	m_4	0.063 (kg)
l_{c3}	0.069 (m)	I_1	0.0017 (kgm^2)
l_{c4}	0.062 (m)	I_2	0.0014 (kgm^2)
m_1	0.126 (kg)	I_4	8.74×10^{-5} (kgm^2)

Table 9.2 Friction parameters of the Five-Bar mechanism

Parameter	Value (units)
f'_{11}	0.01 (Nm/rad s)
f'_{22}	0.01 (Nm/rad s)
f'_{33}	0.00001 (Nm/rad s)
f'_{44}	0.00001 (Nm/rad s)

Table 9.3 Gains and parameters of the finite-time nonlinear PID controller

Gain	Joint 1	Joint 2	Units
k_{p0}	0.2	0.22	Nm/rad
k_{d0}	0.5	0.5	Nms/rad
K_p	0.37	0.36	Nm/rad
K_I	0.1	0.06	Nms/rad
K_d	0.1	0.01	Nm/rad
α_0	0.1	0.1	s^{-1}
α_1	0.5	0.5	(dimensionless)
α_2	0.6666	0.6666	(dimensionless)

Table 9.4 Gains of the nonlinear PID controller

Gain	Joint 1	Joint 2	Units
K_p	1.15	1.15	Nm/rad
K_d	0.5	0.5	Nms/rad
K_i	0.08	0.03	Nm/rads

For comparison purposed, simulations of a PID-like controller inspired in Kelly (1998) applied to the Five-Bar Mechanism were also conducted. The control law of this controller is

$$\tau = -K_p\tilde{q} - K_i \int_0^t \operatorname{Tanh}(\tilde{q}(\sigma))\, d\sigma - K_d\dot{q}$$

The gains used for this controller are shown in Table 9.4. These gains were selected by try and test, in order to obtain the best performance of the controller and avoiding to exceed the maximum torque values.

9.4.2 Simulations Results

The results of the simulations are shown in Figs. 9.2, 9.3, 9.4, 9.5, 9.6 and 9.7. In Fig. 9.2, the position errors at joint 1 from both controllers, the finite-time nonlinear PID controller (FNPID) and the nonlinear PID from Kelly (1998), are shown. In Fig. 9.3, the position errors at joint 2 from both controllers are shown. From these figures, notice that the position errors of the FNPID in steady state are smaller than the position errors of the NPID. In Figs. 9.4 and 9.5, the commanded torques from the FNPID for joint 1 and joint 2, respectively, are shown. In Figs. 9.6 and 9.7, the commanded torques from the NPID for joint 1 and joint 2, respectively, are shown.

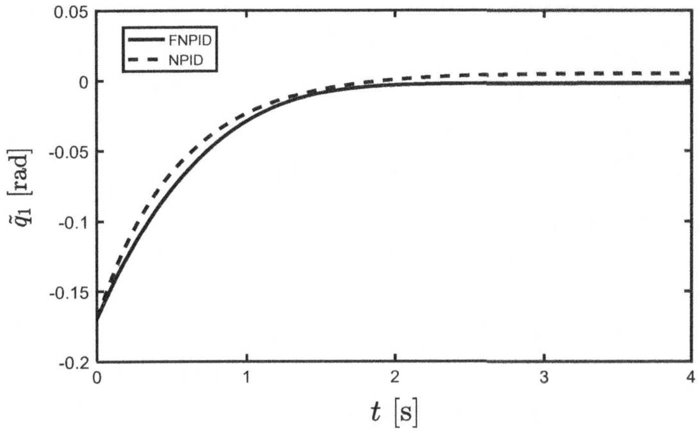

Fig. 9.2 Position errors in joint 1 from both controllers, FNPID and NPID

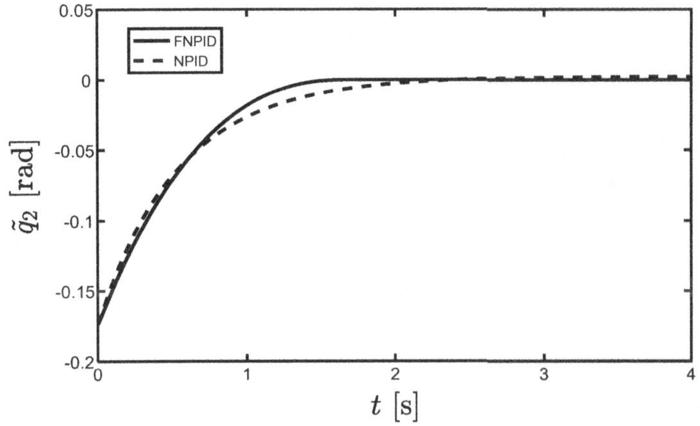

Fig. 9.3 Position errors in joint 2 from both controllers, FNPID and NPID

Notice that the torque signals from the NPID controller for both joints last longer times than the torque signals from the FNPID controller. This may imply smaller and shorter control efforts from the FNPID controller, which may result in improved durability of the drives and motors of the parallel machine. Notice that, as was pointed above, in the simulations we were careful in avoiding exceeding the maximum torque value of 0.2 (Nm).

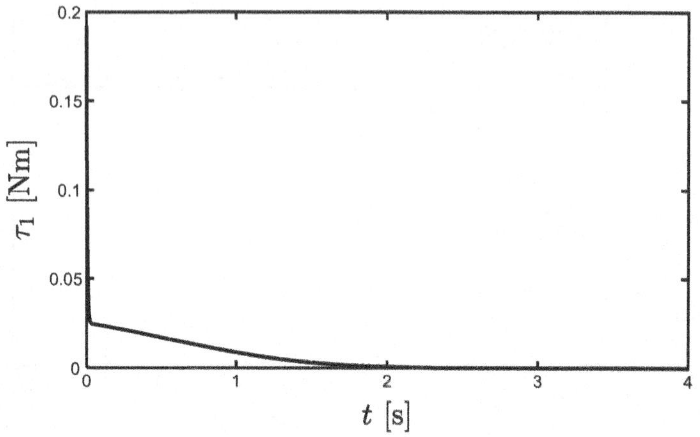

Fig. 9.4 Commanded torque from the FNPID controller, for joint 1

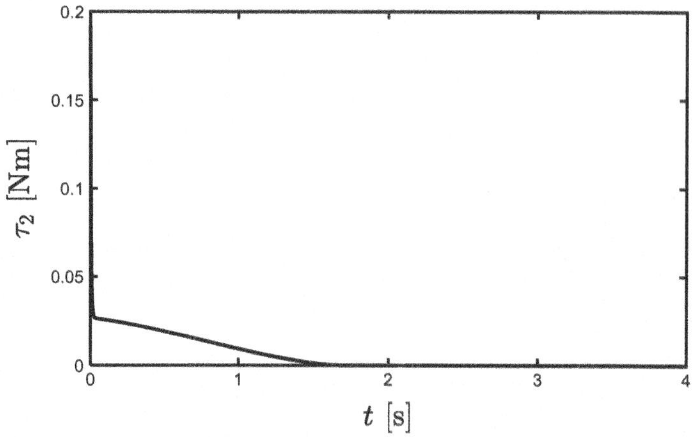

Fig. 9.5 Commanded torque from the FNPID controller, for joint 2

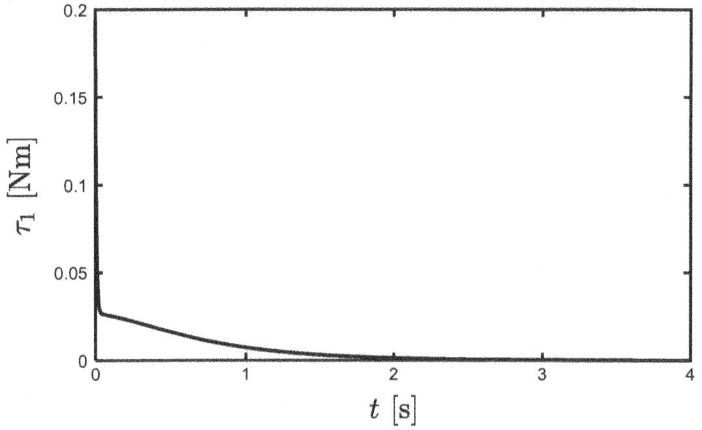

Fig. 9.6 Commanded torque from the NPID controller, for joint 1

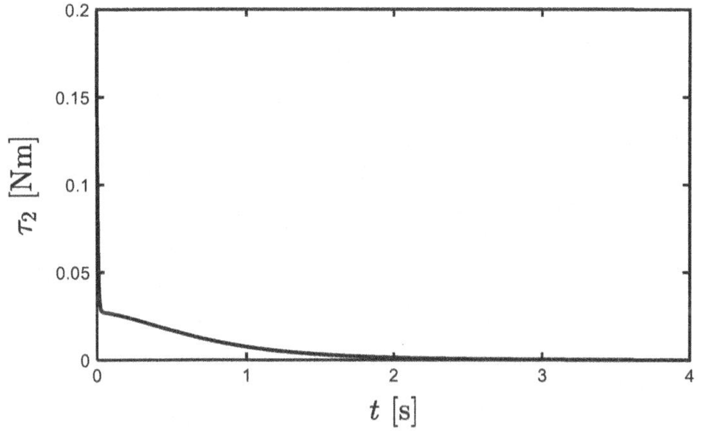

Fig. 9.7 Commanded torque from the NPID controller, for joint 2

9.5 Conclusion

In this work, we have reported the application of a finite-time nonlinear PID regulation controller to a Five-Bar Mechanism. The stability analysis of the system has been carried out, resulting in the global finite-time stability of the closed-loop system.

A dynamic model of a parallel robot, which is subject to mechanical constraints, has been obtained in structure similar to that of a serial robot. This let us analyze the closed-loop system in a similar way to analyzing a system with a serial robot.

Numerical simulations of the proposed controller applied to the model of a Five-Bar Mechanism were conducted. The simulations' results confirm the usefulness of the proposed approach.

References

Amato, F., De Tommasi, G., & Pironti, A. (2013). Necessary and sufficient conditions for finite-time stability of impulsive dynamical linear systems. *Automatica, 49*(8), 2546–2550.

Barnfather, J. D., Goodfellow, M. J., & Abram, T. (2017). Positional capability of a hexapod robot for machining applications. *The International Journal of Advanced Manufacturing Technology, 89*(1–4), 1103–1111. https://doi.org/10.1007/s00170-016-9051-0.

Bhat, S. P., & Bernstein, D. S. (1998). Continuous finite-time stabilization of the translational and rotational double integrators. *IEEE Transactions on Automatic Control, 43*(5), 678–682. https://doi.org/10.1109/9.668834.

Bhat, S. P., & Bernstein, D. S. (2000). Finite-Time stability of continuous autonomous systems. *SIAM Journal on Control and Optimization, 38*(3), 751–766. https://doi.org/10.1137/S0363012997321358.

Bhat, S. P., & Bernstein, D. S. (2005). Geometric homogeneity with applications to finite-time stability. *Mathematics of Control, Signals, and Systems, 17*(2), 101–127. https://doi.org/10.1007/s00498-005-0151-x.

Bourbonnais, F., Bigras, P., & Bonev, I. A. (2015). Minimum-time trajectory planning and control of a pick-and-place Five-Bar parallel robot. *IEEE/ASME Transactions on Mechatronics, 20*(2), 740–749. https://doi.org/10.1109/TMECH.2014.2318999.

Cheng, H., Yiu, Y-K., & Li, Z. (2003). Dynamics and control of redundantly actuated parallel manipulators. *IEEE/ASME Transactions on Mechatronics, 8*(4), 483–491.

Diaz-Rodriguez, M., Valera, A., Mata, V., & Valles, M. (2013). Model-based control of a 3-DOF parallel robot based on identified relevant parameters. *IEEE/ASME Transactions on Mechatronics, 18*(6), 1737–1744. https://doi.org/10.1109/TMECH.2012.2212716.

Dorato, P. (1961). Short time stability in linear time-varying systems. In *IRE International Convention Record* (pp. 83–87). USA: New York.

Enferadi, J., & Shahi, A. (2016). On the position analysis of a new spherical parallel robot with orientation applications. *Robotics and Computer-Integrated Manufacturing, 37,* 151–161. https://doi.org/10.1016/J.RCIM.2015.09.004.

Feng, Y., Yu, X., & Man, Z. (2002). Non-singular terminal sliding mode control of rigid manipulators. *Automatica, 38*(12), 2159–2167. https://doi.org/10.1016/S0005-1098(02)00147-4.

Ghorbel, F. H., Chetelat, O., Gunawardana, R., & Longchamp, R. (2000). Modeling and set point control of closed-chain mechanisms: theory and experiment. *IEEE Transactions on Control Systems Technology, 8*(5), 801–815. https://doi.org/10.1109/87.865853.

Gruyitch, L. T., & Kokosy, A. (1999). Robot control for robust stability with finite reachability time in the whole. *Journal of Robotic Systems, 16*(5), 263–283. http://doi.org/10.1002/(SICI)1097-4563(199905)16:5<263::AID-ROB2>3.0.CO;2-Q.

Hong, Y., Xu, Y., & Huang, J. (2002). Finite-time control for robot manipulators. *Systems & Control Letters, 46*(4), 243–253. https://doi.org/10.1016/S0167-6911(02)00130-5.

Huang, Z., & Cao, Y. (2005). Property identification of the singularity loci of a class of Gough-Stewart manipulators. *The International Journal of Robotics Research, 24*(8), 675–685. https://doi.org/10.1177/0278364905054655.

Kelaiaia, R. (2017). Improving the pose accuracy of the Delta robot in machining operations. *The International Journal of Advanced Manufacturing Technology, 91*(5–8), 2205–2215. https://doi.org/10.1007/s00170-016-9955-8.

Kelly, R. (1998). Global positioning of robot manipulators via PD control plus a class of nonlinear integral actions. *IEEE Transactions on Automatic Control, 43*(7), 934–938. https://doi.org/10.1109/9.701091.

Kelly, R., Santibáñez, V., & Loría, A. (Antonio). (2005). *Control of robot manipulators in joint space*. Berlin: Springer.

Khalil, W. (Wisama), & Dombre, E. (Etienne). (2004). *Modeling, identification and control of robots*. Kogan Page Science.

Khan, W. A., Krovi, V. N., Saha, S. K., & Angeles, J. (2005). Recursive kinematics and inverse dynamics for a planar 3R parallel manipulator. *Journal of Dynamic Systems, Measurement, and Control, 127*(4), 529. https://doi.org/10.1115/1.2098890.

Li, Q., Wu, W., Xiang, J., Li, H., & Wu, C. (2015). A hybrid robot for friction stir welding. *Proceedings of the Institution of Mechanical Engineers, Part C: Journal of Mechanical Engineering Science, 229*(14), 2639–2650. https://doi.org/10.1177/0954406214562848.

Michel, A. (1970). Quantitative analysis of simple and interconnected systems: Stability, boundedness, and trajectory behavior. *IEEE Transactions on Circuit Theory, 17*(3), 292–301. https://doi.org/10.1109/TCT.1970.1083119.

Muller, A. (2005). Internal preload control of redundantly actuated parallel manipulators—Its application to backlash avoiding control. *IEEE Transactions on Robotics, 21*(4), 668–677. https://doi.org/10.1109/TRO.2004.842341.

Nan, R., Li, D., Jin, C., Wang, Q., Zhu, L., Zhu, W., … Qian, L. (2011). The Five-Hundred-Meter Aperture Spherical Radio Telescope (FAST) Project. *International Journal of Modern Physics D, 6*, 989–1024. http://doi.org/10.1142/S0218271811019335.

Pierrot, F., Reynaud, C., & Fournier, A. (1990). DELTA: a simple and efficient parallel robot. *Robotica, 8*(2), 105. https://doi.org/10.1017/S0263574700007669.

Polyakov, A. (2014). Stability notions and Lyapunov functions for sliding mode control systems. *Journal of the Franklin Institute, 351*(4), 1831–1865. https://doi.org/10.1016/J.JFRANKLIN.2014.01.002.

Polyakov, A., & Poznyak, A. (2009). Lyapunov function design for finite-time convergence analysis: "Twisting" controller for second-order sliding mode realization. *Automatica, 45*(2), 444–448. https://doi.org/10.1016/J.AUTOMATICA.2008.07.013.

Ren, L., Mills, J. K., & Sun, D. (2007). Experimental comparison of control approaches on trajectory tracking control of a 3-DOF parallel robot. *IEEE Transactions on Control Systems Technology, 15*(5), 982–988. https://doi.org/10.1109/TCST.2006.890297.

Salinas, A., Moreno-Valenzuela, J., & Kelly, R. (2016). A family of nonlinear PID-like regulators for a class of torque-driven robot manipulators equipped with torque-constrained actuators. *Advances in Mechanical Engineering, 8*(2), 168781401662849. https://doi.org/10.1177/1687814016628492.

Soto, I., & Campa, R. (2014). On dynamic modelling of parallel manipulators: The Five-Bar mechanism as a case study. *International Review on Modelling and Simulations (IREMOS), 7*(3), 531–541. http://doi.org/10.15866/IREMOS.V7I3.1899.

Soto, I., & Campa, R. (2015). Modelling and control of a spherical inverted pendulum on a five-bar mechanism. *International Journal of Advanced Robotic Systems, 12*(7), 95. https://doi.org/10.5772/60027.

Su, Y., & Zheng, C. (2017). PID control for global finite-time regulation of robotic manipulators. *International Journal of Systems Science, 48*(3), 547–558. https://doi.org/10.1080/00207721.2016.1193256.

Weiss, L., & Infante, E. (1967). Finite time stability under perturbing forces and on product spaces. *IEEE Transactions on Automatic Control, 12*(1), 54–59. https://doi.org/10.1109/TAC.1967.1098483.

Wu, H., Handroos, H., & Pessi, P. (2008). Mobile parallel robot for assembly and repair of ITER vacuum vessel. *Industrial Robot: An International Journal, 35*(2), 160–168. https://doi.org/10.1108/01439910810854656.

Xie, F., & Liu, X.-J. (2016). Analysis of the kinematic characteristics of a high-speed parallel robot with Schönflies motion: Mobility, kinematics, and singularity. *Frontiers of Mechanical Engineering, 11*(2), 135–143. https://doi.org/10.1007/s11465-016-0389-7.

Yu, S., Yu, X., Shirinzadeh, B., & Man, Z. (2005). Continuous finite-time control for robotic manipulators with terminal sliding mode. *Automatica, 41*(11), 1957–1964. https://doi.org/10.1016/J.AUTOMATICA.2005.07.001.

Su, Y., & Zheng, C. (2009). A simple nonlinear PID control for finite-time regulation of robot manipulators. In *2009 IEEE International Conference on Robotics and Automation* (pp. 2569–2574). IEEE. http://doi.org/10.1109/ROBOT.2009.5152244.

Su, Y., & Zheng, C. (2010). A simple nonlinear PID control for global finite-time regulation of robot manipulators without velocity measurements. In *2010 IEEE International Conference on Robotics and Automation* (pp. 4651–4656). IEEE. http://doi.org/10.1109/ROBOT.2010.5509163.

Zhao, D., Li, S., Zhu, Q., & Gao, F. (2010). Robust finite-time control approach for robotic manipulators. *IET Control Theory and Applications, 4*(1), 1–15. https://doi.org/10.1049/iet-cta.2008.0014.

Chapter 10
Robust Control of a 3-DOF Helicopter with Input Dead-Zone

Israel U. Ponce, Angel Flores-Abad and Manuel Nandayapa

Abstract This chapter deals with the tracking control problem of a three-degree-of-freedom (3-DOF) helicopter. The system dynamics are given by a mathematical model that considers the existence of a dead-zone phenomenon in the actuators, as well as a first-order dynamic that adds a lag in the system input. This leads to obtain an eighth-order model where the positions are the only available measurements of the system. The control problem is solved using nonlinear $\mathcal{H}_\infty$ synthesis of time-varying systems, the dead-zone is compensated using its inverse model, and a reference model is used to deal with the first-order dynamic in the actuators. Numerical results show the effectiveness of the proposed method, which also considers external perturbations and parametric variations.

Keywords Robust control · Helicopter · Input dead-zone

10.1 Introduction

Recently, there is a lot of attention on unmanned aerial vehicles (UAVs) due to their potential applications. A large list of works can be found in the existing literature. Unmanned helicopters have an advantage over other UAVs because of their unique capabilities to perform tasks such as hover and vertical takeoff and land, needing for that a very limited space (Isidori and Astolfi 1992; Avila et al. 2003; Marconi and Naldi 2007; Gadewadikar et al. 2008).

In this work, we refer to a three-degree-of-freedom (3-DOF) laboratory helicopter developed by Quanser Company that is often used in control research for the design and implementation of control concepts (Quanser 1998). The 3-DOF helicopter system consists of two DC motors mounted at the two ends of a rectangular frame (helicopter frame) that drive two propellers (back and front propellers). There

I. U. Ponce (✉) · A. Flores-Abad · M. Nandayapa
Universidad Autónoma de Ciudad Juárez, Instituto de Ingeniería y Tecnología, Ciudad Juárez, Chihuahua, Mexico
e-mail: Israel.ulises@uacj.mx

O. O. Vergara Villegas et al. (eds.), *Advanced Topics on Computer Vision, Control and Robotics in Mechatronics*,
https://doi.org/10.1007/978-3-319-77770-2_10

are two inputs voltages: one for the front motor and the other for the back motor. The 3-DOF helicopter has three outputs, which correspond to the elevation angle, the pitch angle, and the travel angle. This plant represents a typical underactuated MIMO nonlinear system with large uncertainties which can be utilized as an ideal platform to test effectiveness of control schemes. The system contains various uncertainties such as nonlinearities, coupling effects, unmodeled dynamics, and parametric perturbations which may further increase the difficulties of control. Other important drawback to design a stable controller is the fact that the inputs are aerodynamical forces/torques.

The simplest dynamical model of the 3-DOF helicopter is described in (Quanser 1998) where the friction and other dynamics of the system are neglected. When this model is used to design a position controller, it is difficult to accurately reach the desired position for the closed loop system. To improve the performance of the system, different models are used to control the 3-DOF helicopter such as (Ishutkina 2004; Shan et al. 2005; Andrievsky et al. 2007; Ishitobi et al. 2010). In this work, we consider input dynamics and a dead-zone phenomenon in a model based on those given in (Ishutkina 2004; Andrievsky et al. 2007). These input dynamics relate the input voltages with the torques, adding a lag to the system. These dynamics augment the order of the system and equivalently the degrees of freedom. None of the works referenced in this paper include these dynamics in the controller design. We only can find a reference of this existing dynamic in (Ishutkina 2004).

Attitude control problem (elevation and pitch channels) or position control problem (elevation and travel angles) can be selected as the controlled outputs. The attitude control problem is solved taking into account only a partial dynamic of the system, which simplify the problem (Zheng and Zhong 2011; Wang et al. 2013; Liu et al. 2014) because a fully actuated system is obtained. The position control problem focuses on tracking the references for the elevation and travel angles. To solve the control position problem, different solutions have been proposed, for example, in (Odelga et al. 2012; Liu et al. 2014) a hierarchical control is used. First, the control problem for the travel angle is solved using the reference position for the pitch angle as the control input, and then. the attitude control problem is solved. Another solution for the position control problem is given in (Ishitobi et al. 2010), where a reference model is used. In this paper, we solve the position control problem, considering the pitch angle as dependent of the other desired positions.

Different control techniques have been applied to the 3-DOF helicopter, where robust controllers are more popular, e.g., the PD control with a robust compensator (Zheng and Zhong 2011; Ferreira et al. 2012), the $\mathcal{H}_2, \mathcal{H}_\infty$ or LQR controllers (Li and Shen 2007; Raafat and Akmeliawati 2012; Wang et al. 2013; Liu et al. 2014), and sliding mode controllers (Starkov et al. 2008; Meza-Sanchez et al. 2012a, b; Odelga et al. 2012). Beside these, adaptive controllers are used for this plant (Andrievsky et al. 2007; Gao and Fang 2012). In this work, a $\mathcal{H}_\infty$-synthesis is applied for a nonlinear time-varying system obtained from the model of the plant. Static $\mathcal{H}_\infty$-controllers, which take into account the linearized system, are given by (Ferreira et al. 2012; Wang et al. 2013). The proposed $\mathcal{H}_\infty$-synthesis solves the problem of having only position measurements, obtaining, through a filter, the unknown velocities of the given outputs. Moreover, the $\mathcal{H}_\infty$ ensure a $\mathcal{L}_2$-gain of the chosen output from the disturbances that affect the system.

The aim of this work is to develop a controller capable of ensuring that the position of the system, given by the elevation and travel angles, tracks a desirable smooth enough trajectory. The plant consists of a 3-DOF helicopter including a dynamic and a dead-zone input which generates a system of eight-order equivalent to a system with five degrees of freedom with two degrees of actuation. To simplify the problem, the full system is decomposed, solving first the problem for the plant without considering the input dynamics, and then, a hierarchical control is used to track the reference control, obtained from $\mathcal{H}_\infty$ synthesis, employing the input voltages as the input of the full system.

10.2 Preliminaries

10.2.1 Linear $\mathcal{H}_\infty$ Synthesis of Time-Varying Systems

Let us consider the following state-space representation of a given linear time-varying system

$$\begin{aligned} \dot{x}(t) &= A(t)x(t) + B_1(t)w(t) + B_2(t)u(t), \\ z(t) &= C_1(t)x(t) + D_{12}(t)u(t), \\ y(t) &= C_2(t)x(t) + D_{21}(t)w(t) \end{aligned} \tag{10.1}$$

with the state vector $x(t) \in \mathbb{R}^n$, the control input $u(t) \in \mathbb{R}^m$, the unknown disturbance $w(t) \in \mathbb{R}^r$, the output $z(t) \in \mathbb{R}^l$ to be controlled, and the available measurement $y(t) \in \mathbb{R}^p$, imposed on the system, and with matrices $A(t)$, $B_1(t)$, $B_2(t)$, $C_1(t)$, $C_2(t)$, $D_{12}(t)$, $D_{21}(t)$ of appropriate dimensions.

For convenience for the reader, recall that the system Eq. (10.1) possesses a $\mathcal{L}_2$-gain less than γ if the following inequality holds

$$\int_0^T \|z(t)\|^2 \mathrm{d}t < \gamma^2 \int_0^T \|w(t)\|^2 \mathrm{d}t \tag{10.2}$$

for all $T > 0$, for all the system trajectories initialized at the origin, and for all piecewise continuous functions $w(t) \in \mathcal{L}_2(0,T)$ such that the state trajectories remain in a vicinity of the origin.

The $\mathcal{H}_\infty$-control problem for the system Eq. (10.1) is to find all admissible controllers

$$\begin{aligned} u &= \mathcal{K}(\xi, t), \\ \dot{\xi} &= \mathcal{F}(\xi, y, t), \end{aligned} \tag{10.3}$$

with internal state $\xi \in \mathbb{R}^s$ such that the $\mathcal{L}_2$-gain of the closed-loop system Eq. (10.1), driven by Eq. (10.3), is less than γ. Solving the above problem under γ approaching the infimal achievable level γ^* in Eq. (10.2) yields a (sub)optimal $\mathcal{H}_\infty$-controller with the (sub)optimal disturbance attenuation level γ^* ($\gamma > \gamma^*$).

The following assumptions are imposed on the system Eq. (10.1).

A1. $(A(t), B_1(t))$ is stabilizable, and $(C_1(t), A(t))$ is detectable,
A2. $(A(t), B_2(t))$ is stabilizable, and $(C_2(t), A(t))$ is detectable,
A3. $D_{12}^T(t)C_1(t) \equiv 0$ and $D_{12}^T(t)D_{12}(t) \equiv I$,
A4. $B_1(t)D_{21}^T(t) \equiv 0$ and $D_{21}^T(t)D_{21}(t) \equiv I$.

which are made to simplify the solution to the $\mathcal{H}_\infty$-control problem.

Necessary and sufficient conditions, for the above $\mathcal{H}_\infty$ suboptimal control problem, are formulated in terms of the existence of appropriate solution of certain differential Riccati equations to have a solution with a disturbance attenuation level $\gamma > 0$:

C1. There exists a positive constant ε_0 such that the perturbed differential Riccati equation

$$\begin{aligned}\dot{P}_\varepsilon(t) &= P_\varepsilon(t)A(t) + A^T(t)P_\varepsilon(t) + C_1^T(t)C_1(t)\\ &+ P_\varepsilon(t)\left[\frac{1}{\gamma^2}B_1B_1^T - B_2B_2^T\right](t)P_\varepsilon(t) + \varepsilon I,\end{aligned} \tag{10.4}$$

possesses a positive-definite symmetric solution $P_\varepsilon(t)$ for each $\varepsilon \in (0, \varepsilon_0)$;

C2. While coupled to Eq. (10.4), the perturbed Riccati equation

$$\begin{aligned}\dot{Q}_\varepsilon(t) &= A_\varepsilon(t)Q_\varepsilon(t) + Q_\varepsilon(t)A_\varepsilon^T(t) + B_1(t)B_1^T(t)\\ &+ Q_\varepsilon(t)\left[\frac{1}{\gamma^2 P_\varepsilon B_2 B_2^T P_\varepsilon - C_2^T C_2}\right](t)Q_\varepsilon(t) + \varepsilon I\end{aligned} \tag{10.5}$$

possesses a positive-definite symmetric solution $Q_\varepsilon(t)$ for each $\varepsilon \in (0, \varepsilon_0)$ with $A_\varepsilon(t) = A(t) + \frac{1}{\gamma^2}B_1(t)B_1^T(t)P_\varepsilon(t)$.

Equations (10.4) and (10.5) are utilized to derive a solution of the linear $\mathcal{H}_\infty$-control problem for time-varying systems Eq. (10.1). Under partial state measurements, the $\mathcal{H}_\infty$ synthesis is augmented with a dynamic compensator running in parallel. The compensator is derived by means of the perturbed differential Riccati equations Eqs. (10.4) and (10.5). Summarizing, the following result, extracted from (Orlov and Aguilar 2014), is in force.

Theorem 1 *The $\mathcal{H}_\infty$ suboptimal control problem possesses a solution for the system Eq.* (10.1) *with assumptions* **A1–A4** *and a disturbance attenuation level $\gamma > 0$ if and only if Conditions* **C1** *and* **C2**, *coupled together, are satisfied. Then, the causal dynamic output feedback compensator*

$$\dot{\xi} = A\xi + \left(\frac{1}{\gamma^2}B_1B_1^T - B_2B_2^T\right)P\xi + QC_2^T(y - C_2\xi),$$
$$u = -B_2^T(t)P(t)\xi(t) \tag{10.6}$$

yields a solution to the problem in question.

10.2.2 Nonlinear $\mathcal{H}_\infty$ Synthesis of Time-Varying Systems

We consider a system of the form

$$\begin{aligned}\dot{x}(t) &= f(x(t), t) + g_1(x(t), t)w(t) + g_2(x(t), t)\tilde{u}(t),\\ z(t) &= h_1(x(t), t) + k_{12}(x(t), t)u(t),\\ y(t) &= h_2(x(t), t) + k_{21}(x(t), t)w(t),\end{aligned} \tag{10.7}$$

where $x(t) \in \mathbb{R}^n$ is the state vector, $u(t) \in \mathbb{R}^m$ is the control input, $w(t) \in \mathbb{R}^r$ is the unknown disturbance, $z(t) \in \mathbb{R}^l$ is the unknown output to be controlled, and $y(t) \in \mathbb{R}^p$ is the only available measurement on the system.

The nonlinear $\mathcal{H}_\infty$-control problem for a system Eq. (10.7) is to find a locally stabilizing output feedback controller of the form

$$\begin{aligned}\dot{\tilde{\xi}} &= \tilde{\mathcal{F}}(\tilde{\xi}, y, t),\\ \tilde{u} &= \tilde{\mathcal{K}}(\tilde{\xi}, t)\end{aligned} \tag{10.8}$$

with internal state $\tilde{\xi} \in \mathbb{R}^s$ such that the $\mathcal{L}_2$-gain of the closed-loop system Eq. (10.7) driven by Eq. (10.8) is locally less than γ.

The following assumptions are made on the system Eq. (10.8):

A5. The functions $f(x(t), t)$, $g_1(x(t), t)$, $g_2(x(t), t)$, $h_1(x(t), t)$, $h_2(x(t), t)$, $k_{12}(x(t), t)$, and $k_{21}(x(t), t)$ are of appropriate dimensions and of class C^∞;
A6. $f(0, t) = 0$, $h_1(0, t) = 0$, and $h_2(0, t) = 0$ for all t;
A7. $h_1^T(x(t), t)k_{12}(x(t), t) = 0$, $k_{12}^T(x(t), t)k_{12}(x(t), t) = I$,
$k_{21}(x(t), t)g_1^T(x(t), t) = 0$, $k_{21}(x(t), t)k_{21}^T(x(t), t) = I$.

Assumptions **A5** and **A6** are typical for the nonlinear treatment (Isidori and Astolfi 1992), whereas assumption **A7** contains simplified assumptions related with the linear treatment.

The local synthesis involves the linear $\mathcal{H}_\infty$-control problem of time-varying systems for the linearized system Eq. (10.1), where

$$
\begin{aligned}
A(t) &= \frac{\partial f}{\partial x}(0,t), \quad B_1(t) = g_1(0,t), \quad B_2(t) = g_2(0,t), \\
C_1(t) &= \frac{\partial h_1}{\partial x}(0,t), \quad C_2(t) = \frac{\partial h_2}{\partial x}(0,t), \\
D_{12}(t) &= k_{12}(0,t), \quad D_{21}(t) = k_{21}(0,t).
\end{aligned} \tag{10.9}
$$

A solution for the nonlinear $\mathcal{H}_\infty$-control problem of time-varying system is stated in the following result inspired from (Orlov and Aguilar 2014).

Theorem 2 *Consider system Eq.* (10.7) *with Assumptions* ***A5–A7****. Let conditions* ***C1*** *and* ***C2*** *be satisfied with a certain* $\gamma > 0$ *and with* $(P_\varepsilon(t), Q_\varepsilon(t))$*, being a uniformly bounded positive-definite symmetric solution of Eqs.* (10.4) *and* (10.5)*, corresponding to some* $\varepsilon \in (0, \varepsilon_0)$*. Then, the causal dynamic output feedback compensator*

$$
\begin{aligned}
\dot{\xi} &= f(\xi,t) + \left[\frac{1}{\gamma^2} g_1(\xi,t) g_1^T(\xi,t) - g_2(\xi,t) g_2^T(\xi,t)\right] P_\varepsilon(t)\xi \\
&\quad + Q_\varepsilon(t) C_2^T(t)[y(t) - h_2(\xi,t)], \\
u &= -g_2^T(\xi,t) P_\varepsilon(t)\xi,
\end{aligned} \tag{10.10}
$$

is a local solution of the afore-stated $\mathcal{H}_\infty$*-control problem with the disturbance attenuation level* γ*.*

Proof The validation of Theorem 2 is confined to the specification of Theorem 24 given in Orlov and Aguilar (2014) to the present case.

10.2.3 High Gain Around the Origin Control

We present a controller that globally stabilizes the perturbed system of first order

$$
\dot{x}(t) = \upsilon(x) + w(t) \tag{10.11}
$$

with state $x(t) \in \mathbb{R}$, a feedback law $u(x) \in \mathbb{R}$ referred as high gain around the origin (HGAO) control, and the bounded disturbance $w(t) \in \mathbb{R}$.

The following state feedback

$$
\upsilon(x) = -G(x)x = -\kappa_1(\epsilon + |x|)^{-\alpha} x \tag{10.12}
$$

with parameters $\kappa_1 > 0$, $\epsilon > 0$, and $0 < \alpha < 1$ are proposed to globally stabilize the first-order system Eq. (10.11). If the controller were specified with $\epsilon = 0$, and $\alpha = 1$, the resulting controller would become a discontinuous one $\upsilon(x) = -\kappa_1 \mathrm{sign}(x)$, and the resulting closed-loop system is capable of rejecting bounded disturbances with $|w| < M < \kappa_1$, whereas for $\alpha = 0$, we would obtain the simple control law $\upsilon(x) = -\kappa_1 x$. Therefore, using adequate parameter values of the proposed controller Eq. (10.12), we can get the benefits of a discontinuous controller, given by $\upsilon(x) = -\kappa_1 \text{sign}(x)$, avoiding undesirable chattering effects of the control signal.

The HGAO controller is more robust than the linear controller and can reach a similar performance to the discontinuous one. The following result states the robustness properties for of the HGAO controller.

Theorem 3 *Given $\kappa_1 > 0$, $\epsilon > 0$, and $0 < \alpha < 1$, the continuous closed-loop system Eqs.* (10.11)–(10.12) *is globally asymptotically stable for any disturbance w that satisfy the growth condition*

$$|w(t)| \leq \kappa_0 \frac{|x|}{(\epsilon + |x|)^{\alpha}} = \sigma(x) \tag{10.13}$$

for an arbitrary $\kappa_0 < \kappa_1$.

Proof Satisfying the given conditions of the theorem, the time derivative of the Lyapunov function

$$V(x,t) = \frac{1}{2}x^2 \tag{10.14}$$

computed along the trajectories of Eqs. (10.11)–(10.12) *is given by*

$$\dot{V}(x,t) = -\kappa_1(\epsilon + |x|)^{-\alpha}x^2 + wx \leq -(\kappa_1 - \kappa_0)(\epsilon + |x|)^{-\alpha}|x| \tag{10.15}$$

and since that $\kappa_1 > \kappa_0$ by a condition of the theorem, the global asymptotic stability of Eqs. (10.11)–(10.12) *is then established.*

Figure 10.1 depicts an example of rejected disturbances using the following parameter values: $\kappa_1 = 1$, $\alpha = 0.9$, and $\epsilon = 0.001$. This example shows that the closed-loop system can reject almost disturbances with magnitude less of 1 unit.

The ϵ term is added to avoid singularities in the dynamic model Eqs. (10.11)–(10.12). Furthermore, this term limits the linear gain, $G(x)$, of the control input Eq. (10.12), reaching its maximum value at the origin with $G(0) = \frac{\kappa_1}{\epsilon^{\alpha}}$. The parameter α let us move the border of the rejected disturbances around the origin (see Fig. 10.1).

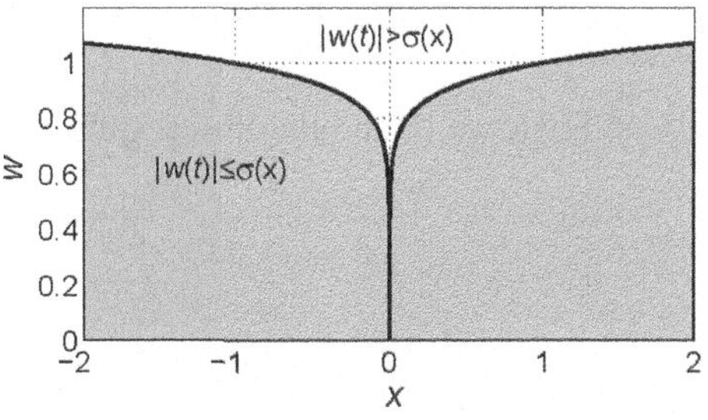

Fig. 10.1 Example of rejected disturbances ($\kappa_1 = 1$, $\alpha = 0.9$, and $\epsilon = 0.001$)

10.2.4 Supertwisting Observer

We consider a system of the form

$$\begin{aligned} \ddot{x}(t) &= f(x, \dot{x}, u, w, t), \\ y(t) &= x(t), \end{aligned} \tag{10.16}$$

which can be expressed in the state space as

$$\begin{aligned} \dot{\chi} &= \begin{bmatrix} \chi_2 \\ f(x, \dot{x}, u, w, t) \end{bmatrix}, \\ y &= \chi_1 \end{aligned} \tag{10.17}$$

where $\chi = \begin{bmatrix} x(t) & \dot{x}(t) \end{bmatrix}^T$ represent the state vector of the system, $t \in \mathbb{R}$ is the time variable, $u(t) \in \mathbb{R}$ and $w(t) \in \mathbb{R}$ are unknown inputs of the given system, $f(x, u, w, t) : \mathbb{R} \times \mathbb{R} \times \mathbb{R} \times \mathbb{R} \to \mathbb{R}$ is a bounded function, and $y(t) \in \mathbb{R}$ represent the output of the system.

For the described system, we state the following problem: using the only available information of the system, $x(t)$, and knowing the bound

$$|f(\cdot)| \leq \Gamma \tag{10.18}$$

we have to construct a robust estimator of velocity. This problem can be solved using the supertwisting observer, first proposed in (Davila et al. 2005), and given by

$$\begin{aligned} \dot{\hat{\chi}}_1 &= -\lambda_1 |\hat{\chi}_1 - \chi_1|^{\frac{1}{2}} \text{sign}(\hat{\chi}_1 - \chi_1) + \hat{\chi}_2 \\ \dot{\hat{\chi}}_2 &= -\lambda_2 \text{sign}(\hat{\chi}_1 - \chi_1) \end{aligned} \tag{10.19}$$

where $\hat{\chi}_1$ and $\hat{\chi}_2$ are the state estimations.

The following result is in force.

Theorem 4 *Suppose that the parameters of the observer Eq.* (10.19) *are selected according to*

$$\lambda_1 > \min\{\Gamma, 2\}, \quad \lambda_2 > 2\Gamma, \tag{10.20}$$

and condition Eq. (10.18) *holds for system Eq.* (10.16). *Then the state estimations* $(\hat{\chi}_1, \hat{\chi}_2)$ *of system Eq.* (10.19) *converge globally asymptotically to the states* (χ_1, χ_2) *of system* (10.17).

Proof First, we define the observation error as

$$\varsigma = \begin{bmatrix} \varsigma_1 \\ \varsigma_2 \end{bmatrix} = \begin{bmatrix} \hat{\chi}_1 - \chi_1 \\ \hat{\chi}_2 - \chi_2 \end{bmatrix}, \tag{10.21}$$

where dynamic defined by (10.17) and (10.19) is governed by the following second-order system:

$$\dot{\varsigma} = \begin{bmatrix} -\lambda_1|\varsigma_1|^{\frac{1}{2}}\text{sign}(\varsigma_1) + \varsigma_2 \\ -\lambda_2\text{sign}(\varsigma_1) - f(\cdot) \end{bmatrix}. \tag{10.22}$$

It is required to demonstrate that given the conditions of the theorem,

$$\lim_{t\to\infty} \varsigma = 0 \tag{10.23}$$

Let us consider the Lyapunov candidate function (Moreno and Osorio 2008)

$$V = 2\lambda_2|\varsigma_1| + \frac{1}{2}\varsigma_2^2 + \frac{1}{2}s^2 \tag{10.24}$$

with

$$s = -\lambda_1|\varsigma_1|^{\frac{1}{2}}\text{sign}(\varsigma_1) + \varsigma_2 \tag{10.25}$$

By computing the time derivative of this function along the trajectories of system Eq. (10.22), we arrive at

$$\begin{aligned} \dot{V} &= -\lambda_1\lambda_2|\varsigma_1|^{\frac{1}{2}} - \frac{1}{2}\lambda_1|\varsigma_1|^{-\frac{1}{2}}s^2 - 2sf(\cdot) - \lambda_1|\varsigma_1|^{\frac{1}{2}}\,\text{sign}(\varsigma_1)f(\cdot), \\ &\le -\lambda_1\lambda_2|\varsigma_1|^{\frac{1}{2}} - \frac{1}{2}\lambda_1|\varsigma_1|^{-\frac{1}{2}}s^2 + 2s|f(\cdot)| + \lambda_1|\varsigma_1|^{\frac{1}{2}}|f(\cdot)|. \end{aligned} \tag{10.26}$$

Employing the well-known inequality $2ab \le a^2 + b^2, a, b \in \mathbb{R}$ and taking into account Eq. (10.18), one arrives at

$$\dot{V} \le -(\lambda_1\lambda_2 - \Gamma^2 - \lambda_1\Gamma)|\varsigma_1|^{\frac{1}{2}} - \frac{1}{2}(\lambda_1 - 2)|\varsigma_1|^{-\frac{1}{2}}s^2 \tag{10.27}$$

which is negative semidefinite when condition Eq. (10.20) holds. By applying the Invariance Principle (Khalil 2002) and since $x \equiv 0$ is the largest invariant set for $\varsigma_1 = 0$, *the origin is globally asymptotically stable.*

In addition to the robustness given by the above theorem, it can be showed that this velocity observer, using adequate parameter values, possesses finite-time convergence (Moreno and Osorio 2008; Orlov et al. 2011), which is a desirable feature.

10.2.5 Inverse Dead-Zone Model

The input dead-zone phenomenon is a non-smooth characteristic that appears commonly in physical components such as mechanical connections and electric servomotors. This phenomenon produces delays and inaccuracies, resulting in poor performance of the system. The inverse model can be used to compensate the effects of the dead-zone on the system (Tao and Kokotovic 1997).

The dead-zone characteristic $\mathcal{D}(\cdot)$ is described by

$$\mathcal{Z}(t) = \mathcal{D}(\mathcal{Y}(t)) = \begin{cases} m_l(\mathcal{Y}(t) + j_l) & \text{if } \mathcal{Y}(t) < -j_l \\ 0 & \text{if } -j_l \leq \mathcal{Y}(t) \leq j_r \\ m_r(\mathcal{Y}(t) - j_r) & \text{if } \mathcal{Y}(t) > j_r \end{cases} \tag{10.28}$$

where j_l and j_r represent the size of the dead-zone, whereas m_l and m_r are the linear ratios between input and output. This model is shown in Fig. 10.2.

The inverse model $\overline{\mathcal{D}}$ of the dead-zone characteristic, depicted in Fig. 10.3, is specified by

$$\overline{\mathcal{Y}}(t) = \overline{\mathcal{D}}(\overline{\mathcal{Z}}(t)) = \begin{cases} -\frac{\overline{\mathcal{Z}}(t) - \bar{m}_l \bar{j}_l}{\bar{m}_l} & \text{if } \overline{\mathcal{Z}}(t) < 0 \\ 0 & \text{if } \overline{\mathcal{Z}}(t) = 0\,. \\ \frac{\overline{\mathcal{Z}}(t) + \bar{m}_r \bar{j}_r}{\bar{m}_r} & \text{if } \overline{\mathcal{Z}}(t) > 0 \end{cases} \tag{10.29}$$

where terms $\bar{j}_l, \bar{j}_r, \bar{m}_l$ and $\bar{m}_r$ are analogous to those of the model Eq. (10.28).

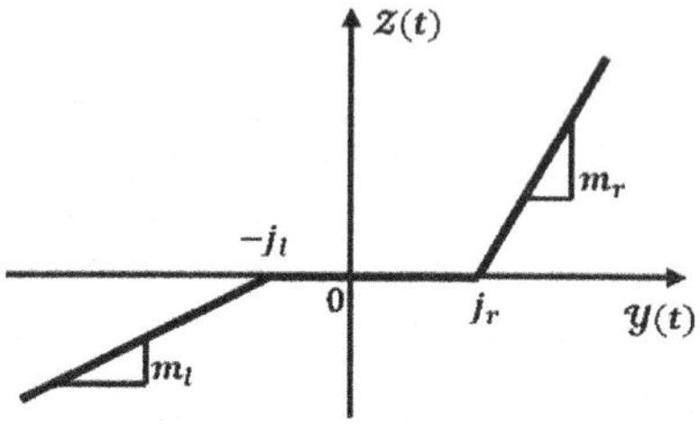

Fig. 10.2 Dead-zone model

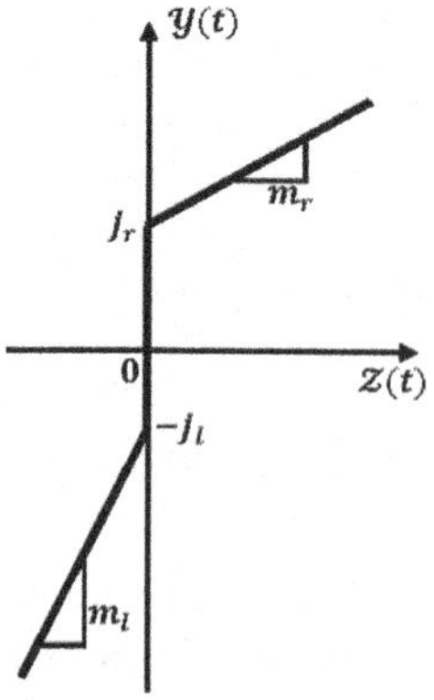

Fig. 10.3 Inverse dead-zone model

When the parameters of the inverse model coincide with those of the dead-zone model, the dead-zone effect is canceled. If this does not happen, the effects of the dead-zone are at least minimized.

10.3 Tracking Control of a 3-DOF Helicopter

The 3-DOF helicopter model, as shown in Fig. 10.4, used in this work is analogous to a tandem rotor helicopter such as the *Boeing CH-47 Chinook* illustrated in Fig. 10.5.

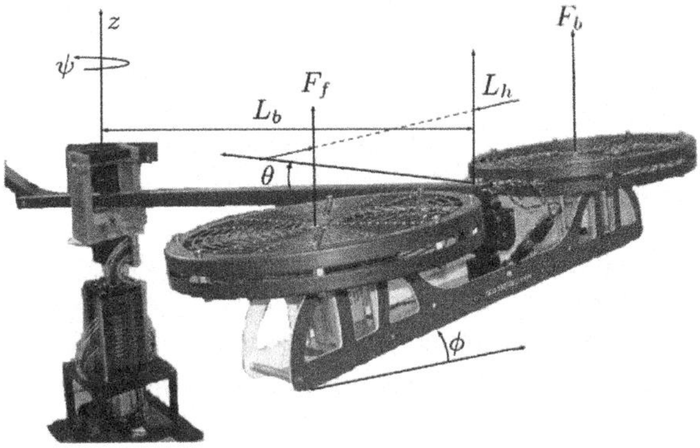

Fig. 10.4 Quanser 3-DOF helicopter prototype

Fig. 10.5 Boeing CH-47 Chinook Helicopter (from Wikipedia)

10.3.1 Dynamic Model

The dynamical model derived from (Quanser 1998; Ishutkina 2004) is given by

$$\begin{aligned} J_e\ddot{\theta} &= -f_e\dot{\theta} - c_e\sin(\theta) + K_f L_b\cos(\phi)(\tau_f + \tau_b) + w_e \\ J_p\ddot{\phi} &= -f_p\dot{\phi} - c_p\sin(\phi) + K_f L_h(\tau_f - \tau_b) + w_p \\ J_t\ddot{\psi} &= -f_t\dot{\psi} - K_p L_b\sin(\phi) + w_t, \end{aligned} \tag{10.30}$$

where θ, ϕ and $\psi \in \mathbb{R}$ are the elevation, pitch, and travel angles. $J_e, J_p, J_t \in \mathbb{R}^+$ and f_e,f_p,f_t are the inertia and viscous friction coefficients of the elevation, pitch, and rotation axis. The terms $c_e\sin(\theta)$ and $c_p\sin(\phi)$ represent the restorative spring torque relative to the elevation and pitch axis, respectively. K_f is the force constant of the motor/propeller combination, and K_p is the force required to maintain the helicopter in flight. L_b is the distance from the pivot point to the helicopter body, and L_h is the distance from the pitch axis to any of the motors. The input torques τ_f and τ_b represent the control action of the front and back DC motors applied to the system. Finally, w_e, w_p and $w_t \in \mathcal{L}_2$ are introduced to take into account different perturbations affecting the system.

The elevation angle, θ, corresponds to the angular displacement of the main sustentation arm with respect to the horizontal axis y. The movement range of the elevation θ is limited between around -1 and 1 rad due to the hardware restriction. The pitch angle, ϕ, defines the movement of the helicopter body and is confined to the domain $\phi \in \left(-\frac{\pi}{2}, \frac{\pi}{2}\right)$. The travel angle corresponds to the rotation of the entire system around the vertical axis. The inertia model of the system is simplified to point masses associated with the two motors and to the counterweight. In addition, friction and aerodynamics drag effects are assumed to be negligible. The force generated by each motor–propeller is assumed to be normal to the propeller plane. The system is controlled by the action of two rotors driven by a corresponding electric DC motor. The collective operation of the two rotors produces two actions acting simultaneously in the system, the lifting system given by the sum of the torques, and the rolling system given by the difference of the torques.

The torques τ_f and τ_b ($\tau_{f,b}$) applied to the system are the result of the torques $\tilde{\tau}_f$ and $\tilde{\tau}_b$ ($\tilde{\tau}_{f,b}$) affected by dead-zone phenomena, which are described as follows

$$\tau_{f,b} = \mathcal{D}(\tilde{\tau}_{f,b}) = \begin{cases} 0 & \text{if } |\tilde{\tau}_{f,b}| \leq j_f, j_b \\ \tilde{\tau}_{f,b} - j_{f,b} & \text{otherwise} \end{cases}, \tag{10.31}$$

where j_f and j_b ($j_{f,b}$) are the size of the dead-zone.

The electric voltages V_f and V_b are the actuation control inputs of the front and back DC motors, respectively. The relation between the voltages (V_f and V_b) and torques ($\tilde{\tau}_f$ and $\tilde{\tau}_b$) that includes an inertial electric lag is given by the following first-order dynamics

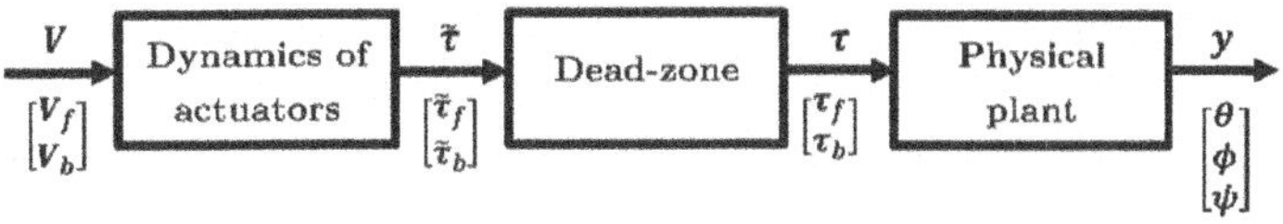

Fig. 10.6 Block diagram of the plant

$$\dot{\tilde{\tau}}_f = -d_1\tilde{\tau}_f + d_2 V_f, \tag{10.32}$$

$$\dot{\tilde{\tau}}_b = -e_1\tilde{\tau}_b + e_2 V_b. \tag{10.33}$$

where d_1, d_2, e_1 and $e_2 \in \mathbb{R}^+$ are some given constants.

The open-loop system described above can be represented using the block diagram given in Fig. 10.6.

10.3.2 Problem Statement

The objective of the control system is achieved such that the system position, (θ, ψ), tracks a desirable smooth enough trajectory, $\mathcal{T}(\theta_d(t), \psi_d(t))$, while also attenuating the system disturbances and the measurement errors w_θ, w_ϕ and w_ψ.

The only available measurements of the system are defined by the positions of the system states corrupted by some measurement errors, defining the system output as

$$y = \begin{bmatrix} \theta + w_\theta \\ \phi + w_\phi \\ \psi + w_\psi \end{bmatrix} \tag{10.34}$$

where w_θ, w_ϕ and $w_\psi \in \mathcal{L}_2$ are measurement errors.

To state the control problem, we introduce the error state vector $x \in \mathbb{R}^6$, defined as

$$x = \begin{bmatrix} x_1 \\ x_2 \\ x_3 \\ x_4 \\ x_5 \\ x_6 \end{bmatrix} = \begin{bmatrix} \theta - \theta_d(t) \\ \dot{\theta} - \dot{\theta}_d(t) \\ \phi - \phi_d(\psi_d(t), t) \\ \dot{\phi} - \dot{\phi}_d(\psi_d(t), t) \\ \psi - \psi_d(t) \\ \dot{\psi} - \dot{\psi}_d(t) \end{bmatrix} \tag{10.35}$$

where θ_d and ψ_d are prespecified trajectories to be tracked, whereas ϕ_d is a trajectory defined by the restrictions of the system.

Rewriting the state equations Eqs. (10.30)–(10.32) in terms of the deviation Eq. (10.35), we have the following time-varying system

$$\dot{x} = \begin{bmatrix} x_2 \\ J_e^{-1}(-f_e(x_2+\dot{\theta}_d) - c_e \sin(x_1+\theta_d) + K_f L_b \cos(x_3+\phi_d)(\mathcal{D}(\tilde{\tau}_f)+\mathcal{D}(\tilde{\tau}_b)) + w_e) - \ddot{\theta}_d \\ x_4 \\ J_p^{-1}(-f_p(x_4+\dot{\phi}_d) - c_p \sin(x_3+\phi_d) + K_f L_h(\mathcal{D}(\tilde{\tau}_f)-\mathcal{D}(\tilde{\tau}_b)) + w_p) - \ddot{\phi}_d \\ x_6 \\ J_t^{-1}(-f_t(x_6+\dot{\psi}_d) - K_b L_b \sin(x_3+\phi_d) + w_t) - \ddot{\psi}_d \end{bmatrix}, \tag{10.36}$$

$$\begin{aligned} \dot{\tilde{\tau}}_f &= -d_1\tilde{\tau}_f + d_2 V_f, \\ \dot{\tilde{\tau}}_b &= -e_1\tilde{\tau}_b + e_2 V_b. \end{aligned} \tag{10.37}$$

where $\mathcal{D}(\tilde{\tau}_{f,b})$ is given by (10.31). The system (10.36)–(10.37) of eighth order represents the dynamics of the entire system, i.e., including the actuator dynamics. The control problem is stated in terms of the dynamics of the 3-DOF helicopter (10.36), as follows.

Let the output to be controlled be given as

$$z = \begin{bmatrix} \tau_f + \tau_b & \tau_f - \tau_b & \rho_1 x_1 & \rho_2 x_2 & \rho_3 x_3 & \rho_4 x_4 & \rho_5 x_5 & \rho_6 x_6 \end{bmatrix}^T, \tag{10.38}$$

Then, the $\mathcal{H}_\infty$ tracking problem is stated in terms of the state deviation vector $x \in \mathbb{R}^6$. Given the error system Eq. (10.36) with the system output Eq. (10.34) and a real number $\gamma > 0$, it is required to find (if any) a causal dynamic output Eq. (10.38) with internal state $\xi \in \mathbb{R}^6$ such that the closed-loop system is internally uniformly asymptotically stable around the origin, whereas its $\mathcal{L}_2$-gain is locally less than γ.

The solution of the above problem only solves the control problem of the plant without considering the actuator dynamics. Therefore, it is necessary to develop a strategy to generate voltages V_f and V_b produced by the torques τ_f and τ_b that are obtained from the proposed $\mathcal{H}_\infty$-synthesis.

10.3.3 Control Strategy

The tracking control problem is solved in two steps. In the first stage, we consider the plant without taking into account the dynamics of the actuator, in this stage is required to find the input torques τ_f and τ_b such that the $\mathcal{H}_\infty$ control problem is solved for the system Eqs. (10.36) and (10.38). The second stage consists of finding the voltages V_f and V_b such that the input torques τ_f and τ_b, derived with $\mathcal{H}_\infty$ synthesis, will be generated by the actuators.

The first stage of the proposed control problem involves a system of six order Eq. (10.36), i.e., considers a system of 3-DOF with two degrees of actuation. To simplifying the control problem of this system, the following control inputs are proposed

$$\begin{aligned} \tau_f + \tau_b = v_1 &= \frac{1}{K_f L_b \cos(\phi)} \left(u_1 + f_e \dot{\theta}_d + c_e \sin(\theta) + J_e \ddot{\theta}_d \right), \\ \tau_f - \tau_b = v_2 &= \frac{1}{K_f L_h} \left(u_2 + f_p \dot{\phi}_d + c_p \sin(\phi) + J_p \ddot{\phi}_d \right), \end{aligned} \tag{10.39}$$

with u_1 and u_2 to be determined from the resulting system obtained by applying Eq. (10.39) on Eq. (10.36)

$$\dot{X}_1 = \begin{bmatrix} x_2 \\ J_e^{-1}(-f_e x_2 + u_1 + w_e) \end{bmatrix}, \tag{10.40}$$

$$\dot{X}_2 = \begin{bmatrix} x_4 \\ J_p^{-1}(-f_p x_4 + u_2 + w_p) \\ x_6 \\ J_t^{-1}(-f_t(x_6 + \dot{\psi}_d) - K_b L_b \sin(x_3 + \phi_d) + w_t) - \ddot{\psi}_d \end{bmatrix}, \tag{10.41}$$

where $X_1 = \begin{bmatrix} x_1 & x_2 \end{bmatrix}^T$ and $X_2 = \begin{bmatrix} x_3 & x_4 & x_5 & x_6 \end{bmatrix}^T$. With the proposed control inputs Eq. (10.39), the system Eq. (10.36) is partially linearized to simplify the $\mathcal{H}_\infty$-control since the system is decomposed into two independent subsystems, one of the linear type.

From dynamics described in Eqs. (10.40)–(10.41), the control objective is focused in two independent subsystems Σ_1 and Σ_2, defined as

$$\Sigma_1 : \begin{cases} \dot{X}_1 \\ Y_1 \\ Z_1 \end{cases}, \qquad \Sigma_2 : \begin{cases} \dot{X}_2 \\ Y_2 \\ Z_2 \end{cases}, \tag{10.42}$$

where the available measurements are represented as

$$Y_1 = \theta - \theta_d + w_\theta, \tag{10.43}$$

$$Y_2 = \begin{bmatrix} \phi - \phi_d + w_\phi \\ \psi - \psi_d + w_\psi \end{bmatrix}, \tag{10.44}$$

and the outputs to be controlled are given as

$$Z_1 = \begin{bmatrix} u_1 & \rho_1 x_1 & \rho_2 x_2 \end{bmatrix}^T, \tag{10.45}$$

$$Z_2 = \begin{bmatrix} u_2 & \rho_3 x_3 & \rho_4 x_4 & \rho_5 x_5 & \rho_6 x_6 \end{bmatrix}^T. \tag{10.46}$$

For each system, Σ_1 and Σ_2, the $\mathcal{H}_\infty$ tracking problem is restated in terms of the state deviation vector X_1 and X_2. Given the error systems Σ_1 and Σ_2 and real numbers $\gamma_1, \gamma_2 > 0$, it is required to find (if any) causal dynamic outputs Eqs. (10.45) and (10.46) with internal states Ξ_1 and Ξ_2 such that the closed-loop systems are internally uniformly asymptotically stable around the origin, whereas their $\mathcal{L}_2$-gains are locally less than γ_1 and γ_2, respectively.

Solving the $\mathcal{H}_\infty$ problem allows us to find the control inputs τ_f and τ_b from Eq. (10.39), where

$$\tau_f = \frac{1}{2}(v_1 + v_2), \tag{10.47}$$

$$\tau_b = \frac{1}{2}(v_1 - v_2), \tag{10.48}$$

correspond to the input torques of the plant.

To achieve the objective of control, it is necessary to find the input voltages V_f and V_b that produce torques Eqs. (10.47) and (10.48). Then, firstly the dead-zone phenomenon is compensated using the inverse dead-zone model, and the input torques $\tilde{\tau}_f$ and $\tilde{\tau}_b$ are obtained. From these obtained torques, $\tilde{\tau}_f$ and $\tilde{\tau}_b$, and using the reference model

$$\dot{\hat{\tau}}_f = -d_1\hat{\tau}_f + d_2V_f, \tag{10.49}$$

$$\dot{\hat{\tau}}_b = -e_1\hat{\tau}_b + e_2V_b. \tag{10.50}$$

the voltages V_f and V_b are obtained such that the torques $\hat{\tau}_f$ and $\hat{\tau}_b$ fulfill the condition

$$\lim_{t\to\infty}\left\|\begin{bmatrix}\hat{\tau}_f - \tilde{\tau}_f\\ \hat{\tau}_b - \tilde{\tau}_b\end{bmatrix}\right\| = 0. \tag{10.51}$$

In this sense, and in terms of the given error vector

$$\zeta = \begin{bmatrix}\zeta_f\\ \zeta_b\end{bmatrix} = \begin{bmatrix}\hat{\tau}_f - \tilde{\tau}_f\\ \hat{\tau}_b - \tilde{\tau}_b\end{bmatrix}, \tag{10.52}$$

the dynamics Eqs. (10.32)–(10.33) are governed by the system

$$\dot{\zeta} = \begin{bmatrix}-d_1\zeta_f - d_1\tilde{\tau}_f - \dot{\tilde{\tau}}_f + d_2V_f\\ -e_1\zeta_b - e_1\tilde{\tau}_b - \dot{\tilde{\tau}}_b + e_2V_b\end{bmatrix}. \tag{10.53}$$

To stabilize the system Eq. (10.53), we use the control law

$$V_f = \frac{1}{d_2}\left(d_1\tilde{\tau}_f + \dot{\hat{\tilde{\tau}}}_f + \upsilon_f\right), \tag{10.54}$$

$$V_b = \frac{1}{e_2}\left(e_1\tilde{\tau}_b + \dot{\hat{\tilde{\tau}}}_b + \upsilon_b\right), \tag{10.55}$$

where $\dot{\hat{\tilde{\tau}}}_f$ and $\dot{\hat{\tilde{\tau}}}_b$ are estimated values of $\dot{\tilde{\tau}}_f$ and $\dot{\tilde{\tau}}_b$, whereas υ_f and υ_b are defined by the control law Eq. (10.12), i.e.,

$$\begin{aligned} \upsilon_f &= -\kappa_{1f}\left(\epsilon_f + |\zeta_f|\right)^{-\alpha_f}\zeta_f \\ \upsilon_b &= -\kappa_{1b}\left(\epsilon_b + |\zeta_b|\right)^{-\alpha_b}\zeta_b. \end{aligned} \tag{10.56}$$

The estimated values $\dot{\hat{\tilde{\tau}}}_f$ and $\dot{\hat{\tilde{\tau}}}_b$ are obtained by applying the supertwisting observer Eq. (10.19), i.e.,

$$\begin{aligned} \dot{\varrho}_f &= -\lambda_{1f}|\varrho_f - \tilde{\tau}_f|^{\frac{1}{2}}\operatorname{sign}(\varrho_f - \tilde{\tau}_f) + \dot{\hat{\tilde{\tau}}}_f \\ \ddot{\hat{\tilde{\tau}}}_f &= -\lambda_{2f}\operatorname{sign}(\varrho_f - \tilde{\tau}_f) \\ \dot{\varrho}_b &= -\lambda_{1b}|\varrho_b - \tilde{\tau}_b|^{\frac{1}{2}}\operatorname{sign}(\varrho_b - \tilde{\tau}_b) + \dot{\hat{\tilde{\tau}}}_b \\ \ddot{\hat{\tilde{\tau}}}_b &= -\lambda_{2b}\operatorname{sign}(\varrho_b - \tilde{\tau}_b) \end{aligned} \tag{10.57}$$

The inverse model Eq. (10.29) is used to compensate the dead-zone Eq. (10.31). We consider a symmetric dead-zone model, i.e., $j_l = j_r$ and $m_l = m_r$. In this case, it is necessary to estimate the gap of the dead-zone for both motors (front and back) which correspond to values j_f and j_b.

The complete system controller is well described in Fig. 10.7.

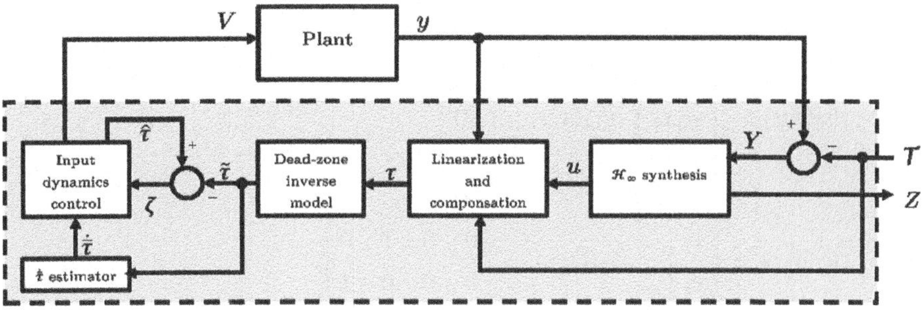

Fig. 10.7 Block diagram of the proposed control

10.3.4 $\mathcal{H}_\infty$ Tracking Control

For subsystems Σ_1 and Σ_2, we apply two different strategies of $\mathcal{H}_\infty$-control. Subsystem Σ_1 corresponds to a linear time-invariant system, and a linear $\mathcal{H}_\infty$-control is applied to stabilize it. The error system Σ_1 is then represented in the form Eq. (10.1) specified by

$$
\begin{aligned}
& A = \begin{bmatrix} 0 & 1 \\ -f_e J_e^{-1} & 0 \end{bmatrix}, \quad B_1 = \begin{bmatrix} 0 & 0 \\ Je^{-1} & 0 \end{bmatrix}, \quad B_2 = \begin{bmatrix} 0 \\ Je^{-1} \end{bmatrix}, \\
& C_1 = \begin{bmatrix} 0 & 0 \\ \rho_1 & 0 \\ & \end{bmatrix}, \quad D_{12} = \begin{bmatrix} 1 \\ 0 \\ & \end{bmatrix}, \\
& C_2 = [[1\,0\,0]\rho_2]\, D_{21} = [0 \quad 1\,0], \\
& u = u_1, \quad w = \begin{bmatrix} w_e & w_\theta \end{bmatrix}^T n
\end{aligned}
\tag{10.58}
$$

and its $\mathcal{H}_\infty$-control solution is obtained by applying Theorem 1.

Subsystem Σ_2 correspond a nonlinear time-variant system, and a time-varying nonlinear $\mathcal{H}_\infty$-control is applied to stabilize it. For this subsystem, we use the form Eq. (10.7), where

$$
\begin{aligned}
& f(x,t) = \begin{bmatrix} x_4 \\ -f_p J_p^{-1} x_4 \\ x_6 \\ J_t^{-1}(-f_t(x_6 + \dot{\psi}_d) - K_b L_b \sin(x_3 + \phi_d)) - \ddot{\psi}_d \end{bmatrix}, \\
& g_1(x,t) = \begin{bmatrix} 0 & 0 & 0 & 0 \\ J_p^{-1} & 0 & 0 & 0 \\ 0 & 0 & 0 & 0 \\ 0 & J_t^{-1} & 0 & 0 \end{bmatrix}, \quad g_2(x,t) = \begin{bmatrix} 0 \\ J_p^{-1} \\ 0 \\ 0 \end{bmatrix}, \\
& h_1(x,t) = \begin{bmatrix} 0 \\ \rho_3 x_3 \\ \rho_4 x_4 \\ \rho_5 x_5 \\ \rho_6 x_6 \end{bmatrix}, k_{12}(x,t) = \begin{bmatrix} 1 \\ 0 \\ 0 \\ 0 \\ 0 \end{bmatrix}, \\
& h_2(x,t) = \begin{bmatrix} x_3 \\ x_5 \end{bmatrix}, \quad k_{21}(x,t) = \begin{bmatrix} 0 & 0 & 1 & 0 \\ 0 & 0 & 0 & 1 \end{bmatrix}, \\
& u(t) = \begin{bmatrix} u_1 & u_2 \end{bmatrix}^T, \quad w(t) = \begin{bmatrix} w_p & w_t & w_\phi & w_\psi \end{bmatrix}^T
\end{aligned}
\tag{10.59}
$$

and to fulfill requirement **A6**, the trajectory ϕ_d and its time derivatives are pre-specified in the form

$$
\begin{aligned}
\phi_d &= \sin^{-1}\left(\frac{J_t}{K_p L_b}\left(-f_t J_t^{-1}\dot{\psi}_d - \ddot{\psi}_d\right)\right),\\
\dot{\phi}_d &= \frac{J_t}{K_p L_b \cos(\phi_d)}\left(-f_t J_t^{-1}\ddot{\psi}_d - \psi_d^{(3)}\right),\\
\ddot{\phi}_d &= \tan(\phi_d)\dot{\phi}_d^2 + \frac{J_t}{K_p L_b \cos(\phi_d)}\left(-f_t J_t^{-1}\psi_d^{(3)} - \psi_d^{(4)}\right).
\end{aligned}
\tag{10.60}
$$

The solution for this nonlinear $\mathcal{H}_\infty$-control problem is obtained by applying Theorem 1.

10.4 Numerical Results

Some numerical simulations were performed to show the efficacy of the proposed method. The parameters of the helicopter, drown from the Quanser 3-DOF helicopter manual (Quanser 1998), are given in Table 10.1. The numerical setup was implemented using *Simulink Version 7.5 (R2010a) from MATLAB 7.10.0.499 (R2010a) for 64-bit (win64)* running on a personal computer with Intel Core i3-3120, 2.50-2.50 GHz, 4 GB processor.

The parameter values for the input dynamics were obtained from (Ishutkina 2004), where $d_1 = 7.3, d_2 = 1, e_1 = 6.2$, and $e_2 = 1$.

We consider a perturbed system with parametric variations. The $\mathcal{H}_\infty$-controller parameters used in the numerical simulations were $\gamma_1 = 320$, $\rho_1 = [300 \quad 0]$,

Table 10.1 Parameter values of the 3-DOF helicopter

Notation	Value	Units
f_e	0.4	N · m · s/rad
f_p	0.013	N · m · s/rad
f_t	0.457	N · m · s/rad
J_e	0.91	kg · m^2
J_p	0.0364	kg · m^2
J_t	0.91	kg · m^2
c_e	0.5	N · m
c_p	0.5	N · m
K_p	0.686	N
K_f	0.5	N/V
L_b	0.66	m
L_h	0.177	m
j_f	0.25	N-m
j_b	0.25	N-m

$\gamma_2 = 800$, $\rho_2 = [200 \quad 0 \quad 280 \quad 0]$, and $\varepsilon_1 = \varepsilon_2 = 0.001$, whereas the HGAO controller parameters were $\kappa_{1f} = \kappa_{1b} = 10$, $\epsilon_f = \epsilon_b = 0.001$ and $\alpha_f = \alpha_b = 0.999$, and the supertwisting observer parameters were $\lambda_{1f} = \lambda_{1b} = 10$ and $\lambda_{2f} = \lambda_{2b} = 15$.

The disturbances affecting the system were given by

$$\begin{aligned} w_e &= 0.05\sin(0.4\pi t), \\ w_p &= 0.02\sin\left(\frac{2}{7}\pi t\right), \\ w_t &= 0.01\sin(0.25\pi t) \end{aligned} \tag{10.61}$$

To design the controller, we consider parametric variations, i.e., considering the following parameter values of the plant: $f_e = 0.44$, $f_p = 0.011$, $f_t = 0.35$, $J_e = 0.8$, $J_p = 0.04$, $J_t = 1.1$, $c_e = 0.53$, $c_p = 0.53$, $K_p = 0.9$, $K_f = 0.53$, $L_b = 0.7$, $L_h = 0.17$, $j_f = 0.3$, $j_b = 0.3$, $d_1 = 7$, $d_2 = 1$,$e_1 = 6$, y $e_2 = 1$.

The trajectory to track was given by

$$\mathcal{T} = \begin{cases} \theta_d(t) = 0.3\sin(0.2\pi t), \\ \psi_d(t) = 0.3\sin(0.1\pi t). \end{cases} \tag{10.62}$$

The position behavior is shown in Figs. 10.8, 10.9, and 10.10, and the control input is shown in Fig. 10.11. These figures demonstrate the effectiveness of the proposed control method.

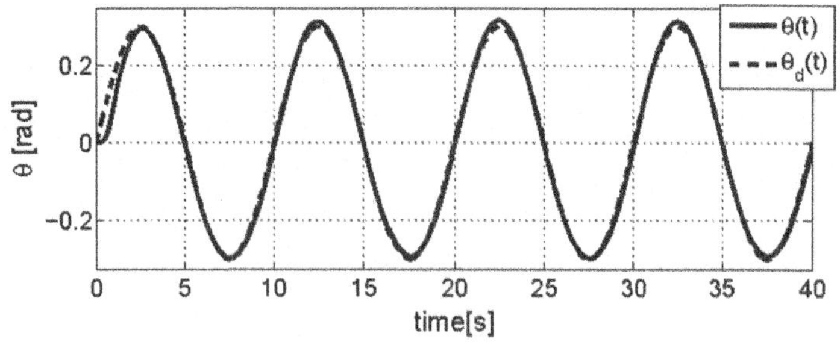

Fig. 10.8 $\theta(t)$ position behavior

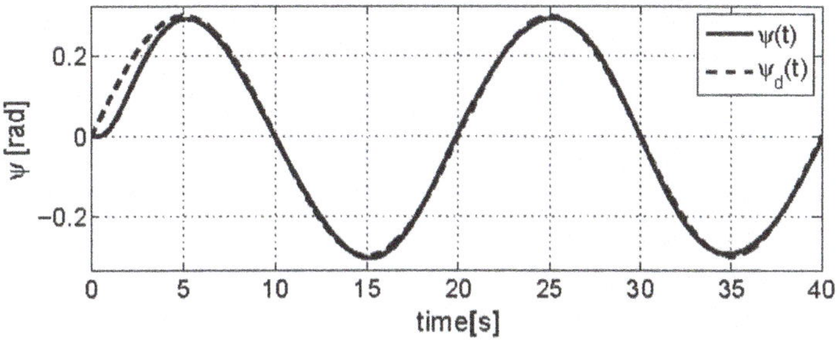

Fig. 10.9 $\psi(t)$ position behavior

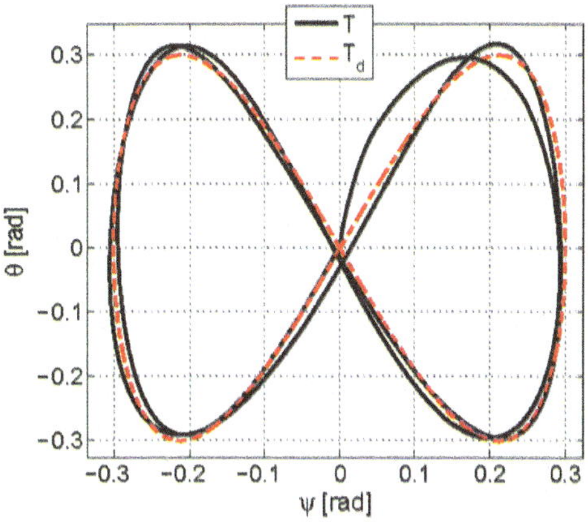

Fig. 10.10 $\psi(t) - \theta(t)$ trajectory behavior

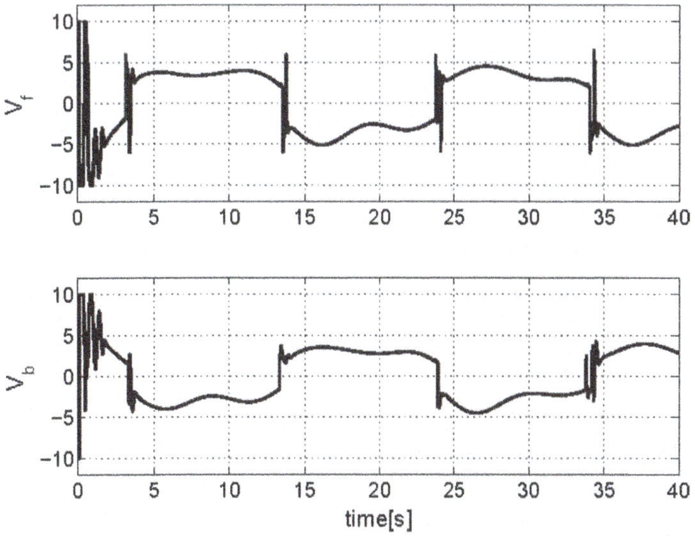

Fig. 10.11 Control signals

10.5 Conclusions

The tracking control problem for a 3-DOF helicopter was solved using $\mathcal{H}_\infty$ synthesis. For this problem, the elevation and travel angles were selected as the outputs to be controlled, while the desired trajectory for the pitch angle was obtained from condition **A6**. The input dead-zone was compensated using its inverse model, and a reference model was used to compensate the first-order dynamics in the actuators, where a first-order controller was developed from a discontinuous one. For this problem, the system positions (θ, ϕ, ψ) were the only available measurements of the system, and external disturbances and parametric variations were considered. Numerical results show the effectiveness of the proposed control.

References

Andrievsky, B., Peaucelle, D., & Fradkov, A. L. (2007, 7). Adaptive Control of 3DOF Motion for LAAS Helicopter Benchmark: Design and Experiments. *American Control Conference, 2007. ACC '07*, (pp. 3312–3317). https://doi.org/10.1109/acc.2007.4282243.

Avila Vilchis, J. C., Brogliato, B., Dzul, A., & Lozano, R. (2003). Nonlinear modelling and control of helicopters. *Automatica, 39,* 1583–1596. https://doi.org/10.1016/S0005-1098(03)00168-7.

Davila, J., Fridman, L., & Levant, A. (2005, 11). Second-order sliding-mode observer for mechanical systems. *Automatic Control, IEEE Transactions on, 50,* 1785–1789. https://doi.org/10.1109/tac.2005.858636.

Ferreira de Loza, A., Rios, H., & Rosales, A. (2012). Robust regulation for a 3-DOF helicopter via sliding-mode observation and identification. *Journal of the Franklin Institute, 349,* 700–718. https://doi.org/10.1016/j.jfranklin.2011.09.006.

Gadewadikar, J., Lewis, F., Subbarao, K., & Chen, B. M. (2008). Structured H-infinity command and Control-loop design for unmanned helicopters. *Journal of Guidance, Control and Dynamics, 31,* 1093–1102. https://doi.org/10.2514/1.31377.

Gao, W.-N., & Fang, Z. (2012, 6). Adaptive integral backstepping control for a 3-DOF helicopter. *Information and Automation (ICIA), 2012 International Conference on*, (pp. 190–195). https://doi.org/10.1109/icinfa.2012.6246806.

Ishitobi, M., Nishi, M., & Nakasaki, K. (2010). Nonlinear adaptive model following control for a 3-DOF tandem-rotor model helicopter. *Control Engineering Practice, 18,* 936–943. https://doi.org/10.1016/j.conengprac.2010.03.017.

Ishutkina, M. A. (2004, 6). *Design and implementation of a supervisory safety controller for a 3DOF helicopter. Ph.D. thesis.* Massachusetts Institute of Technology.

Isidori, A., & Astolfi, A. (1992, 9). Disturbance attenuation and H-∞-control via measurement feedback in nonlinear systems. *IEEE Transactions on Automatic Control, 37,* 1283–1293.

Khalil, H. K. (2002). *Nonlinear Systems.* Prentice Hall.

Li, P.-R., & Shen, T. (2007, 8). The research of 3 DOF helicopter tracking controller. *Machine Learning and Cybernetics, 2007 International Conference on, 1* (pp. 578–582). https://doi.org/10.1109/icmlc.2007.4370211.

Liu, H., Xi, J., & Zhong, Y. (2014). Robust hierarchical control of a laboratory helicopter. *Journal of the Franklin Institute, 351,* 259–276. https://doi.org/10.1016/j.jfranklin.2013.08.020.

Marconi, L., & Naldi, R. (2007). Robust full degree-of-freedom tracking control of a helicopter. *Automatica, 43,* 1909–1920. https://doi.org/10.1016/j.automatica.2007.03.028.

Meza-Sanchez, I. M., Orlov, Y., & Aguilar, L. T. (2012a, 1). Periodic motion stabilization of a virtually constrained 3-DOF underactuated helicopter using second order sliding modes. *Variable Structure Systems (VSS), 2012 12th International Workshop on,* (pp. 422–427). https://doi.org/10.1109/vss.2012.6163539.

Meza-Sanchez, I. M., Orlov, Y., & Aguilar, L. T. (2012b, 1). Stabilization of a 3-DOF underactuated helicopter prototype: Second order sliding mode algorithm synthesis, stability analysis, and numerical verification. *12th International Workshop on Variable Structure Systems (VSS),* (pp. 361–366). https://doi.org/10.1109/vss.2012.6163529.

Moreno, J. A., & Osorio, M. (2008, 12). A Lyapunov approach to second-order sliding mode controllers and observers. *Decision and Control, 2008. CDC 2008. 47th IEEE Conference on,* (pp. 2856–2861). https://doi.org/10.1109/cdc.2008.4739356.

Odelga, M., Chriette, A., & Plestan, F. (2012, 6). Control of 3 DOF helicopter: A novel autopilot scheme based on adaptive sliding mode control. *American Control Conference (ACC)* (Vol. 2012, pp. 2545–2550).

Orlov, Y. V., & Aguilar, L. T. (2014). *Advanced H_∞ control: Towards nonsmooth theory and applications.* New York: Birkhauser.

Orlov, Y., Aoustin, Y., & Chevallereau, C. (2011, 3). Finite Time stabilization of a perturbed double integrator; Part I: Continuous sliding mode-based output feedback synthesis. *Automatic Control, IEEE Transactions on, 56,* 614–618. https://doi.org/10.1109/tac.2010.2090708.

Quanser. (1998). *3D helicopter system with active disturbance. [available]* http://www.quanser.com/choice.asp.

Raafat, S. M., & Akmeliawati, R. (2012). Robust disturbance rejection control of helicopter system using intelligent identification of uncertainties. *Procedia Engineering, 41,* 120–126. https://doi.org/10.1016/j.proeng.2012.07.151.

Shan, J., Liu, H.-T., & Nowotny, S. (2005, 11). Synchronised trajectory-tracking control of multiple 3-DOF experimental helicopters. *Control Theory and Applications, IEE Proceedings, 152,* 683–692. https://doi.org/10.1049/ip-cta:20050008.

Starkov, K. K., Aguilar, L. T., & Orlov, Y. (2008, 6). Sliding mode control synthesis of a 3-DOF helicopter prototype using position feedback. *Variable Structure Systems, 2008. VSS '08. International Workshop on,* (pp. 233–237). https://doi.org/10.1109/vss.2008.4570713.

Tao, G., & Kokotovic, P. V. (1997). Adaptive control of systems with unknown non-smooth non-linearities. *International Journal of Adaptive Control and Signal Processing, 11,* 81–100.

Wang, X., Lu, G., & Zhong, Y. (2013). Robust attitude control of a laboratory helicopter. *Robotics and Autonomous Systems, 61,* 1247–1257. https://doi.org/10.1016/j.robot.2013.09.006.

Zheng, B., & Zhong, Y. (2011, 2). Robust attitude regulation of a 3-DOF helicopter benchmark: Theory and experiments. *Industrial Electronics, IEEE Transactions on, 58,* 660–670. https://doi.org/10.1109/tie.2010.2046579.

Part III
Robotics

Chapter 11
Mechatronic Integral Ankle Rehabilitation System: Ankle Rehabilitation Robot, Serious Game, and Facial Expression Recognition System

Andrea Magadán Salazar, Andrés Blanco Ortega, Karen Gama Velasco and Arturo Abúndez Pliego

Abstract People who have suffered an injury require a rehabilitation process of the affected muscle. Rehabilitation machines have been proposed to recover and strengthen the affected muscle. In this chapter, we propose a novel ankle rehabilitation parallel robot of two degrees of freedom consisting of two linear guides. For the integral rehabilitation, a serious game and a facial expression recognition system are added for entertainment and to improve patient engagement in the rehabilitation process. The serious game has a simple design. This game has three levels and it is controlled with an impedance control, which specific command allowing character game jumps the obstacles. Facial expressions recognition system assists to the serious game. We propose to recognize three different facial expressions to the basic expressions. Based on the experiment results, we concluded that our system is good because it has a performance of 0.95%.

Keywords Parallel robot · Ankle rehabilitation · Facial expression recognition
Serious game

11.1 Introduction

Ligamentous ankle injuries are the most common sports trauma, its frequency approximately represents between 10 and 30% of all sports injuries (Zoch et al. 2003), according to data from different publications. The ankle sprain occurs when the ankle unexpectedly twists or turns in an awkward way beyond what the ligaments can tolerate, being the most common due to an excess of inversion

A. M. Salazar (✉) · A. B. Ortega · K. G. Velasco · A. A. Pliego
National Technologic of Mexico/CENIDET, Cuernavaca, Morelos, Mexico
e-mail: magadan@cenidet.edu.mx

O. O. Vergara Villegas et al. (eds.), *Advanced Topics on Computer Vision, Control and Robotics in Mechatronics*,
https://doi.org/10.1007/978-3-319-77770-2_11

movement, which damages the ankle lateral ligaments. It was observed that sprain occurs when the ankle unexpectedly twists or turns in an awkward way by that when a muscle remains on immobilization tends to weaken, caused stiffness and therefore loses tone and shortened. Salter recommended that all joint affections should be moved continuously through a full range of motion. Salter invented the concept of continuous passive motion, known as CPM (O'Discoll and Giori 2000).

Rehabilitation is the process of restoration of skills by a person who has had an illness or injury so as to regain maximum self-sufficiency and function in a normal or as near normal manner as possible. The rehabilitation is beneficial to reduce spasticity, to increase the muscle mass, and to control the muscle movement.

Robotic ankle rehabilitation devices like CPM machines are used to perform smooth and control motions in rehabilitation therapies to help patients to perform repetitive movements in a well-defined interval and a given speed.

Consequently, there is an increasing research interest in developing rehabilitation machines by technology development companies, institutions, and universities around the world. The main objectives of these machines are (a) rehabilitate the affected part (e.g., knee, ankle, hands, hip), (b) restore mobility, (c) reduce repetitive work of a therapist, (d) increasing the number of therapy services, (e) reduce recovery time and offer a wider range of personalized therapies with precise, and (f) insurance movements (Blanco-Ortega et al. 2012).

The repetitive nature of exercises in therapy sessions is another problem. The patients tend to pull out the rehabilitation process. To solve this problem, serious games have been proposed. Serious games refer to the use of computer games that have a main purpose that is not pure entertainment. These games contribute to increase motivation in rehabilitation sessions (Rego et al. 2010). Our game has simple rules to minimize learning period. Also, it is important the accuracy detecting patient's emotions. Automatic facial expressions recognition provides information nonverbal of patient. Facial expression recognition is a topic widely discussed in Computer Vision. Many researchers have been interested in the analysis of the six basic facial expressions for different applications. We proposed facial expression recognition as interface between serious game and rehabilitator. For this interaction, we recognize three expressions different to basic facial expressions.

In this chapter, we propose a novel comprehensive rehabilitation system that considers an ankle rehabilitation of two degrees of freedom (DOF) (dorsiflexion–plantarflexion and inversion–eversion), a serious game and an artificial vision system for the detection of facial expressions of the patient, which improves the rehabilitation process of the ankle. The movement of dorsiflexion–plantarflexion is considered in this game. Figure 11.1 shows the integration scheme of the proposed integral rehabilitation system.

The main contributions of this chapter are: an ankle rehabilitation system for young people, which allow 25° of dorsiflexion, 45° of plantarflexion, 25° of inversion, 15° degrees of eversion. One serious game with three different levels of difficulty according to the stages of rehabilitation, it helps that the ankle increases his strength and mobility, with a rehabilitation process entertained. Finally, it is also

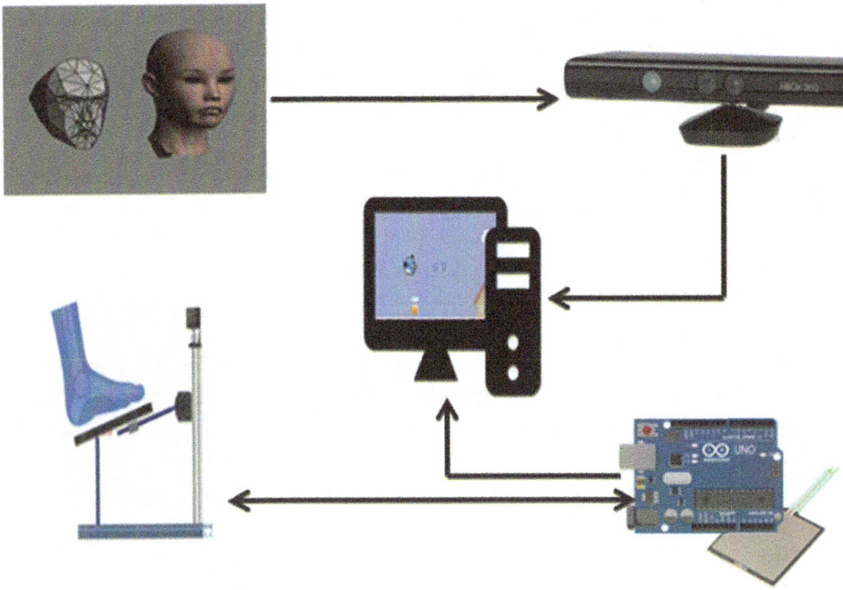

Fig. 11.1 Integration scheme of the proposed rehabilitation system

proposed a system for recognizing facial expressions different from the basic expressions as motivated, unmotivated, and pain, which gives feedback to the serious game by indicating whether to increase or decrease the frequency of obstacles.

The rest of the chapter is organized as follows: Sect. 11.2 reviews the state of the art of ankle rehabilitation systems, serious games, and facial expressions recognition. Section 11.3 explains the design of the proposed rehabilitator. In Sect. 11.4, serious games design is shown; in Sect. 11.5, the proposed facial expressions recognition system is presented. Section 11.6 explains the methodology followed in the experimentation and the results obtained are analyzed. Finally, we provide the conclusions in Sect. 11.7.

11.2 Literature Review

11.2.1 Parallel Robots for Ankle Rehabilitation

Mechatronics has been defined as a synergy of several engineering disciplines. It relates to the design and manufacture of intelligent electromechanical products and devices (Hsu 1997). The adoption of mechatronics or robotized devices for assistance and rehabilitation have given the possibility to improve diagnoses and therapy both in structured places like hospitals or re-education centers as well as at the patient's own home. With respect to manual therapy, the robotized one is more reproducible, more repeatable, and less dependent to the therapist ability; less tiring for the therapist and sometimes may be remotely performed at the patient home. The results are a better quality of life and a reduction of health expenses (Perdereau et al. 2011).

Mechatronic systems for rehabilitation are devices that seek to improve the recovery of a patient after suffering some kind of illness or injury in any part of the body. Some ankle rehabilitation machines proposed are based on configuration of parallel robot whose mechanical structure is formed by a mechanism of closed chain in which the end effector is attached to the fixed base by at least two independent kinematic chains.

In Rutgers University, a rehabilitation ankle device called "The Rutgers Ankle" was proposed, see Fig. 11.2a. This device is a parallel robot with 6-DOF, despite the ankle only has 3-DOF and uses pneumatic actuators. This device includes an interface where the patient interacts virtually through simulation games during their rehabilitation process. This device helps to improve balance, flexibility, and increase muscle strength. It has been used in patients for determining the effectiveness of this rehabilitation device, where they have concluded that require a large capacity compressor to maintain pressure and prevent overheating and low loads in the system (Girone et al. 1999, 2000; Deutsch et al. 2001). A new Rutgers Ankle

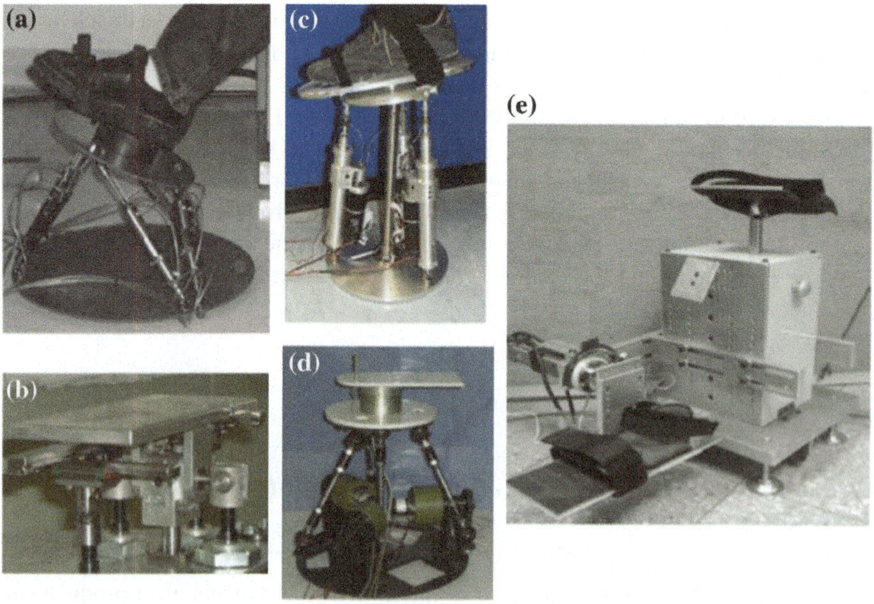

Fig. 11.2 Parallel robots for ankle rehabilitation: **a** 6-DOF (Cioi et al. 2011), **b** 4-DOF (Yoon and Ryu 2005), **c** 2-DOF (Saglia et al. 2009), **d** 3-DOF (Liu et al. 2006), and **e** 1-DOF (Chou-Ching et al. 2008)

device was used to train ankle strength and improved motor control for children with cerebral palsy (CP) (Cioi et al. 2011).

As shown in Fig. 11.2b, the parallel mechanism for ankle rehabilitation comprises two plates for supporting the foot and provides flexion–extension of the fingers. The parallel mechanism is of 4-DOF and uses four pneumatic actuators, which provide dorsi/plantarflexion and inversion/eversion movements for the ankle (Yoon and Ryu 2005). A parallel robot of 2-DOF for ankle rehabilitation is shown in Fig. 11.2c (Saglia et al. 2009). Using a PD control , the parallel robot is operated redundantly to avoid singularities and this way to provide dorsi/plantarflexion and inversion/eversion movements. In Saglia et al. (2010) presents the development of an admittance-based assistive controller for this ankle rehabilitation system. An admittance control technique is used to perform patient-active exercises with and without motion assistance. Electromyography (EMG) signals are used to evaluate patient's effort during training/exercising.

Another parallel robot of 3-DOF (Liu et al. 2006) with a link in the central part to connect the mobile base with fixed base and thus give greater rigidity to the structure and limiting movement is shown in Fig. 11.2d. The authors present simulation results in MSC ADAMS of a virtual prototype and also present the physical prototype.

A robot assistant in ankle rehabilitation of 1-DOF, shown in Fig. 11.2e, provides dorsi/plantarflexion movement to reduce spasticity, increase muscle tone, and improve motor control (Chou-Ching et al. 2008). The authors implemented a proportional-derivative (PD) fuzzy controller combined with a conventional integrated control with feedback of the angular position and torque exerted by the patient through his foot on the robot base.

11.2.2 Serious Games

Ankle rehabilitation consists of several repetitive sessions and intensive activities that become tedious after hundreds of repetitions. On the other hand, serious games are interactive tools, which his goal is to develop a lost ability, and not only fun and entertainment, although they include them. Serious games are a good solution when you want to motive the patient to complete his rehabilitation (Rego et al. 2010).

Several works about serious games for the rehabilitation have been published in the last few years. As can be seen in Table 11.1, the development of this type of games is not new and can be used in multiple areas such as training in a work or learning environment. The development of serious games focused on the rehabilitation area to support the rehabilitation of different parts of the body and specifically the ankle is growing. In this analysis, we found that most of the implemented systems consider the feedback of the patient, used mirror feedback, in which through an avatar the patient can observe the movements made on the screen, facilitating the interaction with the game (Jaume-i-capó and Samčović 2014), while in other works a database stores the number of movements realized in each session.

Table 11.1 Rehabilitation systems with serious game

Author	Area	Rehab	Portability	Development tools	RSM	Feedback
Girone et al. (1999)	Physical therapy	Ankle sprain	Home	World toolkit	Dorsiflexion-plantarflexion eversion-inversion	Yes
Michmizos and Krebs (2012)	Physical therapy	Ankle due to cerebral palsy	Clinic	–	Dorsiflexion-plantarflexion eversion-inversion	Yes
Omelina et al. (2012)	Physical therapy	Cerebral palsy	Home	XML, Kinect SDK, XNA, C++, C#	–	Yes
Burdea et al. (2013)	Physical therapy	Ankle due to cerebral palsy	Clinic and home	–	Dorsiflexion-plantarflexion eversion-inversion	Yes
Goncalves et al. (2014)	Physical therapy	Ankle due hemiparesis	–	C#, XNA	Dorsiflexion-plantarflexion	Yes
Garcia and Navarro (2014)	Physical therapy	Ankle sprain	Home	–	Dorsiflexion-plantarflexion eversion-inversion	Yes
Zhang et al. (2014)	Physical therapy	Ankle sprain	Clinic	–	Dorsiflexion-plantarflexion	Yes
Shah et al. (2014)	Physical therapy	Stroke	Home	Unity 3D, C#	–	Yes
Farjadian et al. (2014)	Physical therapy	Ankle due hemiparesis or cerebral palsy	–	Labview	Dorsiflexion-plantarflexion	Yes
Menezes et al. (2014)	Physical therapy	Stroke	Clinic and home	XNA, Kinect SDK, Unity 3D	Dorsiflexion-plantarflexion	Yes
Jaume-i-capó and Samčović (2014)	Physical therapy	Cerebral palsy	–	C++, OpenNI, OpenCV, Q+	–	Mirror feedback
Tannous et al. (2015)	Physical therapy	Locomotive	Clinic and home	XNA, C#, Visual Studio .Net	–	Mirror feedback
Agas et al. (2015)	Physical therapy	Hand palsy	Clinic	Matlab, RoboPlus, Unity3D	–	Yes
Pasqual et al. (2016)	Physical therapy	Ankle	–	Unity3D	Dorsiflexion-Plantarflexion eversion-inversion	Yes

In ankle rehabilitation, the therapy must be performed through the repetition of specific movements (RSMs) such as movements of dorsiflexion–plantarflexion or eversion–inversion (Girone et al. 1999; Michmizos and Krebs 2012; Burdea et al. 2013; Farjadian et al. 2014; Garcia and Navarro 2014; Goncalves et al. 2014; Menezes et al. 2014; Zhang et al. 2014; Pasqual et al. 2016). The main tools used for the game's implementations are Microsoft XNA Game Studio, Visual Studio . Net, C#, C++, OpenNI by Kinect, OpenCV and Q+.

The objective of the serious game developed here is to propose a rehabilitation of the ankle, in a fun and attractive way to have a better interaction between the patient, the rehabilitation system, and the specialist; proposing an integral system of simple rehabilitation so that it can be used both in the home and in the clinic.

11.2.3 Facial Expression Recognition

According to Fasel and Luettin (2003), a facial expression is a visible manifestation of the affective state, cognitive activity, intention, personality, and psychopathology of a person. Facial expressions in the human are a natural mechanism to express their state of mind that anyone can detect regardless of race, age, or sex. Several factors influence in the formation of facial expressions (Fasel and Luettin 2003), as shown in Fig. 11.3.

According to Ekman and Friesen (1978), the facial expressions are categorized into six emotions: anger, disgust, fear, happiness, sadness, surprise, with 13 action

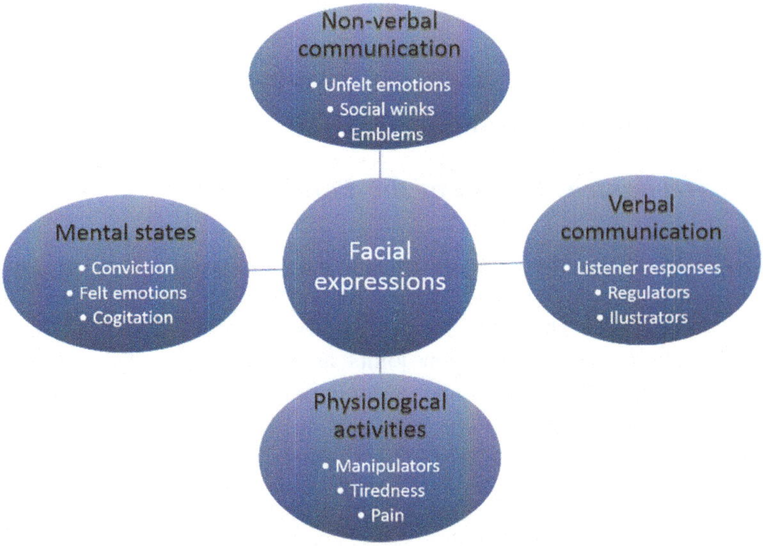

Fig. 11.3 Factors that influence facial expressions formation (Fasel and Luettin 2003)

units, where the automatic detection of the facial components is done through the regions of interest (ROI) of the face, which are mainly the eyes, the mouth, nose, and eyebrows. The development of facial expression recognition systems can be made classifying expressions based on facial action coding system and direct or indirect interpretation of facial expressions.

Facial action coding system (FACS) is a coding system created by Paul Ekman and Freisen (Kumari et al. 2015) for describing facial movements. The FACS identifies the facial muscles that individually or in groups cause changes in facial behaviors. These changes in the face are called action units (AU); then, the FACS is made up of several such action units. This facial action coding system has become a standard for the automatic facial expression recognition (FER).

FER has been a line of research addressed in recent decades, obtaining nonverbal information about the behavior of people. Generally, facial expression automatic systems are carried out by modeling the action units. However, the expression analysis is still complex for current systems of automatic recognition through units of action or specific points, since determining the internal state of a person through their facial muscle movements requires pondering many variables (Porras-Luraschi 2005).

Table 11.2 shows a summary of the state-of-the-art recognition of facial expressions, it can be seen that the most popular techniques for the extraction of these characteristics are: Gabor filters, local binary patterns, principal component analysis (PCA), independent component analysis (ICA) linear discriminant analysis (LDA), local gradient code (LGC), and local directional pattern (LDP).

The most popular classification techniques are, but not limited to support vector machines (SVM), nearest neighbor (NN), artificial neural networks (ANN), and decision trees. We can also observe that in recent years the use of the Kinect sensor for the acquisition of images of facial expressions has increased.

11.3 Design of the Parallel Robot

11.3.1 Mechanical Design of the Parallel Robot

The parallel robot proposed for the ankle rehabilitation consisting of two linear guides actuated with CD geared motors for vertical movements, resulting in a mechanism of 2-DOF. This robot has a movable platform where the foot-ankle is supported. Spherical and translation joints are used to link the movable base through bars. The strut plays an important role in the mechanical design, since it is positioned to counterbalance the foot-leg weight of the patient, and is attached to the movable base by means of a spherical union. The ankle rehabilitation robot provides dorsiflexion/plantarflexion and inversion/eversion movements, as can be observed in Fig. 11.4.

Table 11.2 Summary of automatic facial expressions recognition

Author	Area	Data base	Technique	Precision	Tools	Real time	Expressions
Michel and El Kaliouby (2003)	FER	Cohn-Kanade	SVM	87.9%	–	Yes	Anger, disgust, fear, happiness, sorrow, and surprise
Porras-Luraschi (2005)	FER	Own	Neural networks	–	Visual Studio . Net, Open CV	Yes	Distraction, doubt, interest, happiness, surprise
Tsalakanidou and Malassiotis (2010)	FER	Own	ASM	83.6%	–	Yes	Neutral, disgust, happiness, surprise
Gupta et al. (2012)	Action units recognition	JAFFE	Decision tree	86.66%	Matlab		Neutral, anger, fear, happiness, sadness, surprise
Surbhi (2012)	FER	–	ROI	–	–		–
Arenas et al. (2012)	Feature extraction	Own	PCA, eigenfaces	–	API NUI, Kinect SDK, Matlab		–
Seddik et al. (2013)	Face recognition	JAFFE	PCA, eigenfaces	64%	Open CV, Kinect SDK	Yes	Anger, disgust, fear, happiness, sadness, surprise
Li et al. (2013)	FER	–	AAM, ICP	–	3ds Max, Kinect SDK, Visual Studio	Yes	–
Ijjina and Mohan (2014)	Mapping of facial expressions	EUROCOMM	Convolutional neuronal networks	87.98%	Kinect SDK		Neutral, happiness, jaw drop

(continued)

Table 11.2 (continued)

Author	Area	Data base	Technique	Precision	Tools	Real time	Expressions
Stocchi (2014)	FER	VBN, FaceWarehouse	Facetracking, Zernike moments	87.68%	Kinect SDK	Yes	Anger, disgust, fear, happiness, sadness, surprise
Aly et al. (2014)	FER	FaceWarehouse, own	SVM	95.1% non-frontal, 98% frontal	Kinect SDK	Yes	Anger, disgust, fear, happiness, sadness, surprise
Kakarla and Reddy (2014)	FER	Facetracking	GEMEP-FERA	90%	Kinect SDK	Yes	Anger, disgust, fear, happiness, sadness, surprise
Aly et al. (2015)	FER	Own (VT. KFER)	–	–	Kinect SDK	–	Anger, disgust, fear, happiness, sadness, surprise
Mao et al. (2015)	FER	Own	SVM	87%	Kinect SDK	Yes	Anger, disgust, fear, happiness, sadness, surprise
Li et al. (2013)	FER	JAFFE	LBP, LGC, LGC-HD, LGC-VD, HOG-LDP	89.4231%, 90.3846%, 87.5%, 92.30%, 85.2041%	–		Anger, disgust, fear, happiness, sadness, surprise

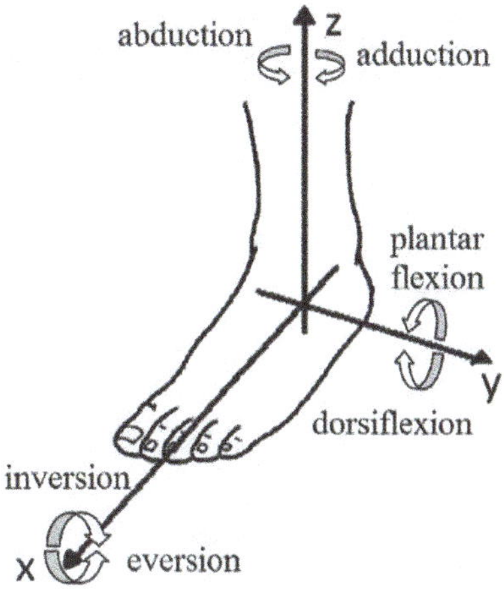

Fig. 11.4 Rotational ankle movements

As a first stage, is a kind of CPM machine, i.e., independent movements will be providedby the movable platform. Only basic and independent movements of dorsiflexion/plantarflexion and inversion/eversion for passive rehabilitation are considered as it is shown in Figs. 11.4 and 11.5. Table 11.3 shows the ankle range of motion.

11.3.2 Kinematic Analysis

Consider the schematic diagram of the parallel robot shown in Fig. 11.6, where r_1, r_2, r_3, and r_4 are the distances of the mobile platform, driven link, mobile base, and

Fig. 11.5 Parallel robot for ankle rehabilitation based on linear guides

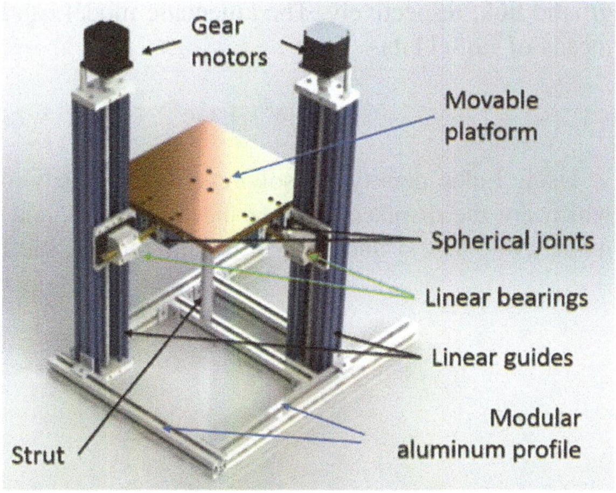

Table 11.3 Range of ankle movements

Type of motion	Max. allowable motion
Dorsiflexion	20.3° a 29.8°
Plantarflexion	37.6° a 45.8°
Inversion	14.5° a 22.0°
Eversion	10.0° a 17.0°
Abduction	15.4° a 25.9°
Adduction	22.0° a 36.0°

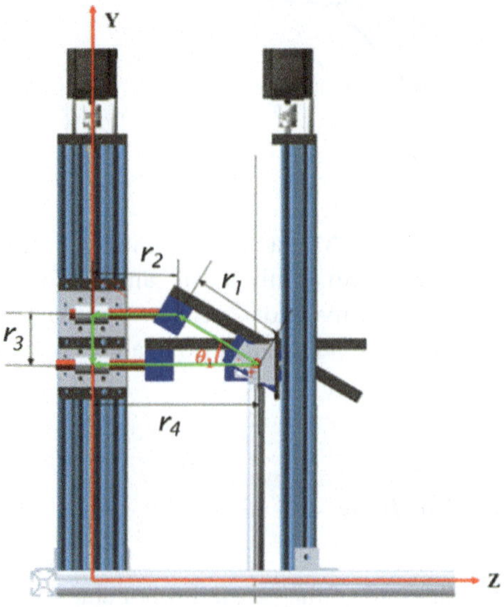

Fig. 11.6 Kinematic analysis based on the vector loop

ground link, respectively. The kinematic model can be expressed in polar form by means of Eq. (11.1).

$$r_1 e^{j\theta_1} + r_2 e^{j\theta_2} + r_3 e^{j\theta_3} = r_4 e^{j\theta_4} \tag{11.1}$$

Using Euler identity and solving simultaneously for the unknown displacements, which are the displacements of the driven link and the mobile base of the linear guide. Note that r_1 and r_4 are constant vectors, see Fig. 11.6.

$$\begin{aligned} r_2 &= r_4 + r_1 \cos\theta_1 \\ r_3 &= r_1 \sin\theta_1 \end{aligned} \tag{11.2}$$

Taking time derivative of the vector loop in Eq. (11.2), using Euler identity and solving for the velocities of the driven link and the mobile base, we obtain

$$\begin{aligned} \dot{r}_2 &= \dot{\theta}\sin\theta_1 \\ \dot{r}_3 &= \mathrm{r}_1\dot{\theta}\cos\theta_1 \end{aligned} \tag{11.3}$$

11.3.3 Dynamic Analysis

Consider Fig. 11.7 where a positive angle θ_1 corresponds to dorsiflexion movement and negative angle θ_1 corresponds to plantarflexion movement, on the y-axis. Similarly, for the other linear guide a positive angle φ represents an eversion movement and negative angle φ an inversion movement.

The mathematical model governing the dynamic for the dorsiflexion–plantarflexion movement (similar for eversion–inversion movement) can be obtained by applying Newton equation, which is given by:

$$J_e\ddot{\theta}_1 = F_1 d_1 - P \tag{11.4}$$

where J_e is the inertia moment of the movable platform, d_1 is the distance between the strut and Acme screw. The control force is F_1 and P is an unknown disturbance (e.g., ankle stiffness, viscous damping, friction forces).

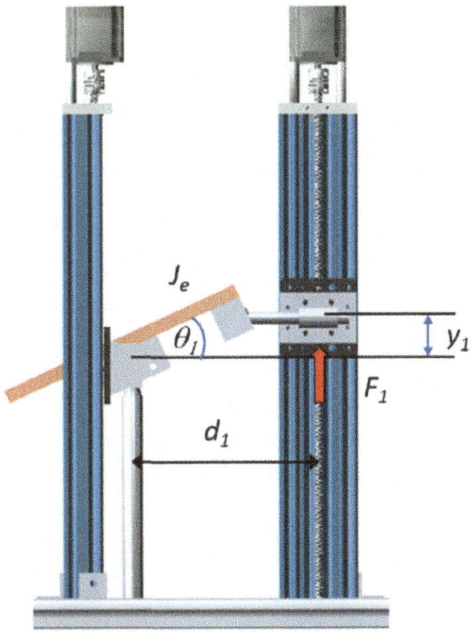

Fig. 11.7 Schematic diagram for a dorsiflexion movement θ_1

The differential Eq. (11.4) is a linear equation, considering P as constant disturbance. A control strategy can be designed to track a time-varying reference trajectory for each axis independently. Gravity effects are neglected due to frictional forces in the screw, the screw is self-locking.

11.3.4 PID-Type Control

For planning trajectory tracking, the following PD-type controller is proposed to obtain the desired position θ_{1d}:

$$v_\theta = \ddot{\theta}_{1d} - \alpha_2\left(\dot{\theta} - \dot{\theta}_{1d}\right) - \alpha_1(\theta - \theta_{1d}) - \alpha_0 \int (\theta - \theta_{1d})\mathrm{d}t \tag{11.5}$$

The control input is given by,

$$F_1 = \frac{J_e}{d_1} v_\theta \tag{11.6}$$

The use of this PID-type controller yields the following closed-loop dynamics for trajectory tracking errors given by $e_\theta = \theta - \theta_{1d}$:

$$\dddot{e}_\theta + \alpha_2 \ddot{e}_\theta + \alpha_1 \dot{e}_\theta + \alpha_0 e_\theta = 0 \tag{11.7}$$

The controller gains α_0, α_1 and α_2 were selected such that the associated characteristic polynomial for the closed-loop system Eq. (11.3) be Hurwitz polynomial (polynomial whose roots are located in the left half-plane of the complex plane), one guarantees that the error dynamics be globally asymptotically stable. The controller gains were set to coincide with those of the desired characteristic polynomial $\left(s^2 + 2\zeta\omega_n s + \omega_n^2\right)(s + b)$ with $\omega_n = 10, \zeta = 0.7, b = 10$.

Desired position trajectory to provide dorsi/plantarflexion movements is given by the following Bézier polynomial:

$$\begin{aligned} \theta_{1d}(t) &= \theta_i + \left(\theta_f - \theta_i\right)\sigma\left(t, t_i, t_f\right)\mu_p^5 \\ \sigma\left(t, t_i, t_f\right) &= \gamma_1 - \gamma_2\mu_p + \gamma_3\mu_p^2 - \cdots + \gamma_6\mu_p^5 \\ \mu_p &= \frac{t - t_i}{t_f - t_i} \end{aligned} \tag{11.8}$$

where $\theta_i = \theta_{1d}(t_i)$ and $\theta_{\mathrm{f}} = \theta_{1d}(t_{\mathrm{f}})$ are the initial and final desired positions, so that the basis of rehabilitation starts from an initial position and goes to a final position with a smooth change, such that:

$$\theta_{1d}(t) = \begin{cases} 0 & 0 \le t < t_i \\ \sigma\left(t, t_i, t_f\right)\theta_f & t_i \le t < t_f \\ \theta_f & t > t_f \end{cases} \tag{11.9}$$

Parameters of the Bézier polynomial function $\theta_{1d}(t)$ (4) are $\gamma_1 = 252$, $\gamma_2 = 1050$, $\gamma_3 = 1800$, $\gamma_4 = 1575$, $\gamma_5 = 700$ and $\gamma_6 = 126$.

11.3.5 Robust GPI Control

Initially, generalized proportional integral (GPI) control was introduced in the context of predictive control of differential flat systems (Fliess and Join 2008). In recent years, GPI controllers have been implemented in power electronics problems and multivariable cases (Franco-González et al. 2007), as well as active vibration control of rotating machinery (Blanco-Ortega et al. 2010). For instance, a GPI controller was proposed in a controlled mass system which was connected to uncertain mass-spring-damper systems in order to reject an unknown disturbance input (Sira-Ramírez et al. 2008).

The unknown disturbance can be included in the rotational equation described by Eq. (11.4) and can be represented as follows:

$$\ddot{\theta}_1 = u_\theta + P_m \tag{11.10}$$

where

$$\begin{aligned} u_\theta &= \frac{F_1 d_1}{J_e} \\ P_m &= -\frac{P}{J_e} \end{aligned} \tag{11.11}$$

Considering the disturbance as a third-order polynomial, given by:

$$P_m = a_3 t^3 + a_2 t^2 + a_1 t + a_0 \tag{11.12}$$

The control strategy is given by the following equation:

$$\begin{aligned} u_\theta = {} & \ddot{\theta}_{1d} - k_5\left(\widehat{\dot{\theta}_1} - \dot{\theta}_{1d}\right) - k_4(\theta_1 - \theta_{1d}) + \\ & - k_3 \int_0^t (\theta_1 - \theta_{1d})\, \mathrm{d}\tau - k_2 \int_0^t \int_0^\tau (\theta_1 - \theta_{1d}) \mathrm{d}\lambda \mathrm{d}\tau \\ & - k_1 \int_0^t \int_0^\tau \int_0^\lambda (\theta_1 - \theta_{1d}) \mathrm{d}\sigma \mathrm{d}\lambda \mathrm{d}\tau \\ & - k_0 \int_0^t \int_0^\tau \int_0^\lambda \int_0^\sigma (\theta_1 - \theta_{1d}) \mathrm{d}\rho \mathrm{d}\sigma \mathrm{d}\lambda \mathrm{d}\tau \end{aligned} \tag{11.13}$$

where

$$\overset{\frown}{\dot{\theta}}_1 = \int_0^t u_x \mathrm{d}\tau$$
$$\dot{\theta}_1 = \overset{\frown}{\dot{\theta}}_1 + \dot{\theta}_1(0) \quad (11.14)$$

Substituting the controller Eq. (11.13) and the disturbance Eq. (11.12) in Eq. (11.4), besides, if we consider the error, $e = \theta_1 - \theta_{1d}$, and its respective derivatives, then we obtain:

$$e^{VI} + k_5 e^{V} + k_4 e^{IV} + k_3 \dddot{e} + k_2 \ddot{e} + k_1 \dot{e} + k_0 e = 0 \quad (11.15)$$

The associated characteristic polynomial for the closed-loop system Eq. (11.15) is given by:

$$s^6 + k_5 s^5 + k_4 s^4 + k_3 s^3 + k_2 s^2 + k_1 s + k_0 = 0 \quad (11.16)$$

The parameters were selected to ensure that the error dynamics was globally asymptotically stable and were set to coincide with those of the desired characteristic polynomial $\left(s^2 + 2\zeta\omega_n + \omega_n^2\right)^3$ with $\omega_n = 10, \zeta = 0.7$.

11.4 Serious Game Design

The serious game needs to be interesting, entertaining, and interactive (Zhang et al. 2014), to motivate the patient. According to the recommendations given in (Michmizos and Krebs 2012), we considered to have a simple visual interface and simple control so that the learning period is short, helping the patient to be autonomous and if possible, that he/she can perform the therapy at home.

The game developed is aimed at young people, ranging from 12 to 20 years old, who need motivation not to abandon their rehabilitation process, having a more attractive and fun therapy. The tool used for its development was Unity3D because it is a flexible development platform that can be implemented in all operating systems (Linux, Mac, Windows, etc.).

The serious game developed is of 1-DOF corresponding to dorsiflexion and plantarflexion movements. The game is proposed to use in active and resistive rehabilitation (strengthening or resistance training). On these rehabilitation modes, the individual realizes all the effort in the exercises. The parallel robot presents an opposing force to the active patient movement, which is gradually increased to improve muscular endurance, see Table 11.4. Contains three levels of difficulty and each level has a game character. The force and angle the patient must apply on the mobile platform increases as level up. The obstacles that must skip the game character also increase its frequency as the game level increases, see Fig. 11.8.

Table 11.4 Serious game specifications

Level	1	2	3
Characteristic game			
Objective	Mobility	Force	Mobility and force
Angle	$8 \leq \theta \leq 10$	$18 \leq \theta \leq 20$	$28 \leq \theta \leq 30$
Force	$F \geq 10$	$F \geq 20$	$F \geq 30$

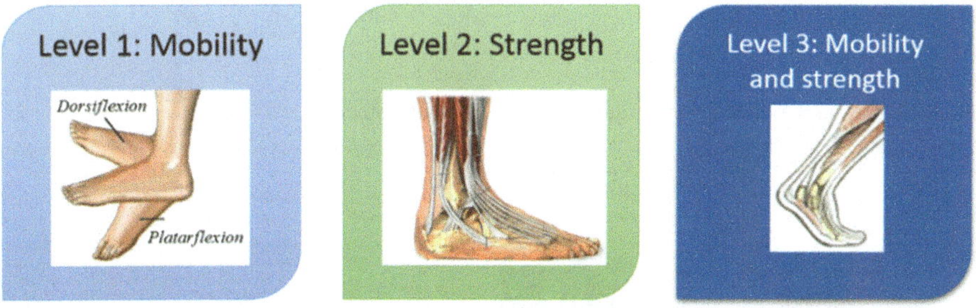

Fig. 11.8 Levels of serious game corresponding to dorsiflexion and plantarflexion movements

The obstacles that must skip the game character also increase its frequency is increasing as the game level. The relations between game character, angle of the movable platform, and force that the patient must apply depending on the game level is shown in Table 11.4.

The purpose of the first game level is for the patient to regain mobility in a small range of motion by applying a small force that does not cause discomfort or pain. In the second level, the patient must apply a greater force to strengthen the affected muscle. Finally, the third level helps both mobility and strengthening the muscles; the range of motion and the force that the patient must apply in the mobile platform are increased. In Fig. 11.9, the relation between angular displacement of the movable platform and the linear displacement of the mobile base of the linear guide is shown.

Another thing is that blue and green tones are used to convey a feeling of well-being and red obstacles, with the intention of capturing the attention of the players, see Fig. 11.10.

Finally, our ankle rehabilitation system contemplates store the aforementioned information about strength, number of movements performed, and therapy time. This information allows for supervision by the therapist. Additionally, improving interaction between rehabilitation prototype and serious game was proposed to

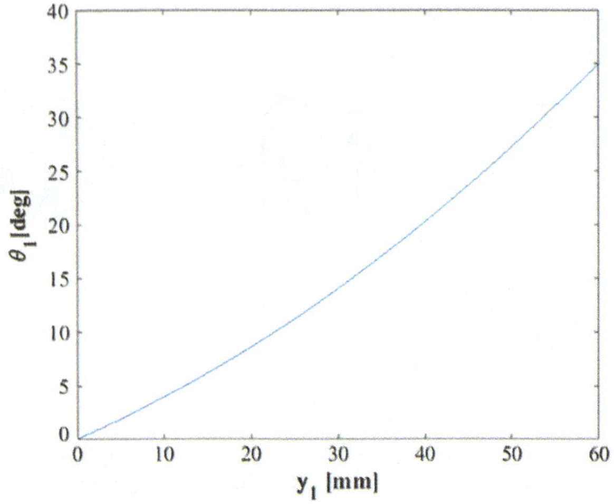

Fig. 11.9 Relationship between angular and linear displacement

Fig. 11.10 Psychology color of game characters

recognize and interpret of facial expressions under naturalistic conditions. Then, if he feels discomfort or pain, the game decreases the strength and frequency of the movement, and if patient is boring, the game increases the frequency of movements and the strength. Of course, we considered the level and supervision of the therapist for these changes.

11.5 Facial Expressions Recognition Proposed

As already mentioned, the system of recognition of facial expressions is a support system for the serious game and the ankle rehabilitator, which is necessary to perform the detection of mental states motivated and unmotivated, and a psychological activity such as the pain. According to Fasel and Luettin (2003), this mental and physical information can be identified visually through facial expressions.

In this paper, the steps considered in facial expression automatic recognition system are (a) Face acquisition, (b) Detection and descriptions of FACS, and finally (c) Facial expression recognition. We used Kinect sensor for the first and second step.

The Kinect sensor was built to revolutionize the way you play video games and the entertainment experience; however, now it has more uses. In recent years, Kinect has been used for other areas such as the recognition of facial expressions, helping to obtain nonverbal information of what a person is feeling.

Kinect provides the Microsoft Face Tracking SDK development tool for the detection, tracking and description of the components of the face and their movements. This software enables us to develop our application for the tracking of the face in real time. This library makes the description of the face using different action units. We proposed to recognize only tree expression: pain, motivated or concentrate, and unmotivated or distracted. For the facial expressions recognition, we used six action units and two movements of the head, these are shown in Fig. 11.11.

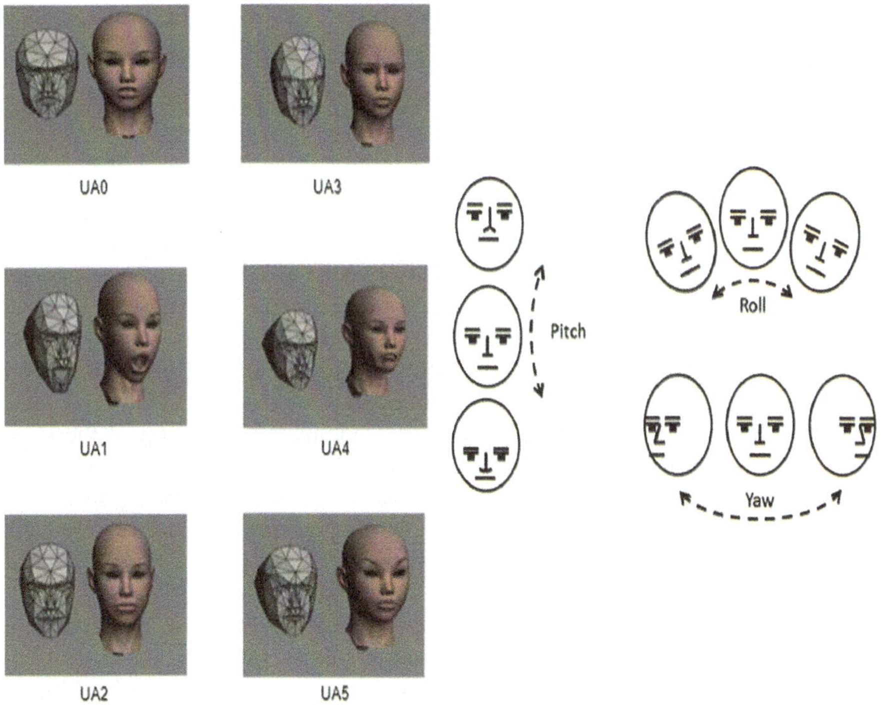

Fig. 11.11 Action units and movements that Kinect sensor recognizes

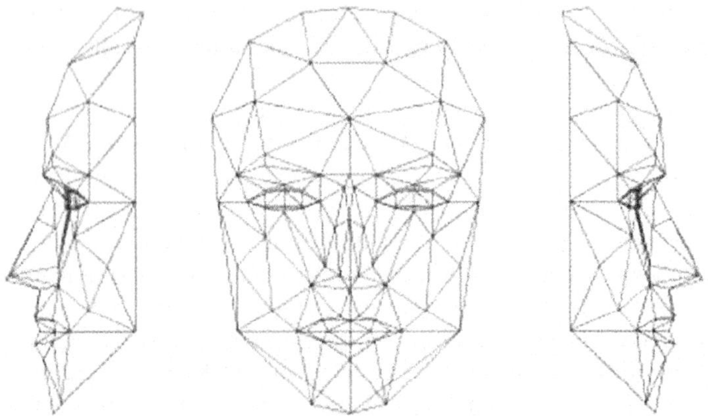

Fig. 11.12 Candide-3 mask

Table 11.5 Equivalent of the action units proposed by Paul Ekman and Kinect sensor for facial expression considered in this paper

Kinect sensor	Ekman	Description
AU0	AU5	Upper lip raiser
AU1	AU26	Jaw drop
AU2	AU20	Lip stretcher
AU3	AU4	Brow lowerer
AU4	AU15	Lip corner depressor
AU5	AU2	Outer brow raiser

The unit of action UA0 refers to lifting the upper lip, the UA1 to lower the jaw, the UA2 to tighten the lips, the UA3 to lower the eyebrows, the UA4 to lower the corners of the lips, and the UA5 to raise the eyebrows. These units of action are measured in a range of -1 to 1. The head movements that Kinect sensor recognized is Pitch (which refers to raising and lowering the head) and Yaw (moving the head from left to right and Roll turning the head), which are measured in a range of $-90°$ to $90°$. For the detection of these movements and action units, Candide-3 is used, which is a parameterized mask of 113 vertices and 168 surfaces (see Fig. 11.12).

The FACS system proposed by Ekman is a complete description of the facial muscles. However, the set of facial actions considered by the Kinect sensor to carry out the recognition of facial expressions is smaller, so it was necessary to make a table of equivalences of both systems. It is important to mention that we only made this table of equivalences for those action units that are part of the three facial expressions considered in this project, see Table 11.5.

Generally, the recognition of basic facial expressions is based on the description made by Ekman (through the FACS) for these emotions. However, we did not find in the literature of the area a description of which action units are involved for the recognition of the three facial expressions considered in this paper. We made an analysis of the data and we propose a description for the three facial expressions

Table 11.6 Proposed description of the facial expressions pain, concentrate or motivated, distraction or unmotivated and transitions

Class	Description
Pain (class 1)	Lip corner depressor (AU4) with values greater than 0.01 Brow lowerer (AU3) and jaw drop (AU1) with values greater than 0.25 Brow lowerer (AU3) and lip stretcher (AU29) with values greater than 0.25
Distraction or unmotivated (class 2)	Pitch with values greater than 35° Pitch with values lower than −10° Yaw with values greater than 35° Yaw with values lower than −25°
Concentrate or motivated (class 3)	Neutral: all units of action between −0.20 and 0.20 Happy: lip corner depressor (AU3) with values lower than −0.35
Transitions (class 4)	Those are not included in any of the previous class

under study from four action units and two head movements. The face is a dynamic object that usually changed by blinking, yawn, move the head from one side to another, lower the head, etc. For these movements, we considered necessary to have another class and it was named transitions. In Table 11.6, you can see this description proposal.

Then, the feature vector of each expression is composed of eight attributes: six action units and two positions of Kinect sensor: x = {AU0, AU1, AU2, AU3, AU4, AU5, Pitch, Yaw}.

11.6 Experimentation

In this section, three tests are presented. First, we can provide simulations results about ankle rehabilitation. The second test is about the facial expression recognition system, and finally, the last test is about integration of all the systems proposed.

11.6.1 Simulation Results

The ankle rehabilitation parallel robot can provide a passive rehabilitation using the PID-type controller or robust GPI controller. In passive exercises, the patient effort is not required, due to the parallel robot moves the ankle-foot in a smooth way. Table 11.7 shows the simulation parameters obtained from the virtual prototype, see Fig. 11.13.

In Fig. 11.13, the real and desired dorsiflexion response and control forces are shown, using the virtual prototype (see Fig. 11.5) and PID-type controller of Eq. (11.6). It shows how smooth movement of 0° to 15° ($\pi/12$ rad) is obtained

Table 11.7 Simulation parameters

$J_e = 0.015$ kg m^2	Inertia moment of movable platform
$P = 0.5$ Nm	Perturbation force due to foot weight
$d_1 = 0.185$ m	Distance from the strut to acme screw

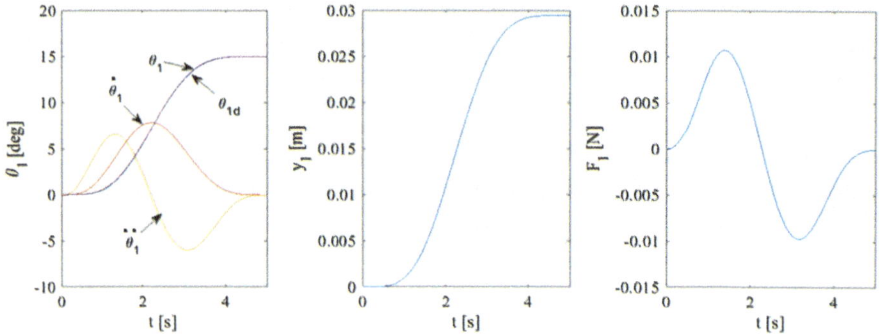

Fig. 11.13 Dorsiflexion response using the PID-type controller for $P = 0$ Nm

using Bézier polynomial of Eq. (11.8). The aim is that the physiotherapist could set the angle interval at the desired time in order that parallel robot provides the required speed based on the process of rehabilitation and improvement of the affected part. It can be seen that for the dorsiflexion movement, the tracking error tends to zero and that the movement is performed smoothly, 15° in 5 s.

The simulation results are shown in Fig. 11.14, corresponding to dorsiflexion movement considering a constant disturbance ($P = 0.5$ Nm). It can be seen that the control force does not compensate the disturbance.

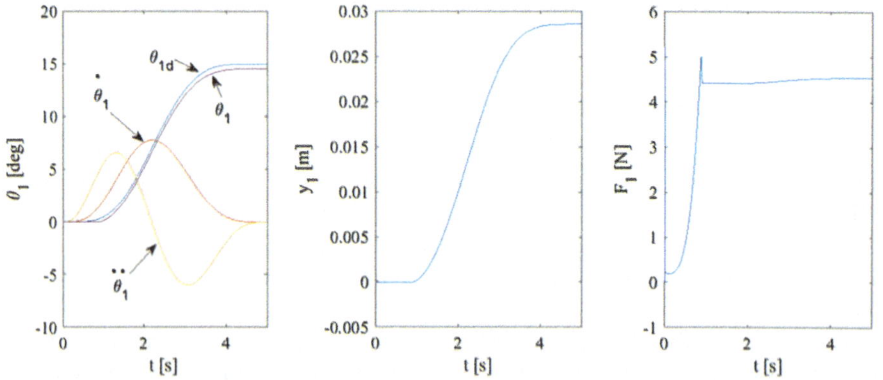

Fig. 11.14 Dorsiflexion response using the PID-type controller for $P = 0.5$ Nm

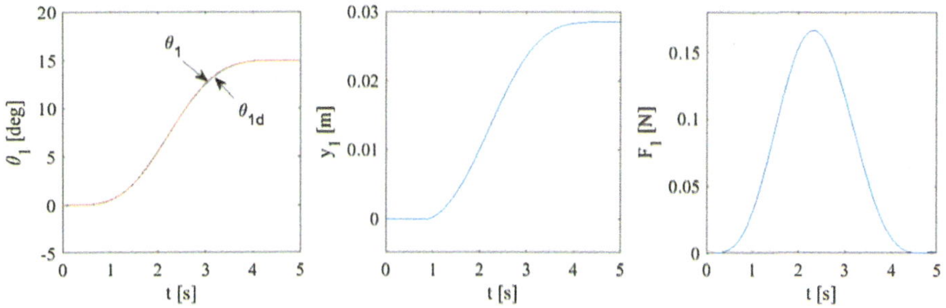

Fig. 11.15 Dorsiflexion response using the robust GPI controller, $P = 0$ N

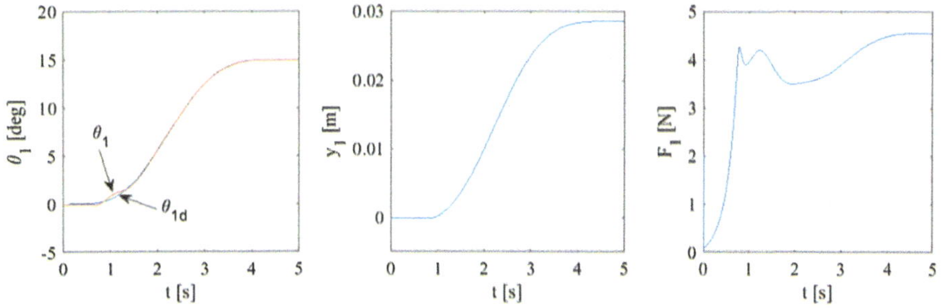

Fig. 11.16 Dorsiflexion response using the robust GPI controller, $P = 0.5$ N

Figure 11.15 corresponds to the dorsiflexion movement. It can be observed how smooth movement of 0° to 15°, with a displacement of 0.03 m on the linear guide, is obtained. The trajectory tracking based on Bézier polynomial Eq. (11.8) smoothly interpolating between zero and a final position located at fifteen degrees from the initial rest position in approximately 5 s.

The simulation results shown in Fig. 11.16 correspond to dorsiflexion movement considering a constant disturbance ($P = 0.5$ Nm). It can be seen that the robust GPI controller compensate the disturbance, the tracking error tends to zero, and that the movement is performed smoothly, 15° in 5 s.

11.6.2 Database

A training database was created with 2303 instances and a database for tests with 306 instances. The images were captured using the Kinect sensor, at 30 frames in one second. We recorded only three expressions for 45 people (15 women and 30 men) between 22 and 30 years old, from México. The facial expression images are true color (24 bits) with measure 640 × 480 pixels. All images have a frontal-view of a single person. The images were acquired under various lighting conditions, in a natural environment.

Kinect sensor has some restrictions for building the database. First, Kinect needs a distance from 60 cm to 1.20 m between volunteers and the Kinect camera. If there is a greater or lesser distance, the Candide-3 mask is not positioned on the face to tracking her. Second, the Kinect sensor does not work correctly with the presence of accessories on the face such as mustache, hair in face, glasses, caps. In this case, our volunteers did not have these accessories.

11.6.3 Facial Expression Recognition

The goal of this test was evaluating the facial expression recognition system. In this work, we provide a comparison with three classification algorithms in the two frameworks in such a way that the best model for the recognition of facial expressions was obtained. Weka (Waikao 2017) is open source software that contains a collection of machine learning algorithms for data mining tasks, Weka 3.8 was used and for all algorithms we considered the option cross validation. Language R (Foundation 2017) is a free software environment for statistical computing and graphics. It compiles and runs on a wide variety of platforms. The algorithms evaluated were Naïve Bayes, C4.5, and Support vector machines (SVMs) using an RBF kernel. We use the default values from frameworks for the cost and gamma.

Database was divided into two parts. For the training, we used a database with captures of 30 people, the test phase was carried out with an integrated database of 15 people different from those considered in the training base. Table 11.8 shows the performance achieved by each of the classification algorithms considered.

Classifier with the lowest performance was the Bayes algorithm. This algorithm was considered for its simplicity to implement it in the integration of the rehabilitation system. The results obtained by Bayes on both platforms are similar.

The algorithm with the best performance, in both Weka and R, is C4.5 reaching in the training phase a value 99% of success in both platforms, but his performance decrease in the test phase up to 90% in the R language. In Weka, the algorithm maintained the good result in the test phase. The tree generated for this model is shown in Fig. 11.17. We decided to consider this decision tree for its integration in

Table 11.8 Comparison of the results obtained with classification algorithms

Classification algorithm	Weka		R	
	Train (%)	CV (%)	Train (%)	Test (%)
Naïve Bayes	83.5866	82.0669	83.2826	83.6601
MSV	89.6657	85.7154	93.0881	91.5032
J48	99.0881	95.7447	99.0881	90.8496

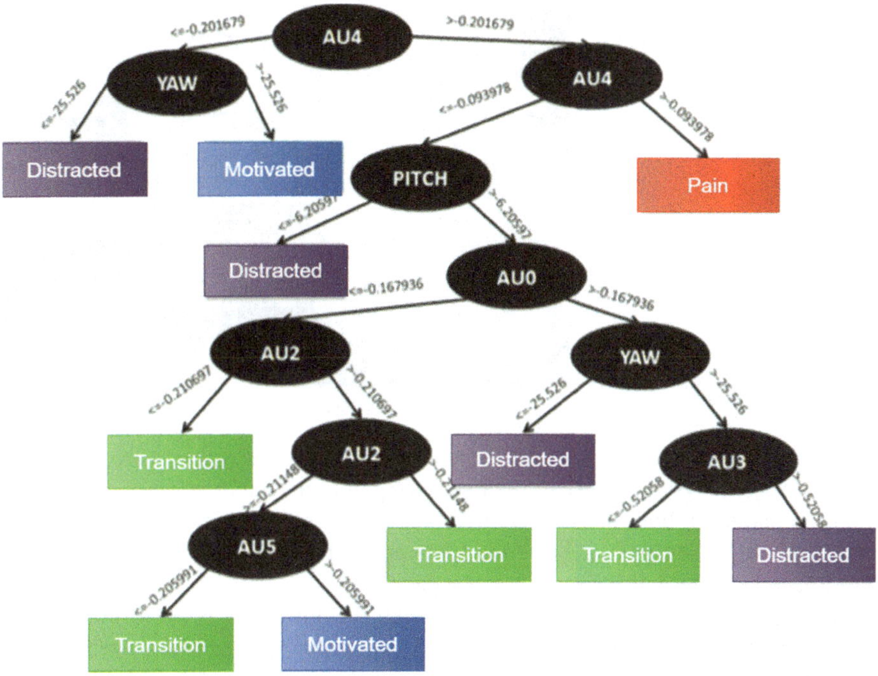

Fig. 11.17 Tree generated for C4.5 in Weka

the rehabilitation system for its simplicity in the rules, its easy represents, for its rapid response and for achieving a close performance to accomplish by the SVM in test phase, in R.

Support vector machines obtain a good result, according the literature; however, it is interesting to note that the implementation carried out in each platform affects the results obtained. SVM achieves a better performance in R. However, this model had not considered because its implementation and integration is more difficult.

Analyzing the confusion matrix of the evaluated algorithms, we conclude that the two main problems in the recognition of facial expressions are due to changes in the luminous intensity and the large number of movements that the face presents continuously, such as blinking, yawn, move the head from one side to another, lower the look.

11.6.4 Integrated Systems

For the development and integration of these three systems, different software and hardware tools were used. We used a laptop with Windows 10 operating system, Kinect sensor, 1-DOF ankle rehabilitation, force sensor, Unity Engine C#, Visual C++, Kinect SDK, and an Arduino.

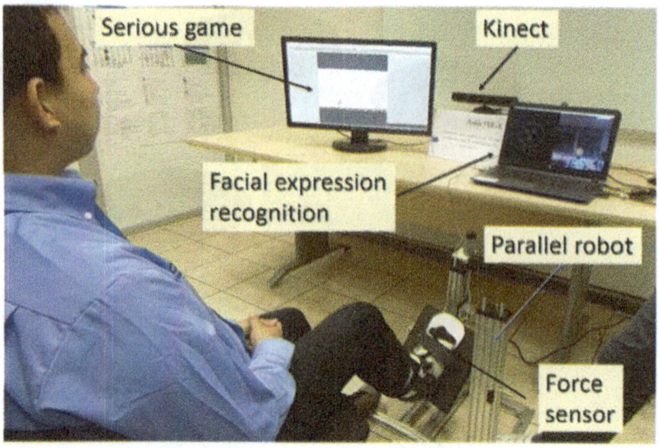

Fig. 11.18 Components of the ankle rehabilitation integral system

The main elements of the ankle rehabilitation integral system are shown in Fig. 11.18. The system consists of a serious game, a Kinect sensor, ankle rehabilitation parallel robot, and the facial expression recognition system. The parallel robot has a force sensor to acquire the force applied by the patient on the movable platform.

The interaction between the patient and the serious game is performed by the Kinect sensor, which by means of the facial expression recognition system detects the patient's condition to increase or decrease the game level based on the frequency of appearance of obstacles, which the serious game character must jump. If the game level is modified, the patient must exert force in the movable platform so that the game character jumps the obstacles. By means of a force sensor placed on the mobile platform, the force magnitude that should be depending on the level of the game is detected. That is, if the required force is not exerted, the angle amplitude will not be achieved and therefore cannot jump the obstacle. On the contrary, if the force is applied, the angle amplitude is achieved, and the character of the game skips the obstacle.

Integral system was evaluated with six healthy people, for two 10-min therapies. Each person used the rehabilitator and play with the developed game, while the automatic recognition of their facial expressions was performed. The results obtained show that the game entertains them while performing therapy. The interface of the game is simple and easy to understand and learn quickly; however, it is necessary to add more challenges to the concentration of the person during a longer period. The communication between the rehabilitator and the game is carried out in time close to the real, with responses less than 10 s. The rehabilitator fulfills its objective by providing the necessary movements for the rehabilitation of the ankle. Finally, the facial expression recognition system makes the recognition of nonverbal information of the patient's face, allowing the levels of the game to be modified depending on their emotional state.

11.7 Conclusions

Mechatronics is a discipline that has had a great impact on the development of technology; product development to facilitate the activities of human beings. In recent years, there has been a great development and interest in the rehabilitation area. In the last decade, many mechatronic systems that contribute to the rehabilitation in both upper and lower extremities have been developed.

An integral rehabilitation system for ankle is proposed, which is based on a serious game and facial expression recognition for entertainment and to improve patient engagement in the rehabilitation process. The tests performance showed that integration of proposed system was successful. In addition, it is proposed and successfully implemented our method for three facial expressions recognition, different to the basic expressions, using FACS and C4.5. Based on the experiment results, we can be concluded that performance is good because it can recognize facial expressions of single person with error between 0 and 10% under naturalistic conditions. As future work, the other two movements of the ankle that the rehabilitator can execute will be included in the serious game.

References

Agas, A., Daitol, A., Shah, U., Fraser, L., Abbruzzese, K., Karunakaran, K., & Foulds, R. (2015). 3-DOF admittance control robotic arm with a 3D virtual game for facilitated training of the hemiparetic hand. In *41st Annual Northeast Biomedical Engineering Conference, (NEBEC)* (pp. 1–2).

Aly, S., Trubanova, A., Abbott, L., White, S., & Youssef, A. (2015). VT-KFER: A kinect-based RGBD + Time dataset for spontaneous and non-spontaneous facial expression recognition. In *International Conference of Biometrics*, Miami (pp. 1–8).

Aly, S., Youssef, A., & Abbott, L. (2014). Adaptive feature selection and data pruning for 3d facial expression recognition using the kinect. In *2014 IEEE International Conference on Image Processing (ICIP)*, Paris, France (pp. 1361–1365).

Arenas, Á.., Cotacio, B., Isaza, E., Garcia, J., Morales, J., & Marín, J. (2012). Sistema de Reconocimiento de Rostros en 3D usando Kinect.

Blanco-Ortega, A., Beltrán, F., Silva, G., & Oliver, M. (2010). Active vibration control of a rotor-bearing system based on dynamic stiffness. *Revista Facultad de Ingeniería Universidad de Antioquia, 55,* 125–133.

Blanco-Ortega, A., Quintero-Mármol, E., Vela-Valdés, G., López-López, G., & Azcaray-Rivera, H. (2012). Control of a virtual prototype for ankle rehabilitation. In *Eighth International Conference on Intelligent Environments, (IE'12)*, Guanajuato, Mexico (pp. 80–86).

Burdea, G., Cioi, D., Kale, A., Janes, W. E., Ross, S. A., & Engsberg, J. R. (2013). Robotics and gaming to improve ankle strength, motor control, and function in children with cerebral palsy —A case study series. *IEEE Transactions on Neural Systems and Rehabilitation Engineering, 21*(2), 165–173.

Chou-Ching, K., Ming-Shaung, J., Shu-Min, C., & Bo-Wei, P. (2008). A specialized robot for ankle rehabilitation and evaluation. *Journal of Medical and Biological Engineering, 28*(2), 79–86.

Cioi, D., Kale, A., Burdea, G., Engsberg, J., Janes, W., & Ross, S. (2011). Ankle control and strength training for children with cerebral palsy using the Rutgers Ankle CP. In *2011 IEEE International Conference on Rehabilitation Robotics*, Zurich, Switzerland (pp. 1–6).

Deutsch, J., Latonio, J., Burdea, G., & Boian, R. (2001). Rehabilitation of musculoskeletal injuries using the rutgers ankle haptic interface: Three case reports. In *Eurohaptics Conference*, Birmingham, UK (pp. 1–4).

Ekman, P., & Friesen, W. (1978). *Facial action coding system: A technique for the measurement of facial movement*. Palo Alto: Consulting Psychologists Press.

Farjadian, A., Nabian, M., Holden, M., & Mavroidis, C. (2014). Development of 2-DOF ankle rehabilitation system. In *40th Annual Northeast Bioengineering Conference, (NEBEC)*, Boston, MA (pp. 1–2).

Fasel, B., & Luettin, J. (2003). Automatic facial expression analysis: A survey. *Pattern Recognition, 36*(1), 259–275.

Fliess, M., & Join, C. (2008). Commande sans modèle et commande à modèle restreint. *e-STA, 5*(4), 1–23.

Foundation, R. (2017, Dec). *The R project for statistical computing*. Retrieved from https://www.r-project.org/.

Franco-González, A., Márquez, R., & Sira-Ramírez, H. (2007). On the generalized-proportional-integral sliding mode control of the "boost-boost" converter. In *4th International Conference on Electrical and Electronics Engineering*, Mexico City (pp. 209–212).

Garcia, J., & Navarro, K. (2014). The mobile RehAppTM: An AR-based mobile game for ankle sprain rehabilitation. In *2014 IEEE 3nd International Conference on Serious Games and Applications for Health (SeGAH)*, Rio de Janeiro (pp. 1–6).

Girone, M., Burdea, G., & Bouzit, M. (1999). The rutgers ankle orthopedic rehabilitation interface. In *Proceedings of the ASME Haptics Symposium*, DSC 67 (pp. 305–312).

Girone, M., Buerdea, G., Bouzit, M., Popescu, V., & Deutsch, J. (2000). Orthopedic rehabilitation using the rutgers ankle interface. In *Proceedings of Medicine Meets Virtual Reality*, IOS Press (pp. 89–95).

Goncalves, A., Dos Santos, W., Consoni, L., & Siqueira, A. (2014). Serious games for assessment and rehabilitation of ankle movements. In *2014 IEEE 3nd International Conference on Serious Games and Applications for Health (SeGAH)*, Rio de Janeiro (pp. 1–6).

Gupta, S., Verma, K., & Perveen, N. (2012). Facial expression recognition system using facial characteristic points and ID3. *International Journal of Computer & Communication Technology (IJCCT), 3*(1), 45–49.

Hsu, T. (1997). Mechatronics. An overview. *IEEE Transactions on Components, Packaging, and Manufacturing Technology: Part C, 20*(1), 4–7.

Ijjina, E., & Mohan, C. (2014). Facial expression recognition using kinect depth sensor and convolutional neural networks. In *13th International Conference in Machine Learning and Applications (ICMLA)*, Detroit, MI (pp. 392–396).

Jaume-i-capó, A., & Samčović, A. (2014). Vision-based interaction as an input of serious game for motor rehabilitation. In *22nd Telecommunications Forum Telfor (TELFOR)*, Belgrade (pp. 854–857).

Kakarla, M., & Reddy, G. (2014). A real time facial emotion recognition using depth sensor and interfacing with second life based virtual 3D avatar. In *International Conference on Recent Advances and Innovations in Engineering (ICRAIE-2014)*, Jaipur (pp. 1–7).

Kumari, J., Rajesh, R., & Pooja, K. (2015). Facial expression recognition: A survey. *Procedia Computer Science, 58*, 486–491.

Li, D., Sun, C., Hu, F., Zang, D., Wang, L., & Zhang, M. (2013). Real-time performance-driven facial animation with 3ds max and kinect. In *3rd International Conference on Consumer Electronics, Communications and Networks, CECNet*, Xianning, China (pp. 473–476).

Liu, G., Gao, J., Yue, H., Zhang, X., & Lu, G. (2006). Design and kinematics simulation of parallel robots for ankle rehabilitation. In *International Conference on Mechatronics and Automation*, Luoyang, Henan (pp. 1109–1113).

Mao, Q., Pan, X., Zhan, Y., & Shen, X. (2015). Sing kinect for real-time emotion recognition via facial expressions. *Frontiers of Information Technology & Electronic Engineering, 16*(4), 272–282.

Menezes, R., Batista, P., Ramos, A., & Medeiros, A. (2014). Development of a complete game based system for physical therapy with kinect. In *IEEE 3nd International Conference on Serious Games and Applications for Health, (SeGAH 2014)*, Rio de Janeiro (pp. 1–6).

Michel, P., & El Kaliouby, R. (2003). Real time facial expression recognition in video using support vector machines. In *5th ACM International Conference on Multimodal Interaction—ICMI '03*, Vancouver, British Columbia (pp. 258–264).

Michmizos, K., & Krebs, H. (2012). Serious games for the pediatric anklebot. In *4th IEEE RAS & EMBS International Conference on Biomedical Robotics and Biomechatronics (BioRob)*, Rome, Italy (pp. 1710–1714).

O'Discoll, S., & Giori, N. (2000). Continuous passive motion (CPM): Theory and principles of applications. *Journal of Rehabilitation Research and Development, 7*(2), 179–188.

Omelina, L., Jansen, B., Bonnechère, B., Jan, S. V., & Cornelis, J. (2012). Serious games for physical rehabilitation: Designing highly configurable and adaptable games. In *Proceeding of 9th International Conference Disability, Virtual Reality & Associated Technologies*, Laval, France (pp. 195–201).

Pasqual, T., Caurin, G., & Siqueira, A. (2016). Serious game development for ankle rehabilitation aiming at user experience. In *6th IEEE International Conference on Biomedical Robotics and Biomechatronics (BioRob)*, Singapore (pp. 1015–1020).

Perdereau, V., Legnani, G., Pasqui, V., Sardini, E., & Visioli, A. (2011). International master program on mechatronic systems for rehabilitation. *Journal sur l'enseignement des sciences et technologies de l'information et des systems, 10*(1006), J3eA.

Porras-Luraschi, J. (2005). *Sistema de reconocimiento de expresiones faciales aplicado a la interacción humano-computadora usando redes neuronales y flujo óptico.* UNAM.

Rego, P., Moreira, P. M., & Reis, L. P. (2010). Serious games for rehabilitation: A survey and a classification towards a taxonomy. In *5th Iberian Conference on Information Systems and Technologies*, Santiago de Compostela (pp. 1–6).

Saglia, J., Tsagarakis, N., Dai, J., & Caldwell, D. (2009). A high performance 2-DOF over-actuated parallel mechanism for ankle rehabilitation. In *IEEE International Conference on Robotics and Automation, (ICRA 2009)*, Kobe, Japan (pp. 2180–2186).

Saglia, J., Tsagarakis, N., Dai, J., & Caldwell, D. (2010). Assessment of the assistive performance of an ankle exerciser using electromyographic signals. In *2010 Annual International Conference of the IEEE Engineering in Medicine and Biology*, Buenos Aires, Argentina (pp. 5854–5858).

Seddik, B., Maamatou, H., Gazzah, S., Chateau, T., & Ben Amara, N. E. (2013). Unsupervised facial expressions recognition and avatar reconstruction from Kinect. In *10th International Multi-Conferences on Systems, Signals & Devices, (SSD13)*, Hammamet, Tunisia (pp. 1–6).

Shah, N., Amirabdollahian, F., & Basteris, A. (2014). Designing motivational games for stroke rehabilitation. In *2014 7th International Conference on Human System Interactions (HSI)*, Costa da Caparica (pp. 166–171).

Sira-Ramírez, H., Beltrán, F., & Blanco, A. (2008). A generalized proportional integral output feedback controller for the robust perturbation rejection in a mechanical system. *eSTA Sciences et Technologies de l'Automotive, 5*, 24–32.

Stocchi, L. (2014). 3D facial expressions recognition using the microsoft kinect. In *18th International Conference on Image Processing (ICIP)*, Dublin, Ireland (pp. 773–776).

Surbhi, V. (2012). ROI segmentation for feature extraction from human facial images. *International Journal of Research in Computer Science*, 61–64.

Tannous, H., Dao, T., Istrate, D., & Tho, M. (2015). Serious game for functional rehabilitation. *Advances in Biomedical Engineering (ICABME)*, Beirut (pp. 242–245).

Tsalakanidou, F., & Malassiotis, S. (2010). Real-time 2D + 3D facial action and expression recognition. *Pattern Recognition, 43*(5), 1763–1775.

Waikao, W. (2017, Dec). *Weka 3: Data mining software in java*. Retrieved from https://www.cs.waikato.ac.nz/ml/weka/.

Yoon, J., & Ryu, J. (2005). A novel reconfigurable ankle/foot rehabilitation robot. In *IEEE International Conference on Robotics and Automation, (ICRA 2005)*, Barcelona, Spain (pp. 2290–2295).

Zhang, M., Zhu, G., Nandakumar, A., Gong, S., & Xie, S. (2014). A virtual-reality tracking game for use in robot-assisted ankle rehabilitation. In *2014 IEEE/ASME 10th International Conference on Mechatronic and Embedded Systems and Applications (MESA)*, Senigallia (pp. 1–4).

Zoch, C., Fialka-Moser, V., & Quittan, M. (2003). Rehabilitation of ligamentous ankle injuries: A review of recent studies. *British Journal of Sports Medicine, 37*(4), 291–295.

Chapter 12
Cognitive Robotics: The New Challenges in Artificial Intelligence

Bruno Lara, Alejandra Ciria, Esau Escobar, Wilmer Gaona and Jorge Hermosillo

Abstract Recent technological advances have provided the manufacturing industry with precise and robust machines that perform better than their human counterparts in tiresome and tedious jobs. Likewise, robots can perform high precision tasks including in hazardous environments. However, a new area of research in robotics has emerged in the last decades, namely cognitive robotics. The main interest in this area is the study of cognitive processes in humans and their implementation and modeling in artificial agents. In cognitive robotics, the use of robots as platforms, in the study of cognition, is the best-suited mechanism as they naturally interact with their environment and learn through this interaction. Following these ideas, in these works, two low-level cognitive tasks are modeled and implemented in an artificial agent. Based on the ecological framework of perception, in the first experiment, an agent learns its body map. In the second experiment, the agent acquires a distance-to-obstacles concept. The agent is let to interact with its environment and allowed to build multimodal representations of its surroundings, known as affordances. Internal models are proposed as a conceptual mechanism which performs associations between different modalities. The results presented here provide the basis for further research on the capabilities of internal models as a constituent cognitive base for higher capabilities in artificial agents.

Keywords Cognitive robotics · Artificial intelligence · Embodied cognition
Embodied robotics · Internal models

B. Lara (✉) · E. Escobar · W. Gaona · J. Hermosillo
Cognitive Robotics Laboratory, Centro de Investigación En Ciencias,
Universidad Autónoma Del Estado de Morelos, Cuernavaca, Morelos, Mexico
e-mail: bruno.lara@uaem.mx

A. Ciria
Pshycology Department, Universidad Nacional Autónoma de México (UNAM),
CDMX, Mexico City, Mexico

O. O. Vergara Villegas et al. (eds.), *Advanced Topics on Computer Vision, Control and Robotics in Mechatronics*,
https://doi.org/10.1007/978-3-319-77770-2_12

12.1 Introduction

The main aim of this chapter is to present a different type of robotics research, namely cognitive robotics. The development of machines for automation has always been inspired by the imitation of human agents. Robot arms, in their attempt to perform tasks traditionally performed by human operators, have a close resemblance to their human counterparts. Moreover, after many years of successful development, industrial arms are capable of performing a wide range of tasks with very high precision, with minimum wear and no boredom.

However, there still exists the issue of imitation, as this is not addressed in depth in all these industrial developments. To a certain extent, the concept of intelligence for industrial robots is, if not irrelevant, very limited. It is in this quest that robotics and artificial intelligence come together. Research in cognitive robotics aims at making use of artificial agents to model, simulate, and understand cognitive processes.

This chapter is organized as follows. Section 12.2 provides a short story of industrial robotics highlighting its limitations in their exploration of human-level intelligence and cognitive processes. It is followed by a quick review of the problems in artificial intelligence, its shortcomings, and changes of paradigm. Section 12.3 then presents the new field of embodied cognition and robotics and their attempt to understand cognition studying low-level cognitive processes. Section 12.4 presents two significant studies that try to model very specific and basic human cognitive abilities. Finally, Sect. 12.5 presents the conclusions of this chapter.

12.2 Robotics and Artificial Intelligence

In one of the first patents registered, the inventor Devol (1967) put forward some of the first ideas for the automation of machinery and manufacturing processes. The first manufacturing robot was sold to the Ford Company, which used it to tend a die-casting machine (Mortimer and Rooks 1987). The company was UNIMATION, creators of the Programmable Universal Machine for Assembly (PUMA) robot developed in 1978.

Since then, robot companies in this field have come out with a variety of automated machines to fulfill manufacturing tasks. Nowadays, robots are sophisticated apparatus that can operate in different environments performing tasks deemed too tiring or too dangerous for human operators such as painting (Graca et al. 2016; Li et al. 2016), assembling of heavy loads (Chuyet al. 2017), or soldering (Draghiciu et al. 2017). They can also perform highly precise work such as in the biosciences (Wu et al. 2016; Zhuanget al. 2018) or medicine (Brown et al. 2017; Rosen et al. 2017).

During all this time and development, a parallel quest has always been present and relates to a universal and far older curiosity: Can we build a machine that acts and thinks as a human being? This has been a central question in what became known as artificial intelligence (AI). The definition of AI has been a topic of debate and analysis since the creation of the field. In general, it is possible to define AI as the field devoted to building tools or agents capable of displaying intelligent behaviors. In 1997, IBM designed Deep Blue, a computer chess system that defeated Garry Kasparov, the then world chess champion. Later on, in 2011, Watson a "question answering machine" defeated the two human champions on the quiz show *Jeopardy*, a game that implies the answering of complex natural language questions very quickly. Intuitively, playing chess or a quiz game is the activities that require intelligence. However, do these computer programs show genuine, intelligent behavior? A deep philosophical question arises in terms of defining what intelligence behavior is, and even more problematic, what it means to have a mind and how the mind is capable of performing intelligent behavior in different contexts and circumstances.

Turing (1950), one of the most influential computer science theoreticians asked the question "Can machines think?" He proposed what became known as the "Turing Test" and claimed that a computer able to pass this test could be considered an intelligent machine. The Turing Test can be described in terms of an imitation game, which is played by three people: a man, a woman, and an interrogator. The aim of the game is for the interrogator to identify which of the two players is the man and which is the woman. The role of the man in the game is to confuse the interrogator to cause a wrong identification. In the other hand, the role of the woman is to help the interrogator in the identification task. During the game, the interrogator stays in a different room and is allowed to pose questions about the identity of the two players. The answers of the two players should be typewritten so that the voice may not help the identification. Now, it is possible to ask the question:

> "What will happen when a machine takes the part of A (the man) in this game?" Will the interrogator decide wrongly as often when the game is played like this as he does when the game is played between a man and a woman? These questions replace our original, "Can machines think?". (Turing1950, p. 443)

The idea of declaring a machine as intelligent by means of a successful Turing Test result is controversial but is irrefutable that linguistic behavior required by the test is at the heart of human cognition (Arkoudas and Bringsjord 2014). Actually, in 1970s some AI researchers argued that computers could understand natural language. Schank and Abelson (1977) developed a technique called "conceptual representation" arguing that understanding language involves causally connecting sentences. In general terms, the conceptual representation technique used scripts, which were defined as a stereotyped sequence of actions and events to represent conceptual relations. Schank and Abelson proposed that the best way to approach the problem of building an intelligent machine is to emulate the human conceptual mechanisms that deal with language. This approach lays in the intersection of psychology and computer science. In this sense, the psychologist Johnson-Laird (1977) suggested a theory of semantics based on the analogy between natural and computer programming languages:

> Artificial languages, which are used to communicate programs of instructions to computers, have both a syntax and a semantics. Their syntax consists of rules for writing well-formed programs that a computer can interpret and execute. Their semantics consists of the procedures that the computer is instructed to execute. (Johnson-Laird 1977, p. 189)

Johnson-Laird (1977) argued that AI researchers were radical by favoring either a purely declarative knowledge based on a set of assertions used in problem-solving or a purely procedural knowledge in terms of procedures. He proposed a procedural semantics methodology, where declarative knowledge is coupled with procedures that can convert its constituents into procedures. Despite the difficulties of whether or not there are semantic primitives into which meanings of words are decomposed, Johnson-Laird believed that this framework can be utterly useful in developing psychological theories about the meanings of words and can be modeled in the form of computer programs. Procedural semantics deal with the meaning of procedures that computers are told to execute (Johnson-Laird 1977, p. 190).

Fodor (1978) made a substantial critique of the procedural semantic methodology. The main critique lies in the argument that procedural semantics confuses semantic theories with theories of sentence comprehension. Mistakenly, the procedural semantics supposed that a theory of how a sentence is understood can be the same as a theory of what the sentence means. In response to Johnson-Laird assumptions about semantics, Fodor wrote:

> The computer models provide no semantic theory at all, if what you mean by a semantic theory is an account of the relation between language and the world (Fodor 1978, p. 229) .... A fortiori, we don't know how to build a robot which can use 'chair' to refer to chairs.... Nobody has the foggiest idea of how to connect this system to the world (how to do the semantics of internal representations).... (Fodor 1978, p. 246–248)

Fodor's statement that computer models provide no meaning to words at all anticipated the Chinese Room argument proposed by the philosopher Searle (1980), which tries to prove that syntax is not sufficient for semantics.

Searle asks the reader to imaging him locked inside a room. The room contains baskets full of symbols and a book of rules. The rules in the book relate certain strings of symbols to other strings of symbols based only on their shapes. Searle is handed a string of these symbols and is expected to find a corresponding string according to the rules. Once he finds it in the book, he is supposed to compose it with the symbols in the baskets and hand it back. The rules relate the symbols in the form of "Take a squiggle-squiggle sign from basket number one and put it next to a squiggle-squiggle sign from basket number two" (Searle 1990, p. 26). Now, Searle explains that the incoming strings of symbols are questions written in the Chinese language and the corresponding handed out strings are the answers to the respective questions. Searle does not speak Chinese; however, for any external Chinese observer outside the room, Searle perfectly understood the questions and the respective answers. Searle proposed that the rule book is the "computer program," the people who wrote the rules are the "programmers," Searle is the "computer," the boxes with symbols are the "database," and the bunches of symbols that are handed to Searle are the "questions," and the bunches handed out are the "answers."

The Chinese Room argument claims that despite the fact that Searle does not understand a word of Chinese the outputs are indistinguishable from those of a native Chinese speaker. Although this "computer program" actually passes the Turing Test, this does not mean that the computer understands the meaning of the symbols. Just manipulating the Chinese symbols is not enough to guarantee cognition, perception, understanding, and thinking (Searle 1990). Human minds have mental contents (semantics), and manipulating the symbols (syntax) is not sufficient for having semantics. Computers would have semantics and not just syntax if their inputs and outputs were put in an appropriate causal relation to the rest of the world (Searle 1990, p. 30).

The Chinese Room was put forward mainly as a response to the work of Schank and Abelson (1977) about "conceptual representation" which claims that computer programs understand the meaning of the words and sentences they are programmed to respond to. However, the main argument also applies to Winograd's SHRDLU (Winograd 1973), Weizenbaum's ELIZA (Weizenbaum 1965), and of course the Turing Test (Turing 1950).

Harnad (1989) defended Searle's argument arguing that symbol meaning is grounded in perceptuomotor categories. Specifically, Harnad (1990) aimed to answer how symbol meaning is to be grounded in something other than just more meaningless symbols. This came to be known as the symbol grounding problem. The standard reply to the symbol grounding problem is that the meaning of the symbols comes from connecting the system with the world (Fodor 1978). However, this assumption underestimates the difficulty of selecting the proper objects, events, and states that symbols refer to (Harnad 1990). He proposes as a possible solution a hybrid nonsymbolic/symbolic system. The nonsymbolic part of the hybrid system refers to the ability to discriminate inputs which depends on the "iconic representations" that are analogs of the proximal sensory projections of distal objects and events. The symbolic part of the system refers to the ability to identify an input reducing the icons to those "invariant features" that will reliably distinguish a member of a category. The output of the category-specific feature detector is the "category representation." With this hybrid system, the match between the words and the world is grounded in perceptual categories or "categorical representations" which are based on the invariants "iconic representations." How does the hybrid system find the invariant features of the sensory projection that makes it possible to categorize and identify objects correctly? The names of the elementary symbols (categorical representations) are connected to nonsymbolic representations (iconic representations) via connectionist networks that extract the invariant features of their analog sensory projections, so that is possible to select the objects to which they refer.

The main focus of AI research turned toward the physical grounding hypothesis, which states that to build a system with intelligent behavior it is necessary to have its representations grounded in the physical world. However, Brooks (1990) suggested that when this approach is implemented, the need for traditional symbolic representations fades entirely.

The main assumption is that the world is the best model of itself as it contains every detail that has to be known. Likewise, an agent must respond continuously to its inputs using its perception of the world instead of a world model (Brooks 1991a). This is the key element of situatedness. Therefore, in this framework, intelligence is determined by the total behavior of the system and how that behavior emerges in relation to the environment. Now, the line between intelligence and environmental interaction disappears.

The idea that intelligence can be conceived as no representational (Brooks 1991b) is often criticized. However, what Brooks suggested relies on the idea that there are representations, but they are partial models of the world. These representations extract only those aspects of the world that are relevant within the context and the specific task. Nevertheless, he highlighted the idea that if the world is the best model of itself, it is necessary to sense it appropriately, so that building a system that is connected to the world via a set of sensors and actuators turns out to be fundamental. In this framework, the agent has a body, sensors, and a motor system, so that it is embodied (Brooks 1991a). Two main reasons make the embodiment of an intelligent system critical. First, only an embodied intelligent agent can deal with the real world. Second, only with a physical grounding framework can any internal symbolic system give meaning to the processing going on within the system.

Pfiefer and Bongard (2007) proposed that only agents that are embodied, whose behavior can be observed as they interact with the environment, are intelligent. Having a body is a prerequisite for any kind of intelligence, and it is necessary for cognition. The embodied cognition framework requires working with real-world physical systems, like robots. Autonomous robots, which are independent of human control, have to be situated by being able to learn about the world through their sensory system during interaction. The ideas around the concept of embodiment produced a major shift in research in AI, which is addressed in the next section.

12.3 Embodied Cognition and Robotics

In the last decades, a new paradigm has started to surface in the sciences concerned with the study of the brain. In this, the body and the environment take an important role in the shaping of the mind. Known globally as embodied cognition, this new way of thinking puts forward the idea that agents are entities that have a body and interact with their environment as they develop (Wilson 2002). It is through this interaction that knowledge arises and forms the basis of cognitive abilities.

In this chapter, two telling examples are presented where an artificial agent acquires specific cognitive abilities through the interaction with the world. In these examples, we explore the concept of *affordances*, a term coined by psychologist Gibson in his seminal (1979) book. According to the ecological approach to perception, an *affordance* is what the environment provides an individual. When we perceive a chair, we perceive its sit-*ability*, all possibilities of interaction with the object are activated to form what we know as the concept *chair*. Furthermore, our actions with the object are also conceived as a modal input to the system, just as is the visual information, the smell and even possible memories of such objects (Barsalou 2008).

In the cognitive sciences, the concept of internal models has been put forward as the means to code affordances. An internal model naturally associates different sensorimotor modalities through the interaction with the world. Here, the work is concerned mainly with the concept of forward models. A forward model is an internal model which incorporates knowledge about sensory changes produced by self-generated actions of an agent. Given a sensory situation St and a motor command Mt (intended or actual action), the forward model predicts the next sensory situation $St + 1$.

These models have been found to be a constituent of the self, giving clear explanations of the functioning of several behaviors and characteristics of the brain. The importance of this type of models stems from the relevance of the prediction of the consequences of our actions for seemingly trivial tasks such as planning and avoiding undesired situations (Blakemore et al. 1998).

Much research has been done on computational forward models for action preparation and movement (Wolpert and Ghahramani 2000; Escobar et al. 2016; Jamone et al. 2016), with highly functional models that account, for example, for hand trajectory planning taking into account different contexts (Wolpert et al. 2003), navigation (Pezzulo and Cisek 2016), and mirror systems computational modeling (Thill et al. 2013). On the other hand, prediction and simulation processes occurring in the brain seem to play a central role in cognition (Barsalou 2009). A thorough review of the importance of these models and their implementations is found in Schillaci et al. (2016).

In this work, two affordances are explored. First, the acquisition of a self body map as an affordance of the agents own body and its basic interaction with the world and second the acquisition of a distance affordance.

12.3.1 Self Body Mapping

The faculties, capabilities, and skills to dynamically interact with the world, which as adult humans we possess, emerge through a long process of tuning and rehearsing of sensorimotor schemes. This idea has taken a central role in research in the cognitive sciences.

A very important example is the acquisition of the sensorimotor schemes that code for the capabilities and reaches of our body. This set of schemes provide us with many cognitive tools, among them the knowledge and coding of our body map, essential for navigating around the environment.

This is exemplified by the telling experiment presented by Warren and Whang (1987), showing that humans have a physically embodied notion of distance when asked to pass through an aperture gap. They found that there is a relation between the subject's shoulder's size and the gap size which translates into an affordance to pass through (*passability affordance* in their own terms). These results point to the use of measures of the world based on the capabilities given to agents by means of the construction of self body-based mappings, in short body mappings. In this construction, forward models as proposed herein play a fundamental role.

In the context of the experiment presented here a robot is provided with a simulation process through a chain of forward models. Chains of forward models have been used in the past (Hoffmann 2007; Möller and Schenk 2008).

A chain of forward models was used by Lara et al. (2007) to avoid collisions and let the robot obtain information on the ownership of its actions. The visual input in these experiments was the data coming from a simulated linear camera. A major drawback of this approach is the need for three time steps to disassociate size and distance of objects in the field of view (Lara and Rendon 2006).

In this work, the experiment goes a step further and uses a stereo vision pair to obtain a disparity map as the input sensory situation and provide the robot with a simulation process in order to evaluate the *affordability* of passing through a gap.

12.3.2 Distance Perception Using Visuomotor Cycles

Distance perception has been studied for long in cognitive sciences and stills a complex problem (Turvey 2004). According to some research hypothesis, the perception of distance is not a geometrical process but an association of multimodal (visual and tactile) sensory information (Braund 2007), influenced by the body, self-motion (Profitt 2006), and the environment (Lappin et al. 2006).

To the best of our knowledge, modeling distance perception without a geometrical framework is an issue that has not been addressed in cognitive robotics. Indeed, the study of spatial cognition in robotics has a long history, and several different techniques have been proposed (see Thrun and Leonard (2008) for an exhaustive review). In the frame of cognitive robotics, the work presented here differs from the brain-anatomical approaches (Tolman 1948; Arleo et al. 2004) in that here the work models basic cognitive functions through internal models (Miall and Wolpert 1996; Wolpert et al. 2001), where the sensorimotor cycle is considered the fundamental unit of cognition and from which, it is hypothesized the modeling of cognitive processes should start (Lungarella et al. 2003).

Instead of giving the robot the explicit means to model the external metric or topology of the free space or to exploit geometrical information from stereo vision techniques (Moons 1998), as in Experiment 1, this experiment goes a step further. Here, the aim is to validate internal models that associate sensorimotor relationships which code distance. The distance affordance is obtained by means of the prediction and reenaction of visuomotor cycles.

Spatial cognitive abilities using internal models and visual sensors have been proposed in the past for particular space features recognition and for localization purposes (Hoffmann and Möller 2004; Möller and Schenck 2008). These works used refined data structures from preprocessed visual information together with forward models. The work presented here differs from these approaches in that here foveated images are used as the sensory situation together with current motor commands as input to a forward model that predicts the next images and a proximal tactile sensory situation. In this way, a notion of distance is coded in robot motor coordinates (i.e., a distance perception in robot's own body-scale units) representing the spatial relationships between the artificial agent and the objects in the environment.

We provide details of the implemented cognitive process in the next section.

12.4 Affordance Acquisition Experiments

In light of the evidence presented above, coherent action is attributable to the existence of basic mechanisms of anticipation in sensorimotor coordination processes. The experiments presented in this section are representative examples in the framework of cognitive robotics and aim at showing how forward models provide an agent with basic capabilities (Escobar et al. 2012; Gaona et al. 2012). In particular, two experimental designs are presented. Firstly, it is shown how forward models constitute a framework for self body mapping allowing the robot to acquire a *passability affordance*. Then, methodological steps are presented for building distance perception capabilities not as a geometrical process but as an association of multimodal (visual and tactile) sensory information during the agent's interaction with its environment.

12.4.1 Experimental Platform

The artificial agent used is a Pioneer 3-DX showed in Fig. 12.1. It has two wheeled-motors, a frontal ring of sonars and a stereoscopic camera. The robot can execute forward and backward movements as well as turn to the right or left with velocities controlled independently for each wheel.

The range sensors are 8 SensComp sonars series 600 with a sensing range of 0.15–5 m with a covering angle of 15°. The sonars are arranged in a ring around the sides and front of the robot, with sonar number 1 pointing to the left of the robot, sonar number 8 to the right, and the remaining six distributed evenly in between.

The stereoscopic camera pair is a STOC-9CM from Videre Design with a resolution of 640 × 480 pixels and a baseline of 9 cms. It has two 1.4f/6.0 mm lenses which give it a 57.3° horizontal field of view (HFOV). The stereoscopic pair is arranged as two digital cameras placed at the same height with parallel optical axis

Fig. 12.1 Robot Pioneer 3D-X

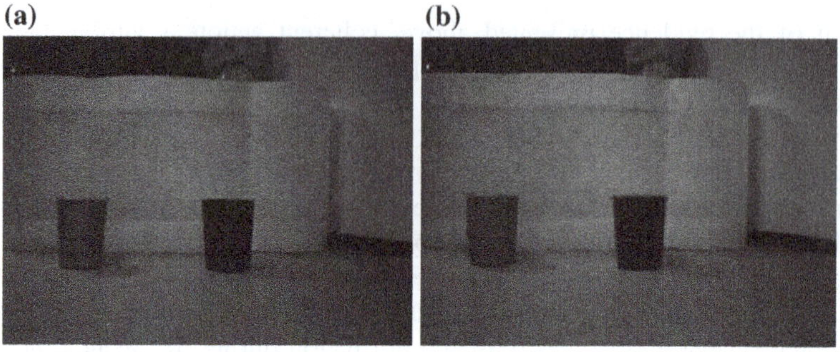

Fig. 12.2 Stereo pair images from the left (**a**), and respectively right (**b**) camera

separated a known distance, and whose intrinsic parameters are irrelevant in our framework. Both cameras provide a monochromatic image (320 × 240) of the scene with values between [0–255]. An example of an acquired image pair on the environment designed for the experiments is shown in Fig. 12.2.

12.4.2 Distance Perception Using Visuomotor Cycles

In stereo vision, the basic principle is that, having two simultaneous images from a scene, a matching is made between features from one image and features from the other. The disparity found between the features from the images is a relative measure of the distance these have to the camera pair or any other pre-defined reference frame. The disparity d of a point X in 3-D with coordinates (x_l,y_l) and (x_r,y_r) in the left and right projections for each of the cameras respectively is found

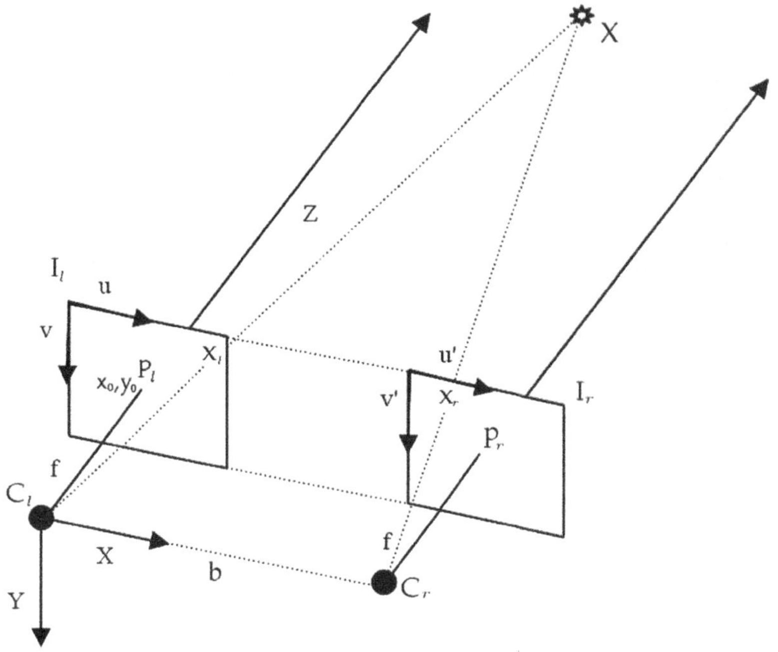

Fig. 12.3 Basic stereo vision geometry

by computing $d = x_l - x_r$. As can be seen from Fig. 12.3, d comes from Eq. 12.1 and is obtained by doing basic geometric correspondence when all other parameters are known, which is the case for a calibrated pair (Hartley and Zisserman 2003).

$$\frac{b - (x_l - x_r)}{Z - f} = \frac{b}{Z} \qquad (12.1)$$

A disparity map is formed by the set of points that overlap in the two images.

Autonomous navigation strategies based on stereo vision have been used with different rates of success (Collins and Kornhauser 2006; Murarka and Kuipers 2009). In particular, the work of Hasan et al. (2009) presents a system that is capable of navigating its way among obstacles. However, the decisions the system takes are strictly based on the values of the disparity map.

All of these works represent a background on the use of traditional computer vision methods to solve autonomous navigation. However, the issue in this work is the use of cognitive models and their applicability. In this exercise, the attempt is made to try to understand their relevance in the search for artificial intelligence. In particular, the aim is providing the robot with the cognitive tools allowing it to anticipate collisions by means of reenacted visuomotor cycles predicting proximal tactile situations without actually moving. To test the proposed model, a look-for-an-output experimental task is constructed.

12.4.2.1 Proposed Model

The model presented here does the learning of a basic body map using a forward model. The model takes as input sensory information and a constant motor command and predicts the next sensory situation. The input sensory data is formed by visual information coming from the disparity map of two images (D_t). The output is formed by the predicted visual information (D_{t+1}) and simulated tactile stimuli (B_{t+1}) coded from threshold capped sonar values (Fig. 12.4).

Visual information and tactile data form a representation of the obstacles in the arena of the robot. Making use of this representation the agent is capable of performing predictions about the sensory changes in the environment. The motor command in Fig. 12.4 can be an executed command or a planned action that is not necessarily executed. This planned action allows the execution of long-term predictions as the output of the forward model can be used as input to a next forward model.

12.4.2.2 Data Preprocessing

Tactile information is obtained by thresholding the values of the sonars. Given the size of the robot and the characteristics of the visual data (see below) a value of 440 or less is defined as a collision, meaning an obstacle is 44 cm away or closer to the robot.

The necessary steps to acquire the visual data from the two images coming from the stereo camera can be seen in Fig. 12.5 and can be described as follows:

Image Acquisition Using a calibrated camera pair STOC-9CM, two simultaneous 320×240 images of the scene are obtained. These images are rectified to correct for the distortion caused by the lenses and sensor geometries.

Disparity Map The disparity map for two images is based on the difference in pixels between the projection of the same point in the left and the right images. The matching of each point in one of the images with its pair in the other image is done using the sum of absolute differences (SAD):

$$\phi_{\mathrm{SAD}}(x, y, d) = \sum_{j=1}^{m} \sum_{i=1}^{n} \mathrm{abs}[V_r(i,j) - V_l(i,j)] \tag{12.2}$$

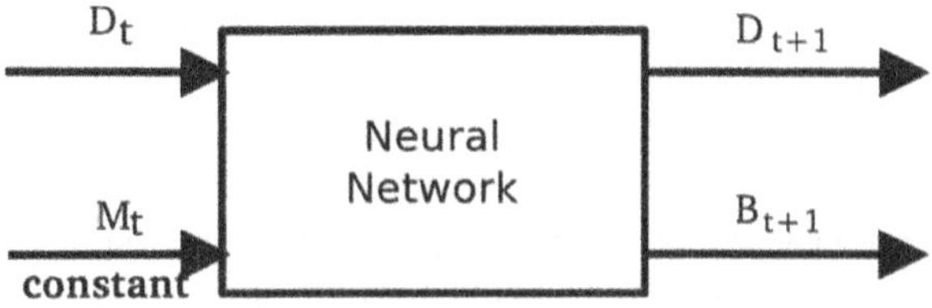

Fig. 12.4 Implemented forward model

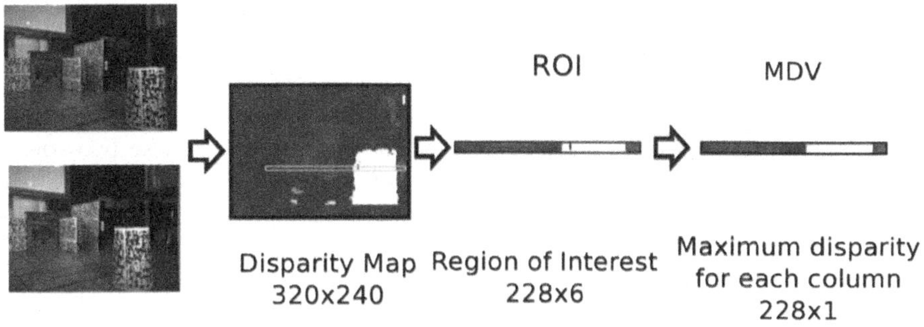

Fig. 12.5 Main steps of visual preprocessing

where $V_l(i,j)$ is the pixel (i,j) in the $n \times m$ window V_l with center in the (x,y) pixel from the left image I_l, likewise, $V_r(i,j)$ is a window on the right image with center at $Ir(x+d,y)$. The central pixels of the two most similar windows are considered to represent the same point in the three-dimensional world.

Several parameters define the maximum and minimum number of disparity values that can be calculated from a pair of images, and this number directly relates to the distance that will be coded in the disparity map. For this work, a good and safe compromise, considering the size of the robot and the parameters of the stereoscopic pair, was set to look for 64 values of disparity. For our system, this means disparities will be found in the range between 34 and 215 cm. (The interested reader can refer to Konolige (1997)).

Region of Interest (ROI) From the disparity map, a 228 × 6 ROI is extracted. The upper limit of the ROI is located at line 152 of the image which, in a scene without obstacles, is located at 2.15 m from the robot; this is the maximum distance for which a disparity value can be calculated. In the horizontal direction, 228 pixels are taken as they are the effective processed area of the image given the size of the masks used for calculating the disparity.

Maximum Disparity Vector (MDV) This vector is formed by taking the maximum disparity for each column of the ROI and represents the closest obstacles in the final 57.3° of the visible field of the camera.

Low Pass Filter Finally, a Gaussian filter with a five-pixel mask is applied to the MDV. This is done primarily to facilitate the learning of the forward model.

12.4.2.3 Network Architecture

To obtain the forward model, we used 57 MLPs which are trained using resilient back-propagation (Riedmiller and Braun 1993). The input is a 228 values vector (VDM) for time t and the output is the VDM for the time t + 1 and the bumper state

B_{t+1}; each of these two vectors has 228 values. Each of the 57 MLP takes as input a 14 value window from the 228 values of the MDV and predicts the central four values of the next time step (Lara et al. 2007). The 57.3° of the MDV is covered by the two front sonars of the Pioneer, which corresponds to sonars 4 and 5, so a vector is composed of 228 binary values depending on whether any of these two sonars present the pre-defined activation. It is important to note that all of the values of the MDV are set to 1 when there is a collision disregarding which of the sonars detected it and to 0 otherwise.

12.4.2.4 Experiments and Results of Experiment 1

The MLPs are trained offline using data collected during walks of the robot in an arena filled with obstacles. The obstacles are texturized to ease the stereo matching problem; they vary in shape and range in height from 30 to 60 cm, which ensures their visibility by the camera pair given that this is mounted on top of the robot.

The robot has a diameter of 38 cm and performs steps of 15 cm. For every step, it takes a snapshot of the scene and does a one-step prediction (OSP) using the forward model. A threshold is set so that if 35 or more neurons are predicting the MVD show activation of 0.45 or higher a long-term prediction (LTP) is triggered. This threshold serves as a warning for a possible future collision. A second threshold is set: four or more neurons predicting the bumper with activation of 0.95 or higher is considered a collision.

The LTP is an internal simulation of the trajectory and consists of using the predicted VMD as input to a next forward model which in turns predict the next VMD. This process can be carried out for a small number of steps as insignificant errors in the OSP accumulate turning the VMD into noise.

Figure 12.6 shows the results of a typical run of the system. The first column shows the VMDs as the agents move in the environment with t = 0 at the top of the image. The next column shows the prediction of the forward model, the network is very accurate, with an average sum squared error of SSE = 0.0043, still, small errors are apparent. After a few steps, the system triggers the LTP. In the example, the LTP is triggered correctly, and a collision is detected after four steps of internal simulation, which is sufficient distance for the agent to take corrective action.

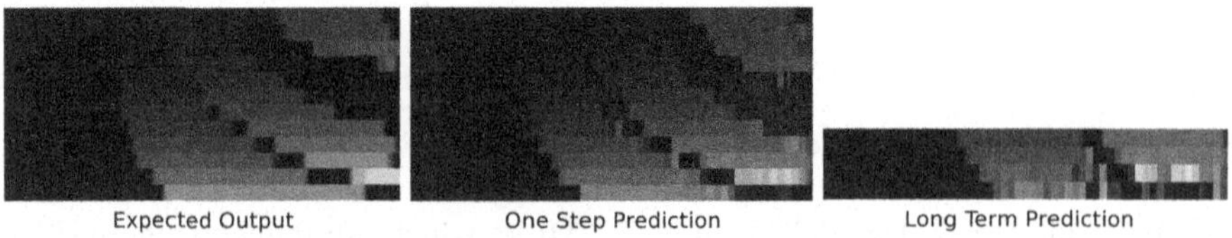

Fig. 12.6 System evaluation with a collision in the right

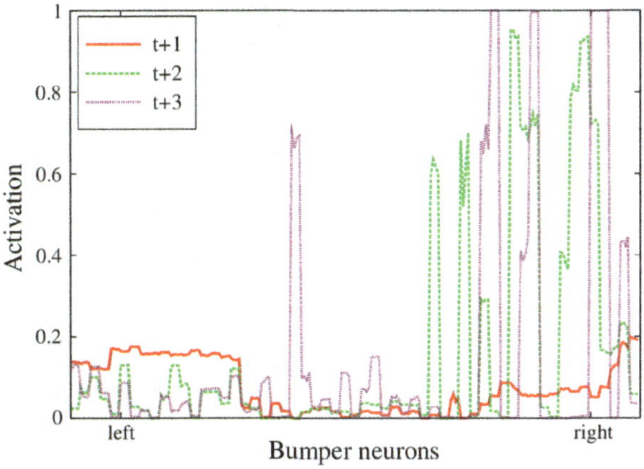

Fig. 12.7 LTP for neurons coding bumper states

The increase in activation of the output neurons coding for the bumper states in the trajectory shown in Fig. 12.6 can be seen in Fig. 12.7. The activation corresponds to the time steps where LTP is performed, which is an internal simulation of the events.

A remarkable emergent property of the system and not predetermined by design is the fact that the activation of the neurons coding for bumper states corresponds to the proximity of the obstacles in the MVD. This is, the activation of right and left bumper corresponds to obstacles in the right and left image regions, respectively. This activation can actually be interpreted as a body map.

Navigation

The robot is programmed to perform a straight trajectory in an arena with obstacles. In case the MVD prediction shows high activation, the LTP is triggered to check whether continuing with that path will lead to a collision. The LTP performs the internal simulation of four steps if a future collision is predicted the robot performs a turn to the opposite side from where the activation is found to be high. It is important to note that the decision on which direction to turn is based on the activation of the neurons coding for bumper activation. As it has been shown, this activation hints to the position of the obstacles in the arena.

The agent is capable of navigating safely in 21 trajectories with different starting points and obstacle configurations. A typical scenario for this experiment can be seen in Fig. 12.8. It is worth noting that the LTP was never triggered falsely; this is, every time it was triggered, there would be a collision in the future. In all cases, the prediction of a collision was made with at least three steps in advance, giving enough room for the agent to make a decision.

Fig. 12.8 Agent facing an obstacle in a typical scenario

Toward a Self-Acquired Body Map

To further evaluate the capabilities of the forward model the agent was set in the center of the arena with obstacles all around. The obstacles had a single passage where the robot could go out. The distances between the obstacles varied from 10 cm to up to approximately 60 cm of the free passage. With an acquired body map, the agent should be able to find the gap in the obstacles where it can pass through.

This experiment allowed us to test the following hypothesis: with an acquired basic body map, the agent should be able to find the gap in the obstacles where it can pass through without the need to move in that direction.

The agent turns around its axis for 360°, every 10° it takes a snapshot of the scene and performs a long-term prediction, recording the number of steps it can predict without registering a collision. Once the robot has completed a whole turn, it heads in the direction where a collision was not detected while performing LTP or where this was detected after the largest number of steps. At this point, the behavior of obstacle avoidance from the previous experiment takes over taking the robot out of the circle of obstacles.

Figure 12.9 shows a typical run of the experiment. In Fig. 12.9a, the agent is turning around in the circle of obstacles, performing the internal simulation of heading in that direction. After completing a whole turn, the robot heads in the direction where the gap between the obstacles is sufficient to pass through. The agent undertakes a small correction in the path as it predicts a possible future collision. This is due to the errors in encoders' measurements and the skidding of the wheels when the agent rotates toward the desired direction. Finally, in Fig. 12.9b, the agent reaches the output.

The system was tried in different obstacle configurations with a 100% success rate. The final path toward the exit was corrected 90% of the times due, again, to the

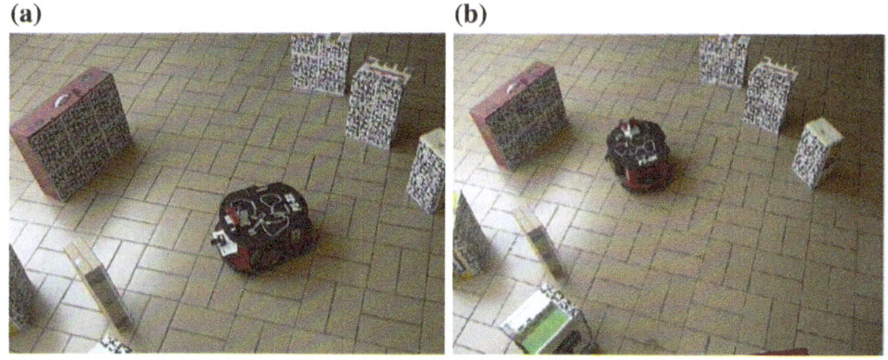

Fig. 12.9 Typical scene for self map experiment

accumulation of errors in encoder readings and the inaccuracy of consecutive motor commands during the exploration phase of the experiments.

12.4.3 *Experiment 2: Distance Perception Using Visuomotor Cycles*

In this second experiment, the methodological steps for building distance perception capabilities are explored, not as a geometrical process but as an association of multimodal (visual and tactile) sensory information during the agent's interaction with its environment.

12.4.3.1 Proposed Forward Model

A schematic view of the proposed forward model is shown in Fig. 12.10. This model receives the two images from the stereo pair (V_L and V_R) and a motor command (M) at time t and produces as output the sensory consequences were that motor command being executed. The consequences of that execution are two resulting sensory states, visual (for both cameras), and tactile (B), at time $t + 1$.

The tactile output is coded as a continuous value in the range 0–1 and represents a measure of the proximity the agent has to objects in its arena.

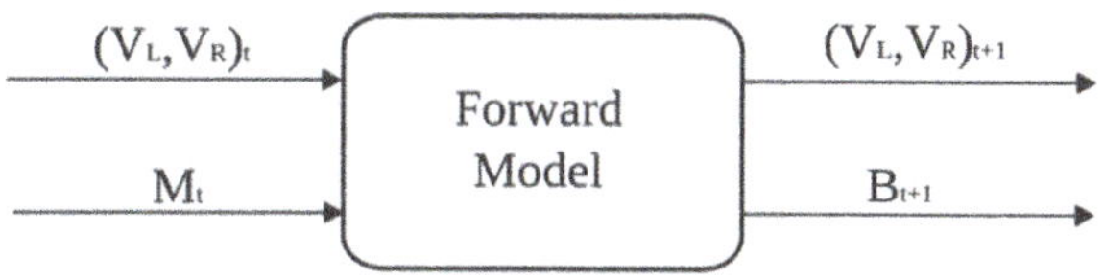

Fig. 12.10 Schematic view of the proposed forward model

The proposed forward model associates future visual and tactile modalities from present visual and motor information. Hence, creating a multimodal sensory representation, the system is relating different sensory modalities around the same perceived situation together with the executed or thought action.

The proposed model produces a notion of distance for navigation through the agent's interaction with the environment. This notion of distance is coded by the multimodal sensory representation, and its units are grounded in the physical capabilities and characteristics of the agent.

12.4.3.2 Data Preprocessing

As a first step, the images were inverted, in the original coding a high pixel value represents white and 0 codes the presence of an obstacle. To reduce the dimensionality of the visual data, we use a foveated imaging technique based on a Gaussian distribution. The fovealisation process is a weighted mask which produces images with high resolution at the center, decreasing it toward the periphery. This technique allowed us to reduce the size and enhance the central region of the images provided by the cameras. The result of applying this process was two final images of size 23 × 24 pixels and is shown in Fig. 12.11.

The motor commands were chosen from three classes, turn 5° to left or to the right, and a forward movement of a step-size, in this case 15 cm: each of these commands was transformed into a vector of values given by three Gaussian functions with the same standard deviation but with different mean according to each type of motor command as can be seen in Fig. 12.12.

Fig. 12.11 **a** Original image (320 × 240) and **b** Foveated image (23 × 24)

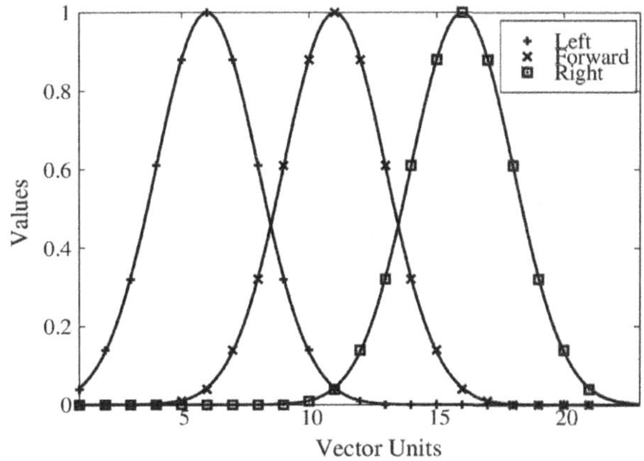

Fig. 12.12 Motor commands transformation

12.4.3.3 Artificial Neural Network

The forward model was coded using artificial neural networks and is made out of 120 local predictors. Each of these network predictors receives two windows of 5×5 pixels from the left and right images and a vector of 21 values for the motor command as input. The output for each network is two windows of 3×3 pixels for the left and right images and one value representing the tactile output. Each input image (left and right) is divided into 120 windows, 10 in the x-direction and 12 in the y-direction and predicts a 3×3 window of the next time step.

The inputs and outputs of the system are overlapped in the vertical and horizontal direction, allowing a prediction of a fully sized output image. This arrangement can be seen in Fig. 12.13, where three different input windows map to their respective output windows. In effect, we have local predictors that take a window of the whole scene and predict a smaller region of the next time step, the system as a whole predicts two full images. The tactile state of the system is represented by a vector of 10 values: a bumpers vector. Each of the ten columns composed of 12 predictors contributes equally to one of the bumper values. The whole vector contains binary values and is 0 when there is no collision and 1 when any of the four front sonars detects a collision.

The training patterns were recollected through random and manual movements executed by the robot with a combination of the previously specified motor commands. As before, the training of the system was done using resilient back-propagation (Riedmiller and Braun 1993).

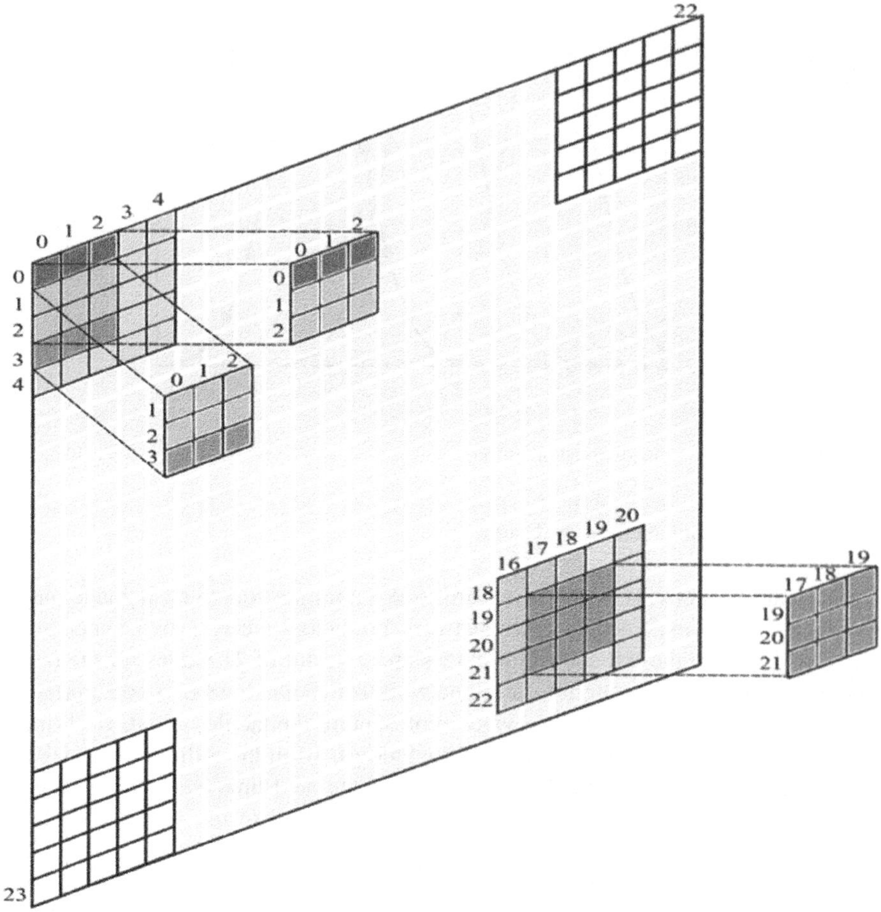

Fig. 12.13 Distribution of input and output windows for the local predictors

12.4.3.4 Results from Experiment 2

Short-Term Prediction

The first test to the trained system was to perform short-term prediction, this is, given a sensory situation (visual) and a motor command, perform a prediction of the next sensory situation (visual and tactile states). A typical result can be seen in Fig. 12.14 where in subfigure (a) we see the initial sensory state. The bright strip running from top to bottom in both images represents an obstacle. It is clear from the images that the obstacle is to the left of the robot.

After a turn to the left, subfigure (b) shows the state of the visual data and subfigure (c) shows the prediction of the system. The prediction for the tactile state, represented by the bumpers vector, can be seen in Fig. 12.15. It shows a clear

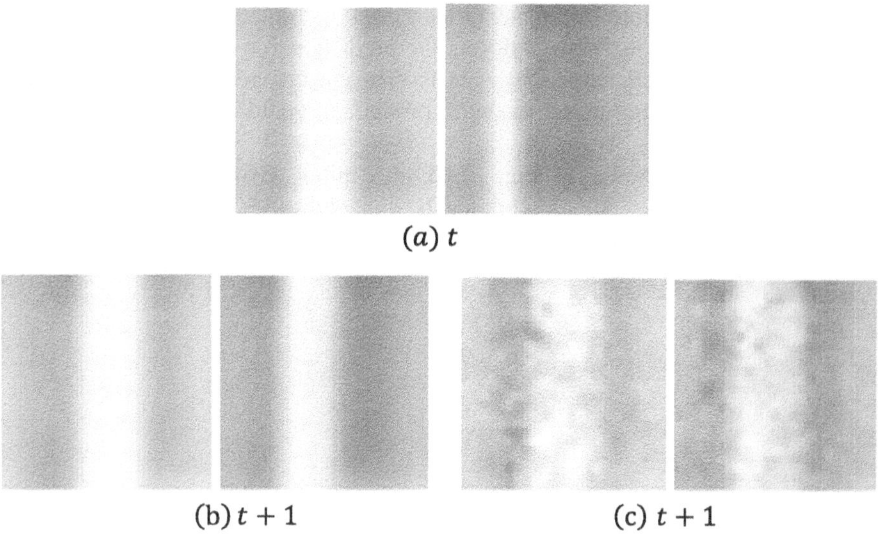

Fig. 12.14 Short-term prediction for the visual state

Fig. 12.15 Short-term prediction for the tactile state

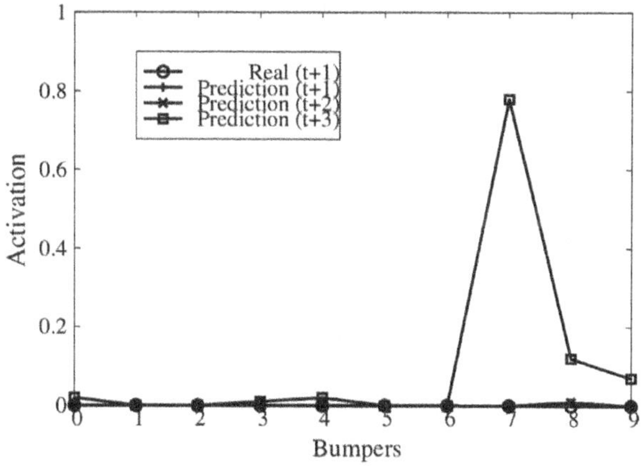

correspondence between the real and predicted state, with a maximum activation value near to 0:05 (5% of the activation range), indicating there is no imminent collision.

Long-Term Prediction

To provide the artificial agent with a long internal simulation process or long-term prediction, the forward model described above can be chained a number of times. This means a consecutive sensorimotor reenaction is performed where the output of the model is used as input for the next time step. For a given initial visual state and a motor command, the system performs the prediction for of the next visual and tactile states. The predicted visual state is then used, together with a respective motor command as input to the model to produce the prediction for the next time step.

In the example shown in Fig. 12.16, the initial visual state is shown in Fig. 12.16a. A chain of motor commands right-right-forward is used as covert actions; i.e., the actions are not executed but internally simulated. The long-term

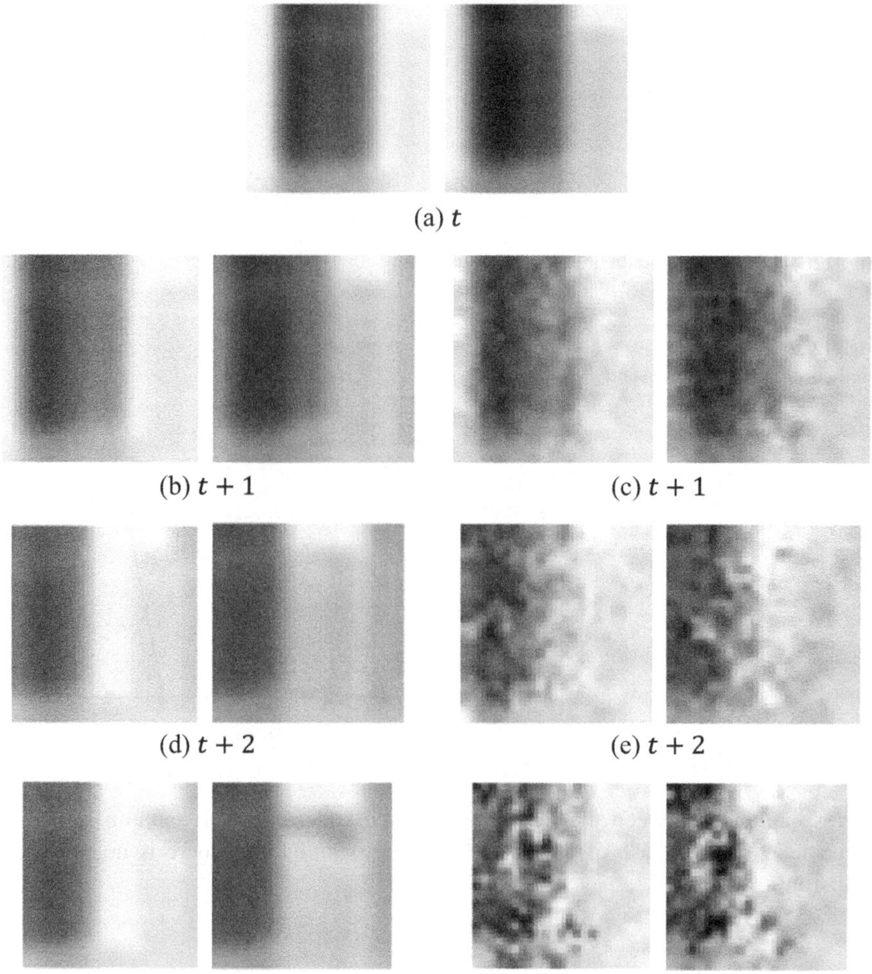

(a) t

(b) $t+1$ (c) $t+1$

(d) $t+2$ (e) $t+2$

(f) $t+3$ (g) $t+3$

Fig. 12.16 Long-term prediction for the visual states

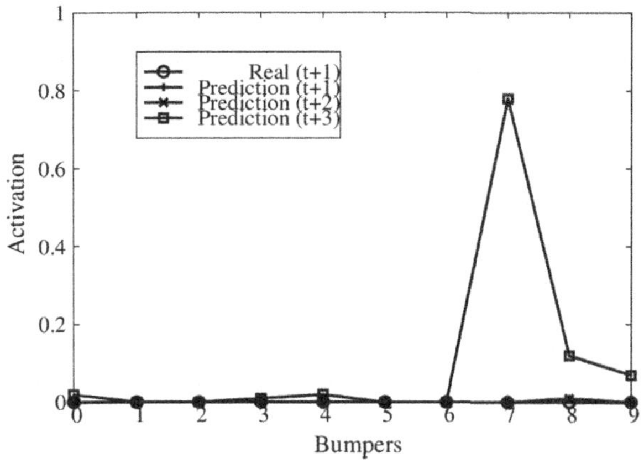

Fig. 12.17 Long-term prediction for the tactile state from time $t + 1$ to $t + 3$

predictions of the system for the next three time steps are shown in the right column of Fig. 12.16; the left column shows the sensory situations once the robot executes the chain of motor commands.

The tactile prediction of the system is shown in Fig. 12.17 for the execution of the three internally simulated motor commands. It is worth noting that the tactile prediction encodes the position of the obstacle in the arena. At time $t + 3$, the neurons of the bumpers vector on the right of the agent produce high activation. As it was the case, if the robot had continued forward on that path, it would have collided with the obstacle.

These results allow us to conclude that the bumpers vector values is coding, at least in an incipient way, the body-reference perception of distance that we are looking for, due to the fact that given the previous sequence of three movements, the system is able to predict a possible collision expressed in terms of the robot's own motor capabilities.

It is worth noting that in no way in the training examples there is a coding for the spatial position of the obstacles related to the tactile data. The whole tactile vector is set to 1 when an obstacle is encountered. As an emergent property, the tactile prediction is an indication of the position of the obstacles in the arena and reproduces previous results (Lara et al. 2007).

It is evident, however, that the visual prediction deteriorates due to the fact that the errors for each step accumulate, distorting the visual input and therefore making more uncertain longer predictions. Nevertheless, the prediction of three time steps allows the robot preventive action. The agent does not need to approach a dangerous or undesired situation, as an internal simulation of its actions allots it knowledge of its surroundings.

12.5 Conclusions

In this work, a different area in the robotics literature is addressed, namely cognitive robotics, which can be considered as the artificial intelligence branch of the cognitive sciences. Cognitive robotics uses artificial autonomous agents to shed light on processes such as perception, learning through sensorimotor interactions with the world, intelligent adaptive behaviour in dynamic environment. The work in the area is framed in the embodied cognition thesis.

Throughout its history, artificial intelligence has suffered a significant number of paradigm turns. This story started in the mid of last century attempting to emulate high-level human cognitive abilities. These turned out to be relatively easy to imitate and sooner rather than later, we had machines capable of defeating chess world champions, general problem solvers and very smart conversationalists.

However, low-level abilities such as walking in uneven terrain, distinguishing a rotten fruit from a ripe one or something as basic as calculating distance to an object, turned out to be very complicated tasks for machines. Artificial intelligence research, for many decades, failed to deliver in this quest.

It is only recently that research in artificial intelligence has turned to the findings and results from the other cognitive sciences searching for a different approach to understand, model and then implement basic behaviors in artificial agents. It is in this framework that two telling examples are presented here.

In the first, an artificial agent making use of a stereo camera and the disparity map calculated from its images learns sensorimotor associations which endow it with safe navigation abilities. The agent of the first example learns the consequences of performing forward movements on the environment that surrounds it. At the same time, the changes in the disparity map are associated with the feeling of crashes with obstacles. The associations are coded by means of a forward model.

The second example brings the capabilities of the forward model a step further by making use of important characteristics of the images coming from the stereo camera. Here, the agent learns associations between visual stimuli, a range of motor commands and the feeling of touching obstacles. These associations allow the agent navigation in complex corridors and sets of obstacles. The predicted tactile values show an important association between the visual data representing the obstacles and their actual position in the space, without this being explicitly coded in the training data.

Both examples have shown that a system of local predictors successfully forms what is known as multimodal sensory representations allotting the agent with a self body map and a notion of distance. Without needing to perform any motor command, the agent is capable of predicting the sensory consequences of its actions. The agent learns these representations by means of its interaction with the environment. Furthermore, the self body knowledge and distance affordances are learned with regards to the agent's own sensorimotor capabilities. To state it plainly, distance to an object is not learned as, for example, centimeters but as number of motor commands to touch the object. It is argued here that this type of knowledge is grounded by the agent providing it with an association between its internal processes and the external world.

The hypothesis put forward in this work is that artificial intelligence can be accounted for by a combination of low-level sensory motor schemes. Furthermore, it is claimed that these primitive sensory motor schemes are coded by means of forward models; forward models represent one of the lowest level requirements in the search for artificial intelligence.

References

Abelson, R., & Schank, R. (1977). *Scripts, plans, goals and understanding* (p. 10). New Jersey: An inquiry into human knowledge structures.

Arkoudas, A. & Bringsjord, S. (2014). Philosophical foundations. In Frankish, K., & Ramsey, W. M. (Eds.), *The Cambridge handbook of artificial intelligence* (pp. 34–63). Cambridge University Press.

Arleo, A., Smeraldi, F., & Gerstner, W. (2004). Cognitive navigation based on nonuniform gabor space sampling, unsupervised growing networks, and reinforcement learning. *IEEE Transactions on Neural Networks, 15*(3), 639–652.

Barsalou, L. (2008). Grounded cognition. *Annual Review of Psychology, 59,* 617–645.

Barsalou, L. (2009). Simulation, situated conceptualization, and prediction. *Philosophical Transactions of the Royal Society of London B: Biological Sciences, 364*(1521), 1281–1289.

Blakemore, S., Goodbody, S., & Wolpert, D. (1998). Predicting the consequences of our own actions: The role of sensorimotor context estimation. *The Journal of Neuroscience, 18*(18), 7511–7518.

Braund, M. (2007). The indirect perception of distance: Interpretive complexities in berkeley's. *Kritike, 1,* 49–64.

Brooks, R. (1990). Elephants don't play chess. *Robotics and autonomous systems, 6*(1–2), 3–15.

Brooks, R. (1991a). Intelligence without reason. *Artificial intelligence: critical concepts, 3,* 107–163.

Brooks, R. (1991b). Intelligence without representation. *Artificial Intelligence, 47*(1–3), 139–159.

Brown, J., O'Brien, C., Leung, S., Dumon, K., Lee, D., & Kuchenbecker, K. (2017). Using contact forces and robot arm accelerations to automatically rate surgeon skill at peg transfer. *IEEE Transactions on Biomedical Engineering, 64*(9), 2263–2275.

Chuy, O., Collins, E., Sharma, A., & Kopinsky, R. (2017). Using dynamics to consider torque constraints in manipulator planning with heavy loads. *Journal of Dynamic Systems, Measurement, and Control, 139*(5), 051001.

Collins, B., & Kornhauser, A. (2006). Stereo vision for obstacle detection in autonomous navigation. *DARPA grand challenge Princeton university technical paper,* 255–264.

Devol, G. (1967). U.S. Patent No. 3,306,471. Washington, DC: U.S. Patent and Trademark Office.

Draghiciu, N., Burca, A., & Galasel, T. (2017). Improving production quality with the help of a robotic soldering arm. *Journal of Computer Science and Control Systems, 10*(1), 11.

Escobar, E., Hermosillo, J., & Lara, B. (2012, November). Self body mapping in mobile robots using vision and forward models. In *2012 IEEE Ninth Electronics, Robotics and Automotive Mechanics Conference (CERMA),* (pp. 72–77). IEEE.

Escobar-Juárez, E., Schillaci, G., Hermosillo-Valadez, J., & Lara-Guzmán, B. (2016). a self-Organized internal Models architecture for coding sensory–Motor schemes. *Frontiers in Robotics and AI, 3,* 22.

Fodor, J. A. (1978). Tom Swift and his procedural grandmother. *Cognition, 6*(3), 229–247.

Gaona, W., Hermosillo, J., & Lara, B. (2012, November). Distance perception in mobile robots as an emergent consequence of visuo-motor cycles using forward models. In *IEEE Ninth Electronics, Robotics and Automotive Mechanics Conference (CERMA),* (pp. 42–47). IEEE.

Gibson, J. (1979). *The Ecological Approach to Visual Perception.* Psychology Press.

Graca, R., Xiao, D., & Cheng, S. (2016). U.S. Patent No. 9,227,322. Washington, DC: U.S. Patent and Trademark Office.

Harnad, S. (1989). Minds, machines and Searle. *Journal of Experimental & Theoretical Artificial Intelligence, 1*(1), 5–25.

Harnad, S. (1990). The symbol grounding problem. *Physica D: Nonlinear Phenomena, 42*(1–3), 335–346.

Hartley, R., & Zisserman, A. (2003). *Multiple view geometry in computer vision.* Cambridge university press.

Hasan, A., Hamzah, R., & Johar, M. (2009, November). Region of interest in disparity mapping for navigation of stereo vision autonomous guided vehicle. In *International Conference on Computer Technology and Development, 2009. ICCTD'09,* (Vol. 1, pp. 98–102). IEEE.

Hoffmann, H. (2007). Perception through visuomotor anticipation in a mobile robot. *Neural Networks, 20*(1), 22–33.

Hoffmann, H., & Möller, R. (2004). Action selection and mental transformation based on a chain of forward models. *From Animals to Animats, 8,* 213–222.

Jamone, L., Ugur, E., Cangelosi, A., Fadiga, L., Bernardino, A., Piater, J., & Santos-Victor, J. (2016). Affordances in psychology, neuroscience and robotics: a survey. *IEEE Transactions on Cognitive and Developmental Systems.*

Johnson-Laird, P. (1977). Procedural semantics. *Cognition, 5*(3), 189–214.

Konolige, K. (1997). Small vision system. hardware and implementation. In *proceedings international symposium on robotics research* (pp. 111–116). ISRR.

Lappin, J., Shelton, A., & Rieser, J. (2006). Environmental context influences visually perceived distance. *Attention, Perception, & Psychophysics, 68*(4), 571–581.

Lara, B., & Rendon, J. (2006, September). Prediction of undesired situations based on multi-modal representations. In *Electronics, Robotics and Automotive Mechanics Conference, 2006* (vol. 1, pp. 131–136). IEEE.

Lara, B., Rendon, J., & Capistran, M. (2007). Prediction of multi-modal sensory situations, a forward model approach. In *Proceedings of the 4th IEEE Latin America Robotics Symposium* (Vol. 1, pp. 504–542).

Li, X., Wang, J., Choi, S., Li, R., Riveland, S., Landsnes, O., & Hara, M. (2016, June). Automatic Gyro Effect Simulation for Robotic Painting Application. In *Proceedings of ISR 2016: 47st International Symposium on Robotics,* (pp. 1–4). VDE.

Lungarella, M., Metta, G., Pfeifer, R., & Sandini, G. (2003). Developmental robotics: A survey. *Connection Science, 15*(4), 151–190.

Miall, R., & Wolpert, D. (1996). Forward models for physiological motor control. *Neural Networks, 9*(8), 1265–1279.

Möller, R., & Schenck, W. (2008). Bootstrapping cognition from behavior—a computerized thought experiment. *Cognitive Science, 32*(3), 504–542.

Moons, T. (1998, June). A guided tour through multiview relations. In *SMILE* (Vol. 98, pp. 304–346).

Mortimer, J., & Rooks, B. (1987). Introduction. In *The International Robot Industry Report* (pp. 1–7). Berlin, Heidelberg: Springer.

Murarka, A., & Kuipers, B. (2009, October). A stereo vision based mapping algorithm for detecting inclines, drop-offs, and obstacles for safe local navigation. In *Intelligent Robots and Systems, 2009. IROS 2009. IEEE/RSJ International Conference on* (pp. 1646–1653). IEEE.

Pezzulo, G., & Cisek, P. (2016). Navigating the affordance landscape: feedback control as a process model of behavior and cognition. *Trends in cognitive sciences, 20*(6), 414–424.

Pfeifer, R., & Bongard, J. (2007). *How the body shapes the way we think: A new view of intelligence.* MIT press.

Proffitt, D. (2006). Distance perception. *Current Directions in Psychological Science, 15*(3), 131–135.

Riedmiller, M., & Braun, H. (1993). A direct adaptive method for faster backpropagation learning: The RPROP algorithm. In *IEEE International Conference on Neural Networks, 1993,* (pp. 586–591). IEEE.

Rosen, J., Sekhar, L., Glozman, D., Miyasaka, M., Dosher, J., Dellon, et al. (2017, May). Roboscope: A flexible and bendable surgical robot for single portal Minimally Invasive Surgery. In *IEEE International Conference on Robotics and Automation (ICRA),* (pp. 2364–2370).

Schillaci, G., Hafner, V., & Lara, B. (2016). Exploration behaviors, body representations, and simulation processes for the development of cognition in artificial agents. *Frontiers in Robotics and AI, 3,* 39.

Searle, J. (1980). Minds, brains, and programs. *Behavioral and Brain Sciences, 3*(3), 417–424.

Searle, J. (1990). Is the brain's mind a computer program? *Scientific American, 262*(1), 26–31.

Thill, S., Caligiore, D., Borghi, A., Ziemke, T., & Baldassarre, G. (2013). Theories and computational models of affordance and mirror systems: An integrative review. *Neuroscience and Biobehavioral Reviews, 37*(3), 491–521.

Thrun, S., & Leonard, J. (2008). Simultaneous localization and mapping. In *Springer handbook of robotics* (pp. 871–889). Berlin, Heidelberg: Springer.

Tolman, E. (1948). Cognitive maps in rats and men. *Psychological Review, 55*(4), 189.

Turing, A. (1950). Computing machinery and intelligence. *Mind, 59*(236), 433–460.

Turvey, M. (2004). Space (and its perception): The first and final frontier. *Ecological Psychology, 16*(1), 25–29.

Warren, W., Jr., & Whang, S. (1987). Visual guidance of walking through apertures: body-scaled information for affordances. *Journal of Experimental Psychology: Human Perception and Performance, 13*(3), 371.

Weizenbaum, J. (1965). ELIZA—A computer program for the study of natural language communication between man and machine. *Communications of the ACM, 9*(1), 36–45.

Wilson, M. (2002). Six views of embodied cognition. *Psychonomic Bulletin & Review, 9*(4), 625–636.

Winograd, T. (1973). A procedural model of language understanding. In R. Schank & K. Colby (Eds.), *Computer models of thought and language* (pp. 152–186). San Francisco: W. H. Freeman.

Wolpert, D., & Ghahramani, Z. (2000). Computational principles of movement neuroscience. *Nature Neuroscience, 3,* 1212–1217.

Wolpert, D., Doya, K., & Kawato, M. (2003). A unifying computational framework for motor control and social interaction. *Philosophical Transactions of the Royal Society of London B: Biological Sciences, 358*(1431), 593–602.

Wolpert, D., Ghahramani, Z., & Flanagan, J. R. (2001). Perspectives and problems in motor learning. *Trends in cognitive sciences, 5*(11), 487–494.

Wu, Y., Chen, H., Chen, Z., He, N., & Liu, B. (2016). Robotic sample preparation system based on magnetic separation. *Journal of Nanoscience and Nanotechnology, 16*(12), 12257–12262.

Zhuang, S., Lin, W., Zhong, J., Zhang, G., Li, L., Qiu, J., et al. (2018). Visual servoed three-dimensional rotation control in Zebrafish Larva heart microinjection system. *IEEE Transactions on Biomedical Engineering, 65*(1), 64–73.

Chapter 13
Applications of Haptic Systems in Virtual Environments: A Brief Review

Alma G. Rodríguez Ramírez, Francesco J. García Luna, Osslan Osiris Vergara Villegas and Manuel Nandayapa

Abstract Haptic systems and virtual environments represent two innovative technologies that have been attractive for the development of applications where the immersion of the user is the main concern. This chapter presents a brief review about applications of haptic systems in virtual environments. Virtual environments will be considered either virtual reality (VR) or augmented reality (AR) by their virtual nature. Even if AR is usually considered an extension of VR, since most of the augmentations of reality are computer graphics, the nature of AR is also virtual and will be taken as a virtual environment. The applications are divided in two main categories, training and assistance. Each category has subsections for the use of haptic systems in virtual environments in education, medicine, and industry. Finally, an alternative category of entertainment is also discussed. Some representative research on each area of application is described to analyze and to discuss which are the trends and challenges related to the applications of haptic systems in virtual environments.

Keywords Haptic systems · Virtual environments · Augmented reality
Virtual reality

13.1 Introduction

Humans are in constant interaction with different environments. The interaction is possible through the sense of sight, touch, taste, hearing, and smell. Basic knowledge is learned through the five senses. Particularly, sight, hearing, and touch are essential to acquire knowledge based on the fact that people learn best when they see, hear, and do (Volunteer Development 4H-CLUB-100 2016).

A. G. Rodríguez Ramírez (✉) · F. J. García Luna · O. O. Vergara Villegas · M. Nandayapa
Instituto de Ingeniería y Tecnología, Universidad Autónoma de Ciudad Juárez,
Ciudad Juárez, Chihuahua, Mexico
e-mail: alma.rodriguez.ram@uacj.mx

O. O. Vergara Villegas et al. (eds.), *Advanced Topics on Computer Vision, Control and Robotics in Mechatronics*,
https://doi.org/10.1007/978-3-319-77770-2_13

Actually, 90% of what is learned by doing is kept by the person who is learning (Volunteer Development 4H-CLUB-100 2016). So, sigh, hearing, and touch are the main senses that allow us to recognize and perceive an environment. When it comes to virtual environments, such as virtual reality (VR) and augmented reality (AR), the user interacts fully or partially with virtual objects only through sight and sometimes also with hearing. The addition of the sense of touch as haptic feedback from a virtual environment could enhance the recognition and perception of the virtual environment. The computational requirements are essential to accomplish this integration.

The evolution of computers in terms of better processing, graphics, and peripherals has allowed the development of virtual environments. Nowadays, virtual environments have been studied and have been applied in different areas such as medicine, education, industry, military, entertainment, aeronautics, among others. All the areas of applications look forward representing a realistic environment, but interaction of users in virtual environments usually misses the feeling of touch which is essential for a realistic experience.

Part of the human experience of interaction with any environment is touching. A common reaction in users, immersed in a virtual environment, is trying to touch the objects in it. Given the importance of senses for interacting in an environment, some robotic systems represent or simulate senses through the integration of different devices and systems like sensors and actuators. For example, the use of vision, acoustic, and haptic systems are commonly used for simulating sight, hearing, and touch, respectively.

The systems, that represent senses in an artificial way, require either unilateral or bilateral transmission of information through an interface as it is shown in Fig. 13.1. The interface works as a means for the exchange of information. For example, a camera as a vision system may represent sight. A camera, just as sight,

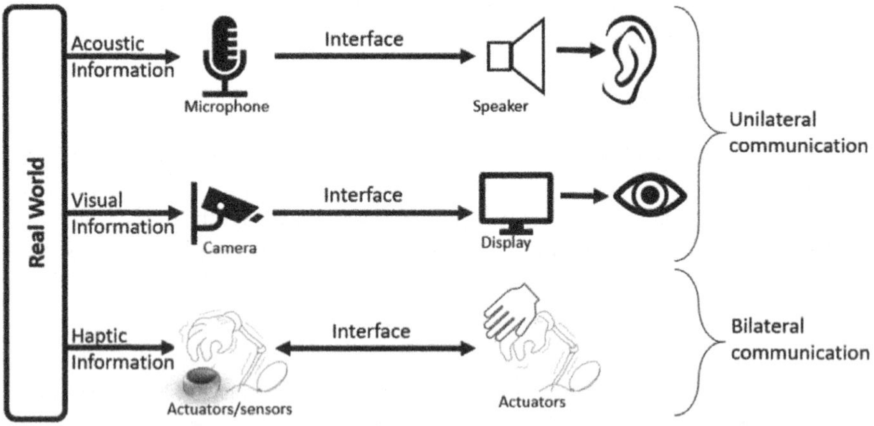

Fig. 13.1 Acoustic, visual, and haptic information acquisition and reproduction

requires a unilateral communication since it captures visual information and displays it on a screen; what was capture does not receive any feedback. When it comes to acoustic systems, the communication is also unilateral. The acoustic system may record an acoustic information and play it through a speaker without having a feedback from the recorder or speaker.

Unlike other senses and systems, touch and haptics require a bilateral communication between the entity and its physical environment. Bilateral communication entails an exchange of information; for example, the feeling a user has when dragging a pencil along a surface involves the exchange of information of the movement of the pencil and the roughness of the surface. Some examples of haptic systems are buttons, vibrators, or complex electromechanical devices that generate reaction forces to emulate the feeling of touch. When it comes to virtual environments such as VR and AR, haptic systems seek to contribute to the immersion of the user by adding information to the interactions with environment. Immersion is a deep mental involvement in something (Oxford University Press 2017). For a user to be involved in an activity, the interaction with the environment is essential.

This chapter is focused on two technologies with the objective of showing how haptic systems have been applied in virtual environments. The interaction of a user in a virtual environment with a haptic system is described in Fig. 13.2. The user will be able to see and touch objects in the virtual environment with a haptic system. So, the immersion is pursued through sight and touch as the user interacts with the environment. The graphic and haptic information is computer rendered so all the information has to be synchronized for the use to feel immersed. Figure 13.2 shows how the graphic rendering generates the virtual object while the haptic rendering generates the haptic information, but both are related; the virtual object generated in the virtual environment is needed to generate the haptic information.

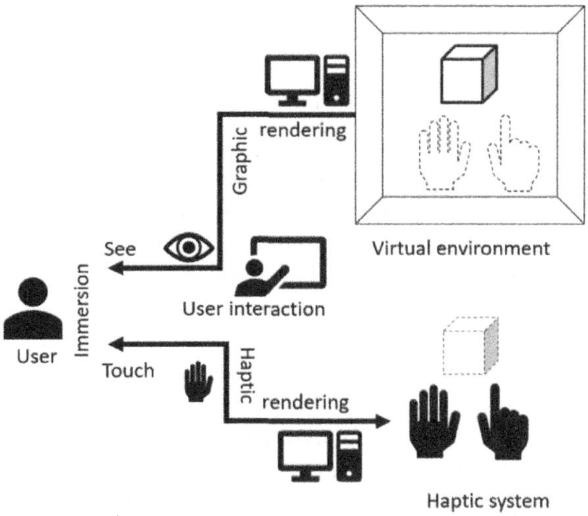

Fig. 13.2 Haptic systems in virtual environments

The virtual environments discussed in this chapter are visual systems based on VR and AR. Haptic systems will be represented as haptic feedback either kinesthetics or cutaneous (discussed in Sect. 13.1.1). Many systems developed by the scientific community, that integrate the feeling of touch, use haptic interfaces. The haptic interface captures information from the environment and sends it to the user to make him/her have the feeling of touch through a device while the device senses the position and force of the user (Lin and Otaduy 2008).

In a virtual environment, users can see and touch rendered objects. Artificial vision systems are carried out by digital image processing techniques, and the generation of images in computer graphics is called graphic rendering. In the case of haptics, haptic rendering is the calculus and simulation of force and/or torque that the user feels when manipulates an object in a virtual environment through a haptic device in real time (Luo and Xiao 2004).

In general, a mathematical model is required to represent the behavior of an object or a system from the real world. A mathematical model of a dynamic system is defined as a set of equations that represent the dynamics of the system (Ogata 1998). When the behavior of an object is mathematically modeled, it is possible to simulate it in a virtual environment. Mathematical models allow to obtain an answer from the interaction between virtual objects and send it to the user through an interface, giving the sensation of touching a virtual object.

There is another way to perceive haptic feedback in a virtual environment; some authors call it pseudo-haptics (Lecuyer et al. 2008; Punpongsanon et al. 2015; Li et al. 2016; Neupert et al. 2016). Pseudo-haptics is the simulation of the haptic sensation when the user interacts in a virtual environment usually through sight. This technique has been proven to enhance the interaction with different materials, helping the user to distinguish them.

The use of haptic systems in virtual environments has increased in the last decades. In Sects. 13.1.1 and 13.1.2, haptic systems and virtual environments are described, respectively.

13.1.1 Haptic Systems

The word haptic refers to the capability to sense a natural or synthetic mechanical environment through touch (Hayward et al. 2004). A haptic system works as a teleoperated system by its bilateral communication nature. Just like teleoperated systems, haptic systems have a master and a slave robot. The master controls the slave's moves; the slave sends feedback to the master in response to the interaction with the remote environment. The objective is that the user, through the master, feels an object even if is not in direct contact with it, the one in contact would be the slave. When it comes to haptic systems in virtual environment, the slave and remote environment are computed, meaning they are virtual (Hayward et al. 2004).

Most of the commercial haptic devices developed are force-feedback based. The lack of realistic touch sensations in the applications is largely due to limitations in traditional force-feedback haptic devices and rendering methods (Aleotti et al. 2016). These are mainly limited in space and transparency.

The fidelity of a system relies on how the feedback is reality like. In haptic systems, the fidelity is called transparency. Some applications require more transparency than others depending on the kind of task carried on. For example, medical assistance applications require more fidelity than entertainment since the user in a medical application interacts with a living patient. In teleoperation applications, if the user is immersed in the remote environment and has the actual feeling of being there, the term of telepresence takes place (Pacchierotti et al. 2014). Telepresence is possible through the feedback that the user receives from the remote environment. The most common feedback in teleoperation is visual and acoustic. To represent completely the human interaction in an environment, all senses should be included. The sense of touch has been taken in count in the last decades though haptic feedback systems.

Haptic feedback may be kinesthetic (position/force feedback) or cutaneous (pressure/temperature feedback). Kinesthetics haptic systems represent the sense of touch through the calculus of contact forces following the physics laws that govern their behavior. Cutaneous feedback is related to the stimuli detected by the skin and relies on measures of the location, intensity, direction, and timing of the contact forces on the fingertips (Pacchierotti et al. 2016).

Ideally, the combination of the two kinds of haptic feedback would represent completely the sense of touch with the constraints of the haptic devices used. For example, Paccierotti et al. (2014) integrated kinesthetic feedback using an Omega 3 (Force Dimension 2017) haptic device and the cutaneous feedback with a non-commercial device developed by the authors. The cutaneous haptic interface consisted of two platforms; one is fixed to the back of the finger and another in the fingertip integrating a force sensor and three 0615S DC micromotors (Faulhaber Group 2017). The system was tested in a virtual environment displaying an object in a computer screen for manipulation. The main objective of this study was to analyze the role of each haptic feedback. Since the stability of a teleoperation system is crucial, it was determined through several experiments that the kinesthetic feedback may bring instability to the system, but it creates more realistic illusion of touch. The cutaneous feedback does not affect the stability of the overall system, but it provides less realism than kinesthetic. The compensation for the feedbacks was not validated for their use in teleoperation tasks because of the performance of cutaneous channel, besides they only considered one degree of freedom (DOF) teleoperation task so it was not realistic for common operations.

13.1.2 Virtual Environments

In this chapter, virtual environments refer to either VR or AR. AR is a mean used to add information to the physical world (Craig 2013) while VR involves only the visual environment without the physical world. Both technologies have in common the generation of virtual information fully or partially in the environment. Typically, virtual objects are computer-generated graphics, so they only represent real objects visually. If the virtual object is programmed to simulate the behavior of its real counterpart, the user will be able to feel it through a haptic interface. We can see virtual objects that are just visual as virtual inert object and the ones programmed with a mathematical model of its behavior as virtual dynamic objects.

On the other hand, motion tracking is an important requirement since its precision, accuracy, speed, and latency affect directly the immersion of the user on its interaction with a virtual environment. AR and VR on their need to place virtual information, either in the real world for AR or keep track of the user in VR, uses image recognition to identify where to place the virtual information and know where the user is. The main techniques are marker and markerless based. Either of these techniques track a mark, and the difference is that marker based uses a digital image processing to identify a physical mark such as an optical square marker, a coated marker with a reflective material (usually a small round object) an array of blinking LEDs, among others. Optical square markers are very popular for their low cost and easy implementation; they consist in a back-square box, of known dimensions, within a white square with an ID mark inside (specific draw, QR code, barcode). Markerless-based technique uses pattern recognition for identify body parts and specific objects to superimpose the virtual information, and also, it could use an external signal such as global positioning system (GPS) to identify where to place the virtual information.

The interaction of a user in a virtual environment requires technology that can accurately measure position and orientation of users and objects of interest as they move in the virtual environment (Rolland et al. 2001). For further information on user tracking technologies and techniques for virtual environments, Rolland et al. (2001) proposed a classification based on the principles of the techniques. The characteristics, limitations, and advantages of each user tracking technique are also described. In Table 13.1, a summary of the classification is shown.

The rest of the chapter is organized as follows. In Sect. 13.2, the objective of the research is presented and so as the databases and selection criteria for the articles described. In Sect. 13.3, the categories of training, assistance, and entertainment for the applications of haptic systems in virtual environments are presented with the description of particular cases. In Sect. 13.4, the discussion about trends and challenges in the integration of haptic systems and virtual environments is presented. Finally, in Sect. 13.5, the conclusions about the issues discussed in the chapter are presented.

13.2 Research Method

The objective of the search presented in this chapter is to show how haptic systems have been applied in virtual environments. When these two technologies are combined, a whole world of application emerges, all seeking for the immersion of

Table 13.1 Classification of user tracking in virtual environments based on (Rolland et al. 2001)

Principle	Classification	Sub-classification
Time of flight (TOF)	Ultrasonic measurements	
	Pulse infrared laser diode	
	GPS	
	Optical gyroscope	
Spatial scan	Outside-in	
	Inside-out	Videometric
		Beam scanning
Inertial sensing	Mechanical gyroscope	
	Accelerometer	
Mechanical linkages		
Phase difference		
Direct field sensing	Magnetic field sensing	Sinusoidal alternating current
		Pulse direct current
		Magnetometer/compass
	Gravitational field sensing	
		Hybrid inertial platforms
		Inside-out inertial
		Magnetic/videometric
	Hybrid systems	TOF/mechanical linkages/videometric position tracker
		TOF/mechanical linkages/videometric 5-DOF tracker

the user in the activity performed. For this research, a virtual environment can be either VR or AR. Four main areas were identified for the application of haptics in virtual environments. Certainly, there are many other areas of applications, but the one described in this chapter circles most of them in a global way.

The databases used for the search were IEEE Xplore (IEEE 2017), ScienceDirect (Elsevier B.V. 2017), ACM Digital Library (ACM Inc. 2017) EMERALD (Emerald Publishing 2017), and Springer (Springer International Publishing AG 2017). These databases are related in general to the areas of computation, technology, engineering, electronics, having also the advantage of having them in the repertory of the UACJ Data Base BIVIR (Biblioteca virtual, virtual library). From the start of the search, the main key words used in the databases were haptics and virtual since this is the technologies of interest. Throughout the investigation, other key words like augmented reality, visuo-haptic, pseudo-haptics, mixed reality, virtual education, virtual training, haptics augmented surgery, and virtual haptics entertainment was used.

The selection criteria for the articles were first based on four areas of application: education, medicine, industry, and entertainment. The researches were not older than 2007 and were taken from journals and conferences related to the two technologies (haptic systems and virtual environments). Finally, other sources used were books, for the fundamental theory and online links, for commercial trends and identification of haptic devices and interfaces.

13.3 Applications of Haptic Systems in Virtual Environments

In this chapter, some works related to haptic systems in virtual environments are described. Three categories are presented: training, assistance, and entertainment. In the first two categories, applications related to education, medicine, and industry will be mentioned. The category of training is based on the application of haptic systems in virtual environments as tools or strategies for acquiring knowledge about a specific task/topic. In contrast, the category of assistance focuses on the application oriented to help during an activity having in count that the user already has the knowledge and experience to do it, but the system is expected to enhance the performance. The final category presented is entertainment. Entertainment industry has played an important role in the development of haptic systems and virtual environments since they have the same objective, the immersion of the final user.

13.3.1 Training

Along their lives, humans are in constant learning. Since humans are born, the process of learning is important, and it starts with the interaction with other human beings and the environment in general. Later, when humans want to acquire an explicit knowledge, they proceed to training activities where the user interacts with a given environment. Haptic systems in virtual environments can be applied in training applications where the user obtains knowledge through interaction and immersion, Fig. 13.3. The user carries on a certain interactive activity; trough sigh and touch, the user is immersed in the activity; the immersion enhances the process of learning; finally, the objective is to acquire certain knowledge. Training is the action of teaching a person or animal a specific skill or type of behavior (Oxford University Press 2017). To enhance the process of training, the integration of different technologies has taken place in the last decades. That is the case of haptic systems and virtual environments.

In general, virtual trainings reproduce an activity and give feedback to the user to get the feeling of doing it on real life. The purpose of training applications is transferring user's knowledge from the virtual experience to real-life operations.

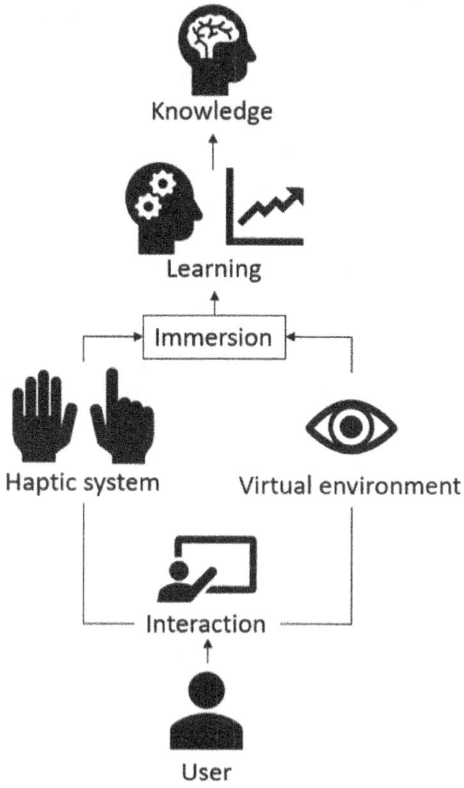

Fig. 13.3 Training applications of haptic systems in virtual environments

This purpose can be achieved through immersion and interaction. If the user has the feeling of reality, he/she will gain certain degree of experience and the ability even if the environment is controlled and safe.

Haptic systems as the feeling of touch in artificial systems work through a haptic interface, while virtual environments simulate the physical surroundings humans interact with trough computer graphics. In combination, these two technologies have tried to enhance different experiences of the user such as the process of training. In the following subsections, some cases of training application are described particularly in the areas of education, medicine, and industry.

13.3.1.1 Education

In general, human machine interfaces (HMIs) are the link between the user and an artificial system, allowing the interaction with computers and machines. Currently, tangible interfaces offer new ways of interaction with virtual objects. Nevertheless, the design of this interface has not been much studied with an educational approach in a spatial learning context (Skulmowski et al. 2016). Lindgren et al. (2016) considered that simulations that use more than one digital technology (mixed reality) help to improve the learning process since they bring with them cognitive and motivational benefits related to immersion and interaction capacity.

Han and Black (2011) incorporated haptic feedback in a physics learning simulator. The study proposed an instructional model for embodied understanding, a theory that suggested that people with perceptual experiences construct multimodal representations to later be able to mentally simulate what is being presented. The instructional model is based on the fact that the first approach to knowledge should be multimodal (visual, auditory, and haptic) and the following instructions had less sensory modalities. The simulator was for experimenting with force transmission through gears. The system included Sidewinder Force-Feedback Joystick II by Microsoft as a haptic interface to generate the feeling of force magnitude needed to spin a gear given an input force. Students could change the configuration of the transmission, composed by four gears, to see how the gears move and the in and out force level (high, medium, and low). The results of recalls, inferences, and transfer tests applied indicated that the addition of haptic feedback had a positive effect in student's comprehension of later instructions with less sensory modalities so as in transfer knowledge to new learning situations.

Potkonjak et al. (2016) discussed about twenty virtual laboratories software based, highlighting the advantages and problems. Some of the advantages detected were the economic saving, the flexibility, the multiple access, the easy adjust of parameters, the resistance to damage, and the capacity to show black boxes. Likewise, some of the problems identified were the computer capacity required, the lack of seriousness from the students caused by the virtual nature of the system, and the fact that the abilities acquired by real experience are different to a simulated experience.

Skulmowski et al. (2016) developed an educational software of 3D (three dimensions) learning that shows a 3D model of a heart with interactive labels of the names of each part. The system consisted in a camera tracking of a plastic heart, and the user held it in one hand while a stylus pen or computer mouse in the other to select a specific part of the heart. The system had two modalities: display and selective. The display mode sets permanent labels in the heart's parts; the user could see all the labels all the time while manipulating the heart's position. In the selective position, the user pointed and clicked the label that wanted to see. The overall system included a Polhemus FASTRAK (Polhemus 2017) motion tracking system and a stylus pen. The comprehension aid of the additional haptic input (plastic heart) is argued to be considerable since it represents a complex structure. In comparison to the subjects who just saw, clicked, and manipulated the heart on a screen, the subjects who had the haptic input required less mental effort according to the tests' results. Also, the selective mode resulted in higher learning performance than permanent display, under certain circumstances.

13.3.1.2 Medicine

Medical applications for training have become popular in the last decades. The benefits that haptic systems and virtual environments bring to the students are mainly based on the possibility of experience a medical procedure without the dangers of treating a living patient. It is certainly difficult to gain the same knowledge through a real experience than through a simulation due to the limitations of the systems in terms of perception of the world and the lack of realistic experiences. With the integration of haptic systems in virtual environments, the user could have a more realistic experience having visual and haptic feedback for medical training.

Rhienmora et al. (2010) developed a dental surgery training simulator. The simulator was developed in two virtual environments using two haptic devices. The system had two modalities, one with VR and the other with AR. In the VR mode, the environment was displayed in a computer screen; the user was able to interact with dental pieces for extraction training. The AR mode used a head-mounted display (HMD); the user manipulated the dental pieces shown also for extraction training. In the second mode, the virtual objects were set using markers. Both modalities required two haptic interfaces Phantom Omni (Sensable Technologies 2016c). It was reported that an experienced dentist confirmed that AR environment had many advantages over VR for dental surgical simulations like the realistic clinical setting.

Lin et al. (2014) developed and validated a surgical training simulator with haptic feedback for safe, repeatable, and cost-effective alternative in learning bone-sawing skill. The system had an Omega.6 as haptic interface and a Display300 as the 3D stereo display. For the haptic rendering, spindle speed, feed velocity, and bone density were considered as variables and multi-point collision detection is method applied. The position and orientation of the virtual tool were continuously updated, according to the position of the end effector of the haptic device. A multi-threading computation environment was applied to maintain, 1000 Hz for haptic rendering and 30 Hz for graphic rendering, update rates. Acoustic feedback was also added. Finally, the validation was based on three experiments: the first proved that the systems were able to differentiate between experimented and novice participants and also improved the performance with repeated practice by decreasing the operative time; the second to prove if the simulator acted as expected; and the third validated the knowledge transfer from training to real procedure in terms of maximal acceleration, and the trained group with lower maximal acceleration suggested that the simulator had positive effects on real sawing.

Chowriappa et al. (2015) developed and tested a training system AR and haptic based for robot-assisted urethrovesical anastomosis (needle driving, needle positioning, and suture placement). The environment called hands-on surgical training (HoST) consisted on a simulator that helps the trainees step by step on the procedures with simultaneous proctoring throughout the training. The experience of the user is visual, auditive, and haptic enable for didactic explanations, annotations, and illustrations in critical steps of the procedure. To evaluate the performance of the training, three tools were used: the Global Evaluative Assessment of Robotic Skills (GEARS) assessment score, the urethrovesical anastomosis evaluation score, and the National Aeronautics and Space Administration (NASA) Task Load Index assessment. Three groups were used for the analysis, the HoST group (received HoST-based training), the control group (video training, not HoST-based), and the cross over group (participants of the control group complete HoST-based training). The urethrovesical anastomosis (UVA) performed in HoST environment was resembled to da Vinci Surgical System. Among all results, according to 70% of all the participants, HoST AR-based environment was as realistic as the actual surgical procedure and 76% felt it was appropriate for learning. In general, HoST and crossover groups had better results in all tests which lead to the conclusion of the improvement in skill acquisition with minimal cognitive demand.

13.3.1.3 Industry

Industry has encounter solutions to train personnel in simulators with haptic feedback. Training personnel is a time- and resource-consuming activity that requires user immersion for it to have an impact on the user's knowledge.

Han et al. (2010) presented the construction of a visuo-haptic training system to satisfy the necessity of training personnel for a production line in manufacturing plants before the actual production begins. The effect of the haptic information for memorize the order of certain positions selected in a plane was studied. The work area covered the reach of an adult person's arm. The haptic interface was a robotic arm, WAM™ Arm. The system was tested in three training modes: visual, visual with haptic feedback, and visual with haptic guide. The visual with haptic feedback had the best results. An observation made was that the haptic guidance annulled the learning performance enhancement under the conditions taken. Passive haptic guidance and alternative guidance algorithms were recommended for better results.

On the other hand, there are industrial training applications for assembly operations. Xia et al. (2012) developed a training system for assembly operations of complex products. The system integrated Phantom and CyberGlove for the haptic feedback in a virtual reality environment. The configuration allowed the user to move freely in a relatively large area. All the data was automatically transferred through a graphic user interface (GUI) from a computer-aided design (CAD) to the virtual environment. The model for the simulation was physics-based taking in count the hierarchical constraints allowing a realistic simulation of the assembly operations. It was proved that the haptic feedback was valuable for the virtual training in the assembly process, using separately Phantom and CyberGlove interfaces. The combination of the two haptic interfaces was not viable for the application since it affected the virtual interaction.

Abidi et al. (2015) identified the assembly operations as the vital process of manufacturing. They developed an assembly training system for the chassis of a blower motor. The development was modular, including the graphic motor module, the physical motor module, and the haptic motor module. In the first module, the virtual environment and objects were developed. In the second module, the physical behavior of the objects was programmed. Finally, the third module was related to the haptic device Phantom Desktop (Sensable Technologies 2016b), now called Geomagic Touch X. In a screen, the user saw the correct order for assembly and then had to do the assembly having the haptic feedback for improving the experience of the virtual assembly task. The case study presented resulted in the identification of the haptic feedback as a beneficial technology for virtual assembly tasks. The importance of the physical features was also identified for the realistic simulation of the assembly task, features like restitution coefficient, control spring stiffness, and other should have been included; so, as a stereoscopic visualization for enhancing the user's immersion.

Carlson et al. (2016) evaluated a virtual assembly task using different combinations of interfaces used by the user. The task consisted in manipulating two different pieces at the same time with two haptic interfaces for insert one in the other. The haptic interfaces used for the first experiment were a Phantom Omni (Sensable Technologies 2016c) and a 5DT Data Glove (Virtual Realities, LLC 2017) and for the second one two Phantom Omni. Several combinations related to the device used by the dominant hand were also made, but it was not proved that there was any significant difference in this combination. The insertion in assembly tasks is a difficult operation for training simulation for the complexity in synchronizing the instruments. In general, it was reported that participants performed equally well in all treatment conditions. The tests did not include gravity force in the objects; the haptic feedback was limited to object's collisions. Either way, the participants showed interest in the experiments.

13.3.2 Assistance

The combination of haptic systems and virtual environments, such as VR and AR, has attracted the attention of assistance applications, mainly in the areas of education, medicine, and industry. Assistance is the action of helping someone by sharing work (Oxford University Press 2017). In this subsection, the cases presented are related to systems developed with the purpose of assisting in different tasks for enhancing the performance, productivity, and/or precision of the user.

Figure 13.4 described how the user interacts with a system through sight and touch with the integration of haptic systems and virtual environments, respectively. The integration of these two technologies contributes to the user's immersion in the task carried on. The overall system has the purpose of aiding the user in a specific task, taking in count that the user already has the knowledge and skills to do it. For example, if the task involves a dangerous procedure, the system could help by warning the user, visually and tangibly, if danger is close. The final objective of this application, and mentioned before, is to enhance the performance, productivity,

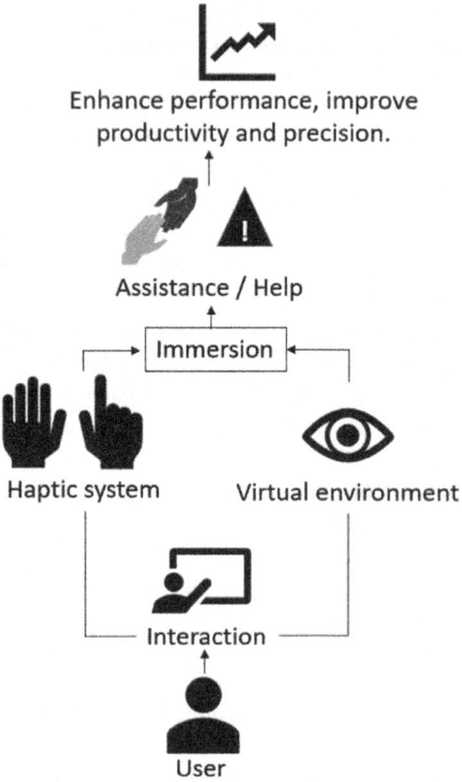

Fig. 13.4 Assistance applications of haptic systems in virtual environments

and/or precision of the user. Unlike training applications, assistance applications assume the user already have the ability or skill to carry on the task and the systems are developed to help the user have a better result.

13.3.2.1 Education

In education, the teachers usually use different tools and strategies to improve the process of teaching. Unlike training applications for education, assistance applications are focused on facilitating the teaching–learning process more than teaching a specific topic. For example, Csongei et al. (2012) developed a system called ClonAR that allowed the user to clone and to edit objects from real world. First, the real object was scanned by a Kinect Fusion (Microsoft 2017). Then, the object was rendered and could be edited in a visuo-haptic AR environment. The information for the rendering was not managed in meshes, instead signed distance fields (SDF) were used because of the Kinect Fusion. They assured the information flow was faster than meshes. The system was tested as a didactic tool, but other possible applications were identified such as medical training a medical education.

Eck and Sandor (2013) defined the term visuo-haptic augmented reality (VHAR) as a technology that allows the user to see and to touch virtual objects. The authors presented a software development platform called HARP that allowed to program VHAR applications. The platform worked either as an educational tool or just for application development. The authors used H3DAPI (SenseGraphics AB 2012), a haptic software development platform that is open source and they complemented it with the Phantom Omni haptic device (Sensable Technologies 2016c). The platform developed was tested and validated by undergraduate students who used it for making different projects. The applications developed in HARP were limited to 30 FPS (frames per second) so the image was reported as looking shaky. Another limitation was that it did not allow the use of more efficient rendering techniques.

Murphy and Darrah (2015) created a set of twenty applications for teaching math and science to students with visual impairments. Certainly, the objective is to teach the students, but the haptic feedback was taken as tool/strategy to improve the teaching–learning process in students with visual difficulties. The haptic device used was the Novint Falcon (Novint 2017). The applications were developed with the haptics software developers kit (HSDK) and the game engine GameStudio. The students were able to select the application of interest and interact with the virtual objects on the simulator through the haptic device having the feeling of touching the objects, and also, acoustic and visual feedback was included. Some of the applications were a plant cell nucleus, volume of shapes, gravity of planets, and exploration of atoms. Six applications were tested in classroom with pre- and post-tests for each application. The results showed, in general, a significant learning gain for all the applications tested, and most of the teachers agreed in the easy to use characteristic of the whole system.

13.3.2.2 Medicine

The rehabilitation of patients is a motivation for the development of systems with haptic feedback in virtual environments. Other important motivation for integrating robotic systems in medical assistance, particularly in surgery application, is having the capacity to perform minimally invasive procedures. The precision of a machine combined with the intuition and ability of a medical doctor is a powerful combination when it comes to surgery. The following researches described are examples of assistance application in medicine.

Atif and Saddik (2010) developed a rehabilitation system based on AR technology and adopted the concept of tangible objects. When a person has a partial paralysis in the body, usually they require rehabilitation assistance. In the system proposed, the user had to manipulate objects in such way that gradually could increase the complexity of the movements based on daily life activities such as handling and moving a cup. The user obtained the haptic feedback through the haptic interfaces Phantom (Sensable Technologies 2016b) and CyberGrasp (CyberGlove Systems Inc. 2017); the visual feedback was obtained from a head-mounted display (HMD) iWear VR920 (Vuzix 2017). The results of the tests coincide with the motivation of the subjects to do the exercises besides the difficulty in depth perception (overcame with practice of the participants). Some of the participants felt arm fatigue because of the weight of the tangible object, but this could be changed in customized exercises depending on the subject's capabilities. In general, the study showed efficiently motivation in patients and the capability of the system to measure important performance factors for the assess of the patient's treatment progress such as task completion time.

Unilateral spatial neglect (USN) is a post-stroke neurological disorder that causes a failure in stimuli response of the brain hemisphere damaged. The patients who have USN present spatial deficits such as stepping into objects when walking and only can dress on side of their body. Tsirlin et al. (2010) studied a therapy application based on a string haptic workbench. The technique for rehabilitation included a space interface device for artificial reality (SPIDAR) and a Fastrack stylus attached. SPIDAR is a device that has a ring suspended by wires, a pair of red and green glasses, and a large screen. An object was displayed on the screen and perceived as a 3D object by the user. Then, the user moved the ring with the finger and had the feeling of touching the object. This occurred when the position of the ring and the object was the same. This illusion was possible because the motion of the string was restricted when the collision occurred. The study revealed that spatial biases could be induced when the user was in a scenario where he/she should avoid a perturbed sensorimotor experience in one side of space. The tests were made with subjects who had to draw a trajectory with the Fastrak stylus and felt a disturbance on one side of the space. For example, when the user traced a line from left to right and the right hemispace as disturbed, they induced a significant bias to the left.

Yamamoto et al. (2012) presented a system for surgical robotic assistance tested in artificial tissue. The system had a pair of haptic devices Phantom Premium (Sensable Technologies 2016a) communicated with a master–slave control and a Bumblebee2 IEEE-1394 stereo-vision camera (FLIR Integrated Imaging Solutions, Inc. 2017). The authors made a prohibited-region user-defined to make sure the procedure was minimally invasive, and the healthy tissues stayed safe. The region of interest was augmented so the user could carry out the task easily and reliability. For the tests, the artificial prostate tissue was reconstructed as the user interacted with it. The task consisted in a teleoperated palpation of tissue to differentiate soft and hardener surfaces, in real time. The forbidden region virtual fixture was found to be useful in the procedure and so as the haptic feedback during the experiments. The force feedback resulted in discontinuities; this could be fixed with modification of impedance and edge geometry of the virtual fixtures according to the authors.

Haptic devices allow users to interact with a remote or virtual environment through the sense of touch (Díaz et al. 2014). Since some surgery devices are manipulated by pedal, Díaz et al. (2014) proposed the use of a pedal with a double haptic channel. The double haptic channel was referred to the hand and foot haptic feedback received during a procedure. The haptic feedback would help the surgeon perform the necessary task not only visual based but also tactile since the surgeon cannot feel what the instrument is touching. The one DOF pedal system proposed consisted in a Maxon RE40 DC motor and a transmission cable (26.66:1), with a

Quantum Devices QD145 encoder. The pedal had a peak torque of 10.72 Nm, a continuous torque of 5.36 Nm, and 15° of workspace. The performance of the haptic pedal was validated on a user-study, with warning signals and resistance to tool's penetration, during a drilling procedure with a double haptic channel. The hand haptic feedback was acquired through a PHANToM 1.0 device with a micro-vibrating electric motor attached at the tip of the PHANToM's stylus. The haptic pedal controlled the speed of the drill, and the resistance torque of the tool's penetration was emulated back to the pedal, so the user could feel that resistance. In general, during the experiments, the users with haptic feedback had a faster reaction to warning signals. The results indicated that the haptic information is helpful during a drilling procedure, and it improves the surgeon accuracy.

Also, haptic systems in combination with VR have helped in diagnosis task and medical analysis. This is the case of the cephalometric diagnosis and analysis, and the current 2D and 3D tools are often complicated, impractical, and not intuitive (Medellín-Castillo et al. 2016). Medellín-Castillo et al. (2016) presented a solution to the disadvantages of the cephalography analysis based on a haptic approach. The proposed system required a haptic device, either a Phantom Omni (Sensable Technologies 2016c) or a Falcon (Novint 2017). Since they used the platform H3DAPI (SenseGraphics AB 2012) (open source haptics software development platform that uses the open standards OpenGL and X3D), the system could interact with any of the two haptic devices. The 2D and 3D crane models were imported in the interface where the haptic interaction was integrated. The user manipulated the crane and had the feeling of touch through a pencil/pen easing the processes of diagnose and surgery planning.

13.3.2.3 Industry

The assistance in industrial applications, by virtual environments and haptic systems, is presented in the cases of assistance for path planning in maintenance assembly/disassembly, motion-impaired operators, and welding. Certainly, the combination of virtual environments and haptic systems could be used in other assistance cases, for example design, manufacturing, and data analysis.

Hassan and Yoon (2010) presented a haptic-based approach for path planning in virtual maintenance assembly/disassembly (MAD). This approach consisted on an automatic system that processed MAD starting by loading the CAD models and then assemble them to their final position to get a sequence table and afterward apply a path planning algorithm; finally, the haptic control mode is applied in the virtual environment. The haptic mode guides the user throughout the maintenance for revision. The 3D algorithm for MAD optimization was based on 3D potential field and genetic algorithm. Computational time was considerable but, the expected results were optimal for the MAD. The haptic feedback was reported as very useful for assisting the user in the simulation process.

Asque et al. (2014), looking to reduce of error rates and targeting times in GUIs for motion-impaired operators in human-computer interaction, presented two haptic assistive techniques using the Phantom Omni. The reduction of this error rates and time targeting in industrial applications could improve productivity and efficiency in human–computer interaction operations. The techniques are based on a virtual plane designed with deformable cones and deformable switches to develop a haptic virtual switch for implementation on existing GUIs. For the experimentation of the techniques, six measurements were defined in terms of characteristics of the clicking operation. Gravity wells and haptic cones were implemented: the first, based on a bounding volume with a spring force toward the center of that volume (Asque et al. 2014); the second, based on the cursor clamping to the apex at the target center by extracting the button position to embed the cones correctly into the mesh of a virtual plane. Finally, deformable virtual switches were developed to help people with physical disabilities target and operate accurately different devices and interfaces. The first experiment of cursor analysis of the haptic assistance proved significant improvements in the measures. The second experiment of the effect of target size (small, medium, and large) and shape showed that only in small and medium, the haptic condition has a significant effect and that target shape has less significant effect on the participant's performance than the haptic condition and the target size.

Ni et al. (2017) noticed that programming remote robots for welding manipulation becomes difficult when the only feedback, from the remote site, is visual information. They proposed an AR application with a haptic device integrated. They used a display for showing the real robot and augment a virtual arm and the end effector. The robot was manipulated by moving the haptic device (PHANToM) remotely. The user got the feedback from the remote robot to the haptic device before the end effector reached the welding surface. This helped keeping a constant distance while the user defined the welding path. The workpiece was captured by a Kinect camera for a 3D point cloud data acquisition. The virtual robotic arm was placed in scene using a marker in the physical workspace of the real robot. The system was tested by ten users with no background in welding or robot programming. The test consisted in recording the path followed to weld two workpieces. The user could choose the welding points as they moved the remote virtual arm as they saw the real scene with the augmented robot on a display. The user-defined paths from the actual welding path were within ± 15 mm of the actual path.

13.3.3 Entertainment

Entertainment has become such an important part of our lives that new studies have pursuit the understanding of the "Psychology of Entertainment" (Invitto et al. 2016). Entertainment has been studied with a multidisciplinary approach having in mind that is related to learning, perception, emotions, communication, marketing, science, therapy, and others (Ricciardi and Paolis 2014). That might be the reason why entertainment is an attractive area of application for the developers of haptic systems in virtual environments. In general, the applications of entertainment seek

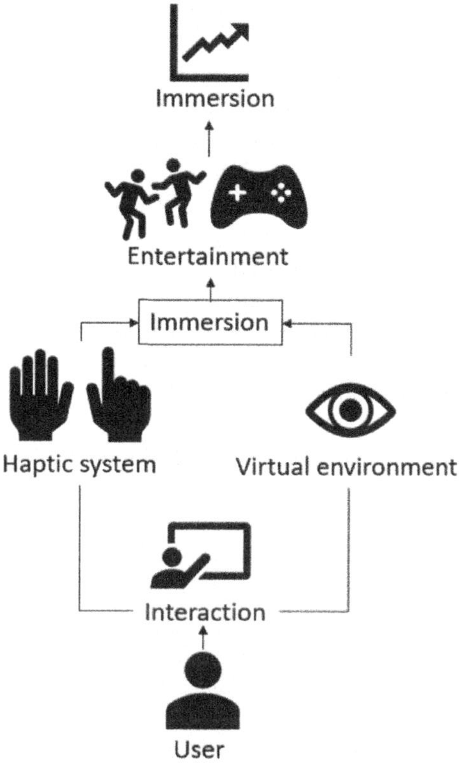

Fig. 13.5 Entertainment applications of haptic systems in virtual environments

for the user to feel comfortable and immerse in the environment given as described in Fig. 13.5. In the case of video games, immersion is usually accomplished through audio, graphics, and simple haptic feedback like vibration.

Other application of entertainment includes haptic systems like the one developed by Magnenat-Thalmann et al. (2007). The system consisted in an interactive virtual environment where the user could fix the hair of a virtual character. The user had the feeling of manipulating the hair by using different virtual tools to comb, wet, dry, and cut. A SpaceBall 5000 (Spacemice 2017) and a Phantom Desktop (Sensable Technologies 2016b) were used, so as an algorithm called virtual coupling based on physics modeling for the haptic representation. The algorithm was also used to link the haptic device with the virtual tools in a stable way.

From Disney Research Laboratories, Bau and Poupyrev (2012) developed a system they called REVEL. The system was based on AR combining visual and haptic feedback to virtual and real objects inserted in the reality. The visual feedback was delivered through a display that allowed the user to see the reality with virtual objects inserted. The haptic feedback allowed the user feeling the virtual object through reverse electrovibration (induce AC signal in the user instead of in the object). The system maintained a constant tactile sensation by adjusting the signal amplitude dynamically to compensate all the varied impedances. The signals generated and applied to the user were safe since the current applied to the user was in the microampere range (max. 150 μA). When the user touched a real object, a capacitive sensing of the touch occurred, and the haptic augmented feedback was delivered from a database. The touch sensing of virtual objects was optical, a user finger tracking through a Kinect. The virtual objects were inserted with markers. The system required an infrastructure previously prepared for the tactile augmentation when touching an object.

Sodhi et al. (2013), also from Disney Research Laboratories, developed the project AIREAL. AIREAL was a device that gave the user the feeling of free air textures. The device consisted of a servo actuated flexible nozzle that generated an air vortex and a camera to measure the target's distance, and the camera was mounted over a gimbal structure. The vortex control was based on four dimensions mainly: pulse frequency, intensity, location, and multiplicity. The experiences of the users consisted on having a projection, for example, over the hand, of an object and the air haptic feedback should coincide with it in space and time. The system could synchronize with others of the same type to create a whole atmosphere. The authors tested the systems simulating an environment where the user felt seagulls flying around while seeing it on a computer game. The system presented by Sodhi et al. (2013) made considerable sound considered as noisy and could make the user feel uncomfortable because of the position of the devices.

On the other hand, Ouarti et al. (2014) developed a test platform to differentiate between a visual, haptic, and visuo-haptic experience of a user in a virtual world. The experiment of interest was the visuo-haptic one where the system had the capacity to make the user feel like being inside of an accelerating car. The user could see on a screen a video generated in a graphic engine for simulating the movement. The system had a Virtuose (Haption 2017) haptic device connected to a mechanism to simulate the movement when the car accelerated. The authors concluded about the importance of the haptic feedback synchronized with the video for the user to be immersed in the game.

Israr et al. (2014) presented a story-telling application. Just like any other entertainment application, immersion was important for the user to have a satisfying experience. The application was kids oriented. The system was capable of making the user felt like it was raining, something walked around, started a motor, and others. Each feeling was classified within a haptic vocabulary list, and it could be intensified as wanted. The effects were also visual and auditory. The system had two modules. The first module was an arrangement of tactile vibrators called tactors C-2 (Engineering Acoustics Inc. 2017) aligned in the back and waist (arranged in a vest). The second module was a graphic interface for the manipulation of parameters of vibrators (mainly time and intensity), this could be shown in a computer or a mobile device. When someone said a phrase of the haptic vocabulary, a vibration corresponding to it was produced.

Punpongsanon et al. (2015) developed a system called SoftAR which was an application of AR where the user could feel the softness of an object. The user could see a projection of a surface over a real object and when the user touched it, he/she saw the deformation of the material projected. The projection created the haptic illusion. The author also identified marketing and design applications of the system, to show clients or designers different options of materials to select based on the softness simulated.

13.4 Trends and Challenges

In Tables 13.2, 13.3, 13.4 and 13.5, a summary of the researches, described in representation of each area of application, is shown. From Tables 13.2, 13.3, 13.4 and 13.5 it should be noted that most of the applications use commercial haptic interfaces, so there are opportunities in the development of customized haptic devices. The design of customized devices could help in the development of more complex, cheapest, specialized, and/or precise applications to enhance the immersion of the user given a specific application. In a future, people will be able to touch virtual information and do it in a natural way just like interacting with the natural environment.

Nowadays, the educational applications, of haptic systems in virtual environments, have impact in society since currently digital technologies are been more incorporated as strategies for teaching. Nevertheless, the access to some technologies may not be at hand for everyone yet one day the use of haptic systems and virtual environments will be used as didactic materials like today we use books and computers. Certainly, with the advantage of access to intelligent mobile devices, it seems feasible to scale different applications in the classroom or for the common use. When it comes to apply haptic systems in virtual environment in training for education, it seemed to be difficult. The difficulty resided in the fact that a virtual experience might not be as good as gaining real experience. Multisensorial feedback has recently been taken as a solution to this problem, integrating different technologies such as computer vision, haptic feedback, and audio effects; it is possible to have a close approach to real tasks. On the other hand, assistance applications for education have big potential for simplifying the teaching–learning process. Teachers will be benefited with the flexibility of educational tool based on haptic systems in virtual environments. The use of multisensorial experiences trough haptic systems and virtual environments has also taken place in medical applications.

On one hand, training medical applications have become very popular in the last decades. The main limitation is the use of commercial haptic devices. The next step in haptic systems development for medical applications is to develop customized devices. The transparency is the main factor of a haptic device; it could be improved by implementing low friction actuators. The development of realistic simulators in virtual environments with haptic feedback has become an important trend given the fact that user in this field requires the feeling of touch to gain the experience needed. On the other hand, medical assistance applications focus on enhancing the performance of the user with multiple feedback of sensory information such as visual and haptic.

Table 13.2 Researches of applications for haptic systems in virtual environments

References	Category	Area of application	Haptic feedback	Haptic interface	VE	Latency/ real time/ frequency	Graphic/haptic Rendering	Subjects	Tracking	Display device
Magnenat-Thalmann et al. (2007)	Entertainment	Entertainment	Kinesthetic	Phantom Desktop and SpaceBall 5000	VR	Real time	Haptic and visual sensory channels are processed in separate threads	–	Simulation	Computer screen
Rhienmora et al. (2010)	Training	Medicine	Kinesthetic	Phantom Omni	VR/ AR	Real time	PolyVox library	Students and two instructors	Marker-based	HMD
Han et al. (2010)	Training	Industry	Kinesthetic	7-DOFWAM Arm	VR	–	Physics-based modeling UDP comms	36	Teleoperation UDP comms	42-inch LCD TV (1920 1080 pixels; LG Electronics Inc., Korea)
Atif and Saddik (2010)	Assistance	Medicine	Kinesthetic and cutaneous	Phantom; CyberGrasp	AR	Real time	ARToolKit; CHAI3D; Open Dynamics Engine; VirtualHand SDK	15	Marker-based	Display or HMD
Tsirlin et al. (2010)	Assistance	Medicine	Kinesthetic	SPIDAR	VR	–	–	15	Fastrack stylus	Stereo google

Table 13.3 Cont. Researches of applications for haptic systems in virtual environments

References	Category	Area of application	Haptic feedback	Haptic interface	VE	Latency/real time/ frequency	Graphic/ haptic rendering	Subjects	Tracking	Display device
Bau and Poupyrev (2012)	Entertainment	Entertainment	Cutaneous	Reverse electrovibration system	AR	Real time; ~150 ms latency	ARToolkit library	2 applications	Kinect Fusion	Proyector; mobile device
Eck and Sandor (2013)	Assistance	Education	Kinesthetic	Phantom Omni	AR	1000 FPS in the haptic loop and 30 FPS for graphical rendering	Parallel graphic render rate	9 projects	Marker-based	Canon VH-2007 HWD
Sodhi et al. (2013)	Entertainment	Entertainment	Cutaneous	AIREAL	VR	~139 ms latency of a vortex	–	5 applications	Depth sensor PMD Camboard Nano; Kinect Fusion	Computer screen/ tablet/ proyector
Díaz et al. (2014)	Assistance	Medicine	Kinesthetic and cutaneous	Phantom Premium and pedal	VR	Real time; 1 kHz sampling rate	dSPACE 1104; OpenGL; virtual springs model	12	Simulation	Computer screen
Ouarti et al. (2014)	Entertainment	Entertainment	Kinesthetic	Virtuose	VR	–	OpenGL	17	Simulation	Computer screen
Israr et al. (2014)	Entertainment	Entertainment	Cutaneous	Tractors tipo C-2	VR	–	–	85	Button to enable the feeling described	iPad

Table 13.4 Cont. Researches of applications for haptic systems in virtual environments

References	Category	Area of application	Haptic feedback	Haptic interface	VE	Latency/real time/ frequency	Graphic/haptic rendering	Subjects	Tracking	Display device
Lin (2014)	Training	Medicine	Kinesthetic	Omega.6	VR	Real time; update rates of 1000 Hz for haptic rendering and 30 Hz for graphic rendering	CHAI3D; OpenGL; multi-threading computation environment	25 × 6	Polaris, NDI Canada; simulation; markerless	Display300
C.T.Asque et al. (2014)	Assistance	Industry	Kinesthetic	Phantom Omni	VR	Real time	CHAI3D API	6	–	Computer screen
Abidi et al. (2015)	Training	Industry	Kinesthetic	Phantom Desktop	VR	RT comms ph-hap eng	OpenGL; GLUT libraries; PhysX; OpenHaptics	Case study: a blower house assembly	Simulation	Computer screen
Parinya and Kosuke (2015)	Entertainment	Entertainment	Visual	Pseudo-haptic	AR	Real time	NVIDIA GeForce GT520 2 GB	17 3 Elastic objects simulated	Marker-based	NEC NP-L51 WD 1280 × 800 70 Hz ANSI lumen
Chowriappa et al. (2015)	Training	Medicine	Kinesthetic	Do not say	AR	Real time	–	52	–	–
Murphy and Darrah (2015)	Assistance	Education	Kinesthetic	Novint Falcon	VR	–	HSDK; GameStudio	32	Simulation	Computer screen

Table 13.5 Cont. Researches of applications for haptic systems in virtual environments

References	Category	Area of application	Haptic feedback	Haptic interface	VE	Latency/ real time/ frequency	Graphic/haptic rendering	Subjects	Tracking	Display device
Skulmowski et al. (2016)	Training	Education	Cutaneous	Stylus	VR	~4 ms latency	The tracked position and rotation was smoothed over 5 frames (approx. 83 ms)	96	Polhemus FASTRAK motion tracking system (six degrees-of-freedom, 60 Hz, 4 ms latency)	24" iiyama ProLite E2473HS screen (1920 1080 pixels)
Carlson et al. (2016)	Training	Industry	Kinesthetic and cutaneous	Phantom Omni/ 5DT data glove	VR	IS-900 at around 4 ms and the patriot at around 17 ms	OpenSceneGraph; VR JuggLua	52	Polhemus patriot magnetic tracker InterSense IS-900 hybrid inertial and ultrasonic tracking system.	Computer screen
Medellín-Castillo et al. (2016)	Assistance	Medicine	Kinesthetic	Phantom omni/ falcon	VR	Haptic device latency	Microsoft foundation classes; visualization toolkit library; H3DAPI	5 2D and 1 3D cephalometric radiographs, 21 dental surgeons	Simulation	Computer screen
Ni et al. (2017)	Assistance	Industry	Kinesthetic	Phantom device	AR	Real time	Point cloud data; Implicit surface of workpieces	10	Marker-based	Computer screen

Finally, the entertainment industry has a considerable contribution in the development of haptic systems in virtual environment. These applications have worked as development platform since their customers demand more immersive experiences in every way, every day. The game developers have proposed themselves to make systems with more quality in graphics, sound, haptic feedback, and more.

If researchers keep looking for new ways to make more realistic environments, more transparent haptic systems and find a way to combine these two technologies in a natural way, the possibilities to have all these applications in our common life grow.

13.5 Conclusions

The technologies of VR, AR, and haptics are in fast growing and have been of great interest of technology innovators. There are infinite possibilities of application in the use of this technology. The researches described were classified in training, assistance, and entertainment with a sub-classification for the first and second one, according to their area of application in educational, medical, and industrial. Nevertheless, some of the authors coincide with the fact that the development of a haptic system in a virtual environment may have a multidisciplinary impact.

Users in all areas demand immersive experiences. The lack of the feeling of touch in virtual environment limits the user immersion and could lead to a low-interest response from the user. Besides the enhancement of immersive experiences, haptic virtual environment-based trainings can improve safe acquisition of technical and basic skills. On the other hand, assistance application meets the benefits of haptic systems in virtual environments by the enhancements of operations in all areas of application.

References

Abidi, M., Ahmad, A., Darmoul, S., & Al-Ahmari, A. (2015). Haptics assisted virtual assembly. *IFAC-PapersOnLine*, *48*(3), 100–105.

ACM, Inc. (2017). *ACM digital library*. Retrieved from http://dl.acm.org/.

Aleotti, J., Micconi, G., & Caselli, S. (2016). Object interaction and task programming by demonstration visuo-haptic augmented reality. *Multimedia Systems, 22*(6), 675–691.

Asque, C., Day, A., & Laycock, S. (2014). Augmenting graphical user interfaces with haptic assistance for motion-impaired operators. *International Journal of Human-Computer Studies, 72,* 689–703.

Atif, A., & Saddik, A. E. (2010). AR-REHAB: An augmented reality framework for poststroke-patient rehabilitation. *IEEE Transactions on Instrumentation and Measurement, 59*(10), 1–10.

Bau, O., & Poupyrev, I. (2012). REVEL: Tactile feedback technology for augmented reality. *ACM Transactions on Graphics, 89,* 1–11.

Carlson, P., Vance, J., & BergNee, M. (2016). An evaluation of asymmetric interfaces for bimanual virtual assembly with haptics. *Virtual Reality, 20*(4), 193–201.

Chowriappa, A., Raza, S., Fazili, A., Field, E., Malito, C., Samarasekera, D., et al. (2015). Augmented-reality-based skills training for robot-assisted urethrovesical anastomosis: A multi-institutional randomised controlled trial. *BJU International, 115*(2), 336–345.

Craig, A. B. (2013). *Understanding augmented reality: Concepts and applications*. Newnes.

Csongei, M., Hoang, L., Eck, U., & Sandor, C. (2012). ClonAR: Rapid redesign of real-world objects. *IEEE International Symposium on Mixed and Augmented Reality*, 277–278.

CyberGlove Systems Inc. (2017). *Overview*. Retrieved from http://www.cyberglovesystems.com/cybergrasp/.

Díaz, I., Gil, J., & Louredo, M. (2014). A haptic pedal for surgery assistance. *Computer Methods and Programs in Biomedicine, 116*(2), 97–104.

Eck, U., & Sandor, C. (2013). HARP: A framework for visuo-haptic augmented reality. *IEEE Virtual Reality*, 145–146.

Elsevier B.V. (2017). *Explore scientific, technical, and medical research on sciencedirect*. Retrieved from http://www.sciencedirect.com/.

Emerald Publishing. (2017). *Discover new things*. Retrieved from http://www.emeraldinsight.com/.

Engineering Acoustics Inc. (2017). *C2-HDLF*. Retrieved from https://www.eaiinfo.com/product/c2-lf/.

Faulhaber Group. (2017). *DC-micromotors series 0615…S*. Retrieved from https://www.faulhaber.com/en/products/series/0615s/.

FLIR Integrated Imaging Solutions, Inc. (2017). *Bumblebee2 1394a*. Retrieved from https://www.ptgrey.com/bumblebee2-firewire-stereo-vision-camera-systems.

Force Dimension. (2017). *Omega.3*. Retrieved from http://www.forcedimension.com/products/omega-3/overview.

Han, G., Lee, J., Lee, I., & Choi, S. (2010). Effects of kinesthetic information on working memory for 2D sequential selection task. *IEEE Haptics Symposium*, 43–46.

Han, I., & Black, J. (2011). Incorporating haptic feedback in simulation for learning physics. *Computers and Education*, 2281–2290.

Haption SA. (2017). *Virtuose 6D*. Retrieved from https://www.haption.com/site/index.php/en/products-menu-en/hardware-menu-en/virtuose-6d-menu-en.

Hassan, S., & Yoon, J. (2010). Haptic based optimized path planning approach to virtual maintenance assembly/disassembly (MAD). In *The 2010 IEEE/RSJ International Conference on Intelligent Robots and Systems* (pp. 1310–1315). Taipei, Taiwan: IEEE.

Hayward, V., Astley, O., Cruz-Hernandez, M., Grant, D., & Robles-De-La-Torre, G. (2004). Haptic interfaces and devices. *Sensor Review, 24*, 16–29.

IEEE. (2017). *IEEE Xplore digital library*. Retrieved from http://ieeexplore.ieee.org/Xplore/home.jsp.

Invitto, S., Faggiano, C., Sammarco, S., Luca, V., & Paolis, L. (2016). Haptic, virtual interaction and motor imagery: Entertainment tools and psychophysiological testing. *Sensors, 16*(3), 1–17.

Israr, A., Zhao, S., Schwalje, K., Klatzky, R., & Lehman, J. (2014). Feel effects: Enriching storytelling with haptic feedback. *ACM Transactions on Applied Perception, 11*(3), 1–14.

Lecuyer, A., Burkhardt, J.-M., & Tan, C.-H. (2008). A study of the modification of the speed and size of the cursor for simulating pseudo-haptic bumps and holes. *ACM Transactions on Applied Perception, 5*(13), 1–32.

Li, M., Sareh, S., Xu, G., Ridzuan, M., Luo, S., Xie, J., et al. (2016). Evaluation of pseudo-haptic interactions with soft objects in virtual environments. *PLoS One, 11*(6), 1–17.

Lin, Y., Wang, X., Wu, F., Chen, X., Wang, C., & Shen, G. (2014). Development and validation of a surgical training simulator with haptic feedback for learning bone-sawing skill. *Journal of Biomedical Informatics, 48*, 122–129.

Lin, M., & Otaduy, M. (2008). *Haptic rendering foundations, algorithms, and applications*. A K Peters.

Lindgren, R., Tscholl, M., Wang, S., & Johnson, E. (2016). Enhancing learning and engagement through embodied interaction within a mixed reality simulation. *Computers & Education, 95*, 174–187.

Luo, Q., & Xiao, J. (2004). Physically accurate haptic rendering with dynamic effects. *IEEE Computer Graphics and Applications, 24*(6), 60–69.

Magnenat-Thalmann, N., Montagnol, M., Bonanni, U., & Gupta, R. (2007). Visuo-haptic interface for hair. In *International Conference on Cyberworlds*, 3–12.

Medellín-Castillo, H., Govea-Valladare, E., Pérez-Guerrero, C., Gil-Valladaresc, J., Limd, T., & Ritchie, J. (2016). The evaluation of a novel haptic-enabled virtual reality approach for computer-aided cephalometry. *Computer methods and programs in biomedicine, 130*(C), 46–53.

Microsoft. (2017, March). *Kinect fusion*. Retrieved from https://msdn.microsoft.com/en-us/library/dn188670.aspx.

Murphy, K., & Darrah, M. (2015). Haptics-based apps for middle school students with visual impairments. *IEEE Transactions on Haptics, 8*(3), 318–326.

Ni, D., Yew, A., Ong, S., & Nee, A. (2017). Haptic and visual augmented reality interface for programming welding robots. *Advanced Manufacturing, 5*(3), 191–198.

Neupert, C., Matich, S., Scherping, N., Kupnik, M., Werthscheutzky, R., & Hatzfeld, C. (2016). Pseudo-haptic feedback in teleoperation. *IEEE Transactions on Haptics, 9*(3), 397–408.

Novint. (2017, March). *Falcon technical specifications*. Retrieved from http://www.novint.com/index.php/novintxio/41.

Ogata, K. (1998). *Ingeniería de Control Moderna*. Pearson Educación.

Ouarti, N., Lécuyery, A., & Berthozz, A. (2014). Haptic motion: Improving sensation of self-motion in virtual worlds with force feedback. *IEEE Haptics Symposium*, 167–174.

Oxford University Press. (2017). *English oxford living dictionaries*. Retrieved from https://en.oxforddictionaries.com/.

Pacchierotti, C., Prattichizzo, D., & Kuchenbecker, K. (2016, February). Cutaneous feedback of fingertip deformation and vibration for palpation in robotic surgery. *IEEE Transactions on Biomedical Engineering, 63*(2), 278–287.

Pacchierotti, C., Tirmizi, A., & Prattichizzo, D. (2014). Improving transparency in teleoperation by means of cutaneous tactile force feedback. *ACM Transactions on Applied Perception, 11*(1), 1–16.

Punpongsanon, P., & Kosuke, S. (2015). SoftAR: Visually manipulating haptic softness perception in spatial augmented reality. *IEEE Transactions on Visualization and Computer Graphics, 21*(11), 1279–1288.

Polhemus. (2017). *FASTRAK*. Retrieved from http://polhemus.com/motion-tracking/all-trackers/fastrak.

Potkonjak, V., Gardner, M., Callaghan, V., Mattila, P., Guetl, C., Petrovic, V., et al. (2016). Virtual laboratories for education in science, technology, and engineering: A review. *Computers & Education, 95*, 309–327.

Rhienmora, P., Gajananan, K., Haddawy, P., Dailey, M., & Suebnukarn, S. (2010). Augmented reality haptics system for dental surgical skills training. In *VRST'10 Proceedings of the 17th ACM Symposium on Virtual Reality Software and Technology* (pp. 97–98).

Ricciardi, F., & Paolis, L. (2014). A comprehensive review of serious games in health professions. *International Journal of Computer Games Technology*, 1–14.

Rolland, J., Davis, L., & Baillot, Y. (2001). Survey of tracking technology for virtual environments. In W. Barfield, & T. Caudell (Eds.), *Fundamentals of wearable computers and augmented reality* (p. 836). CRC Press.

Sensable Technologies. (2016a). *Geomagic phantom premium haptic devices*. (Geomagic, Editor) Retrieved from http://www.geomagic.com/es/products/phantom-premium/overview/.

Sensable Technologies. (2016b). *Phantom desktop haptic device*. Retrieved from http://www.geomagic.com/archives/phantom-desktop/specifications/.

Sensable Technologies. (2016c). *Phantom omni haptic device*. (Geomagic, Editor) Retrieved from http://www.geomagic.com/archives/phantom-omni/specifications/.

SenseGraphics AB. (2012). *What is H3DAPI*. Retrieved from http://www.h3dapi.org/.

Skulmowski, A., Pradel, S., Kühnert, T., Brunnett, G., & Rey, G. (2016). Embodied learning using a tangible user interface: The effects of haptic perception and selective pointing on a spatial learning task. *Computers and Education, 92*(C), 64–75.

Sodhi, R., Poupyrev, I., Glisson, M., & Israr, A. (2013). AIREAL: Interactive tactile experiences in free air. *ACM Transactions on Graphics, 134*(1–134), 10.

Spacemice. (2017). *Spaceball 5000*. Retrieved from http://spacemice.org/index.php?title=Spaceball_5000.

Tsirlin, I., Dupierrix, E., Chokron, S., Ohlmann, T., & Coquillart, S. (2010). Multimodal virtual reality application for the study of unilateral spatial neglect. *IEEE Virtual Reality*, 127–130.

Virtual Realities, LLC. (2017). *5DT data glove 5 ultra*. Retrieved from https://www.vrealities.com/products/data-gloves/5dt-data-glove-5-ultra-2-2.

Volunteer Development 4H-CLUB-100. (2016). 4H.VOL.115 learning Styles_2016: Retrieved from http://4h.okstate.edu/literature-links/ lit-online/others/volunteer/4H.VOl.115%20learning%20Styles_08.pdf/.

VUZIX. (2017). *VUZIX, view the future*. Retrieved from https://www.vuzix.com/.

Xia, P., Lopes, A., Restivo, M., & Yao, Y. (2012). A new type haptics-based virtual environment system for assembly training of complex products. *International Journal of Advanced Manufacturing and Technology, 58*(1–4), 379–396.

Yamamoto, T., Abolhassani, N., Jung, S., Okamura, A., & Judkins, T. (2012). Augmented reality and haptic interfaces for robot-assisted surgery. *The International Journal of Medical Robotics and Computer Assisted Surgery, 8*(1), 45–56.

Chapter 14
Experimental Analysis of a 3-DOF Articulated Flat Empennage

Miguel Angel García-Terán, Ernesto Olguín-Díaz, Mauricio Gamboa-Marrufo, Angel Flores-Abad and Fidencio Tapia-Rodríguez

Abstract This paper presents the aerodynamic analysis of a bio-inspired empennage, which mimics the way that the tail of some birds moves. To avoid a kinematic chain in the proposed design, the three axes of rotation intersect in a single point. To define the relationship between the attitude of the empennage and the aerodynamic coefficients, different tests were conducted in an open-circuit wind tunnel at low velocity, where the attitude of the empennage was changed, and the aerodynamic effects were measured by means of a force sensor and expressed with six dimensional wrenches, including forces and torques. Wrenches were transformed from the sensor's reference frame to a frame located at the aerodynamic center of the empennage to compute the aerodynamic coefficients. A multiple regression analysis revealed a coupling effect between the aerodynamic coefficients and the attitude of the proposed empennage, as changes in any of the three Euler angles modify the aerodynamic coefficients. Besides, it is shown that both the longitudinal and the translational motion of the vehicle can be controlled by the proposed bio-inspired empennage.

M. A. García-Terán (✉)
UACJ Department of Manufacturing and Industrial Engineering, Ciudad Juárez, Mexico
e-mail: angel.garcia@uacj.mx

M. A. García-Terán
CINVESTAV Ramos Arizpe, Coahuila, Mexico

E. Olguín-Díaz
Department of Robotics and Advanced Manufacturing, CINVESTAV, Ramos Arizpe, Coahuila, Mexico

M. Gamboa-Marrufo
Department of Structures and Materials, UADY, Mérida, Yucatán, Mexico

A. Flores-Abad
University of Texas at El Paso, El Paso TX, USA

F. Tapia-Rodríguez
Department of Engineering, Universidad Panamericana de Guadalajara, Zapopan, Jalisco, Mexico

O. O. Vergara Villegas et al. (eds.), *Advanced Topics on Computer Vision, Control and Robotics in Mechatronics*,
https://doi.org/10.1007/978-3-319-77770-2_14

Keywords Flat empennage · Bio-inspired empennage · Aerodynamic sectional approach

14.1 Introduction

Birds have the ability to modify position, attitude, and shape of both the wings and the tail independently, as well as the shape of their bodies in order to develop a specific flight mode. Furthermore, birds change the attitude of their feathers during the flapping wings. It is noteworthy that flight modes of birds depend on the species of bird and flight technique is particular of each individual bird no matter if they belong to the same species, (Gatesy and Dial 1993; Alexander 2002; Biewener 2003; Gottfried 2007; Tobalske 2007). After this, the definition of a bio-inspired morphing unmanned aerial vehicle (UAV) arises naturally as a UAV that changes its external shape during flight to adapt to the environment (Valasek 2011).

There are two opinions concerning the function of the tail during bird flight. Pennycuick in (Pennycuick 2008) affirms that the effects of the bird's tail are not significant during the flight; then, flight modes of the birds are developed by means of the wings. On the other hand, in Tucker (1992), Gatesy and Dial (1993), Thomas (1993), Gottfried (2007), Su et al. (2012) show that this element is very important during different locomotion movements. In the same sense, these establish that it is necessary to develop studies about the physic contributions because of the complexity for understanding its functions during the flight (Tucker 1992; Kirmse 1998; Biewener 2003; Videler 2005; Shyy et al. 2008).

The work proposed by (Alexander 2002) established that birds used the tilt movement of the tail to counteract the tilting of the wings. It is noteworthy that results showed that the effects of the wings were predominant and the effects of the tail were not important. In the same sense, in Pennycuick (2008) it was stated that the tail of the birds produces neither a lift force nor a moment; then the tail does not improve the stability of the birds. Notwithstanding, the results suggested that birds modify the angle of attack of each wing for controlling their movements. On the other hand, in Tucker (1992), Gatesy and Dial (1993), Su et al. (2012), it has been established that the tail of the birds is an important element for taking off, for landing, for developing acrobatic movements, and different flight modes. Last works concluded that the interaction between the tail and wings is not clear because of the problems which represent the process of measurement. Then, in Su et al. (2012), the hover flight was studied because this flight mode was considered appropriate to analyze the interactions between tail and wings. In accordance with Su et al. (2012), the birds changed both: attitude and area of the tail (spread and folded tail) for modifying the aerodynamic forces. These changes improved the stability and maneuverability of the birds, because the lift and drag forces were related to the area and the angle of attack of the tail. Then, the birds synchronized the tilt, folded, and spread of the tail with the wings to recover the body posture.

On the other hand, in Gottfried (2007) it was defined that the contribution of the tail to the directional stability is a function of the sideslip angle and the lift coefficient.

The electrical activity of the tail muscles of pigeons for different locomotion movements and flight modes were presented in Gatesy and Dial (1993). The results showed that the electrical activity changed during the transition phase from the take-off to flight, the flight to landing and between flight modes. It is noteworthy that in accordance with the results the flapping flight requires more contributions from the tail. Tucker (1992) analyzed the pitching equilibrium of the Harris hawk where the primary wing feathers were clipped. The hawks presented problems to achieve the equilibrium when gliding while increasing the percentage of clipped feathers. Results showed that the hawks spread the tail and changed the position of the wings to achieve the longitudinal stability. On the other hand, in Thomas (1993) the aerodynamic properties of a tail of bird were determined. The work proposed a model based on slender lifting surface theory, and it concluded that there are two situations where the birds need the forces produced by the tail. The first is at slow velocities where the longitudinal instability is presented; the second during acrobatic movements and hover flight where high forces are required to control. These effects were produced by both the upward and downward movements of the tail to cancel the longitudinal unbalance and the banking motion.

14.2 State of the Art

The study and development of bio-inspired morphing UAVs are focused on new designs, materials, mechanism, dynamic modeling, and controllers, all of them implemented typically on the wings, (Valasek 2011; Paranjape et al. 2011a, b, 2012a, b). However, the studies about the bio-inspired empennage or bio-inspired tail are limited because most of the times, the aerodynamic effects of the empennage are considered negligible. However, the main contributions of the empennage to a fixed-wing aerial vehicle are given by the moments produced by the aerodynamic forces, which are important for both the longitudinal and lateral–directional stability.

The analysis of stability of soaring birds using a RC model airplane with similar dimensions and weight of a Raven is presented in Hoey (1992). The vehicle included an articulated empennage, which can develop both the bank motion and the tilt motion, as well as flaps on the lower surface of the wings. The results showed that the direction of the lateral force was defined by the attitude of the empennage. Moreover, there were at least two combinations that produce the same sign. Bank motion of the empennage affects not only the lateral–directional stability but also the airplane pitch motion. Results suggested that soaring birds control their lateral stability by means of the adverse-yaw effects, and also that they use the dihedral angles and the motion of the tail when performing rapid turns. The work affirms that soaring birds used more the dihedral angle to stabilize the flight when soaring than when gliding.

In Leveron (2005), Higgs (2005), Rivera-Parga et al. (2007) the design, test, and dynamic model of a micro-aerial vehicle (MAV) with a 2-DOF articulate empennage, to develop a portable aerial vehicle, were presented. The work determined the behavior of the MAV and the aerodynamic effects produced by the attitude changes of the empennage. Tail included both bank and tilt motions with respect to the MAV body. Tests were developed in a wind tunnel at low velocity, where the forces and moments on the vehicle were measured for different empennages and different attitude settings. Results showed that both longitudinal stability and lateral–directional stability of the vehicle were affected by the empennage. Results suggested that the empennage acted as a spoiler under specific conditions because the lift did not increase despite the attitude changes; however, the longitudinal stability was not compromised because the lift, drag, and pitching moment were mildly affected. On the other hand, both the direction of the lateral force as well as the yaw moment were affected by the empennage tilt movement, whose values were similar to those of a typical airplane. Notice that these results were consistent with the results of Hoey (1992). In the same sense, the attitude of the empennage produces slight changes on the roll moment of the vehicle, but these effects can improve the orientation. The research showed that the empennage motions modify the attitude of the MAV; nevertheless, both the longitudinal and lateral–directional stability are coupled, and it is necessary to define a control strategy.

The longitudinal stability and the controllability of an ornithopter, that included a variable tail with one DOF, were presented in Han et al. (2008). The work proposed the dynamic model of the vehicle and using a path following control the flapping frequency and the tilt angle of the tail are adjusted. The results suggested that the synchronization of the wings and the tail guarantees the longitudinal stability, it is important to comment that the lateral stability was not analyzed. In the same sense, the design, the perching control, and the experimental test of MAVs having articulated wings and variable tilt horizontal flat tail (lacking the vertical stabilizer) (Paranjape et al. 2011a, b, 2012a, b). The dynamic model included the inertial effects produced by the attitude changes of wings; furthermore, a perching control strategy was proposed based on the variable dihedral angles of the wings and the tilt motion of the tail. Experimental results showed that while the lateral–directional motion was stabilized, the system required large control effort and the yaw dynamics become slow because of the lack of the vertical stabilizer.

The design of a bio-inspired autonomous aircraft with a rotatable empennage was presented in (Muller et al. 2015). The vehicle includes five different airfoils per wing and an empennage with two DOF, which consist of an open chain, and a set of ailerons as control surfaces. The vehicle was tested in a wind tunnel, and the results were consistent with the results presented previously in Hoey (1992), Rivera-Parga et al. (2007). The previous works consider that the tail of birds has two degrees of freedom (the bank and tilt motions). However, in accordance with the observation process, there are at least two more DOF which correspond to the pan motion and the capability of both folding and spreading the tail. Then in this work, the design of a bio-inspired empennage which includes the bank, tilt, and pan motion is presented in order to mimic the main movements on the tail of the birds and to analyze their effects.

14.3 Background

The aerodynamic forces and moments arise from the interaction between a body and the air (fluid) flow. These effects are defined to be function of a set of dimensionless coefficients which depend on the shape of the airfoil, and they are weighted by some geometric parameters, the relative velocity, the angle of attack, the sideslip angle, and their rate of change (Stevens and Lewis 2003; Stengel 2004; Cook 2007). The aerodynamic coefficients of an aerial vehicle include the aerodynamic contribution of the vehicle components (the wings, the vertical stabilizer, the horizontal stabilizer, and the fuselage) and the control surfaces. It is noteworthy that the coefficients are affected by the position and attitude of that vehicle elements; then, there is a set of coefficients for each configuration over the same vehicle. This is one of the reasons why the analysis of bio-inspired aerial vehicles is complicated. However, an aerodynamic sectional approach can recover the qualitative behavior of an aerial vehicle, which defines the aerodynamic forces and moment as the sum of the contribution acting on each vehicle component, (Noth 2008; Roscam 2003; Olguín-Díaz and García-Terán 2014). The aerodynamic effects produced either by an articulated wing or an empennage can be expressed as a function of its attitude using the appropriated transformations. Furthermore, the aerodynamic effects can be expressed by means of either a pair of 3D vectors (force and moment vectors) or a single 6D vector (which is known as wrench) that includes both force and moment vectors.

14.3.1 The Rotation Matrix

The set of variables that define the position and attitude of a rigid body with respect to either an inertial reference frame or a local reference frame are known as the pose of the rigid body. Let Σ_0 be an inertial reference frame (Earth-fixed frame) and Σ_1 a local reference frame rigidly attached to a body at point 1 as shown in Fig. 14.1. Vector $\mathbf{r}^{(0)} = (x, y, z)^T \in \mathbb{R}^3$ is the position of the local reference Σ_1 with respect to the inertial reference frame Σ_0 expressed with coordinates of the inertial reference frame Σ_0. Both reference frames can be related by means of a rotation matrix $\mathbf{R}_0^1(\boldsymbol{\theta}) \in SO(3)$ that is parameterized by an attitude vector $\boldsymbol{\theta} \in \mathbb{R}^m$ (for $m = \{3, 4\}$). Notice that the dimension of the vector $\boldsymbol{\theta}$ depends on the attitude representation, then there are multiple ways to parameterize the matrix $\mathbf{R}_0^1(\boldsymbol{\theta})$. In aeronautic, the attitude is represented typically by the "roll-pitch-yaw" angles vector $\boldsymbol{\theta} = (\phi, \theta, \psi)^T$. Therefore, for this attitude representation the rotation matrix $\mathbf{R}_0^1(\boldsymbol{\theta})$ is given by (1), where the trigonometric functions sine(x) and cosine(x) are abbreviated for simplification as $s_x = \sin(x)$ and $c_x = \cos(x)$, (Stevens and Lewis 2003; Stengel 2004; Olguín-Díaz and García-Terán 2014).

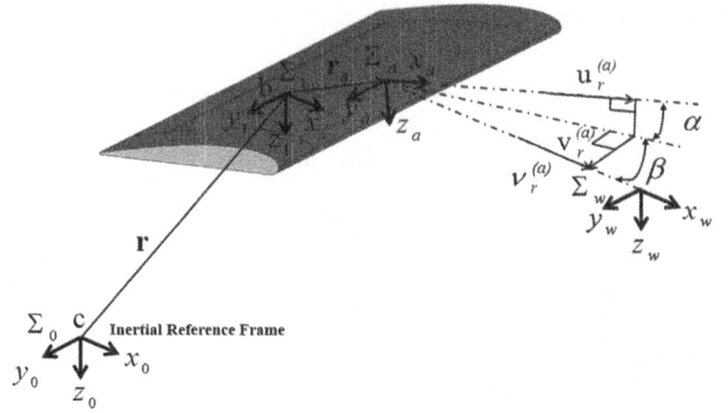

Fig. 14.1 General schematic of a wing and its reference frames

$$\mathbf{R}(\boldsymbol{\theta}) = \begin{bmatrix} c_\psi c_\theta & -s_\psi c_\phi + c_\psi s_\theta s_\phi & s_\psi s_\phi + c_\psi s_\theta c_\phi \\ s_\psi c_\theta & c_\psi c_\phi + s_\psi s_\theta s_\phi & -c_\psi s_\phi + s_\psi s_\theta c_\phi \\ -s_\theta & c_\theta s_\phi & c_\theta c_\phi \end{bmatrix} \tag{14.1}$$

Since the above-mentioned reference frame Σ_1 may be used as the root frame of the aerial vehicle, additional frames for each aerodynamic section may be of some important use. Let the reference frame Σ_a be the aerodynamic frame as shown in Fig. 14.1 which is placed at the aerodynamic center of the wing (or section) and oriented along the chord line of the wing. This reference frame may be parameterized by means of the dihedral angle φ, the incidence angle $\bar{\alpha}$, and the sweep angle Λ_i of the section, (Olguín-Díaz and García-Terán 2014). In accordance with the aerodynamic sectional modeling if the wing is hinged, the matrix $\mathbf{R}_1^a$ is parameterized by means of the additional movements; the rotation matrix $\mathbf{R}_1^a(\varphi, \bar{\alpha}, \Lambda) \in SO(3)$ is constant for constant parameters: $\varphi, \bar{\alpha}, \Lambda$.

The absolute rotation matrix of the aerodynamic frame with respect to the inertial one can be made by composed rotations in the appropriate order: $\mathbf{R}_0^a = \mathbf{R}_0^1\mathbf{R}_1^a$ (where arguments are excluded only for simplification purposes in the notation). Moreover, since the rotation matrix is an orthogonal matrix it arises $(\mathbf{R}_0^a)^{-1} = (\mathbf{R}_0^a)^T$. Notice that the position of Σ_a respects to Σ_0 expressed in the inertial reference frame Σ_0 is the addition of the vector $\mathbf{r}^{(0)}$ and the vector $\mathbf{r}_a^{(a)}$ in accordance with Eq. (14.2). Then, the position and attitude of any point that belongs to the body can be determined by means of the last procedure (Siciliano et al. 2009).

$$\mathbf{r}_a^{(0)} = \mathbf{r}^{(0)} + \mathbf{R}_0^1\mathbf{R}_1^a\mathbf{r}_a^{(a)} = \mathbf{r}^{(0)} + \mathbf{R}_0^a\mathbf{r}_a^{(a)} \tag{14.2}$$

14.3.2 The Extended Vectors

An extended vector is given by the pair $(\mathbf{a}, \mathbf{b})$, where $\mathbf{a} \in \mathbb{R}^3$ is a linear vector and $\mathbf{b} \in \mathbb{R}^3$ is a free vector, both belonging to the Euclidian space, and expressed in the same reference frame. It is noteworthy that there are different extended vectors and each one has a specific meaning and mathematical properties (Featherstone 2010a, b). Let $v_b^{(1)} \in R^6$ be the extended velocity vector called twist, and $\mathbf{F}_b^{(1)} \in \mathbb{R}^6$ the extended force vector or wrench, both in the reference frame Σ_1 (Fig. 14.1) and expressed by Eqs. (14.3) and (14.4):

$$\mathbf{v}_b^{(1)} = \begin{pmatrix} \mathbf{v}_b^{(1)} \\ \boldsymbol{\omega}_b^{(1)} \end{pmatrix} \in \mathcal{M} \subset \mathbb{R}^6 \tag{14.3}$$

$$\mathbf{F}_b^{(1)} = \begin{pmatrix} \mathbf{f}_b^{(1)} \\ \mathbf{n}_b^{(1)} \end{pmatrix} \in \mathcal{F} \subset \mathbb{R}^6 \tag{14.4}$$

where $\mathbf{v}_b^{(1)}$, which belongs to the motion space $\mathcal{M} \subset \mathbb{R}^6$, contains the linear velocity $\mathbf{v}_b^{(1)} \in \mathbb{R}^3$ of the point b and the angular velocity of the body to which the point belongs; notice that both vectors describe the motion of the body. On the other hand, the wrench $\mathbf{F}_b^{(1)}$, which belongs to the force space $\mathcal{F} \subset \mathbb{R}^6$, contains the force vector $\mathbf{f}_b^{(1)} \in \mathbb{R}^3$ that acts on the point b and the moment vector $\mathbf{n}_b^{(1)} \in \mathbb{R}^3$ acting on the body, all of them expressed in the local reference frame Σ_1.

The extended vectors can be expressed between any two reference frames, i.e., Σ_0 and Σ_1 by means of an extended rotation $\mathcal{R}_0^1$ and an extended translation $\mathcal{T}\left(\mathbf{r}_{c/b}^{(1)}\right)$ that are presented by Eqs. (14.5) and (14.6), respectively:

$$\mathcal{R}_0^1 = \begin{bmatrix} \mathbf{R}_0^1 & 0 \\ 0 & \mathbf{R}_0^1 \end{bmatrix} \tag{14.5}$$

$$\mathcal{T}\left(\mathbf{r}_{c/b}^{(1)}\right) = \begin{bmatrix} I & -\left[\mathbf{r}_{c/b}^{(1)}\times\right] \\ 0 & I \end{bmatrix} \tag{14.6}$$

where the operator $[\mathbf{a}\times] \in so(3)$ is a third order skew-symmetric matrix (i.e.:$[\mathbf{a}\times]^T = -[\mathbf{a}\times] = [-\mathbf{a}\times]$) that expresses the 3D vector cross product $[\mathbf{a}\times]\mathbf{b} = \mathbf{a} \times \mathbf{b}$, for all $(\mathbf{a}, \mathbf{b}) \in \mathbb{R}^3$. Notice that the vector $\mathbf{r}_{c/b}^{(1)}$ is the distance from Σ_1 to Σ_0, expressed with frame Σ_1 coordinates, but it can be expressed in the opposite way as $\mathbf{r}^{(0)}$. The term $\mathbf{R}_0^1(\boldsymbol{\theta}) \in SO(3)$ in Eq. (14.5) for this case is the rotation matrix that is expressed in Eq. (14.1).

Using the operators in Eqs. (14.5) and (14.6) both twists and wrenches can be transformed between any two reference frames. For instance, between Σ_0 and Σ_1 it follows, (Olguín-Díaz and García-Terán 2014):

$$\begin{array}{ll}
\mathbf{v}_c^{(1)} = \mathcal{T}\left(\mathbf{r}_{c/b}^{(1)}\right)\mathbf{v}_b^{(1)} & \mathbf{v}_b^{(1)} = \mathcal{T}^{-1}(\mathbf{r}_{c/b}^{(1)})\,\mathbf{v}_c^{(1)} = \mathcal{T}(-\mathbf{r}_{c/b}^{(1)})\mathbf{v}_c^{(1)} \\
\mathbf{v}_b^{(0)} = \mathcal{R}_0^1\,\mathbf{v}_b^{(1)} & \mathbf{v}_b^{(1)} = \mathcal{R}_0^{1T}\,\mathbf{v}_b^{(0)} \\
\mathbf{F}_c^{(1)} = \mathcal{T}^T\left(\mathbf{r}_{c/b}^{(1)}\right)\mathbf{F}_b^{(1)} & \mathbf{F}_b^{(1)} = \mathcal{T}^{-T}(\mathbf{r}_{c/b}^{(1)})\,\mathbf{F}_c^{(1)} = \mathcal{T}^T(-\mathbf{r}_{c/b}^{(1)})\mathbf{F}_c^{(1)} \\
\mathbf{F}_b^{(0)} = \mathcal{R}_0^1\,\mathbf{F}_b^{(1)} & \mathbf{F}_b^{(1)} = \mathcal{R}_0^{1T}\,\mathbf{F}_b^{(0)}
\end{array}$$

It is noteworthy that the extended translation operator needs to be transposed to translate the wrench. Most of the times, it is necessary to include both operations (translation and rotation) to express an extended vector with respect to a different reference frame. Extended motion operation (EMO) arises as a new operator that includes both rotation and translation. EMO operator can be defined by different ways; thus, a general expression is presented as follows, (Olguín-Díaz and García-Terán 2014):

$$\mathcal{X}(\mathbf{d}_{a/b}, R_j^i) \triangleq \mathcal{R}_j^{iT}\mathcal{T}(\mathbf{d}_{a/b}^{(j)}) = \mathcal{T}(\mathbf{d}_{a/b}^{(i)})\mathcal{R}_j^{iT} \tag{14.7}$$

The first term of Eq. (14.7) is defined by rotating an extended vector that was translated previously, and the second one is obtained by translating the previous rotated vector, then the twist and wrench are directly transformed as shown in the following expression:

$$\mathbf{v}_a^{(i)} = \mathcal{X}(\mathbf{d}_{a/b}, R_j^i)\mathbf{v}_b^{(j)} \tag{14.8}$$

$$\mathbf{F}_a^{(i)} = \mathcal{X}^{-T}(\mathbf{d}_{a/b}, R_j^i)\mathbf{F}_b^{(j)} \tag{14.9}$$

14.3.3 The Aerodynamic Wrench

Aerodynamic wrench is expressed as: $\mathbf{F}_A^{(a)} = (\mathbf{f}_A^{(a)T}, \mathbf{n}_A^{(a)T})^T \in \mathbb{R}^6$, which includes both the force vector $\mathbf{f}_A^{(a)} \in \mathbb{R}^3$ and the moment vector $\mathbf{n}_A^{(a)} \in \mathbb{R}^3$ that are expressed in the local reference frame Σ_a (Olguín-Díaz and García-Terán 2014). It is noteworthy that in aerodynamic bodies, while the aerodynamic moment vector $\mathbf{n}_A^{(a)} \in \mathbb{R}^3$ is referred to the reference frame Σ_a [see Eq. (14.10)], the aerodynamic force vector coordinates (and in consequence their aerodynamic coefficients) are computed with respect to the relative wind velocity [see Eq. (14.11)]. Then, there must exist a fourth reference frame Σ_w placed at some point in the body (normally the neutral pressure point) such that the x-axis of this frame is aligned with the relative incidence velocity, (Stengel 2004):

$$\mathbf{f}_A^{(w)} = \bar{q}\, S \begin{pmatrix} -C_D(\cdot) \\ C_Y(\cdot) \\ -C_L(\cdot) \end{pmatrix} \tag{14.10}$$

$$\mathbf{n}_A^{(a)} = \bar{q}\, S \begin{pmatrix} b\, C_l(\cdot) \\ \bar{c}\, C_m(\cdot) \\ b\, C_n(\cdot) \end{pmatrix} \tag{14.11}$$

In Eqs. (14.10) and (14.11) $\bar{q} = \frac{1}{2}\rho\|\mathbf{v}_r\|^2$ is the dynamic pressure that is a function of the air density ρ and the relative wind velocity $\mathbf{v}_r$. The terms b, $\bar{c}$, and S are the span, the mean aerodynamic chord and the area of the wing. Finally, $C_D(\cdot)$, $C_Y(\cdot)$, $C_L(\cdot)$, $C_l(\cdot)$, $C_m(\cdot)$ and $C_n(\cdot)$ are the aerodynamic coefficients which are determined either by means of experimental procedures or computed fluid dynamics (CFDs) software. These coefficients are functions of the aerodynamic angles (the angle of attack α and the side slip angle β) and their time derivatives $(\dot{\alpha}, \dot{\beta})$; the angular velocity coordinates $\boldsymbol{\omega}^{(a)} = (p, q, r)^{\mathrm{T}} \in \mathbb{R}^3$; the span b; the mean aerodynamic chord $\bar{c}$; the aspect ratio AR; the cross-coupling effects representing by some specific aerodynamic coefficients $C_{xz}(\cdot)$; and the control surfaces, such as ailerons δ_a, elevator δ_e, or rudder δ_r (Roscam 2003; Stevens and Lewis 2003; Cook 2007; Noth 2008).

The aerodynamic force vector needs a frame coordinates transformation $\mathbf{f}_A^{(a)} = \mathbf{R}_w^{aT}(\alpha, \beta)\mathbf{f}_A^{(w)}$, by some rotation matrix $\mathbf{R}_A \triangleq \mathbf{R}_w^a \in SO(3)$ that represents the relative attitude of the aerodynamic frame Σ_a with respect to the wind frame Σ_w, and which can be represented by the next matrix:

$$\mathbf{R}_{A_i}(\alpha, \beta) = \mathbf{R}_w^a = \begin{bmatrix} c_\alpha c_\beta & s_\beta & c_\beta s_\alpha \\ -c_\alpha s_\beta & c_\beta & -s_\alpha s_\beta \\ -s_\alpha & 0 & c_\alpha \end{bmatrix} \in SO(3) \tag{14.12}$$

being parameterized by means of the angle of attack $\alpha = \arctan(\mathrm{v}_{r_z}/\mathrm{v}_{r_x})$ and the side slip angle $\beta = \arcsin(\mathrm{v}_{r_x}/\left\|\mathbf{v}_r^{(a)}\right\|)$ that are a function of the relative wind velocity vector $\mathbf{v}_r^{(a)}$, (Stevens and Lewis 2003).

The relative velocity $\mathbf{v}_r^{(a)}$ is computed by the difference between the section (wing) velocity of the wing $\mathbf{v}_a^{(a)}$ and the wind velocity $\mathbf{v}_w^{(0)}$. For proper addition both velocities need to be expressed with respect to the same reference frame; for instance, the aerodynamic frame of the section Σ_a:

$$\mathbf{v}_r^{(a)} = \mathbf{v}_a^{(a)} - \mathbf{R}_0^{aT}(\boldsymbol{\theta})\mathbf{v}_w^{(0)} \tag{14.13}$$

Based on the concept of the EMO operator, the aerodynamic wrench $\mathbf{F}_A^{(a)}$ can be expressed in any reference frame by means of Eq. (14.14), where $\mathcal{X}_A$ is parameterized by the distance between Σ_a and Σ_1 and the rotation matrix that related both reference frames. Notice that the wrench is a function of the relative wind twist $\mathbf{v}_r^{(a)}$ that is computed by means of Eq. (14.15).

$$\mathbf{F}_A^{(1)} = \mathcal{X}_A^T \, \mathbf{F}_A^{(a)}(\mathbf{v}_r^{(a)}) \tag{14.14}$$

$$\mathbf{v}_r^{(a)} = \mathcal{X}_A \, \mathbf{v}_1^{(1)} - \mathcal{R}_0^{aT} \, \mathbf{v}_w^{(0)} \tag{14.15}$$

The term $\mathcal{R}_0^a$ in Eq. (14.15) is the extended rotation which relates both Σ_0 and Σ_a reference frames. In the same expression, the term $\mathbf{v}_w^{(0)}$ is the wind twist which represents the environmental effects that is referred to the inertial reference frame. Then, in accordance with Eq. (14.14), it is possible to express the aerodynamic effects of an aerodynamic body with respect to any reference frame using the appropriated transformations.

14.4 The Experimental Setup

The design of bio-inspired UAVs with articulated empennage typically considers that this aerodynamic body includes two DOF; however, in accordance with the analysis of the flight of birds, these living beings have the capability to develop at least four movements. The design of a bio-inspired empennage presents different problems that are related with the size, the weight, and the number of DOF, as well as the undesirable effects that are related with the shape of the empennage. In these sections, the design of a bio-inspired empennage is presented, where the PT-40 RC model from *Great Planes®* was considered to define the dimension and the shape of the tail. Moreover, in order to analyze the static stability of the bio-inspired empennage an experimental testbed was built to measure the forces and moments for different attitudes.

14.4.1 Design of the Bio-inspired Empennage

Based on the dynamic flight of the birds and the movements of its tail, an empennage with three DOF was designed. The empennage consisted of a flat single surface, which corresponds with the horizontal stabilizer of the PT-40 RC model.

Figure 14.2 presents a 3D view of the mechanism that consists in a set of conical gears, which allows to develop both bank (Φ) and tilt (Θ) movements by the combination of the rotation of two servomotors. The third movement, which corresponds to the pan motion (Ψ), is produced by a pair of conical gears that change the rotation axis in order to produce a rotation along the vertical axis. This arrangement reduces the size of the mechanism and allows that all of the axes of rotation intersect at the same point, and then the complexity of the extended vectors transformation is reduced.

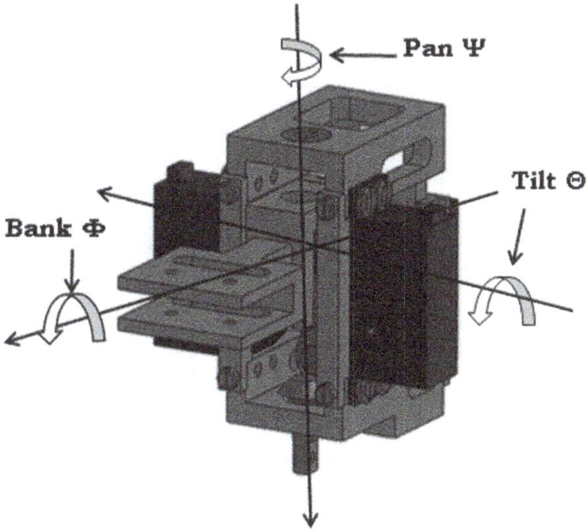

Fig. 14.2 Design of the mechanism for the bio-inspired empennage

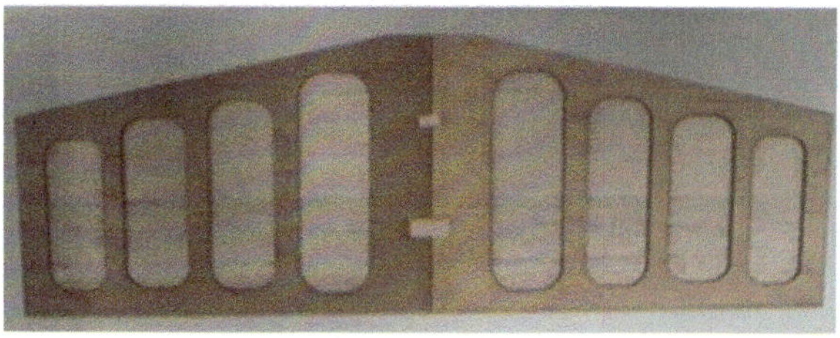

Fig. 14.3 Design of the mechanism for the tail

Figure 14.3 shows the design and manufacture of the tail that was made of two wood plates, and in order to reduce weight some material was removed. Plates were covered with polyester film.

14.4.2 Experimental Setup

In order to compute the static aerodynamic coefficients using a wind tunnel, an experimental setup was built (see Fig. 14.4). The moving tail mechanism is mounted on a column with a circular hollow section to reduce the wind flow perturbations and undesirable aerodynamic forces and moments.

A 6D force sensor was mounted under this arrangement, with the use of steel plates, which were manufactured in accordance with the specifications. In order to avoid the perturbations produced by the boundary layers of the walls, an additional

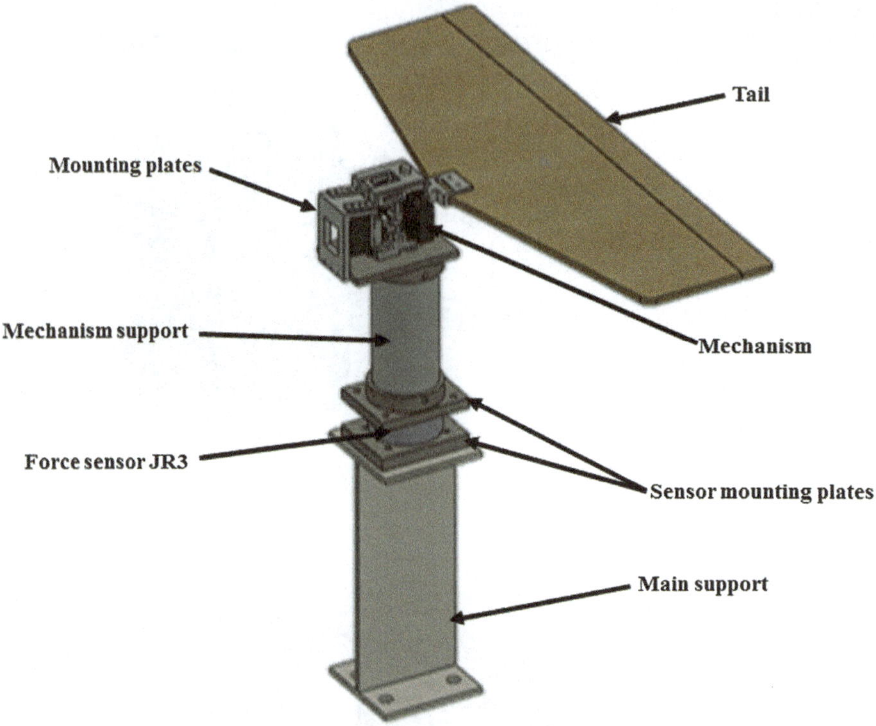

Fig. 14.4 Experimental test bed

I-shaped beam was included to firmly fix and locate the bio-inspired empennage at the center of the wind tunnel test section, as shown in Fig. 14.5. The pitot tube was aligned with the wind flow in order to measure the wind velocity.

Reaction forces and moments were measured with a 6D force sensor JR3 (mod. 67M25A3-140-DH) and a DSP-based receiver PCI card that works at 33 MHz and 32-Bit. In Table 14.1, the most relevant characteristics of the force sensor are presented.

Figure 14.6 presents a block diagram that exemplifies the instrumentation of the system. The attitude of the bio-inspired empennage was controlled by an open source electronic board, which receives the angular set points through the computer serial port from MATLAB. The MATLAB program selects the direction, the angular displacement, and the sequence of the rotations in order to achieve the desired attitude. The wind velocity was measured by a manometer PCE-P01 of PCE Instruments and a pitot tube. The reaction forces and moments and wind velocity were recorded in a text file.

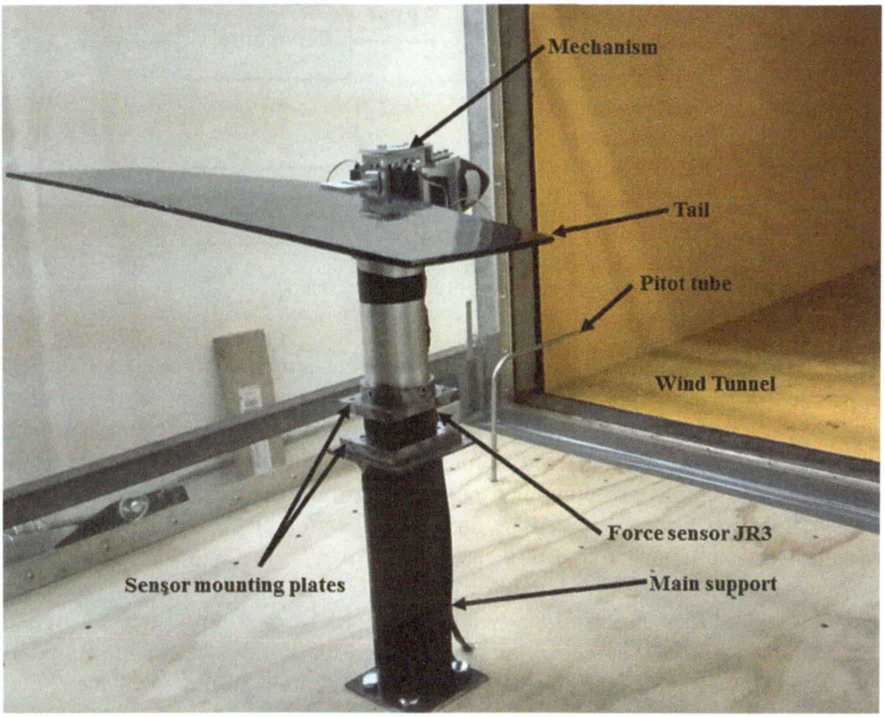

Fig. 14.5 Mounting of the experimental setup in the wind tunnel

Table 14.1 Characteristic of the 6D force sensor

Parameter	
Measurement range of force along *x*-axis	±100 N
Measurement range of force along *y*-axis	±100 N
Measurement range of force along *z*-axis	±200 N
Digital resolution of the force along *x* and *y*-axes	0.025 N
Digital resolution of the force along *z*-axis	0.050 N
Measurement range of moment along *x*-axis	±6.3 Nm
Measurement range of moment along *y*-axis	±6.3 Nm
Measurement range of moment along *z*-axis	±6.3 Nm
Digital resolution of moment along *x*, *y*, and *z*-axes	0.0016 Nm

14.5 Test and Results

The objective of the experiments is to define the relationship between the attitude of the bio-inspired empennage and the aerodynamic forces and moments. The tests consist on measuring the reaction forces and moments produced by the

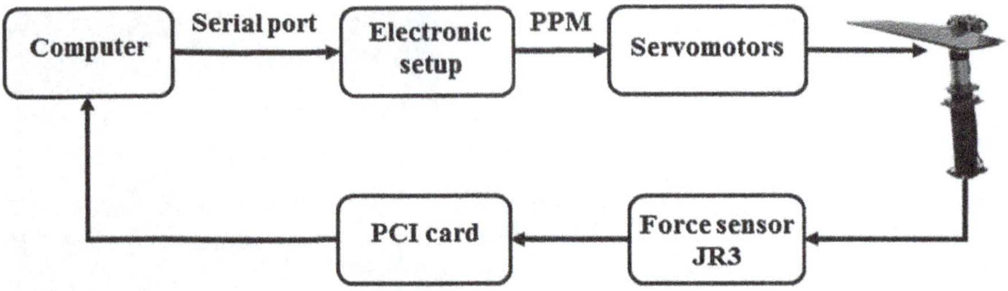

Fig. 14.6 Block diagram of electronic and instrumentation setup

aerodynamic effects for the given range of different attitudes combinations of the empennage. A proper transformation from the center of the sensor, corresponding to the experimental setup, must be performed to produce the appropriate coordinate force measurements at the aerodynamic center of the bio-inspired empennage. These transformations are computed considering the reference frames defined in Fig. 14.7, which were computed by an EMO operator given by Eq. (14.7). The force sensor's reference frame is located at its own center and the aerodynamic reference frame is at the aerodynamic center at 25% of the chord line.

Three tests at the same constant wind velocity of $V_w = 20$ km/h were designed. By combining two movements, which are termed basic movement and secondary movement, the attitude is changed. Table 14.2 presents a summary of the movements, ranges, and increments (Δ_B for basic and Δ_S for secondary movement respectively) for each test. The first test was developed considering the pan motion Ψ as basic movement and the bank Φ motion as secondary movement,

Fig. 14.7 Relative position and attitude of the empennage relative to the sensor reference frame

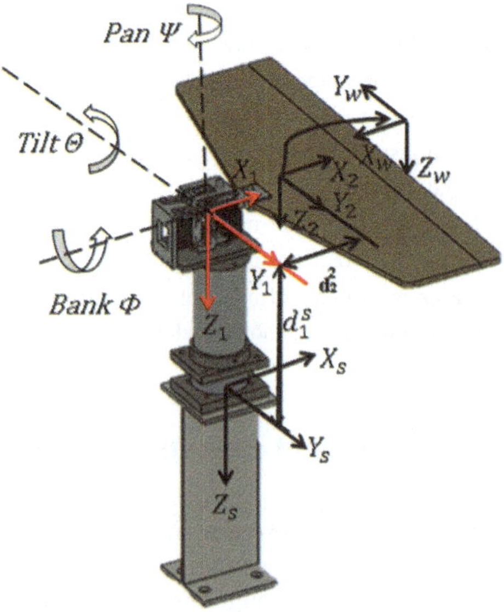

Table 14.2 Ranges of motion for the tests

Test	Basic movement range	Secondary movement range	$\Delta_{\mathbf{B}}$and$\Delta_{\mathbf{S}}$
First test	$-40° \leq \Psi \leq 40°$	$-40° \leq \Phi \leq 40°$	$\Delta\Psi = \Delta\Phi = 8°$
Second test	$-40° \leq \Psi \leq 40°$	$-40° \leq \Theta \leq 40°$	$\Delta\Psi = \Delta\Theta = 8°$
Third test	$-30° \leq \Theta \leq 30°$	$-20° \leq \Phi \leq 20°$	$\Delta\Theta = \Delta\Phi = 5°$

Table 14.3 Sequence of movements for each test

	0	$\Delta_{\mathbf{S}}$	$2\Delta_{\mathbf{S}}$	$\cdots$	**Max**	**Max** $-\Delta_{\mathbf{S}}$	$\cdots$	**Min**	**Min** $+\Delta_{\mathbf{S}}$	$\cdots$	0
0	$\mathbf{F}^R_{1,1}$	$\mathbf{F}^R_{1,2}$	$\mathbf{F}^R_{1,3}$	$\cdots$	$\cdots$	$\cdots$	$\cdots$	$\cdots$	$\cdots$	$\cdots$	$\mathbf{F}^R_{1,j}$
$\Delta_{\mathbf{B}}$	$\mathbf{F}^R_{2,1}$	$\mathbf{F}^R_{2,2}$	$\mathbf{F}^R_{2,3}$	$\cdots$	$\cdots$	$\cdots$	$\cdots$	$\cdots$	$\cdots$	$\cdots$	$\mathbf{F}^R_{2,j}$
$2\Delta_{\mathbf{B}}$	$\mathbf{F}^R_{3,1}$	$\mathbf{F}^R_{3,2}$	$\mathbf{F}^R_{3,3}$	$\cdots$	$\cdots$	$\cdots$	$\cdots$	$\cdots$	$\cdots$	$\cdots$	$\mathbf{F}^R_{3,j}$
$\vdots$	$\vdots$	$\vdots$	$\vdots$	$\ddots$							$\vdots$
Max	$\vdots$	$\vdots$	$\vdots$		$\ddots$						$\vdots$
Max $-\Delta_{\mathbf{B}}$	$\vdots$	$\vdots$	$\vdots$			$\ddots$					$\vdots$
$\vdots$	$\vdots$	$\vdots$	$\vdots$				$\ddots$				$\vdots$
Min	$\vdots$	$\vdots$	$\vdots$					$\ddots$			$\vdots$
Min $+\Delta_{\mathbf{B}}$	$\vdots$	$\vdots$	$\vdots$						$\ddots$		$\vdots$
$\vdots$	$\vdots$	$\vdots$	$\vdots$							$\ddots$	$\vdots$
0	$\vdots$	$\vdots$	$\vdots$								$\mathbf{F}^R_{i,j}$

the increments for both movements were $\Delta_\Psi = \Delta_\Theta = 8°$. The second and third tests combine the pan-tilt ($\Psi - \Theta$) and tilt-bank ($\Theta - \Phi$) motions, respectively. Both movements basic and secondary were either increased or decreased using stepwise angular position.

Table 14.3 presents the sequence and the values for each step, where the first column corresponds to the values for the basic movement and the first row contains the values for the secondary movement, so that for each combination the forces and moments were measured.

The next pseudocode summarizes the procedure that was followed for each test:

```
for i = 1 : 1 : n
        Δ_B(i) ;
        for j = 1 : 1 : m
                Δ_S(j) ;
                k = 1;
                 for t =;0 : Δ_t : 20
                          F^(S)(i,j,k) = read(Sensor JR3) ;
```

```
            k ++
        end
```
$$F_A^{(S)}(i,j) = \text{mean}\left(F^{(S)}(i,j,k)\right)$$
```
    end
end
```

where i and j correspond to the rows (basic movement) and columns (secondary movement) position, respectively. In accordance with the pseudocode, the test consisted in establishing the value of the basic movement when selecting the ith row. For each position of the basic movement, a sweep of all of the values of the secondary movement was made when selecting the jth column. Forces and moments were measured at a sampling period of 0.001 s during 20 s for each combination, and the average of this measurements was recorded at the position (i,j) in Table 14.3. This process defines a complete measurement and to guaranty stable results, the average of six cycles was computed. Figure 14.8 presents the attitude of the bio-inspired empennage for the first (left) and third (right) tests.

The data measured by the sensor were transformed in order to define the aerodynamic forces and moments in the aerodynamic center of the bio-inspired empennage. The corresponding transformation was computed in accordance with (7), which is a function of the attitude of the empennage and the position of the aerodynamic center. The relationship between the attitude of the empennage and the aerodynamic coefficients was defined by a multiple regression analysis, which is based on the least square theory. The analysis of the results showed that the data can be approximated to a high-order polynomial function (five order) where the model includes interaction between the movements.

Each observed value is a function of the independent variables (Φ, Θ, and Ψ) and a set of regression coefficients in accordance with the next expression:

Fig. 14.8 Test two and three. Figure shows the attitude of the stabilizer due to the bank (left) and tilt (right) motions

$$\mathbf{C}_{x_i} = \mathbf{M}_{\mathrm{p}}\mathbf{B}_{\mathrm{x_i}} \tag{14.16}$$

where $\mathbf{M}_p$ is the predictors matrix and expresses the relationship between the independent variables, and it is represented by Eq. (14.17) where n is the nth value of the independent variables. In Eq. (14.16), the term $\mathbf{B}_{x_i}$ represents the regression coefficients vector and it is expressed by Eq. (14.18).

$$\mathbf{M}_p = \begin{pmatrix} 1 & \Phi_1 & \Theta_1 & \Psi_1 & \Phi_1^2 & \Phi_1\Theta_1 & \Phi_1\Psi_1 & \Theta_1^2 & \dots & \Psi_1^5 \\ 1 & \Phi_2 & \Theta_2 & \Psi_2 & \Phi_2^2 & \Phi_2\Theta_2 & \Phi_2\Psi_2 & \Theta_2^2 & \dots & \Psi_2^5 \\ 1 & \Phi_3 & \Theta_3 & \Psi_3 & \Phi_3^2 & \Phi_3\Theta_3 & \Phi_3\Psi_3 & \Theta_3^2 & \dots & \Psi_3^5 \\ \vdots & \vdots & \vdots & \vdots & \vdots & \vdots & \vdots & \vdots & \vdots & \vdots \\ 1 & \Phi_n & \Theta_n & \Psi_n & \Phi_n^2 & \Phi_n\Theta_n & \Phi_n\Psi_n & \Theta_n^2 & \dots & \Psi_n^5 \end{pmatrix} \tag{14.17}$$

$$\mathbf{B}_{x_i} = \begin{pmatrix} b_0(\cdot) \\ b_1(\cdot) \\ b_2(\cdot) \\ \vdots \\ b_n(\cdot) \end{pmatrix} \tag{14.18}$$

Finally, $\mathbf{C}_{x_i}$ is the vector of observed values, which correspond with the experimental data; see Eq. (14.19). Notice that both Eqs. (14.18) and (14.19) are general expressions, then there is a pair of vector $\mathbf{B}_{x_i}$ and $\mathbf{C}_{x_i}$ for each aerodynamic coefficient.

$$\mathbf{C}_{x_i} = \begin{pmatrix} c_0(\cdot) \\ c_1(\cdot) \\ c_2(\cdot) \\ \vdots \\ c_n(\cdot) \end{pmatrix} \tag{14.19}$$

Figure 14.9 shows the results from raw data of the experimental tests.

To facilitate the analysis of the experimental data, a multiple regression was applied according to Eq. (14.16), which leads to the definition of the six coefficient vectors given by Eq. (14.18). Numerical results are presented in Tables 14.3–14.8 of the appendix. It is noteworthy that the regression model has a correlation greater than 95% for all of the cases. Using the obtained coefficients, the tests were reproduced and the results are presented in Fig. 14.10.

Figure 14.11 presents the results for the first test, where one can observe that a combination of both movements produces changes on all of the aerodynamic coefficients. In accordance with the graph of drag coefficient C_D, there are different

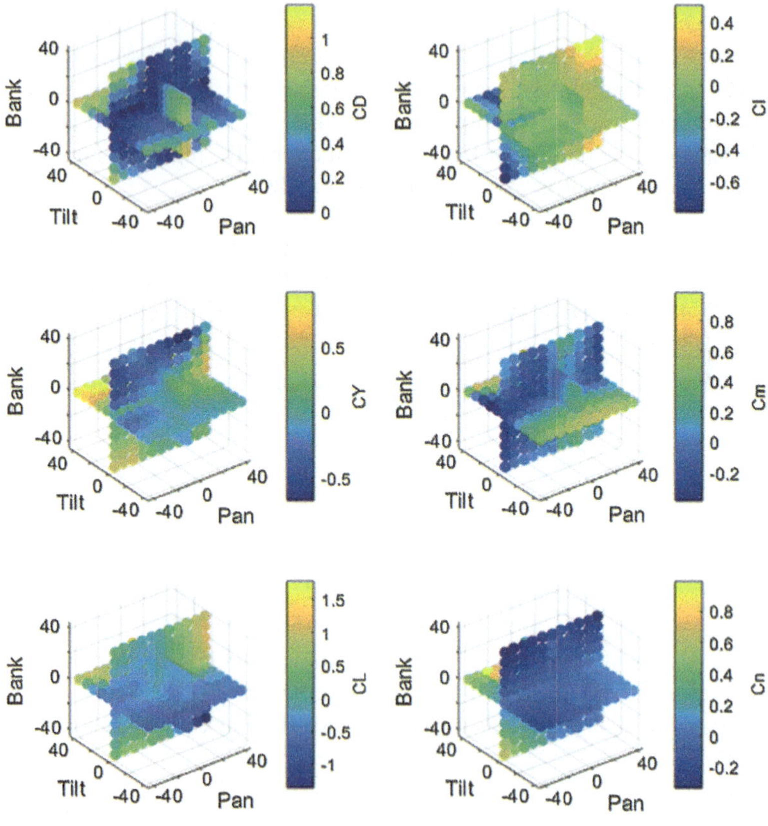

Fig. 14.9 Experimental data

combinations that produce the same output value. Notice that for the side force coefficient C_Y, which affects the lateral–directional stability, its value changes in magnitude and sign as a function of the pan motion. That means that it is possible to change the direction of the lateral force by this movement. On the other hand, the magnitude of the yawing moment coefficient C_n changes due to the pan motion and for a specific value of the bank motion the pan motion changes the sign of the coefficient. It is noteworthy that the combination of the maximum and minimum values of these movements produces changes in both sign and amplitude of rolling moment coefficient C_l. Bank motion changes the magnitude and direction of the lift coefficient C_L, and these effects are enhanced by the pan motion. Pitch moment coefficient C_m exhibits both sign and amplitude changes and there are different combinations that produce the same value. Notice that these combinations do not produce the same values for the rest of the coefficients.

Figure 14.12 shows the results of the second test, where it is possible to observe that the tilt motion affects principally the longitudinal stability (C_D, C_L, C_m) if the

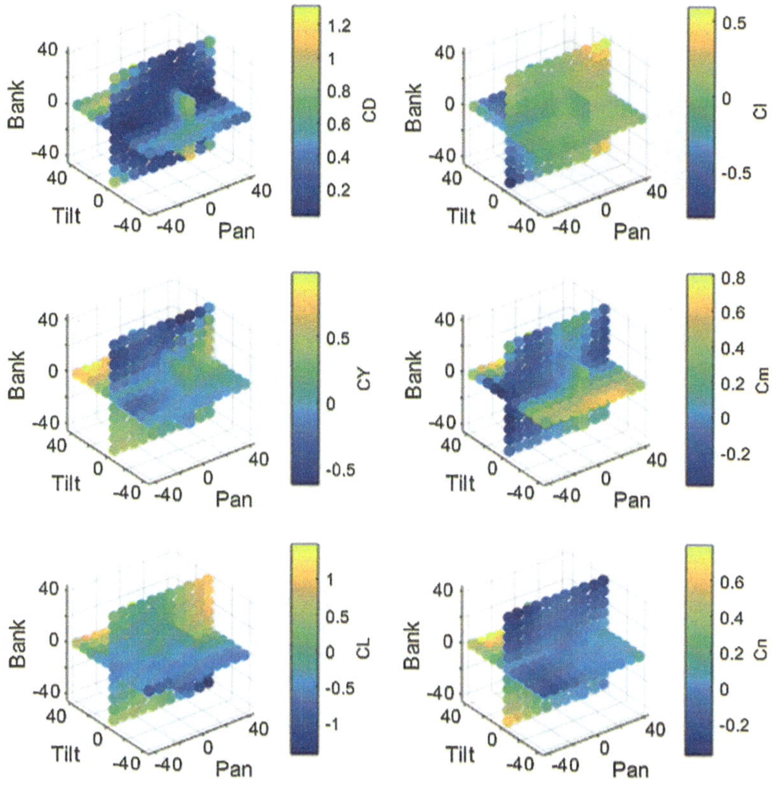

Fig. 14.10 Results from the multiple regression model

pan motion is null; however, when this condition is not fulfilled, the lateral–directional stability is modified. Coefficients (C_D, C_L, C_m) are maximized due to the pan motion in accordance with the graphs, where the maximum values are produced by the combination of the maximum values of both motions, this behavior is common due to the increase of the angle of attack. Notice that the aerodynamic coefficients which are related with the lateral–directional stability (C_Y, C_l, C_n) exhibit slight changes for positive values of both movements; however, for negative values of the movements the coefficients exhibit significant changes respect to the rest of the possible combinations. Results of the third test are depicted in Fig. 14.13, and these suggest that the drag coefficient C_D is symmetrical with respect to the tilt movement. Nevertheless, with respect to the bank motion, it seems that there are no significant changes produced by this movement. Notice that as the rest of the tests there are different combinations which produce the same drag value.

Lift coefficient C_L exhibits significant changes due to Θ, this behavior is common because of the dependence of the lift with the angle of attack. Furthermore, for the lift coefficient, the bank motion Φ produces minimum changes. However,

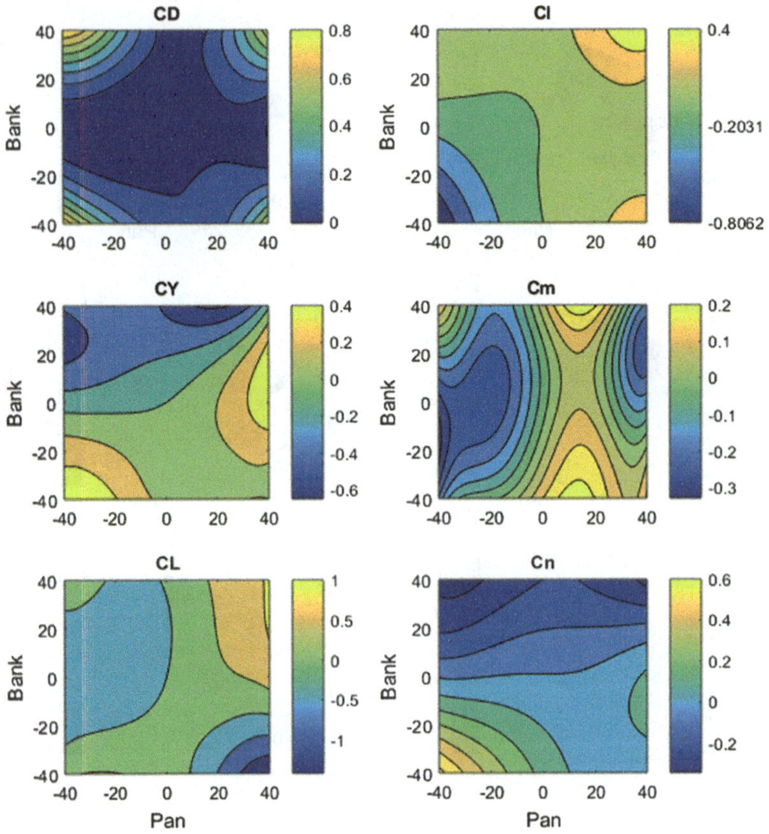

Fig. 14.11 Results pan (Ψ)-bank (Φ)

the main contribution of Φ is seen in the lateral coefficient C_Y, where the direction of the lateral force can be changed by the bank position of the empennage.

Lateral coefficient exhibits more changes for positive values of both Θ and Φ. Graph of the roll moment coefficient shows that C_l varies its value mainly as a response to positive values of the tilt motion Θ. In the same graph, it is observed that there is no symmetry with respect to any of the movements. Nevertheless, the effects are more significant for positive values of Θ. Pitch moment coefficient C_m is symmetrical respect to Θ; however, the coefficient exhibits more changes when combining positive values of Θ with the range of values of Φ. Notice that the bank motions seem to produce no main effects on C_m; furthermore, it is important to comment that there are more than two combinations that produce the same value. The corresponding results to the yaw moment coefficient C_n suggest that the significant changes are produced by positive values of Θ, in the same sense the bank movement, for any value of Θ, affects the magnitude of C_n.

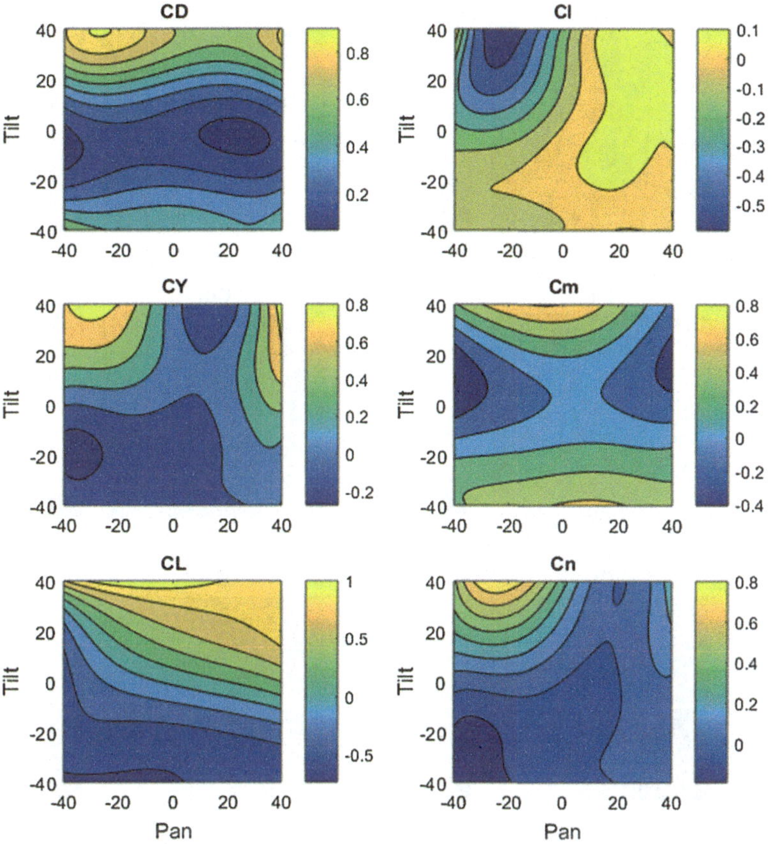

Fig. 14.12 Results pan (Ψ)-tilt (Θ)

14.6 Conclusions

In this work, a 3-DOF empennage with the capability of aiding in the flight control of aerial vehicles was introduced. The proposed design was bio-inspired on the way that some birds move their tail to control their flight. The document focused on the experimental aerodynamic analysis of the system to determine the aerodynamic coefficients and their relationship with the attitude change of the empennage. The spatial transformations of the six-axis force sensor from inertial to a local frame located at the aerodynamic center of the empennage are performed using extended vectors such that torque and force are treated in a single expression. A number of experiments were conducted at low velocity in a wind tunnel. Experimental results were fitted using a multiple regression method base on a least square approximation. Four-dimensional plots and contours were used to facilitate the visualization of the connection between the attitude change and the variation on the aerodynamic

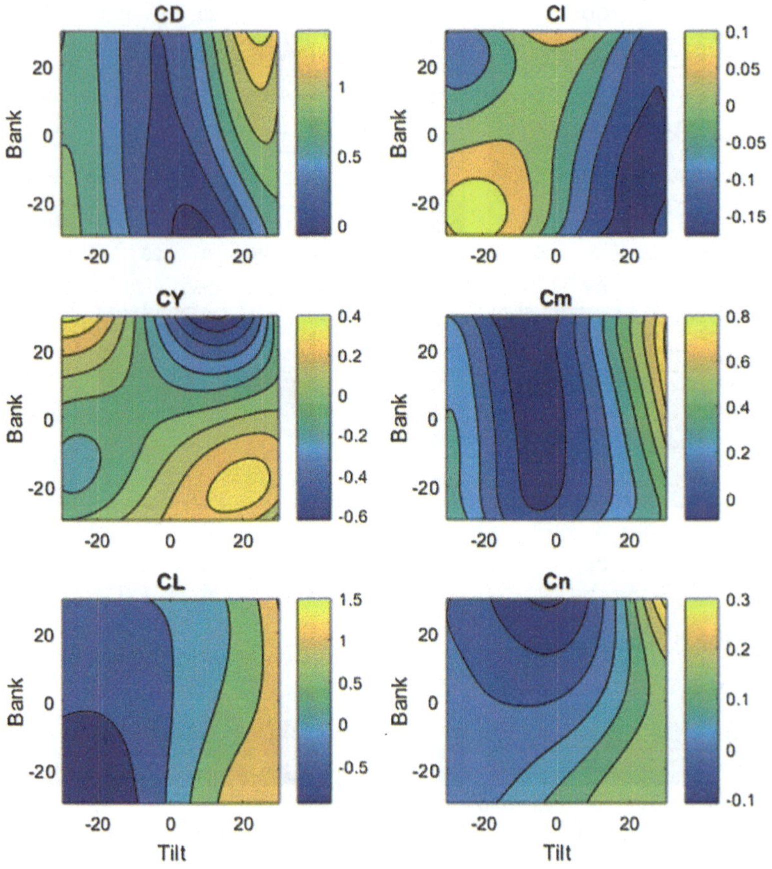

Fig. 14.13 Results tilt (Θ)-bank (Φ)

coefficients. The obtained coefficients have potential application on the development of control algorithms and will be the next assessment of the proposed bio-inspired empennage.

Appendix

See Tables 14.4, 14.5, 14.6, 14.7, 14.8 and 14.9.

Table 14.4 Regression coefficients for C_D

Coef.	Valor	Coef.	Valor	Coef.	Valor	Coef.	Valor	Coef.	Valor
b_0	1.35E−01	b_{11}	−1.72E−06	b_{22}	3.75E−08	b_{33}	3.40E−08	b_{44}	−6.45E−11
b_1	4.23E−04	b_{12}	3.66E−06	b_{23}	−2.24E−08	b_{34}	3.93E−08	b_{45}	−5.49E−10
b_2	8.06E−03	b_{13}	−3.43E−06	b_{24}	0.00E + 00	b_{35}	4.45E−10	b_{46}	0.00E + 00
b_3	7.25E−04	b_{14}	0.00E + 00	b_{25}	2.91E−07	b_{36}	3.53E−10	b_{47}	0.00E + 00
b_4	−1.63E−04	b_{15}	−2.17E−06	b_{26}	3.80E−08	b_{37}	−1.74E−09	b_{48}	0.00E + 00
b_5	−1.48E−04	b_{16}	−3.98E−07	b_{27}	0.00E + 00	b_{38}	2.68E−09	b_{49}	−1.29E−09
b_6	−4.48E−05	b_{17}	3.72E−05	b_{28}	0.00E + 00	b_{39}	0.00E + 00	b_{50}	−1.75E−09
b_7	6.78E−04	b_{18}	−3.64E−05	b_{29}	−4.78E−09	b_{40}	1.74E−09	b_{51}	−4.27E−08
b_8	1.19E−03	b_{19}	−1.20E−06	b_{30}	−5.48E−07	b_{41}	1.10E−09	b_{52}	9.04E−09
b_9	−7.60E−05	b_{20}	7.14E−08	b_{31}	−5.29E−07	b_{42}	0.00E + 00	b_{53}	3.04E−09
b_{10}	−1.04E−06	b_{21}	7.41E−08	b_{32}	8.83E−07	b_{43}	0.00E + 00	b_{54}	5.73E−08
								b_{55}	2.56E−10

Table 14.5 Regression coefficients for C_Y

Coef.	Valor	Coef.	Valor	Coef.	Valor	Coef.	Valor	Coef.	Valor
b_0	5.86E−03	b_{11}	1.09E−05	b_{22}	1.04E−07	b_{33}	2.50E−07	b_{44}	−5.63E−10
b_1	8.51E−04	b_{12}	−6.11E−06	b_{23}	6.38E−08	b_{34}	4.06E−08	b_{45}	1.62E−09
b_2	6.43E−03	b_{13}	−1.38E−05	b_{24}	0.00E+00	b_{35}	−1.88E−09	b_{46}	0.00E+00
b_3	−8.21E−03	b_{14}	0.00E+00	b_{25}	−3.04E−08	b_{36}	−8.69E−10	b_{47}	0.00E+00
b_4	1.95E−04	b_{15}	−4.33E−06	b_{26}	6.58E−08	b_{37}	4.26E−09	b_{48}	0.00E+00
b_5	−4.78E−04	b_{16}	−5.82E−06	b_{27}	0.00E+00	b_{38}	5.58E−09	b_{49}	−2.89E−09
b_6	2.22E−04	b_{17}	2.53E−07	b_{28}	0.00E+00	b_{39}	0.00E+00	b_{50}	2.13E−09
b_7	−9.18E−04	b_{18}	−2.26E−05	b_{29}	−1.88E−07	b_{40}	3.33E−09	b_{51}	9.40E−09
b_8	6.27E−05	b_{19}	2.79E−06	b_{30}	−2.83E−08	b_{41}	−2.76E−09	b_{52}	1.88E−08
b_9	−1.48E−04	b_{20}	−2.90E−08	b_{31}	6.26E−07	b_{42}	0.00E+00	b_{53}	1.05E−09
b_{10}	5.88E−06	b_{21}	2.52E−07	b_{32}	3.08E−07	b_{43}	0.00E+00	b_{54}	3.05E−09
								b_{55}	−1.06E−09

Table 14.6 Regression coefficients for C_L

Coef.	Valor	Coef.	Valor	Coef.	Valor	Coef.	Valor	Coef.	Valor
b_0	−8.17E−03	b_{11}	−3.90E−06	b_{22}	−9.97E−08	b_{33}	1.45E−07	b_{44}	−1.03E−09
b_1	1.70E−02	b_{12}	1.90E−05	b_{23}	−8.54E−08	b_{34}	−5.73E−09	b_{45}	6.11E−09
b_2	3.02E−02	b_{13}	−2.06E−05	b_{24}	0.00E+00	b_{35}	2.22E−09	b_{46}	0.00E+00
b_3	−2.90E−03	b_{14}	0.00E+00	b_{25}	−3.04E−08	b_{36}	−1.95E−09	b_{47}	0.00E+00
b_4	5.44E−05	b_{15}	−1.57E−05	b_{26}	−2.66E−07	b_{37}	−3.95E−09	b_{48}	0.00E+00
b_5	3.05E−04	b_{16}	−1.51E−05	b_{27}	0.00E+00	b_{38}	2.62E−09	b_{49}	2.16E−09
b_6	5.38E−04	b_{17}	8.32E−06	b_{28}	0.00E+00	b_{39}	0.00E+00	b_{50}	5.26E−09
b_7	−1.08E−03	b_{18}	8.09E−05	b_{29}	2.22E−08	b_{40}	−5.45E−10	b_{51}	1.88E−09
b_8	1.92E−04	b_{19}	−8.71E−07	b_{30}	−2.49E−08	b_{41}	4.12E−09	b_{52}	2.12E−09
b_9	9.85E−05	b_{20}	−2.15E−08	b_{31}	9.55E−07	b_{42}	0.00E+00	b_{53}	2.44E−09
b_{10}	−5.46E−06	b_{21}	7.11E−08	b_{32}	9.48E−07	b_{43}	0.00E+00	b_{54}	−1.45E−07
								b_{55}	5.14E−10

Table 14.7 Regression coefficients for C_l

Coef.	Valor	Coef.	Valor	Coef.	Valor	Coef.	Valor	Coef.	Valor
b_0	−6.18E−03	b_{11}	−1.14E−06	b_{22}	−6.87E−08	b_{33}	−8.27E−08	b_{44}	−5.97E−10
b_1	7.15E−03	b_{12}	5.96E−06	b_{23}	5.41E−08	b_{34}	−2.79E−08	b_{45}	−7.25E−11
b_2	−7.35E−03	b_{13}	5.04E−06	b_{24}	0.00E+00	b_{35}	−2.61E−10	b_{46}	0.00E+00
b_3	7.70E−04	b_{14}	0.00E+00	b_{25}	5.06E−08	b_{36}	1.92E−09	b_{47}	0.00E+00
b_4	−3.47E−05	b_{15}	−2.82E−06	b_{26}	−8.65E−08	b_{37}	−7.40E−10	b_{48}	0.00E+00
b_5	3.91E−04	b_{16}	6.22E−06	b_{27}	0.00E+00	b_{38}	−4.34E−09	b_{49}	2.15E−09
b_6	−3.30E−05	b_{17}	−7.91E−06	b_{28}	0.00E+00	b_{39}	0.00E+00	b_{50}	−2.14E−09
b_7	2.46E−04	b_{18}	8.38E−06	b_{29}	1.31E−08	b_{40}	2.12E−09	b_{51}	5.29E−09
b_8	−9.47E−05	b_{19}	5.95E−07	b_{30}	2.63E−08	b_{41}	1.29E−10	b_{52}	−2.39E−09
b_9	6.39E−05	b_{20}	−2.90E−09	b_{31}	−1.57E−07	b_{42}	0.00E+00	b_{53}	−9.50E−10
b_{10}	−1.66E−06	b_{21}	−1.68E−07	b_{32}	−1.14E−07	b_{43}	0.00E+00	b_{54}	−1.19E−08
								b_{55}	−2.41E−10

Table 14.8 Regression coefficients for C_m

Coef.	Valor	Coef.	Valor	Coef.	Valor	Coef.	Valor	Coef.	Valor
b_0	−4.78E−03	b_{11}	−1.53E−05	b_{22}	−1.22E−07	b_{33}	−3.73E−08	b_{44}	6.03E−10
b_1	6.19E−03	b_{12}	−3.25E−06	b_{23}	−7.43E−08	b_{34}	2.02E−08	b_{45}	6.59E−10
b_2	1.96E−03	b_{13}	−4.63E−06	b_{24}	0.00E+00	b_{35}	3.89E−09	b_{46}	0.00E+00
b_3	−2.58E−04	b_{14}	0.00E+00	b_{25}	7.07E−08	b_{36}	2.23E−09	b_{47}	0.00E+00
b_4	−1.89E−04	b_{15}	6.44E−06	b_{26}	−5.68E−08	b_{37}	1.57E−09	b_{48}	0.00E+00
b_5	−3.38E−05	b_{16}	1.02E−06	b_{27}	0.00E+00	b_{38}	9.52E−10	b_{49}	−1.52E−09
b_6	2.55E−06	b_{17}	5.24E−06	b_{28}	0.00E+00	b_{39}	0.00E+00	b_{50}	−4.26E−10
b_7	1.91E−04	b_{18}	3.80E−05	b_{29}	5.65E−08	b_{40}	−3.51E−09	b_{51}	−9.03E−10
b_8	7.13E−04	b_{19}	8.05E−07	b_{30}	−1.78E−07	b_{41}	5.49E−09	b_{52}	−5.91E−09
b_9	1.37E−05	b_{20}	3.07E−08	b_{31}	−1.69E−08	b_{42}	0.00E+00	b_{53}	−1.40E−09
b_{10}	−8.97E−06	b_{21}	2.53E−08	b_{32}	−5.28E−08	b_{43}	0.00E+00	b_{54}	−6.50E−08
								b_{55}	−4.11E−10

Table 14.9 Regression coefficients for C_n

Coef.	Valor	Coef.	Valor	Coef.	Valor	Coef.	Valor	Coef.	Valor
b_0	3.63E−03	b_{11}	6.88E−06	b_{22}	−1.75E−08	b_{33}	4.89E−08	b_{44}	5.86E−10
b_1	−1.38E−03	b_{12}	−1.03E−05	b_{23}	−9.15E−09	b_{34}	1.11E−08	b_{45}	1.55E−09
b_2	4.01E−03	b_{13}	−9.27E−06	b_{24}	0.00E+00	b_{35}	−1.36E−09	b_{46}	0.00E+00
b_3	−3.09E−03	b_{14}	0.00E+00	b_{25}	−3.07E−08	b_{36}	−2.56E−09	b_{47}	0.00E+00
b_4	4.25E−05	b_{15}	−2.10E−06	b_{26}	8.03E−09	b_{37}	3.73E−09	b_{48}	0.00E+00
b_5	−3.64E−04	b_{16}	−1.82E−06	b_{27}	0.00E+00	b_{38}	4.15E−09	b_{49}	3.85E−10
b_6	1.28E−04	b_{17}	1.62E−06	b_{28}	0.00E+00	b_{39}	0.00E+00	b_{50}	1.16E−09
b_7	−7.92E−05	b_{18}	−6.15E−06	b_{29}	−2.07E−09	b_{40}	−1.05E−09	b_{51}	3.13E−09
b_8	1.47E−04	b_{19}	2.04E−06	b_{30}	−3.66E−08	b_{41}	−1.42E−09	b_{52}	2.09E−09
b_9	−2.31E−05	b_{20}	−3.18E−09	b_{31}	1.05E−07	b_{42}	0.00E+00	b_{53}	−1.72E−09
b_{10}	3.69E−06	b_{21}	1.74E−07	b_{32}	4.85E−08	b_{43}	0.00E+00	b_{54}	1.28E−08
								b_{55}	−1.05E−09

References

Alexander, D.-E. (2002). *Nature's flyers: Birds, insects, and the biomechanics of flight*. Marryland: The Johns Hopkins University Press.

Biewener, A.-A. (2003). *Animal locomotion (Oxford animal biology series)*. Oxford: Oxford University Press.

Cook, M.-V. (2007). *Flight dynamics principles* (2nd ed.). Amsterdam: Elsevier.

Featherstone, R. (2010a). A beginner's guide to 6-D vectors (part 1) what they are, how they work, and how to use them. *IEEE Robotics and Automation Magazine, 17*(3), 83–94.

Featherstone, R. (2010b). A beginner's guide to 6-D vectors (part 2) from equations to software. *IEEE Robotics and Automation Magazine, 17*(4), 88–99.

Gatesy, S.-M., & Dial, K. P. (March de 1993). Tail muscle patterns in walking and flying pigeons Columa Livia. *The Journal of Experimental Biology, 176*, 55–76.

Gottfried, S. (2007). Tail effects on yaw stability in birds. *Journal of Theoretical Biology, 249*(3), 464–472.

Han, J.-H., Lee, J.-Y., & Kim, D.-K. (2008). Ornithopter modeling for flight simulation. *International Conference on Control, Automation and Systems, in COEX, Seoul, Korea*.

Higgs, T. J. (2005). *Modeling, stability, and control of a rotatable tail on a micro air vehicle*. Department of Aeronautics and Astronautics. Air force institute of technology. Air university.

Hoey, R. G. (1992). Research on the stability and control of soaring birds. In *28th National Heat Transfer Conference, AIAA*, 393–401.

Kirmse, W. (1998). Morphometric features characterizing flight properties of Palearctic eagles. In R. D. Chancellor, B.-U. Meyburg & J. J. Ferrero (Eds.), *Holarctic birds of prey ADENEX-WWGBP* (pp. 339–348).

Leveron, T. A. (2005). *Characterization of a rotary flat tail as a spoiler and parametric analysis of improving directional stability in a portable UAV*. Department of Aeronautics and Astronautics. Air force institute of technology. Air university.

Muller, B., Clothier, R., Watkins, S., & Fisher, A. (2015). Design of bio-inspired autonomous aircraft for bird management. In *Proceedings of the 16th Australian International Aerospace Congress (AIAC16)*, pp. 370–377.

Noth, A. (2008). *Design of solar powered airplanes for continuous flight (Ph.D. thesis)*. Swiss Federal Institute of technology Zurich.

Olguín-Díaz, E., & García-Terán, M. A. (2014). Aerodynamic sectional modeling with the use of extended vectors. In *Unmanned Aircraft Systems (ICUAS), 2014 International Conference on*, (pp. 459–469).

Paranjape, A., Kim, J., Gandhi, N., & Chung, S.-J. (2011a). Experimental demonstration of perching by an articulated wing MAV. In *AIAA Guidance, Navigation, and Control Conference*, August.

Paranjape, A.-A., Chung, S.-J., & Selig, M. (2011b). Flight mechanics of a tailless articulated wing aircraft. *Bioinspiration & Biomimetic, 6*(2), 1–20.

Paranjape, A. A., Chung, S.-J., Hilton, H. H., & Chakravarthy, A. (2012a). Dynamics and performance of tailless micro aerial vehicle with flexible articulated wings. *AIAA Journal, 50*(5), 1177–1188.

Paranjape, A.-A., Kim, J., & Chung, S.-J. (2012b). Closed-loop perching and spatial guidance laws for bio-inspired articulated wing MAV. In *AIAA Guidance, Navigation, and Control Conference*, 21.

Pennycuick, C.-J. (2008). *Modelling the flying bird (theoretical ecology series)*. Amsterdam: Elsevier.

Rivera-Parga, J., Reeder, M. F., Leveron, T., & Blackburn, K. (November–December de 2007). Experimental study of a micro air vehicle with a rotatable tail. *Journal of Aircraft, AIAA, 44*(6), 1761–1768.

Roscam, J. (2003). *Airplane flight dynamic and automatic flight control* (6th ed.). DAR corporation.

Shyy, W., Yongsheng, L., Tang, J., Viieru, D., & Liu, H. (2008). *Aerodynamics of low reynolds number flyers*. Cambridge: Cambridge Aerospace Series.
Siciliano, B., Sciavicco, L., Villani, L., & Oriolo, G. (2009). *Robotics, modelling, planning and control* (2nd ed.). Berlin: Springer.
Stengel, R. (2004). *Flight dynamics*. Princeton University press.
Stevens, B., & Lewis, F. L. (2003). *Aircraft control and simulation* (2nd ed.). Hoboken: Wiley.
Su, J.-Y., Ting, S.-C., Chang, Y.-H., & Yang, J.-T. (2012). A passerine spreads its tail to facilitate a rapid recovery of its body posture during hovering. *Journal of the Royal Society, 9*(72), 1674–1684.
Thomas, A.-L. (1993). On the aerodynamics of birds' tails. *Philosophical Transactions of the Royal Society B, 340*(1294), 361–380.
Tobalske, B. (2007). Biomechanics of bird flight. *The Company of Biologists Ltd, 210*(18), 3135–3146.
Tucker, V.-A. (1992). Pitching equilibrium, wing span and tail span in a gliding Harris Hawk, Parabuteo Unicinctus. *The Journal of Experimental Biology, 165*, 21–41.
Valasek, J. (2011). *Morphing aerospace vehicles and structures* (Primera Edición ed.). Aerospace Engineering Department Texas A&M University USA. Hoboken: Wiley.
Videler, J.-J. (2005). *Avian flight (Oxford ornithology series)*. Oxford: Oxford University Press.

Chapter 15
Consensus Strategy Applied to Differential Mobile Robots with Regulation Control and Trajectory Tracking

Flabio Mirelez-Delgado

Abstract In this article, the problem of performing different tasks with a group of mobile robots is addressed. To cope with issues like regulation to a point or trajectory tracking, a consensus scheme is considered. Three topologies were tested in simulation. The first goal was to make consensus in the group of robots, after the consensus point was relocated to achieve a regulation control. The last objective was to follow a desired trajectory moving the consensus point along the predefined path. The proposal was validated through experimental test with a group of three differential mobile robots.

Keywords Consensus strategy · Mobile robot · Trajectory tracking

15.1 Introduction

Nowadays, the abundance of resources in autonomous vehicles allows to increase the effectiveness of tasks through cooperative work. Greater effectiveness and operational capacity can be achieved by using autonomous coordinated vehicles. The use of multiple coordinated robots has several advantages over single robot systems. The most important are: The complexity of the task to be carried out may be greater. The task can be distributed to the elements of the group in an equitable manner. Building several simple robots is usually less expensive than building a large and complex one. Multiple robots can solve problems faster by solving tasks in parallel. The introduction of multiple robots adds robustness to the system through redundancy (Chung and Slotine 2009).

F. Mirelez-Delgado (✉)
Centro de Investigación y de Estudios Avanzados del Instituto Politécnico Nacional Unidad Saltillo, Av. Industria Metalúrgica N° 1062, Parque Industrial Saltillo-Ramos Arizpe, C.P, 25900 Ramos Arizpe, Coahuila, Mexico
e-mail: flabiodariomirelezdelgado@gmail.com

O. O. Vergara Villegas et al. (eds.), *Advanced Topics on Computer Vision, Control and Robotics in Mechatronics*,
https://doi.org/10.1007/978-3-319-77770-2_15

Historically, some of the earliest work in multiple robots grappled with the idea of swarming robots to make formations (Desai et al. 2001; Yamaguchi et al. 2001; Sun and Mills 2002; Takahashi et al. 2004; Sun and Mills 2007; Antonelli et al. 2009). Regulation to a fixed point is another research topic widely studied in mobile robotics (Huijberts et al. 2000) as the trajectory tracking, with a single robot (Nijmeijer and Rodríguez-Angeles 2004) or with a swarm (Siméon et al. 2002). Interest in this area is due to the ability of biological societies to complete tasks together faster than individually. One of the initial problems in the control of cooperative robots comes from the need to share information. Sharing information is a necessary condition for cooperation. For example, the relative position of the robots among themselves, the speed of each vehicle, etc. The exchange of information becomes a crucial part of the problem.

The structure of this chapter is as follows: Sect. 15.2 is related to the main element in the group of robots, a differential mobile robot. In this section the kinematic model is explained. Section 15.3 is about consensus strategy used in this paper and the three different topologies. The control algorithms used to perform consensus, regulation, and tracking are explained in Sect. 15.4. Section 15.5 presents the simulation results meanwhile Sect. 15.6 shows the experimental results. Finally, Sect. 15.7 provides a conclusion for this work.

15.2 Differential Mobile Robot

Mobile robotic platforms are increasingly common at the industry and as service robots. The most common are wheel robots with differential control (DMR). The tasks in a general way for this class of mobile robots are:

- Movements from Point to Point: The robot is given a desired configuration and it must reach it from an initial position.
- Trajectory Tracking: A reference point in the robot must follow a certain desired trajectory in a Cartesian plane starting from a certain initial position.

Be $q \in Q$ the n-vector of generalized coordinates for an DMR. The simplest model is that of the unicycle. It means a single tire rolling on a plane. The generalized coordinates are $q = (x, y, \theta) \in R^2 \times SO^1(n = 3)$. The non-holonomic restriction which means that the tire cannot move laterally is given by:

$$A(q)\dot{q} = \dot{x}\,\sin\theta - \dot{y}\,\cos\theta = 0 \tag{15.1}$$

Kinematic model for a DMR, based on Fig. 15.1, is given by:

$$\begin{pmatrix} \dot{x} \\ \dot{y} \\ \dot{\theta} \end{pmatrix} = g_1(q)v + g_2(q)\omega = \begin{pmatrix} \cos\theta \\ \sin\theta \\ 0 \end{pmatrix} v + \begin{pmatrix} 0 \\ 0 \\ 1 \end{pmatrix} \omega \tag{15.2}$$

where v and ω are the linear and angular velocities, respectively.

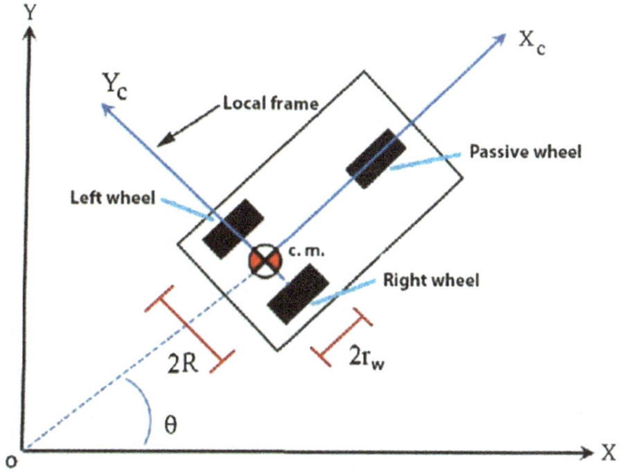

Fig. 15.1 Differential mobile robot on a plane

15.3 Consensus Algorithm

When multiple vehicles agree on the value of a variable of interest, it is said that the robots reached consensus. To reach consensus, there must be a variable of interest which is being shared by all the robots involved. Examples include a representation of the center of the figure of the formation, time of arrival at the desired point, the direction of the movement, the size of the perimeter being monitored, among others. By necessity the consensus is designed to be distributed, assuming only neighboring neighbor interaction between the robots. The objective is to design an updating law so that the status of each value of each vehicle converges to a common point. If a n number of vehicles in the group are assumed, the topology of the communication can be represented through a direct graph.

$$G_n \triangleq (v_n, \xi_n) \tag{15.3}$$

where $v_n = 1, 2, \ldots, n$ is the set of nodes, and $\xi_n \leq v_n \times v_n$ is the set of corners. The most common algorithm of continuous dynamic consensus is:

$$\dot{x}_i(t) = -\sum_{j=1}^{n} a_{ij}(t)\left(x_i(t) - x_{ij}\right), i = 1, \ldots, n \tag{15.4}$$

where a_{ij} is the input (*ij*) of the adjacent matrix $a_n \in R^{n \times n}$ associated with G_n at time *t*. x_i is the information state of the vehicle '*i*'. If $a_{ij} = 0$, the vehicle *i* does not receive information from *j*. A consequence of Eq. (15.4) is that $x_i(t)$ is taken to the information of its neighbors.

Equation (15.4) can be re-written as:

$$\dot{x}_i(t) = L_n(t)x(t) \tag{15.5}$$

where $x = [x_1, \ldots, x_n]^{\mathrm{T}}$ is the state. $L_n(t) = [l_{ij}] \in R^{n\times n}$ is the non-symmetric Laplacian matrix associated with G_n. The consensus is reached by the vehicle group if $\forall x_i(0)$, $\forall v_{i,j} = 1, \ldots, n$, $[x_i(t) - x_j(t)] \to 0, t \to \infty$.

15.3.1 Communication Topologies

The topology of communication is the name given to the configuration or the way in which the robot members of the team communicate or exchange information. For this project, various topologies seen in Ren and Beard (2008) were used.

For the topology presented in Fig. 15.2a, we have the following system.

$$\begin{aligned} \dot{x}_1 &= a_{12}(x_1 - x_2) \\ \dot{x}_2 &= a_{23}(x_2 - x_3) \\ \dot{x}_3 &= 0 \end{aligned} \tag{15.6}$$

The Laplacian matrix is given by:

$$L = \begin{bmatrix} 1 & -1 & 0 \\ 0 & 1.5 & 1.5 \\ 0 & 0 & 0 \end{bmatrix} \tag{15.7}$$

The vector $\upsilon(t)$ is obtained by SVD:

$$\upsilon(t) = \begin{bmatrix} 0 \\ 0 \\ 1 \end{bmatrix} \tag{15.8}$$

For the topology presented in Fig. 15.2b, we have the following system.

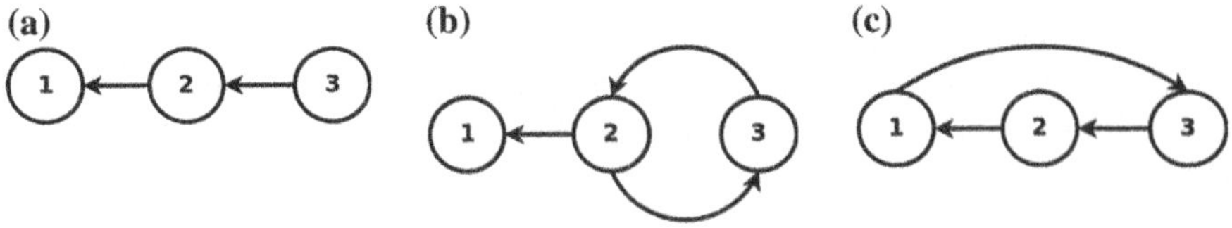

Fig. 15.2 Communication topologies

$$\begin{aligned}\dot{x}_1 &= a_{12}(x_1 - x_2)\\ \dot{x}_2 &= a_{23}(x_2 - x_3)\\ \dot{x}_3 &= a_{32}(x_3 - x_2)\end{aligned} \tag{15.9}$$

The Laplacian matrix is given by:

$$L = \begin{bmatrix} 1 & -1 & 0 \\ 0 & 1.5 & 1.5 \\ 0 & -2 & 2 \end{bmatrix} \tag{15.10}$$

The vector $\upsilon(t)$ is obtained by SVD:

$$\upsilon(t) = \begin{bmatrix} 0 \\ 0.5714 \\ 0.4286 \end{bmatrix} \tag{15.11}$$

For the topology presented in Fig. 15.2c, we have the following system.

$$\begin{aligned}\dot{x}_1 &= a_{12}(x_1 - x_2)\\ \dot{x}_2 &= a_{23}(x_2 - x_3)\\ \dot{x}_3 &= a_{31}(x_3 - x_1)\end{aligned} \tag{15.12}$$

The Laplacian matrix is given by:

$$L = \begin{bmatrix} 1 & -1 & 0 \\ 0 & 1.5 & 1.5 \\ -2 & 0 & 2 \end{bmatrix} \tag{15.13}$$

The vector $\upsilon(t)$ is obtained by SVD:

$$\upsilon(t) = \begin{bmatrix} 0.4615 \\ 0.3077 \\ 0.2308 \end{bmatrix} \tag{15.14}$$

15.4 Control Algorithms

15.4.1 Consensus

As previously mentioned, the consensus process is achieved when all the vehicles agree and reach a variable of interest. Based on the topologies shown in the previous figures, the necessary control is shown so that each topology reaches consensus.

15.4.1.1 Topology 1

The Laplacian matrix for topology 1 is constructed according to the group connections as shown in Eq. (15.7).

$$L = \begin{bmatrix} 1 & -1 & 0 \\ 0 & 1.5 & 1.5 \\ 0 & 0 & 0 \end{bmatrix}$$

The control law needed to achieve consensus in the group is given by:

$$\begin{aligned} \upsilon_1 &= -((-(x_2 - x_1))\cos(\theta_1) + (-(y_2 - y_1))\sin(\theta_1)) \\ \omega_1 &= -(\theta_1 - \arctan 2((y_2 - y_1), (x_2 - x_1))) \\ \upsilon_2 &= -1.5((-(x_3 - x_2))\cos(\theta_2) + (-(y_3 - y_2))\sin(\theta_2)) \\ \omega_1 &= -1.5(\theta_2 - \arctan 2((y_3 - y_2), (x_3 - x_2))) \\ \upsilon_3 &= 0 \\ \omega_3 &= 0 \end{aligned} \tag{15.15}$$

15.4.1.2 Topology 2

For the second topology, we have that the Laplacian matrix is as Eq. (15.10):

$$L = \begin{bmatrix} 1 & -1 & 0 \\ 0 & 1.5 & -1.5 \\ 0 & -2 & 2 \end{bmatrix}$$

That means the control law needed to achieve consensus in the group is given by:

$$\begin{aligned} \upsilon_1 &= -((-(x_2 - x_1))\cos(\theta_1) + (-(y_2 - y_1))\sin(\theta_1)) \\ \omega_1 &= -(\theta_1 - \arctan 2((y_2 - y_1), (x_2 - x_1))) \\ \upsilon_2 &= -1.5((-(x_3 - x_2))\cos(\theta_2) + (-(y_3 - y_2))\sin(\theta_2)) \\ \omega_1 &= -1.5(\theta_2 - \arctan 2((y_3 - y_2), (x_3 - x_2))) \\ \upsilon_3 &= -2((-(x_2 - x_3))\cos(\theta_2) + (-(y_2 - y_3))\sin(\theta_2)) \\ \omega_3 &= -2(\theta_3 - \arctan 2((y_2 - y_3), (x_2 - x_3))) \end{aligned} \tag{15.16}$$

15.4.1.3 Topology 3

Last, the Laplacian matrix for third topology is as depicted in Eq. (15.13):

$$L = \begin{bmatrix} 1 & -1 & 0 \\ 0 & 1.5 & -1.5 \\ -2 & 0 & 2 \end{bmatrix}$$

and the control law needed to achieve consensus in the group is given by:

$$\begin{aligned}
\upsilon_1 &= -((-(x_2 - x_1))\cos(\theta_1) + (-(y_2 - y_1))\sin(\theta_1)) \\
\omega_1 &= -(\theta_1 - \arctan 2((y_2 - y_1), (x_2 - x_1))) \\
\upsilon_2 &= -1.5((-(x_3 - x_2))\cos(\theta_2) + (-(y_3 - y_2))\sin(\theta_2)) \\
\omega_1 &= -1.5(\theta_2 - \arctan 2((y_3 - y_2), (x_3 - x_2))) \\
\upsilon_3 &= -2((-(x_2 - x_3))\cos(\theta_2) + (-(y_2 - y_3))\sin(\theta_2)) \\
\omega_3 &= -2(\theta_3 - \arctan 2((y_2 - y_3), (x_2 - x_3)))
\end{aligned} \tag{15.17}$$

15.4.2 Regulation Control

The kinematic model presented in Eq. (15.2) cannot be transformed into a linear controllable system using static state feedback. However, the system can be transformed via feedback into simple integrators (De Luca et al. 2001).

$$\begin{aligned}
\xi_1 &= \theta \\
\xi_2 &= x\cos\theta + y\sin\theta \\
\xi_3 &= x\sin\theta + y\cos\theta
\end{aligned} \tag{15.18}$$

The existence of a canonical form for the dynamic model of DMR allows a general and systematic development of control strategies of open loop and closed loop. The most useful structure is the so-called chain shape, which is obtained by deriving the previous system:

$$\begin{aligned}
\dot{\xi}_1 &= \theta = u_1 \\
\dot{\xi}_2 &= \dot{x}\cos\theta - x\sin(\theta)\dot{\theta} + \dot{y}\sin\theta + y\cos(\theta)\dot{\theta} = u_2 \\
\dot{\xi}_3 &= \dot{x}\sin\theta - x\cos(\theta)\dot{\theta} - \dot{y}\cos\theta + y\sin(\theta)\dot{\theta} = \xi_2 u_1
\end{aligned} \tag{15.19}$$

This can be written as:

$$\begin{aligned}
\dot{\xi}_1 &= u_1 \\
\dot{\xi}_2 &= u_2 \\
\dot{\xi}_3 &= \xi_2 u_1
\end{aligned} \tag{15.20}$$

Thus,

$$\upsilon = u_2 + \xi_3 u_1 \tag{15.21}$$

$$\omega = u_1 \tag{15.22}$$

15.4.3 Trajectory Tracking

For track tracking, it is assumed that the DMR is represented by a point (x, y). Which must follow a trajectory in the Cartesian plane represented by $(x_d(t), y_d(t))$ where $t \in [0, T]$, and possibly $T \to \infty$. The reference trajectory parameterized in the time used in this work is given by Eq. (15.23).

$$\begin{aligned} x_d &= -0.5 + \frac{a}{3\sin(2(2\pi(t)/n)} \\ y_d &= \frac{a\ \sin(2(2\pi(t)))}{n/2} \\ \theta_d &= \arctan 2(\dot{y}, \dot{x}) \end{aligned} \tag{15.23}$$

where a is the width of the trajectory, t is the current time, and n the time in which it is desired to complete the cycle; therefore, the commands of reference speeds are given by:

$$\upsilon_d = \pm\sqrt{\dot{x}_d^2(t) + \dot{y}_d^2(t)} \tag{15.24}$$

$$\omega_d = \frac{\ddot{y}_d(t)\dot{x}_d(t) - \ddot{x}_d(t)\dot{y}_d(t)}{\dot{x}_d^2(t) + \dot{y}_d^2(t)} \tag{15.25}$$

State tracking errors are defined as Eq. (15.26).

$$\begin{bmatrix} e_1 \\ e_2 \\ e_3 \end{bmatrix} = \begin{bmatrix} \cos\theta & \sin\theta & 0 \\ -\sin\theta & \cos\theta & 0 \\ 0 & 0 & 1 \end{bmatrix} \begin{bmatrix} x_d - x \\ y_d - y \\ \theta_d - \theta \end{bmatrix} \tag{15.26}$$

Using the following nonlinear transformation of velocity inputs:

$$\upsilon = \upsilon_d \cos e_3 - u_1 \tag{15.27}$$

$$\omega = \omega_d - u_2 \tag{15.28}$$

the error dynamics becomes:

$$\dot{e} = \begin{bmatrix} 0 & \omega_d & 0 \\ -\omega_d & 0 & 0 \\ 0 & 0 & 0 \end{bmatrix} e + \begin{bmatrix} 0 \\ \sin e_3 \\ 0 \end{bmatrix} \upsilon_d + \begin{bmatrix} 1 & 0 \\ 0 & 0 \\ 0 & 1 \end{bmatrix} \begin{bmatrix} u_1 \\ u_2 \end{bmatrix} \tag{15.29}$$

The law of feedback is given by:

$$u_1 = -k_1 e_1 \tag{15.30}$$

$$u_2 = -k_2 \operatorname{sign}(\upsilon_d(t)) e_2 - k_3 e_3 \tag{15.31}$$

In terms of the original inputs, the design leads to the nonlinear controller variant in time (De Luca et al. 2001):

$$\upsilon = \upsilon_d \cos(\theta_d - \theta) + k_1 [\cos\theta(x_d - x) + \sin\theta(y_d - y)] \tag{15.32}$$

$$\omega = \omega_d + k_2 sign(\upsilon_d) [\cos\theta(x_d - x) - \sin\theta(y_d - y) + k_3(\theta_d - \theta)] \tag{15.33}$$

15.5 Simulations

Using the different topologies shown in Fig. 15.2 and the control laws from Sect. 15.4, the following was achieved: Consensus, regulation, and trajectory tracking for all robot members of the team.

15.5.1 Topology 1

15.5.1.1 Consensus

The consensus in position and orientation for the first topology on a plane was simulated and Fig. 15.3 shows each robot behavior.

The circles denote where each robot begins, and the pentagon indicates where the robots finish their movements. The (*) mark is used to represent the front of the robot.

Figure 15.4 shows the orientation for the robots. At the end of the graph, it is clear how the heading angles converge to the same value as the robot 1 has. This is due to the connections made on topology 1.

The robots reached consensus as shown in the Figs. 15.3 and 15.4. The linear and angular speeds of each robot to achieve the position and orientation consensus are shown in Figs. 15.5 and 15.6.

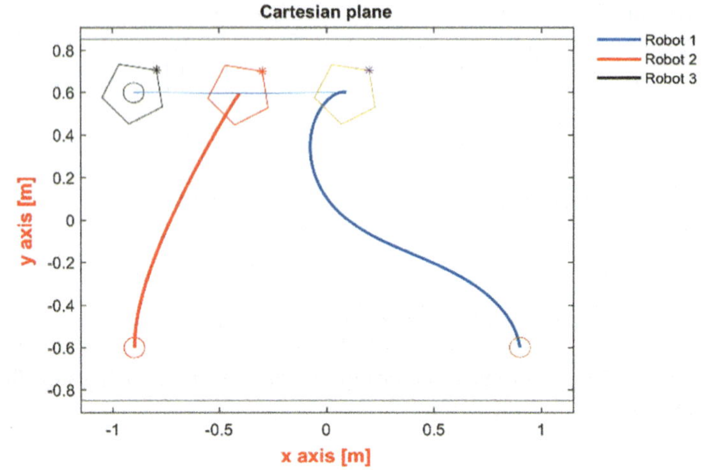

Fig. 15.3 Position consensus for robots in topology 1

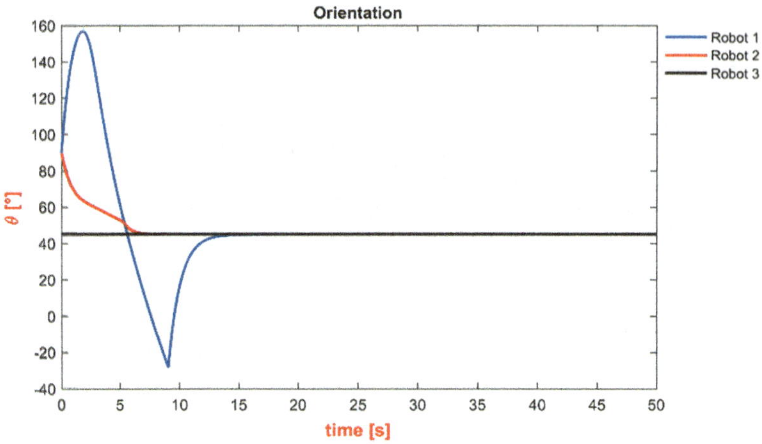

Fig. 15.4 Orientation consensus for robot in topology 1

15.5.1.2 Regulation

Once the consensus process is over, the regulation stage continues, in which the regulation control at a point leads to the states of the robots being modified in such a way that they reach a desired position and orientation. Figure 15.7 shows how the robots reach a desired position and orientation.

The evolution for orientation angles for each member of the group is depicted in Fig. 15.8.

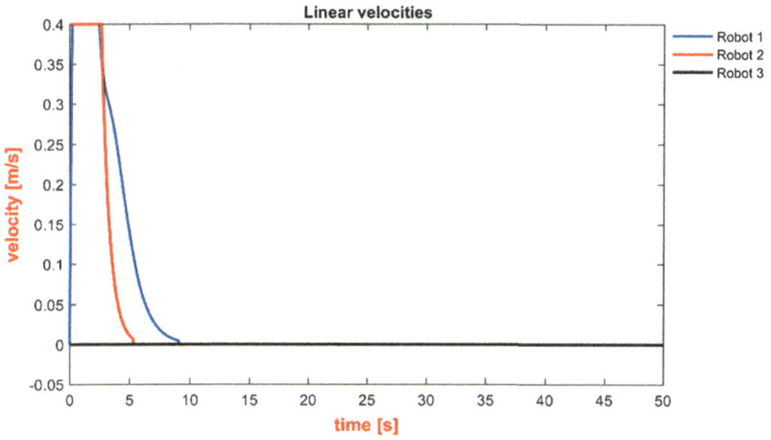

Fig. 15.5 Linear velocities for robots consensus in topology 1

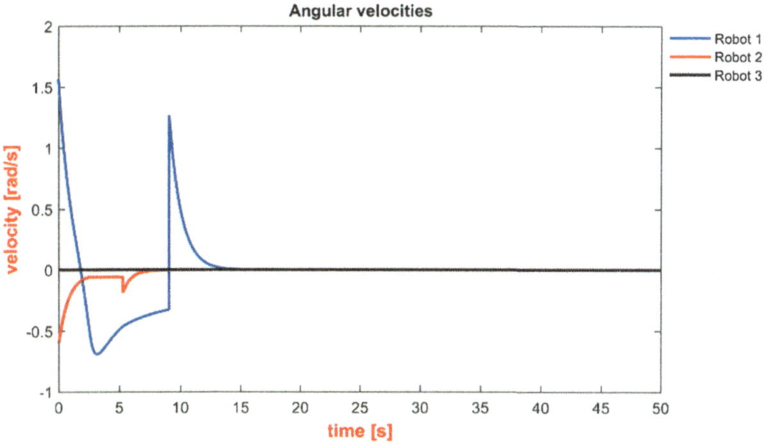

Fig. 15.6 Angular velocities for robots consensus in topology 1

The robots arrived at the desired position as shown in Figs. 15.7 and 15.8. The linear and angular speeds of each robot to achieve this are shown in Figs. 15.9 and 15.10.

15.5.1.3 Trajectory Tracking

Once the robots reach a desired point on Cartesian plane, the next step is to apply a tracking control that will guide the robots to follow a predetermined trajectory. In this case, the desired trajectory is an 8 shape, also known as Lemniscata.

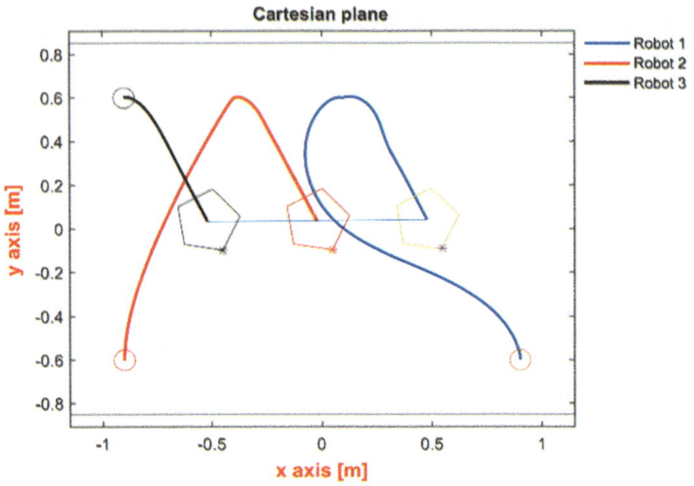

Fig. 15.7 Robots movements for regulation control on consensus for topology 1

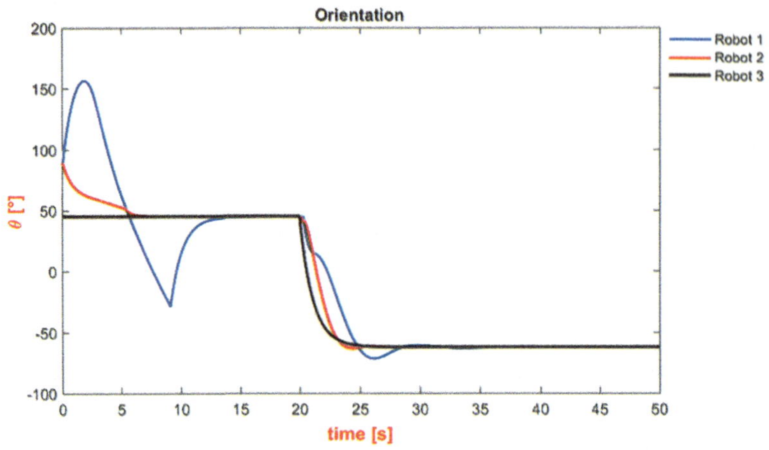

Fig. 15.8 Robots orientation for regulation control on consensus for topology 1

The results of the simulation are shown in Fig. 15.11. Figure 15.12 represents the orientation for each robot along the trajectory, and Figs. 15.13 and 15.14 show the linear and angular velocity, respectively.

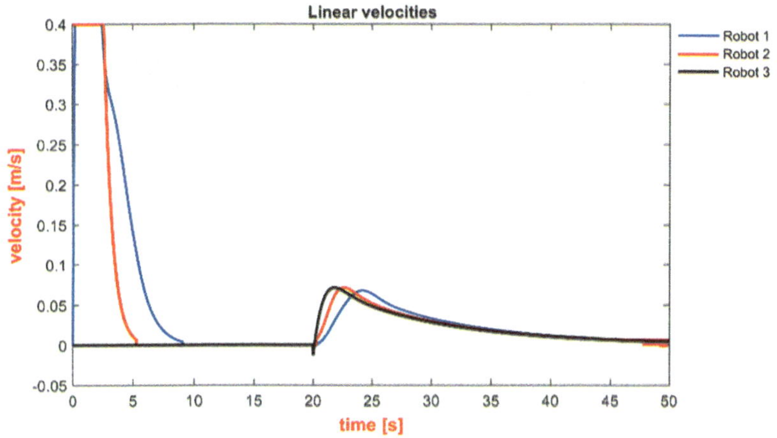

Fig. 15.9 Linear velocities for robots, regulation on consensus for topology 1

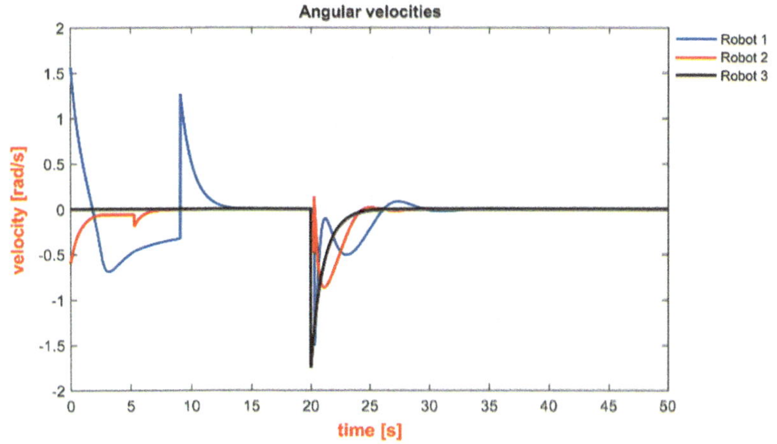

Fig. 15.10 Angular velocities for robots, regulation on consensus for topology 1

15.5.2 Simulation Results for Topologies 2 and 3

The simulations were performed for the topologies 2 and 3 to compare the behavior for the group of robots. In Tables 15.1 and 15.2, the comparison between topology 2 and 3 is depicted. According to the procedure done for topology 1, the main aspects to analyze are Cartesian plane movements, orientation, linear, and angular velocity. These four points are presented in three scenarios; consensus, regulation, and tracking.

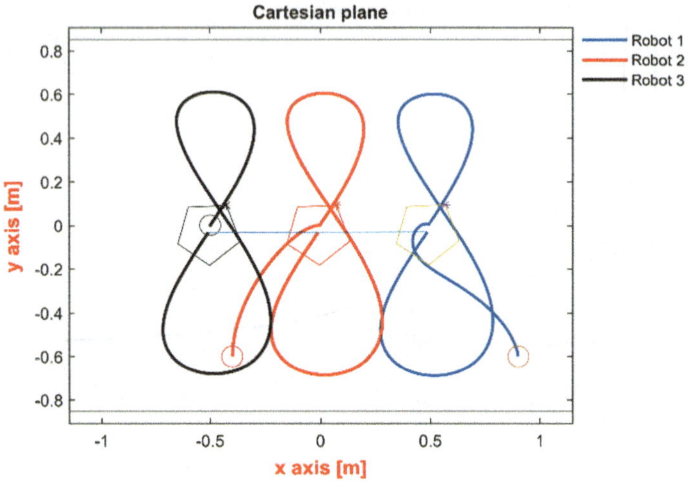

Fig. 15.11 Trajectory tracking for robots in consensus, topology 1

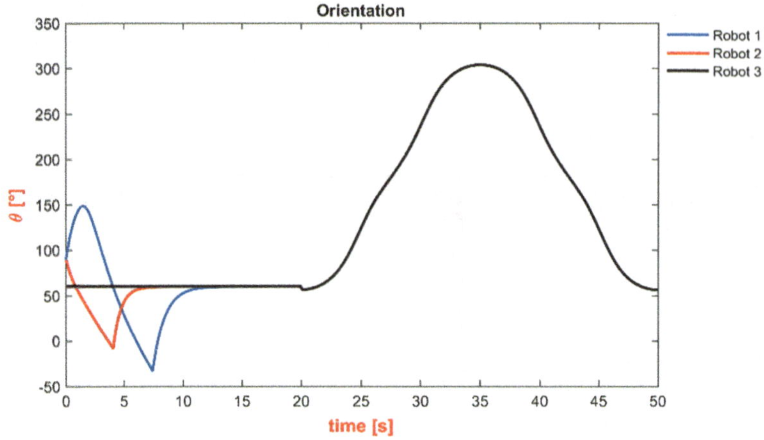

Fig. 15.12 Robots orientation for trajectory tracking in consensus, topology 1

15.6 Experimental Results

The experimental results were obtained using the following equipment:

- Three differential mobile robots iRobot Create.

  0.2605 [m] between wheels.
  0.045 [m] wheel radius.

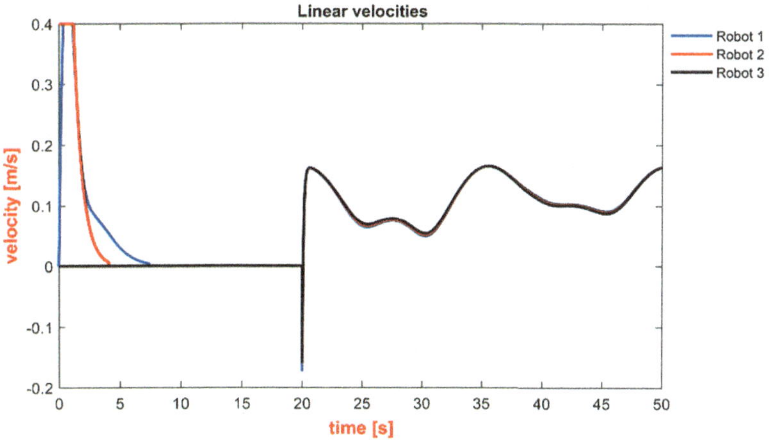

Fig. 15.13 Linear velocities for robots on trajectory tracking, topology 1

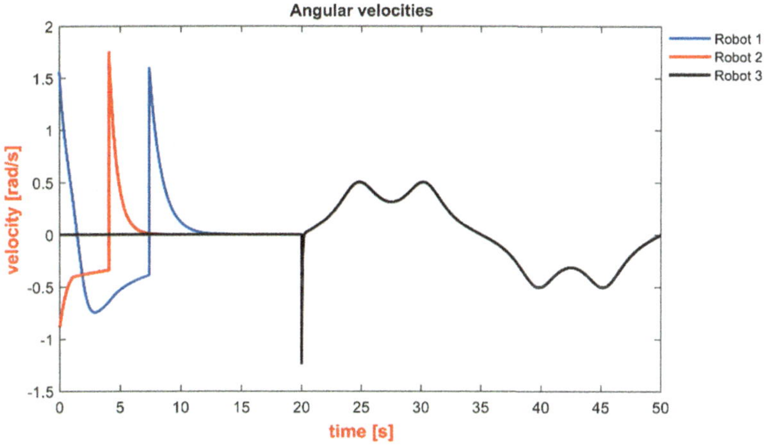

Fig. 15.14 Angular velocities for robots on trajectory tracking, topology 1

- Camera uEye-1220SE-M-CL (monocromatic).
- Field of view of 2.3 [m] × 1.7 [m].
- SO Ubuntu 12.04.
- C++ programming.
- Bluetooth centralized communication.

For implementation, topology 3 was selected with the following results.

Figure 15.15 shows the three stages (consensus, regulation, and tracking) for three DMR. The circles denote the initial conditions and the pentagons are used to mark where the robots finish their trajectories. The green line depicts the desired trajectory which must be followed by the consensus point.

Table 15.1 Simulations for topologies 2

Modality		Topology 2
Consensus	Cartesian plane	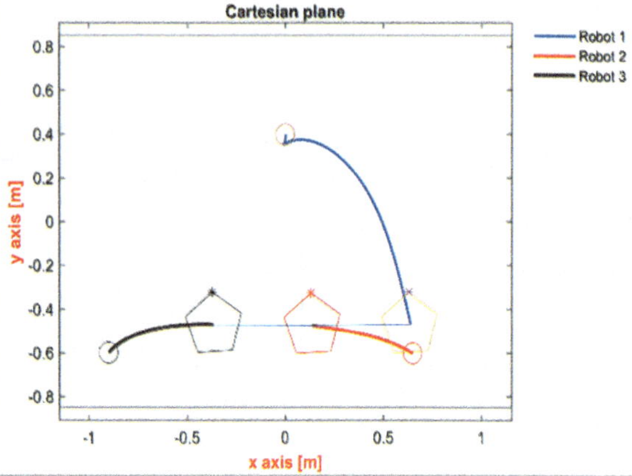
	Orientations	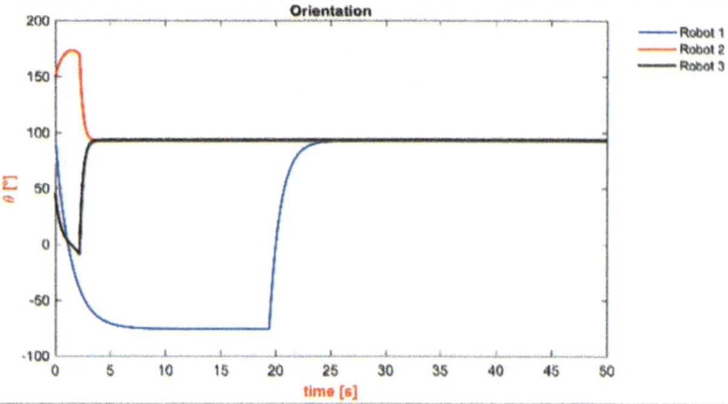
	Linear velocities	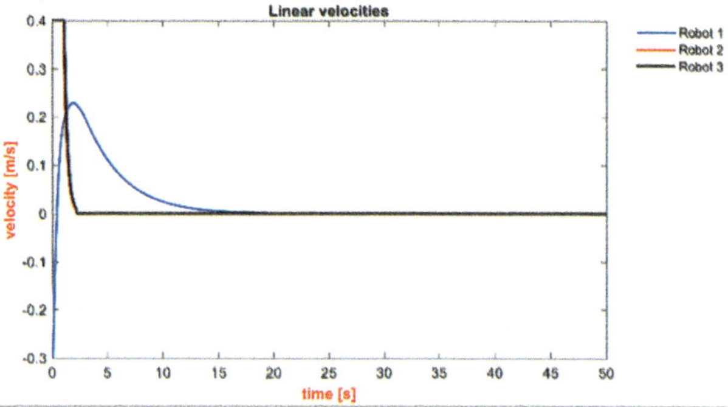
	Angular velocities	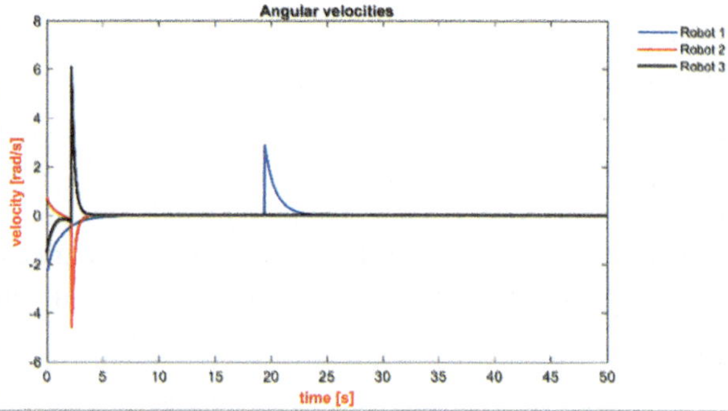

(continued)

Table 15.1 (continued)

Modality		Topology 2
Regulation	Cartesian plane	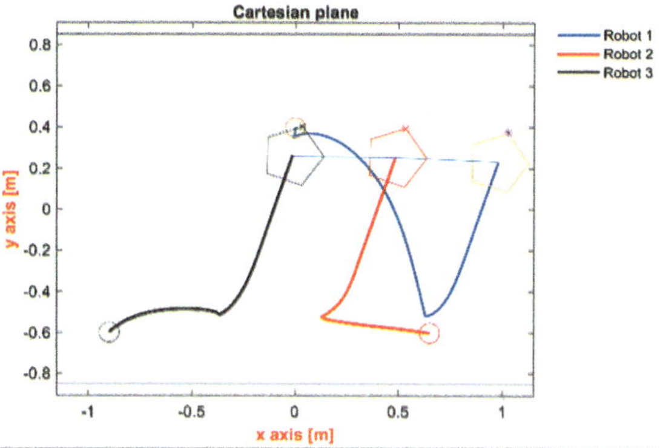
	Orientations	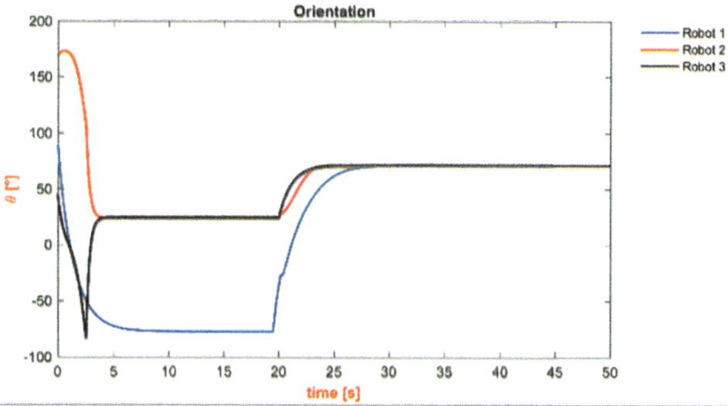
	Linear velocities	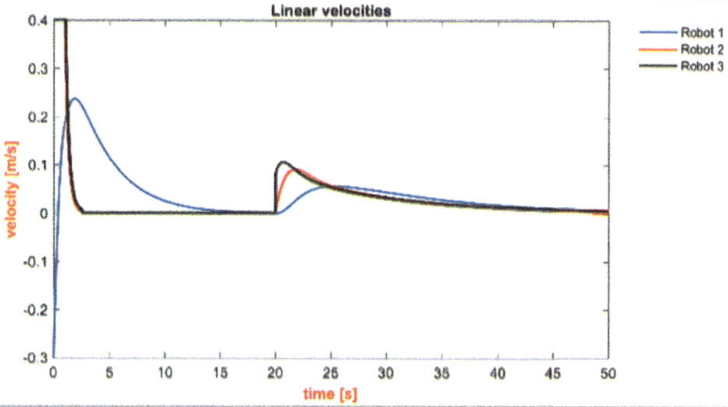
	Angular velocities	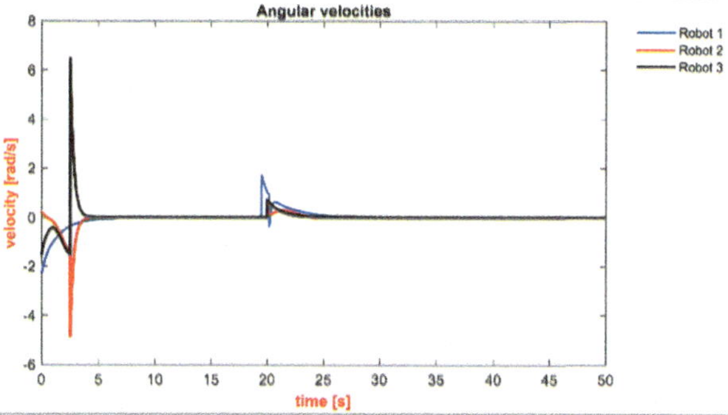

(continued)

Table 15.1 (continued)

<table>
<tr><th colspan="2">Modality</th><th>Topology 2</th></tr>
<tr><td rowspan="4">Tracking</td><td>Cartesian plane</td><td>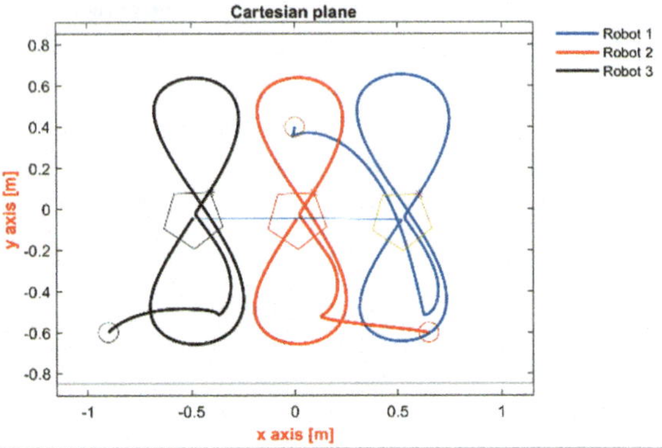
</td></tr>
<tr><td>Orientations</td><td>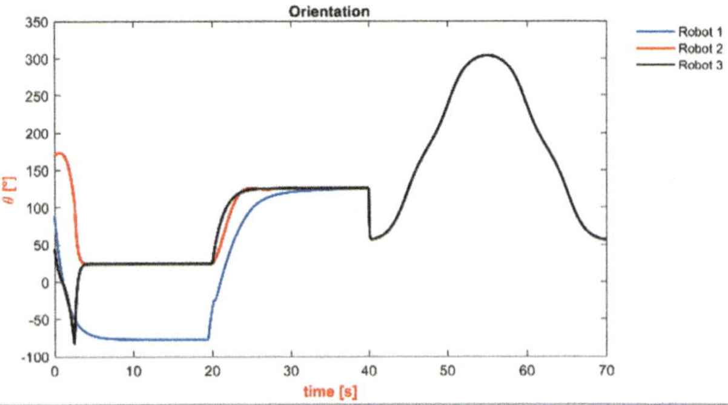
</td></tr>
<tr><td>Linear velocities</td><td>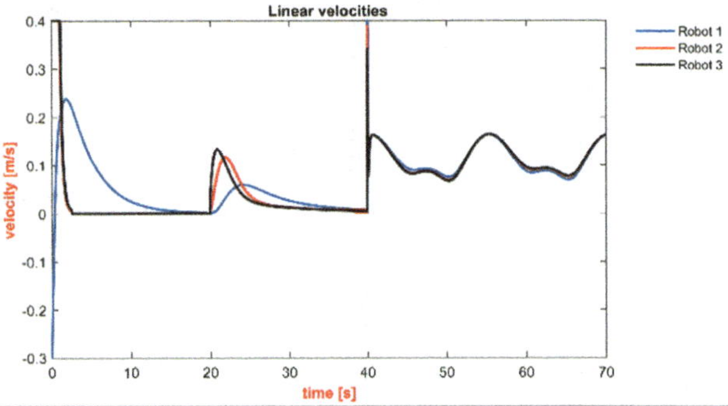
</td></tr>
<tr><td>Angular velocities</td><td>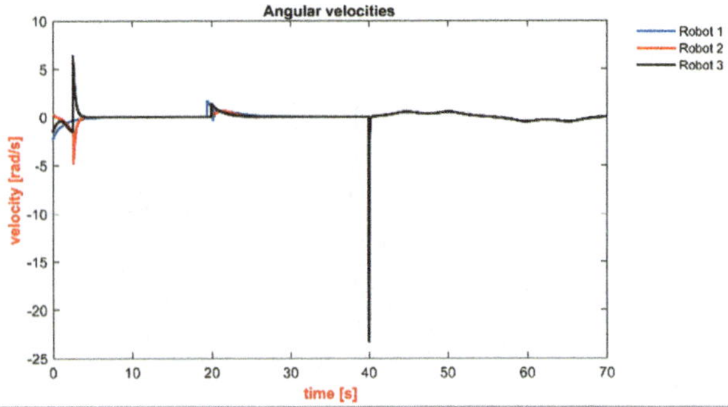
</td></tr>
</table>

Table 15.2 Simulations for topologies 3

Modality		Topology 3
Consensus	Cartesian plane	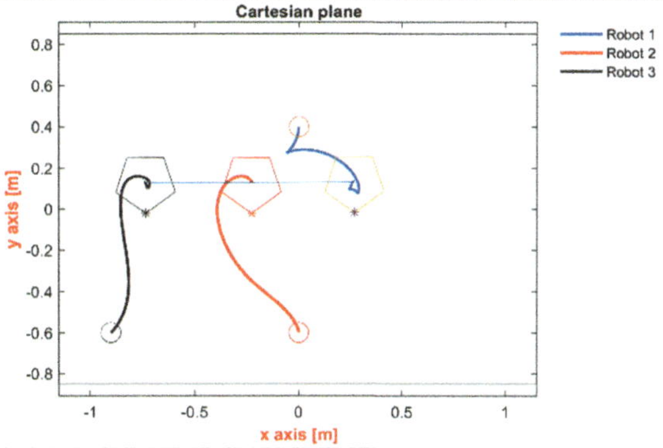
	Orientations	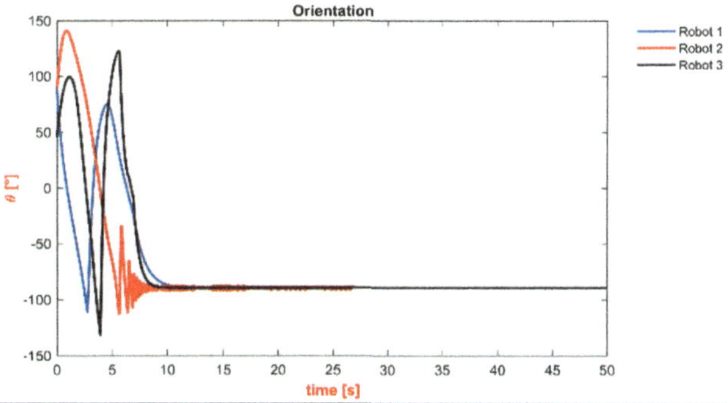
	Linear velocities	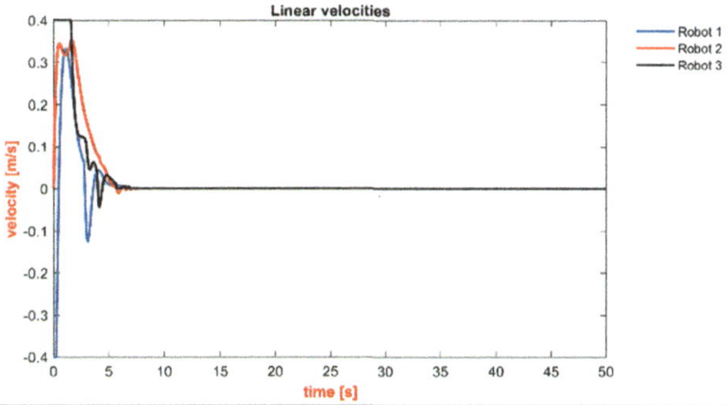
	Angular velocities	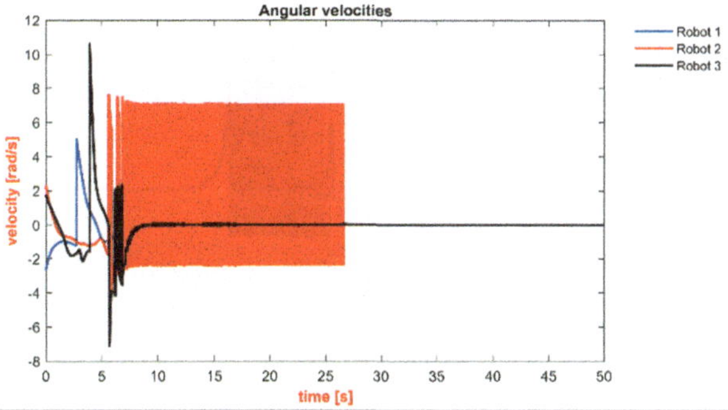

(continued)

Table 15.2 (continued)

Modality		Topology 3
Regulation	Cartesian plane	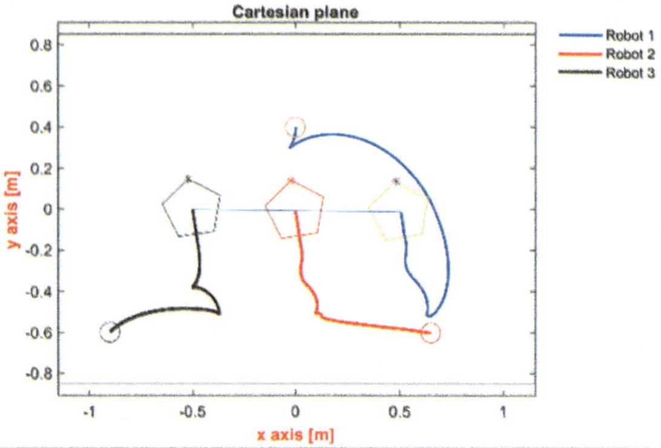
	Orientations	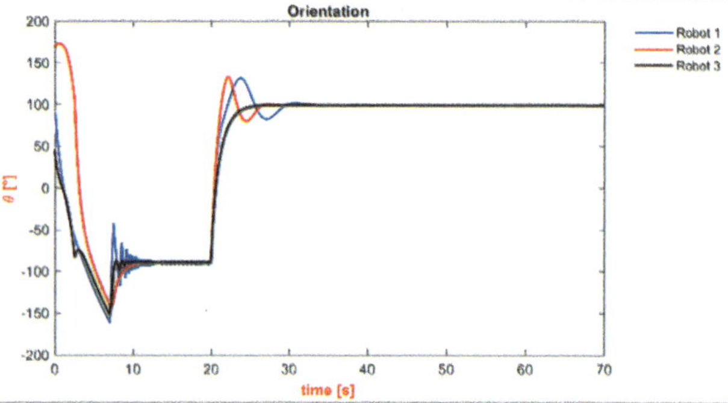
	Linear velocities	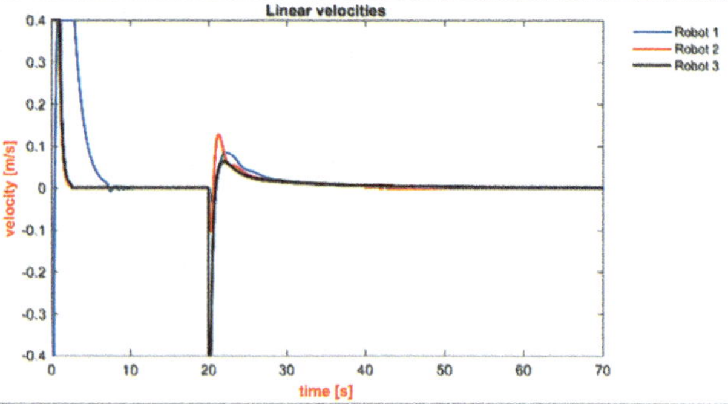
	Angular velocities	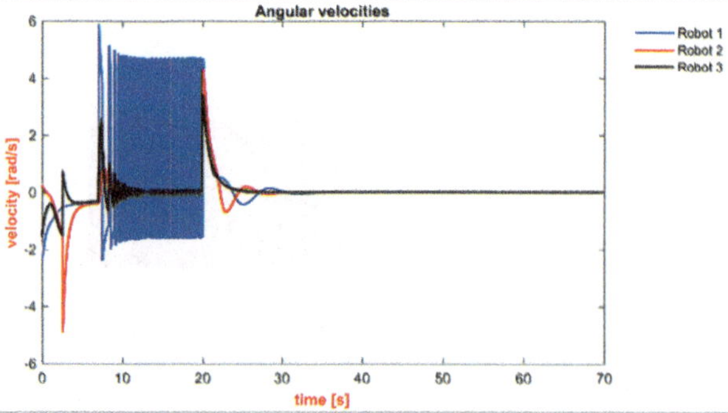

(continued)

Table 15.2 (continued)

Modality		Topology 3
Tracking	Cartesian plane	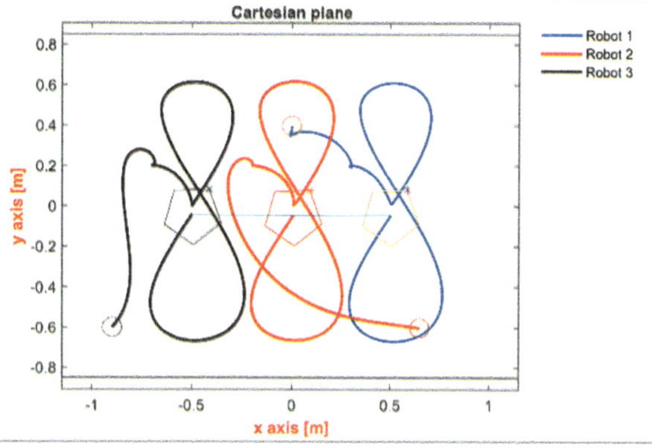
	Orientations	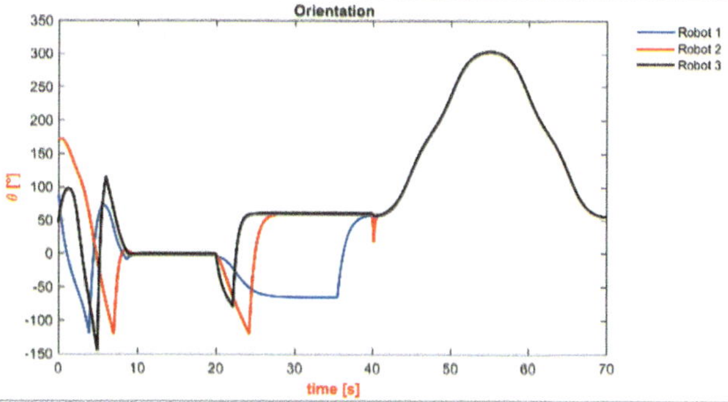
	Linear velocities	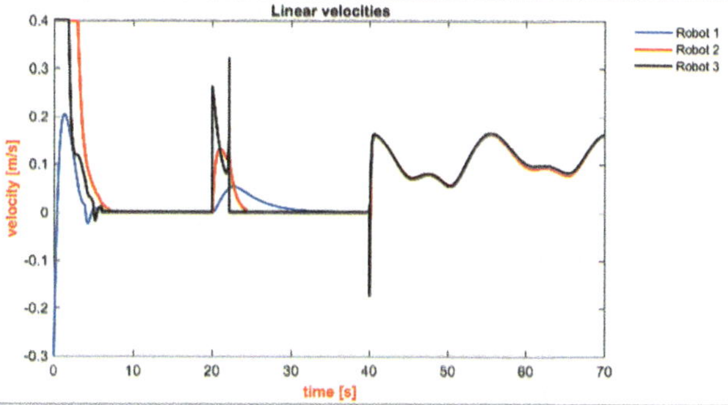
	Angular velocities	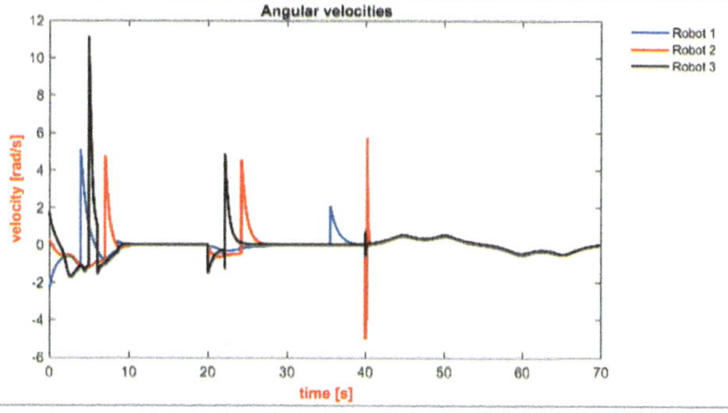

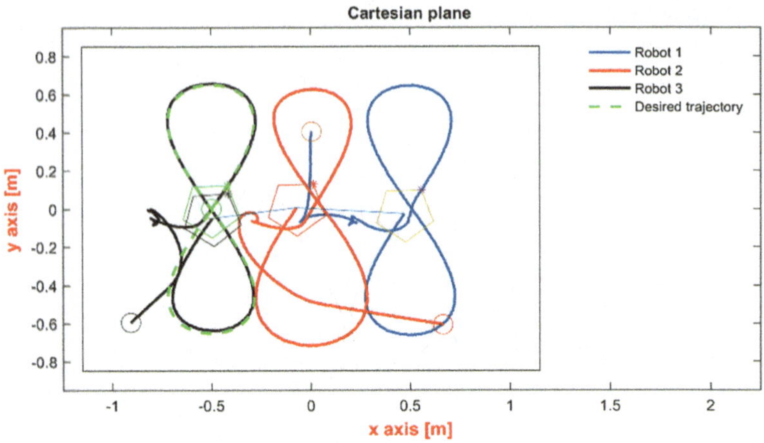

Fig. 15.15 Experimental result using topology 3 for consensus, regulation, and trajectory tracking

Figure 15.16 shows the behavior for the heading angles of each robot during the experiment. At the end of this graph, we can see how the robot has the same orientation as they are following the desired path.

In Figs. 15.17 and 15.18, we can see the evolution for linear and angular velocities in the robots during the experiment.

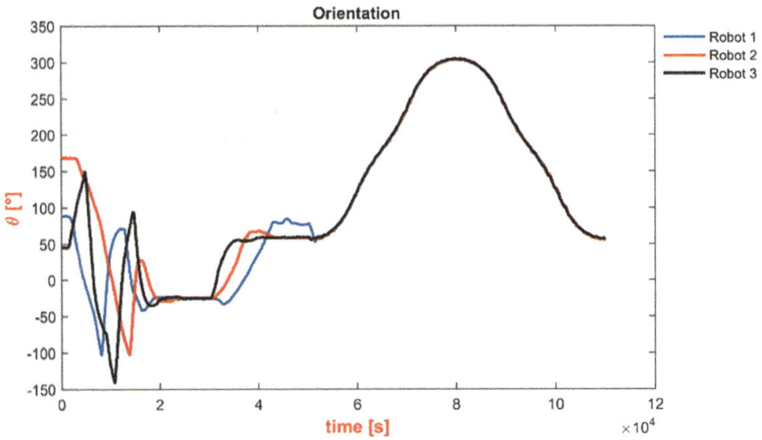

Fig. 15.16 Robots orientation for consensus, regulation and trajectory tracking with topology 3

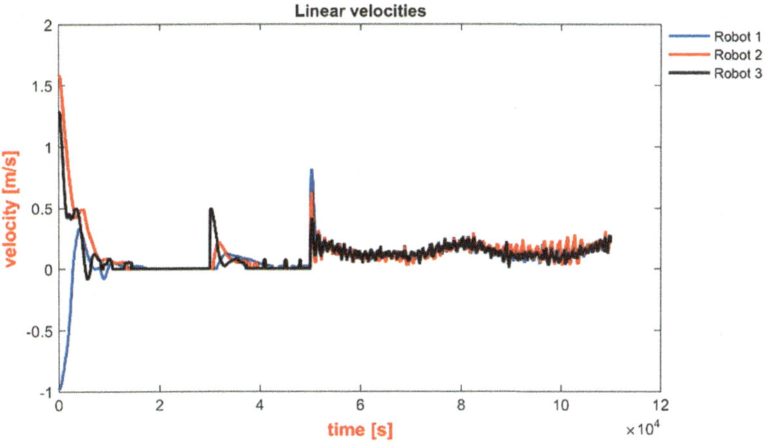

Fig. 15.17 Linear velocities during the experiment

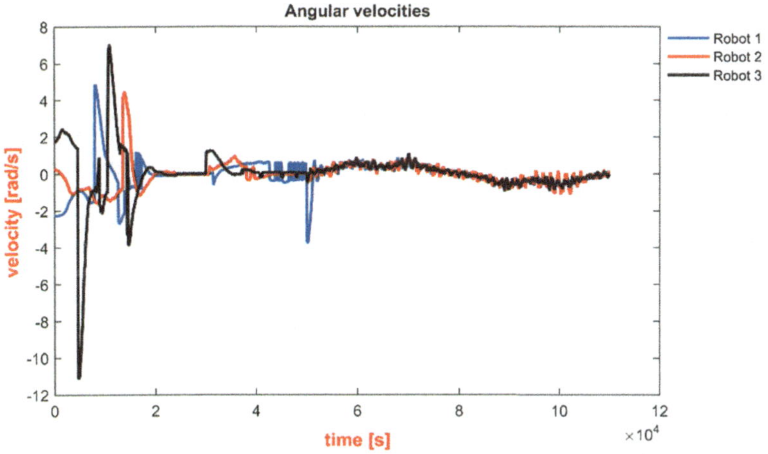

Fig. 15.18 Angular velocities during the experiment

15.7 Conclusions

It was shown that three different mobile robots can achieve consensus in their three states can perform regulation to a fixed point with consensus and follow a path with only displacing the consensus point.

The weights or values of the coefficients of the Laplacian matrix influence not only the value of the consensus point, but also in the robot's behavior on regulation and trajectory tracking. This aspect must be carefully handled at topology design.

The results of the implementation differ from the simulations due to factors such as lighting, physical limitations of the robots and other factors inherent to the experimental platform. The experimental validation demonstrates that through consensus cooperation techniques in mobile robots can be established.

References

Antonelli, G., Arrichiello, F., & Chiaverini, S. (2009). Experiments of formation control with multirobot systems using the null-space-based behavioral control. *IEEE Transactions on Control Systems Technology, 17*(5), 1173–1182.

Chung, S., & Slotine, J. (2009). Cooperative robot control and concurrent synchronization of Lagrangian systems. *IEEE Transactions on Robotics, 25*(3), 686–700.

De Luca, A., Oriolo, G., & Vendittelli, M. (2001). Control of wheeled mobile robots: An experimental overview. In S. Nicosia, B. Siciliano, A. Bicchi, & P. Valigi (Eds.), *Lecture notes in control and information sciences* (Vol. 270). Berlin, Heidelberg: Springer.

Desai, J., Ostrowski, J., & Kumar, V. (2001). Modeling and control of formations of non-holonomic mobile robots. *IEEE Transactions on Robotics and Automation, 17*(6), 905–908.

Huijberts, H., Nijmeijer, H., & Willems, R. (2000). Regulation and controlled synchronization for complex dy-namical systems. *International Journal of Robust and Nonlinear Control, 10*(5), 336–377.

Nijmeijer, H., & Rodríguez-Angeles, A. (2004). Control synchronization of differential mobile robots. In *6th IFAC Symposium on Nonlinear Control Systems*, California, USA, pp. 579–584.

Ren, W., & Beard, R. (2008). *Distributed consensus in multi-vehicle cooperative control: Theory and application*. London: Springer.

Siméon, T., Leroy, S., & Laumond, J. (2002). Path coordination for multiple mobile robots: A resolution-complete algorithm. *IEEE Transactions on Robotics and Automation, 18*(1), 42–49.

Sun, D., & Mills, J. (2002). Adaptive synchronized control for coordination of multi-robot assembly tasks. *IEEE Transactions on Robotics and Automation, 18*(4), 498–510.

Sun, D., & Mills, J. K. (2007). Controlling swarms of mobile robots for switching between formations using synchronization Concept. In *IEEE International Conference on Robotics and Automation*, Roma, Italy, pp. 2300–2305.

Takahashi, H., Nishi, H., & Ohnishi, K. (2004). Autonomous decentralized control for formation of multiple mobile robots considering ability of robot. *IEEE Transactions on Industrial Electronics, 51*(6), 1272–1279.

Yamaguchi, H., Arai, T., & Beni, G. (2001). A distributed control scheme for multiple robotic vehicles to make group for-mations. *Robotics and Autonomous Systems, 36*(4), 125–147.

The manufacturer's authorised representative in the EU is Springer Nature Customer Service Centre GmbH, Europaplatz 3, 69115 Heidelberg, Germany. If you have any concerns regarding our products, please contact ProductSafety@springernature.com

Printed and bound by CPI Group (UK) Ltd, Croydon, CR0 4YY
30/09/2026
02225230-0015